Recent Advances in Mathematics, Statistics and Computer Science

Recent Advances in Mathematics, Statistics and Computer Science

Bihar, India, 29–31 May 2015

Editors

Arun Kumar Sinha

R Rajesh

Prabhat Ranjan

Rajesh Pratap Singh

Central University of South Bihar, India

World Scientific

NEW JERSEY • LONDON • SINGAPORE • BEIJING • SHANGHAI • HONG KONG • TAIPEI • CHENNAI • TOKYO

Published by

World Scientific Publishing Co. Pte. Ltd.
5 Toh Tuck Link, Singapore 596224
USA office: 27 Warren Street, Suite 401-402, Hackensack, NJ 07601
UK office: 57 Shelton Street, Covent Garden, London WC2H 9HE

British Library Cataloguing-in-Publication Data
A catalogue record for this book is available from the British Library.

RECENT ADVANCES IN MATHEMATICS, STATISTICS AND COMPUTER SCIENCE

ISBN 978-981-4696-16-6

Printed in Singapore

Preface

The technological impacts of the advancements in the fields of Mathematics, Statistics and Computer Science have been immense. In this scenario, International Conference on Recent Advances in Mathematics, Statistics and Computer Science (ICRAMSCS, May 29th – 31st, 2015) was organised by the School of Mathematics, Statistics and Computer Science to provide a common platform for the presentation of high quality and latest research trends in the global area of Mathematics, Statistics and Computer Science. It will continue to provide valuable information related to recent advances in the fields of Mathematics, Statistics and Computer Science. It will serve to foster an effective communication among researchers, engineers, and practitioners working under a common interest area involving Mathematics, Statistics and Computer Science with the aim of holistic development of mankind.

The major highlights of this book includes comprehensive research papers and articles on real life problems of research and applications in Mathematics, Statistics and Computer Science and particularly the new approaches, models, algorithms in these areas to stimulate the young researchers. This book presents a cutting-edge research in the areas like high performance computing (which focuses on aggregating computing power to solve big problems efficiently), data mining (which focuses on analysing data and summarizing it into useful information for efficient storage and processing), digital image/signal processing, software engineering, computer networks, operations research, survey sampling, distribution theory, reliability, fluid dynamics, automata theory, spatial statistics, etc. All the submissions underwent strenuous peer review process which was comprised of expert reviewers. Each contribution has been carefully selected for inclusion based on the significance of the research in the field.

Contents

Databased Intrinsic Weights of Indicators of Multi-Indicator Systems and Performance Measures of Multivariate Rankings of Systemic Objects*

G. P. Patil

Center for Statistical Ecology and Environmental Statistics
Department of Statistics, Penn State University
University Park, Pennsylvania16820, USA
E-mail: gpp@psu.edu, gpp2234@gmail.com

S. W. Joshi

Department of Computer Science, Slippery Rock University
Slippery Rock, Pennsylvania 16057 USA
E-mail: sharadchandra.joshi@sru.edu

In this paper, we discuss concepts, methods, and tools of partial order based multivariate ranking of objects leading to novel and innovative measures of performance of ranking methods for given data sets/data matrices of objects and features (indicators). We also develop novel and innovative intrinsic differential weights of relative importance of indicators with implications on their prioritization and subsequent selection status.We also provide illustrative examples using 25x3 data matrix with 25 objects and 3 indicators, giving intrinsic relative weights of the indicators indicating their databased relative importance. Further, we derive the rankings of the objects using different ranking methods constructing multi-indicator object rank scores, given by the weighted composite index, comparability weighted net superiority index, MCMC-based weighted indicator cumulative rank frequency distribution index, and MCMC-based average rank index. Finally, the ranking performance measures of these ranking methods are computed for the illustrative data matrices/data sets. We conclude the paper with selected references and extended bibliography.

Keywords: Multi-Indicator Ranking; Databased Weights; Performance Measures.

1. Introduction

Multivariate Nonparametric Statistics for purposes of inference on multivariate median, multivariate order statistics, and multivariate image reconstruction and enhancement is presently occupied with issues of multivariate ranking involving

* Based on the initial part of the inaugural keynote lecture given by the first author at the International Conference on Recent Advances in Mathematics, Statistics, and Computer Science held at the Central University of Bihar, Patna, Bihar, India, during May 29–31, 2015.

data depth measures and affine invariance. See [5], [8], [21], [9], [28], [29], [6], [7], [22].

Genome Wide Association Studies are presently occupied with issues of gene discoveries and variables selection with oracle properties among others. These issues involve multivariate ranking of objects and variables/indicators. See [3], [20], [24], [25], [26], [27].

Problems of this comparative nature are arising in various areas of inferential and exploratory importance, such as, surveillance geo-informatics, bio-geo-informatics, networks assessments, banking predictive analytics, drug development research, etc. [1], [2], [10], [11], [12], [13], [23].

As a result, an exciting field of study is emerging within the discipline of knowledge discovery, data mining, and statistical learning. It is the comparative knowledge discovery in the multi-indicator information fusion systems. See [18], [10], [11], [2].

In this paper, we discuss concepts, methods, and tools of partial order based multivariate ranking of objects leading to novel and innovative measures of performance of ranking methods for data sets (datamatrices) of objects and their features (indicators). We also develop novel and innovative intrinsic differential weights of relative importance of indicators with implications on their prioritization and subsequent selection status.

2. Preliminaries

To set the stage, we consider a multivariate data set in the form of an nxm data matrix $[x_{ij}]$, i= 1, 2, …, n; j= 1, 2, …, m, where n rows correspond to n objects $a_1, a_2, \ldots, a_i, \ldots, a_n$ and m columns correspond to m indicators $I_1, I_2, \ldots, I_j, \ldots, I_m$.

Objects may be entities, such as, individuals, units, pixels, areas, regions, patients, genes, drugs, documents, clients, products, tools with relevant characteristics/features as potential indicators for some single or multiple outcomes, endpoints, concepts, domains.

Indicators may be measurable characteristics/features of objects with common orientation indicative of some un-measurable abstract/latent concept for objects. For example, a larger magnitude of an indicator will be indicative of a correspondingly larger magnitude of the latent concept/trait of an object. And vice versa.

As a simple example, consider size of an individual as the abstract concept. Consider height, weight, and volume of the individual as indicators of size with assumed common orientation of positive monotonicity/positive correlations.

Generally speaking, larger the size, larger the indicator, and larger the indicator, larger the size.

The three indicators/indicator measurements may have three-dimensional elliptical distribution with pairwise positive correlations.

We note that we thus have an m-dimensional data set consisting of n data points, with no measurement column available for any response variable y.

The multivariate data set is usually a nonlinear partially ordered set (poset). Not all pairs of objects are comparable. For a two-indicator set up, the following diagram in Figure 1 may be suggestive. Interestingly, every object here induces its four quadrants defined by the horizontal and vertical lines passing through it. Every object in its first quadrant, where both indicators are larger, is comparable to it, and together they are defined to make its "UpSet". Every object in its third quadrant, where both indicators are smaller, is comparable to it, and together they are defined to make its "DownSet". Clearly, every object in its second and fourth quadrants, where the two indicators are in conflict, is incomparable to it.

Ranking amounts to linearizing the poset by ranking the objects with appropriate scalar rank-scores consistent with the comparability in the data matrix. Rank-scores need to inherit the comparabilities in the data set. Incomparable pairs are expected to become comparable in either direction. We will see later that the UpSet and DownSet of an object help define a rank-score for it.

3. Indicator Relative Importance Weight Vector

On which line is the linearized set to lie? Without loss of generality, which axis passing through the origin is to be selected? What can be said of the separations between successive objects when ranked? Projections on a ray through the origin have been popular. The ray is determined by $w = (w_1, \ldots, w_m)$, where $w_j > 0$, with summation of w_j being unity, a differential weight vector, measuring relative importance of indicators for the abstract concept. Projection is a fixed scalar multiple of what is popularly called weighted composite index with weight vector w.

Choice of w involves subjective trade off/compensation among indicators. It becomes a sensitive political issue between stakeholders. Reconciliation in view of data matrix evidence becomes a practical challenge and scientific/statistical opportunity.

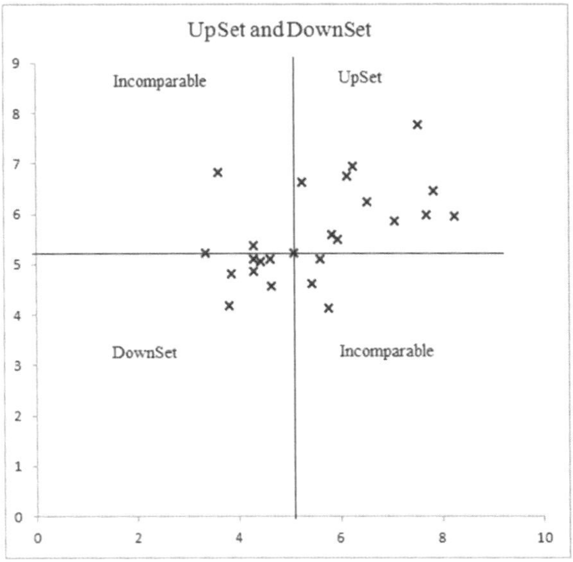

Fig. 1. Four quadrants showing UpSet, DownSets, and Incomparable objects with respect to a particular object in a two-indicator set up.

Can we think of a data based w intrinsic to the data matrix? And relative to such a w, and its corresponding ray, can we think of alternative ways of computing appropriate rank-scores, which do not involve indicator tradeoffs? And if we can think of several methods of rank-scores and resultant rankings, is it possible to measure their individual performance level to help find a best method among them for the given data set? Interestingly, all of these questions are frontier questions that we should wish to address. And fortunately, we now have some initial answers that we wish to share on the challenging issues of multivariate ranking over the past several decades.

4. Data Matrix Based Intrinsic Differential Weight Vector w^I for the Indicator Set to Measure Relative Importance of Indicators

In this paper, we will discuss, "Pairwise Object Comparisons, and Indicator Agreements among Object Comparison Disagreements," as a basis for the formulation.

Consider Multivariate Zeta Matrix: Object x Object Comparability n×n Matrix. Cell Entry: m-variate bit, binary digit: 111…000… 101100…01, where 1 if $a_i \geq a_j$, and 0, otherwise.

- Comparability cell has all 1's, or, all 0's in its bit, indicating collective agreement among indicators.
- Incomparability cell has some 1's and some 0's in its bit, indicating collective disagreement among indicators.
- For each incomparability cell, count for each indicator the number of agreements with the collectivity of indicators. Add up for each indicator over all of the incomparability cells. Normalize/unitize to give the intrinsic w^I we are looking for.
- Incidentally, and importantly, this intrinsic w^I also provides a powerful basis for comparison and selection of indicators.

5. Conceptualizing and Computing Performance Measure of a Ranking Method

Consider Multivariate Zeta Matrix as before: But, this time, Cell entry: (m+1)-variate bit with the first m variates as before, and the $(m+1)^{th}$ variate corresponding to the Ranking.

- For each incomparability cell, count for each indicator the agreement with the Ranking. Add up for each indicator over all the incomparability cells.
- Normalize/unitize to give the w^R induced by the Ranking R.

Define its performance measure PMR by corr/gen. corr(w^I, w^R).

6. Some Comparability Invariant/Partial Order Based Ranking Methods

Method 1: Weighted Composite Index for Rank-score: WCI. See [1], [18].

Given a differential weight vector w = (w1, w2, ..., w_m), and the indicator values of the object to be x = (x_1, x_2, …, x_m), the correspondingly weighted composite index for the object rank-score is given by the weighted average of the values of the indicators for the object, $w_1x_1 + w_2x_2 + \ldots + w_mx_m$, which is equivalent to the inner product

$$w.x = |w|\ |x| \cos(w,x) = |w| \text{ x projection of x on w.}$$

And we get w.d = 0, w.d > 0, w.d < 0, where $d = x_1 - x_2$, where now, x_1 and x_2 are indicator value vectors of some objects 1 and 2. Figure 2 illustrates the concept with a two-indicator space, where $w.x_1$ = OP and $w.x_2$ = OQ.

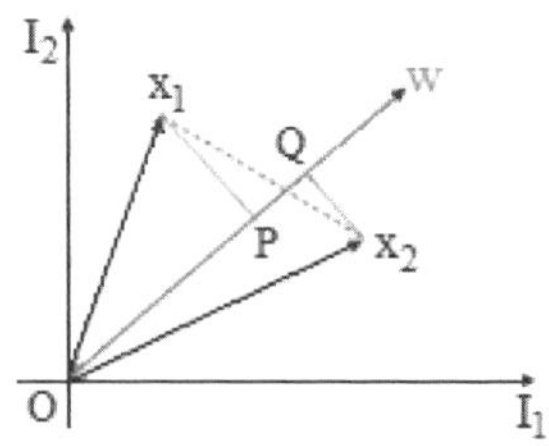

Fig. 2. Comparison of objects 1 and 2 using Method 1.

Method 2: Comparability Weighted Net Superiority Index for Rank-score: CWNSI. See [10], [11], [16].

With larger the better, and smaller the worse protocol, its DownSet provides a measure of superiority for the object x and its UpSet provides a measure of its inferiority.

Let us define O(x) to be the cardinal size of the DownSet of x, and F(x) to be the cardinal size of its UpSet, leading us to define

$$\begin{aligned}\text{Rank-score }(x) &= (O(x)- F(x))(O(x) + F(x))/(n-1)\\ &= \text{Net Superiority} \times \text{Comparability.}\end{aligned}$$

For a two-indicator example illustrated in Figure 1 before, n = 25, O(x) = 7, and F(x) = 11.

Method 3: MCMC based Average Rank for Rank-score: ARI. See [1].

This method attempts to construct a comparability invariant population of indicators/voters relevant to the data matrix indicators/voters as a random sample from it. For this purpose, the method engages in an MCMC, sequentially producing comparability invariant permutations of the objects, yielding what are called linear extensions that specify ranks to the objects.

With the sequence of linear extensions, each object receives a sequence of ranks, giving a sequence of its cumulative rank averages. The MCMC stops when all of the sequences of the cumulative rank averages converge. The limiting rank averages of individual objects are then defined to be their rank scores, providing the ranking of the objects due to this method.

Method 4: MCMC based Weighted indicator Cumulative Rank Frequency Distribution for Stochastic Rank-score: WICRFDI. See [19], [18].

This method starts as Method 3, but, instead of computing cumulative rank averages in the process of MCMC, it computes cumulative rank frequency distributions for each object, and the MCMC stops, when these cumulative rank frequency distributions converge for all objects. They are comparability invariant under stochastic ranking, and in multiple steps, produce ranking of objects.

7. Ranking Performance Measure: An Illustrative Example

We will show calculation for a 25 by 3 data matrix in some detail. First we need to compute the data-based intrinsic weights.

Consider the data matrix shown below in Table 1. Its Hasse diagram is in Figure 3.

Multivariate zeta matrix in its entirety is in Table 2. The multivariate zeta matrix has 25 main columns for 25 objects. Each main column is a group of four sub-columns. Of these to compute intrinsic weights we use three sub-columns with headings I_1, I_2, and I_3 for three indicators at this time. The fourth column with the heading of r (for ranking) will be used in a later step to compute performance measure of a particular ranking of the objects. The matrix has 25 rows for the 25 objects as identified in the leftmost column. Entries in sub-columns are bits: 0 (zero) or 1 (one). The entry in a sub-cell formed by the row corresponding to object a_i and the sub-column corresponding to indicator I_k for object a_j is one if and only if the value x_{ik} of I_k for object a_i is greater than or equal to the value x_{jk} of I_k for object $a_{j,}$ otherwise it is zero. Object a_i is comparable with object a_j if all entries in the group of three sub-columns are identical. If the entries are not all identical and no indicator values for the two objects are equal then the two objects are incomparable. Cells corresponding to all pairs of incomparable objects are shaded in the multivariate zeta matrix in Table 2. Only these cells are used in computation of intrinsic weights.

As the next step, for each group of three sub-cells corresponding to a pair of incomparable objects we count the number of times the bit-value in a sub-cell occurs within the group of the three sub-cells. For the current data matrix with three indicators this count is either 1 or 2. For each indicator such counts are added up over all sub-cells corresponding to the indicator. The total for indicator I_j is denoted by A_j for agreement totals. For this illustrative example, we present individual agreement counts in Table 3. This table has 25 rows for 25 objects and 25 main columns for 25 objects. Each main column has three sub-columns.

Table 1. Data matrix.

Object	I_1	I_2	I_3
a_1	4.273	5.139	4.766
a_2	4.630	4.601	4.645
a_3	8.226	5.983	7.500
a_4	7.044	5.885	7.247
a_5	3.586	6.848	6.165
a_6	5.245	6.643	4.699
a_7	5.928	5.524	6.926
a_8	6.237	6.966	8.543
a_9	4.275	5.391	5.161
a_{10}	6.109	6.775	7.527
a_{11}	5.756	4.151	6.216
a_{12}	5.410	4.639	3.901
a_{13}	4.285	4.879	3.296
a_{14}	4.602	5.137	4.369
a_{15}	7.498	7.789	7.654
a_{16}	5.065	5.243	6.806
a_{17}	7.676	6.010	9.181
a_{18}	6.503	6.261	4.105
a_{19}	4.430	5.090	3.657
a_{20}	3.783	4.204	3.280
a_{21}	3.840	4.849	3.171
a_{22}	3.335	5.249	2.570
a_{23}	7.801	6.488	7.544
a_{24}	5.580	5.140	4.371
a_{25}	5.823	5.623	7.018

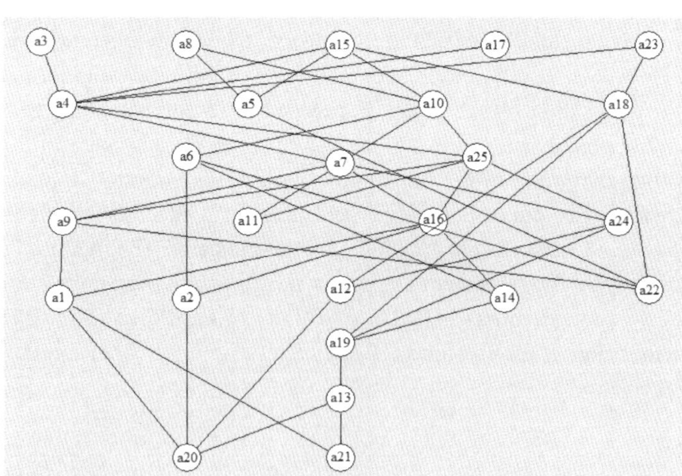

Fig. 3. Hasse diagram for data matrix in Table 1.

Table 2. Multivariate zeta matrix for data in Table 1.

Obj	a1				a2				a3				a4				a5				a6				a7				a8				a9				a10				a11				a12				a13			
	I1	I2	I3	r	I1	I2	I3	r	I1	I2	I3	r	I1	I2	I3	r	I1	I2	I3	r	I1	I2	I3	r	I1	I2	I3	r	I1	I2	I3	r	I1	I2	I3	r	I1	I2	I3	r	I1	I2	I3	r	I1	I2	I3	r	I1	I2	I3	r
a1	1	1	1	1	0	1	1	1	0	0	0	0	0	0	0	0	1	0	0	0	0	0	1	0	0	0	0	0	0	0	0	0	0	0	0	0	0	0	0	0	0	1	0	0	0	1	1	1	0	1	1	1
a2	1	0	0	0	1	1	1	1	0	0	0	0	0	0	0	0	1	0	0	0	0	0	0	0	0	0	0	0	0	0	0	0	1	0	0	0	0	0	0	0	0	1	0	0	0	0	1	1	1	0	1	1
a3	1	1	1	1	1	1	1	1	1	1	1	1	1	1	1	1	1	0	1	1	1	0	1	1	1	1	1	1	1	0	0	0	1	1	1	1	1	0	0	1	1	1	1	1	1	1	1	1	1	1	1	1
a4	1	1	1	1	1	1	1	1	0	0	0	0	1	1	1	1	1	0	1	1	1	0	1	1	1	1	1	1	1	0	0	0	1	1	1	1	1	0	0	0	1	1	1	1	1	1	1	1	1	1	1	1
a5	0	1	1	1	0	1	1	1	0	1	0	0	0	1	0	0	1	1	1	1	0	1	1	1	0	1	0	0	0	0	0	0	0	1	1	1	0	1	0	0	0	1	0	1	0	1	1	1	0	1	1	1
a6	1	1	0	1	1	1	1	1	0	1	0	0	0	1	0	0	1	0	0	0	1	1	1	1	0	1	0	0	0	0	0	0	1	1	0	1	0	0	0	0	0	1	0	1	0	1	1	1	1	1	1	1
a7	1	1	1	1	1	1	1	1	0	0	0	0	0	0	0	0	1	0	1	1	1	0	1	1	1	1	1	1	0	0	0	0	1	1	1	1	0	0	0	0	1	1	1	1	1	1	1	1	1	1	1	1
a8	1	1	1	1	1	1	1	1	0	1	1	1	0	1	1	1	1	1	1	1	1	1	1	1	1	1	1	1	1	1	1	1	1	1	1	1	1	1	1	1	1	1	1	1	1	1	1	1	1	1	1	1
a9	1	1	1	1	0	1	1	1	0	0	0	0	0	0	0	0	1	0	0	0	0	0	1	0	0	0	0	0	0	0	0	0	1	1	1	1	0	0	0	0	0	1	0	0	0	1	1	1	0	1	1	1
a10	1	1	1	1	1	1	1	1	0	1	1	0	0	1	1	1	1	0	1	1	1	1	1	1	1	1	1	1	0	0	0	0	1	1	1	1	1	1	1	1	1	1	1	1	1	1	1	1	1	1	1	1
a11	1	0	1	1	1	0	1	1	0	0	0	0	0	0	0	0	1	0	1	0	1	0	1	0	0	0	0	0	0	0	0	0	1	0	1	1	0	0	0	0	1	1	1	1	1	0	1	1	1	0	1	1
a12	1	0	0	0	1	1	0	0	0	0	0	0	0	0	0	0	1	0	0	0	1	0	0	0	0	0	0	0	0	0	0	0	1	0	0	0	0	0	0	0	0	1	0	0	1	1	1	1	1	0	1	1
a13	1	0	0	0	0	1	0	0	0	0	0	0	0	0	0	0	1	0	0	0	0	0	0	0	0	0	0	0	0	0	0	0	1	0	0	0	0	0	0	0	0	1	0	0	0	1	0	0	1	1	1	1
a14	1	0	0	0	0	1	0	1	0	0	0	0	0	0	0	0	1	0	0	0	0	0	0	0	0	0	0	0	0	0	0	0	1	0	0	0	0	0	0	0	0	1	0	0	0	1	1	1	1	1	1	1
a15	1	1	1	1	1	1	1	1	0	1	1	1	1	1	1	1	1	1	1	1	1	1	1	1	1	1	1	1	1	1	0	1	1	1	1	1	1	1	1	1	1	1	1	1	1	1	1	1	1	1	1	1
a16	1	1	1	1	1	1	1	1	0	0	0	0	0	0	0	0	1	0	1	1	0	0	1	1	0	0	0	0	0	0	0	0	1	0	1	1	0	0	0	0	0	1	1	1	0	1	1	1	1	1	1	1
a17	1	1	1	1	1	1	1	1	0	1	1	1	1	1	1	1	1	0	1	1	1	0	1	1	1	1	1	1	1	0	1	1	1	1	1	1	1	0	1	1	1	1	1	1	1	1	1	1	1	1	1	1
a18	1	1	0	1	1	1	0	1	0	1	0	0	0	1	0	0	1	0	0	1	1	0	0	1	1	1	0	0	1	0	0	0	1	1	0	1	1	0	0	0	1	1	0	1	1	1	1	1	1	1	1	1
a19	1	0	0	0	0	1	0	0	0	0	0	0	0	0	0	0	1	0	0	0	0	0	0	0	0	0	0	0	0	0	0	0	1	0	0	0	0	0	0	0	0	1	0	0	0	1	0	0	1	1	1	1
a20	0	0	0	0	0	0	0	0	0	0	0	0	0	0	0	0	1	0	0	0	0	0	0	0	0	0	0	0	0	0	0	0	0	0	0	0	0	0	0	0	0	1	0	0	0	0	0	0	0	0	0	0
a21	0	0	0	0	0	1	0	0	0	0	0	0	0	0	0	0	1	0	0	0	0	0	0	0	0	0	0	0	0	0	0	0	0	0	0	0	0	0	0	0	0	1	0	0	0	1	0	0	0	0	0	0
a22	0	1	0	0	0	1	0	0	0	0	0	0	0	0	0	0	0	0	0	0	0	0	0	0	0	0	0	0	0	0	0	0	0	0	0	0	0	0	0	0	0	1	0	0	0	1	0	0	0	1	0	0
a23	1	1	1	1	1	1	1	1	0	1	1	1	1	1	1	1	1	0	1	1	1	0	1	1	1	1	1	1	1	0	0	0	1	1	1	1	1	0	1	1	1	1	1	1	1	1	1	1	1	1	1	1
a24	1	1	0	1	1	1	0	1	0	0	0	0	0	0	0	0	1	0	0	0	1	0	0	0	0	0	0	0	0	0	0	0	1	0	0	1	0	0	0	0	0	1	0	0	1	1	1	1	1	1	1	1
a25	1	1	1	1	1	1	1	1	0	0	0	0	0	0	0	0	1	0	1	1	1	0	1	1	0	1	1	1	0	0	0	0	1	1	1	1	0	0	0	0	1	1	1	1	1	1	1	1	1	1	1	1

Table 2. (*Continued*)

Obj	a13				a14				a15				a16				a17				a18				a19			
	I1	I2	I3	r	I1	I2	I3	r	I1	I2	I3	r	I1	I2	I3	r	I1	I2	I3	r	I1	I2	I3	r	I1	I2	I3	r
a1	0	1	1	1	0	1	1	1	0	0	0	0	0	0	0	0	0	0	0	0	0	0	1	0	0	1	1	1
a2	1	0	1	1	1	0	1	0	0	0	0	0	0	0	0	0	0	0	0	0	0	0	1	0	1	0	1	1
a3	1	1	1	1	1	1	1	1	1	0	0	0	1	1	1	1	1	0	0	0	1	0	1	1	1	1	1	1
a4	1	1	1	1	1	1	1	1	0	0	0	0	1	1	1	1	0	0	0	0	1	0	1	1	1	1	1	1
a5	0	1	1	1	0	1	1	1	0	0	0	0	0	1	0	0	0	1	0	0	0	1	1	0	0	1	1	1
a6	1	1	1	1	1	1	1	1	0	0	0	0	1	1	0	0	0	1	0	0	0	1	1	0	1	1	1	1
a7	1	1	1	1	1	1	1	1	0	0	0	0	1	1	1	1	0	0	0	0	0	0	1	1	1	1	1	1
a8	1	1	1	1	1	1	1	1	0	0	1	0	1	1	1	1	0	1	0	0	0	1	1	1	1	1	1	1
a9	0	1	1	1	0	1	1	1	0	0	0	0	0	1	0	0	0	0	0	0	0	0	1	0	0	1	1	1
a10	1	1	1	1	1	1	1	1	0	0	0	0	1	1	1	1	0	1	0	0	0	1	1	1	1	1	1	1
a11	1	0	1	1	1	0	1	1	0	0	0	0	1	0	0	0	0	0	0	0	0	0	1	0	1	0	1	1
a12	1	0	1	1	1	0	0	0	0	0	0	0	1	0	0	0	0	0	0	0	0	0	0	0	1	0	1	1
a13	1	1	1	1	0	0	0	0	0	0	0	0	0	0	0	0	0	0	0	0	0	0	0	0	0	0	0	0
a14	1	1	1	1	1	1	1	1	0	0	0	0	0	0	0	0	0	0	0	0	0	0	1	0	1	1	1	1
a15	1	1	1	1	1	1	1	1	1	1	1	1	1	1	1	1	0	1	0	0	1	1	1	1	1	1	1	1
a16	1	1	1	1	1	1	1	1	0	0	0	0	1	1	1	1	0	0	0	0	0	0	1	1	1	1	1	1
a17	1	1	1	1	1	1	1	1	1	0	1	1	1	1	1	1	1	1	1	1	1	0	1	1	1	1	1	1
a18	1	1	1	1	1	1	0	1	0	0	0	0	1	1	0	0	0	1	0	0	1	1	1	1	1	1	1	1
a19	1	1	1	1	0	0	0	0	0	0	0	0	0	0	0	0	0	0	0	0	0	0	0	0	1	1	1	1
a20	0	0	0	0	0	0	0	0	0	0	0	0	0	0	0	0	0	0	0	0	0	0	0	0	0	0	0	0
a21	0	0	0	0	0	0	0	0	0	0	0	0	0	0	0	0	0	0	0	0	0	0	0	0	0	0	0	0
a22	0	1	0	0	0	1	0	0	0	0	0	0	0	1	0	0	0	0	0	0	0	0	0	0	0	1	0	0
a23	1	1	1	1	1	1	1	1	1	0	0	0	1	1	1	1	1	1	0	0	1	1	1	1	1	1	1	1
a24	1	1	1	1	1	1	1	1	0	0	0	0	1	0	0	0	0	0	0	0	0	0	1	0	1	1	1	1
a25	1	1	1	1	1	1	1	1	0	0	0	0	1	1	1	1	0	0	0	0	0	0	1	1	1	1	1	1

Obj	a20				a21				a22				a23				a24				a25			
	I1	I2	I3	r	I1	I2	I3	r	I1	I2	I3	r	I1	I2	I3	r	I1	I2	I3	r	I1	I2	I3	r
a1	1	1	1	1	1	1	1	1	1	0	1	1	0	0	0	0	0	0	1	0	0	0	0	0
a2	1	1	1	1	1	0	1	1	1	0	1	1	0	0	0	0	0	0	1	0	0	0	0	0
a3	1	1	1	1	1	1	1	1	1	1	1	1	1	0	0	0	1	1	1	1	1	1	1	1
a4	1	1	1	1	1	1	1	1	1	1	1	1	0	0	0	0	1	1	1	1	1	1	1	1
a5	0	1	1	1	0	1	1	1	1	1	1	1	0	1	0	0	0	1	1	1	0	1	0	0
a6	1	1	1	1	1	1	1	1	1	1	1	1	0	1	0	0	0	1	1	1	0	1	0	0
a7	1	1	1	1	1	1	1	1	1	1	1	1	0	0	0	0	1	1	1	1	1	0	0	0
a8	1	1	1	1	1	1	1	1	1	1	1	1	0	1	1	1	1	1	1	1	1	1	1	1
a9	1	1	1	1	1	1	1	1	1	1	1	1	0	0	0	0	0	1	1	0	0	0	0	0
a10	1	1	1	1	1	1	1	1	1	1	1	1	0	1	0	0	1	1	1	1	1	1	1	1
a11	1	0	1	1	1	0	1	1	1	0	1	1	0	0	0	0	1	0	1	1	0	0	0	0
a12	1	1	1	1	1	0	1	1	1	0	1	1	0	0	0	0	0	0	0	0	0	0	0	0
a13	1	1	1	1	1	1	1	1	1	0	1	1	0	0	0	0	0	0	0	0	0	0	0	0
a14	1	1	1	1	1	1	1	1	1	0	1	1	0	0	0	0	0	0	0	0	0	0	0	0
a15	1	1	1	1	1	1	1	1	1	1	1	1	0	1	1	1	1	1	1	1	1	1	1	1
a16	1	1	1	1	1	1	1	1	1	0	1	1	0	0	0	0	0	1	1	1	0	0	0	0
a17	1	1	1	1	1	1	1	1	1	1	1	1	0	0	1	1	1	1	1	1	1	1	1	1
a18	1	1	1	1	1	1	1	1	1	1	1	1	0	0	0	0	1	1	0	1	1	1	0	0
a19	1	1	1	1	1	1	1	1	1	0	1	1	0	0	0	0	0	0	0	0	0	0	0	0
a20	1	1	1	1	0	0	1	0	1	0	1	1	0	0	0	0	0	0	0	0	0	0	0	0
a21	1	1	0	1	1	1	1	1	1	0	1	1	0	0	0	0	0	0	0	0	0	0	0	0
a22	0	1	0	0	0	1	0	0	1	1	1	1	0	0	0	0	0	1	0	0	0	0	0	0
a23	1	1	1	1	1	1	1	1	1	1	1	1	1	1	1	1	1	1	1	1	1	1	1	1
a24	1	1	1	1	1	1	1	1	1	0	1	1	0	0	0	0	1	1	1	1	0	0	0	0
a25	1	1	1	1	1	1	1	1	1	1	1	1	0	0	0	0	1	1	1	1	1	1	1	1

For example, entries in the three sub-columns corresponding to pair (a_{10}, a_3) are 1, 2, 2. This is because the three entries in corresponding sub-columns in Table 2 are 0 1, 1. Here I_1 bit agrees with itself only and thus its agreement count is 1. I_2-bit agrees with I_3-bit. Thus agreement counts for I_2 and I_3 are 2 each. The matrix in Table 3 is symmetric and so we only could have done with either the upper or the lower triangular matrix. In fact, it may be mentioned here that in a computer program much of the computation can be carried out without maintaining matrices which are used here for clarity and conceptual purposes. Here A_1, A_2, and A_3 are, respectively, 364, 344, and 412 adding to 1120 so that intrinsic weights which are proportional to A_1, A_2, and A_3 are 0.325000, 0.307143, and 0.367857.†

Remark: Multivariate zeta matrix (also the traditional zeta matrix) can be used to compute UpSets and DownSets that are needed to compute ranking using CWNSI since for a given object in the column corresponding to the object the number of groups of sub-cells (except the one diagonally located) with all 1's is the size of its UpSet and in the row corresponding to the object the number of groups of sub-cells (except the ones diagonally located) with all 1's is the size of its DownSet.

To compute the performance measure of arbitrary ranking of objects we need to see how closely individual indicators agree with the ranking of the objects. For a given pair (a_i, a_j) if both the indicator and the ranking rank a_i higher than a_j then they agree or if both the indicator and the ranking rank $a_i$lower than a_j then they agree. Otherwise the indicator and the ranking disagree. We keep the count of agreements in a matrix shown in Table 4. For the specific example here,we use the ranking defined by the weighted composite index based on the intrinsic weights (0.325000, 0.307143, 0.367857) and assign higher ranks (closer to 1) to objects with larger index score.

These ranks are shown in Table 4.

We use the fourth sub-column with the heading r in the group of four sub-columns in the matrix shown in Table 2 to show how the ranking rates object a_i compared a_j for all combinations of pairs (a_i, a_j). If the ranking assigns a_i a rank higher than it assigns a_j then the entry is 1 else it is 0. The mutual agreement counts between ranking method and individual indicators are in Table 5.

† Calculations for this example were carried out using MicroSoft™ Excel 2013© and were verified with a C++ program in MicroSoft Visual Studio 2010©. Calculated weights are displayed here to six decimal places after rounding off. However, computation of all index scores that determined various rankings used weights computed with full default precision of Excel and double in Visual C++. Values in the data matrix were generated to three decimal places.

Table 3. Consensus table.

Obj	a1			a2			a3			a4			a5			a6			a7			a8			a9			a10			a11			a12			a13		
	I1	I2	I3	I1	I2	I3	I1	I2	I3	I1	I2	I3	I1	I2	I3	I1	I2	I3	I1	I2	I3	I1	I2	I3	I1	I2	I3	I1	I2	I3	I1	I2	I3	I1	I2	I3	I1	I2	I3
a1				1	2	2							1	2	2	2	2	1													2	1	2	1	2	2	1	2	2
a2	1	2	2										1	2	2										1	2	2				2	1	2	2	2	1	2	1	2
a3													2	1	2	2	1	2				1	2	2				1	2	2									
a4													2	1	2	2	1	2				1	2	2				1	2	2									
a5	1	2	2	1	2	2	2	1	2	2	1	2				1	2	2	2	1	2				1	2	2	2	1	2	2	1	2	1	2	2	1	2	2
a6	2	2	1				2	1	2	2	1	2	1	2	2				2	1	2				2	2	1				2	1	2	1	2	2			
a7													2	1	2	2	1	2																					
a8							1	2	2	1	2	2																											
a9				1	2	2							1	2	2	2	2	1													2	1	2	1	2	2	1	2	2
a10							1	2	2	1	2	2	2	1	2																								
a11	2	1	2	2	1	2							2	1	2	2	1	2							2	1	2							2	1	2	2	1	2
a12	1	2	2	2	2	1							1	2	2	1	2	2							1	2	2				2	1	2				2	1	2
a13	1	2	2	2	1	2							1	2	2										1	2	2				2	1	2	2	1	2			
a14	1	2	2	2	1	2							1	2	2										1	2	2				2	1	2	1	2	2			
a15							1	2	2													2	2	1															
a16													2	1	2	2	2	1							2	1	2				1	2	2	1	2	2			
a17							1	2	2				2	1	2	2	1	2				2	1	2				2	1	2									
a18	2	2	1	2	2	1	2	1	2	2	1	2	1	2	2	1	2	2	2	2	1	1	2	2	2	2	1	1	2	2	2	2	1						
a19	1	2	2	2	1	2							1	2	2										1	2	2				2	1	2	2	1	2			
a20													1	2	2																2	1	2						
a21				2	1	2							1	2	2																2	1	2	2	1	2			
a22	2	1	2	2	1	2																									2	1	2	2	1	2	2	1	2
a23							1	2	2				2	1	2	2	1	2				1	2	2				2	1	2									
a24	2	2	1	2	2	1							1	2	2	1	2	2							1	2	2				2	1	2						
a25													2	1	2	2	1	2	1	2	2																		
Totals	16			21			11			8			30			24			7			8			15			9			29			18			11		
		20			18			13			7			33			21			6			11			20			9			17			19			10	
			19			21			16			10			42			25			7			11			20			12			29			23			14

Table 3. (*Continued*)

Obj	a14			a15			a16			a17			a18			a19		
	I1	I2	I3	I1	I2	I3	I1	I2	I3	I1	I2	I3	I1	I2	I3	I1	I2	I3
a1	1	2	2										2	2	1	1	2	2
a2	2	1	2										2	2	1	2	1	2
a3				1	2	2				1	2	2	2	1	2			
a4										0	0	0	2	1	2			
a5	1	2	2				2	1	2	2	1	2	1	2	2	1	2	2
a6							2	2	1	2	1	2	1	2	2			
a7													2	2	1			
a8				2	2	1				2	1	2	1	2	2			
a9	1	2	2				2	1	2	0	0	0	2	2	1	1	2	2
a10										2	1	2	1	2	2			
a11	2	1	2				1	2	2				2	2	1	2	1	2
a12	1	2	2				1	2	2							2	1	2
a13																		
a14													2	2	1			
a15										2	1	2						
a16													2	2	1			
a17				2	1	2							2	1	2			
a18	2	2	1				2	2	1	2	1	2						
a19																		
a20																		
a21																		
a22	2	1	2				2	1	2							2	1	2
a23				1	2	2				2	2	1						
a24							1	2	2				2	2	1			
a25													2	2	1			
Totals	12			6			13			15			28			11		
		13			7			13			10			29			10	
			15			7			14			15			23			14

Obj	a20			a21			a22			a23			a24			a25			Totals		
	I1	I2	I3	I1	I2	I3	I1	I2	I3	I1	I2	I3	I1	I2	I3	I1	I2	I3			
a1							2	1	2				2	2	1						
a2				2	1	2	2	1	2				2	2	1						
a3										1	2	2									
a4																					
a5	1	2	2	1	2	2				2	1	2	1	2	2	2	1	2			
a6										2	1	2	1	2	2	2	1	2			
a7																1	2	2			
a8										1	2	2									
a9													1	2	2						
a10										2	1	2									
a11	2	1	2	2	1	2	2	1	2				2	1	2						
a12				2	1	2	2	1	2												
a13							2	1	2												
a14							2	1	2												
a15										1	2	2									
a16							2	1	2				1	2	2						
a17										2	2	1									
a18													2	2	1	2	2	1			
a19							2	1	2												
a20				2	2	1	2	1	2												
a21	2	2	1				2	1	2												
a22	2	1	2	2	1	2							2	1	2						
a23																					
a24							2	1	2												
a25																					
Totals	7			11			22			11			14			7			364		
		6			8			11			11			16			6			344	
			7			11			22			13			15			7			412

Table 4. Ranks induced by intrinsic weights.

Object	I_1	I_2	I_3	r
a_1	4.273	5.139	4.766	17
a_2	4.630	4.601	4.645	19
a_3	8.226	5.983	7.500	5
a_4	7.044	5.885	7.247	7
a_5	3.586	6.848	6.165	12
a_6	5.245	6.643	4.699	13
a_7	5.928	5.524	6.926	9
a_8	6.237	6.966	8.543	3
a_9	4.275	5.391	5.161	16
a_{10}	6.109	6.775	7.527	6
a_{11}	5.756	4.151	6.216	14
a_{12}	5.410	4.639	3.901	20
a_{13}	4.285	4.879	3.296	22
a_{14}	4.602	5.137	4.369	18
a_{15}	7.498	7.789	7.654	2
a_{16}	5.065	5.243	6.806	10
a_{17}	7.676	6.010	9.181	1
a_{18}	6.503	6.261	4.105	11
a_{19}	4.430	5.090	3.657	21
a_{20}	3.783	4.204	3.280	24
a_{21}	3.840	4.849	3.171	23
a_{22}	3.335	5.249	2.570	25
a_{23}	7.801	6.488	7.544	4
a_{24}	5.580	5.140	4.371	15
a_{25}	5.823	5.623	7.018	8

The vector W^R of the ranking method obtained from the totals in the rightmost column is (0.308057, 0.251185, 0.440758) and the corresponding PMR is corr((0.325000, 0.307143, 0.367857), (0.308057, 0.251185, 0.440758)) = 0.9999793.

It is of interest to use W^R itself to construct a new composite index and measure PMR of the new ranking obtained this way. We can continue this process iteratively until the two successive rankings are identical. Without reproducing our calculations, we find that this iterative process soon terminates with stable ranking. Table 6 shows the iterative rankings.

It is also of interest to see progression of the iterative sequence of W^R graphically. Patil and Joshi investigated (see [18]) equivalence classes of weight vectors with respect to ranking using composite index. For data matrices with three indicators, these equivalence classes can be shown as a partition of the triangular plane $w_1 + w_2 + w_3 = 1$ in the three dimensional Euclidean space. For the present data matrix, the sequence of the converging weight vectors are labeled by letters P, Q, R, S on the 'weight triangle' S, P being the initial and S being the final vectors (Figure 4). Points R and S are very close to each other.

Table 5. Agreements between ranking and indicators.

Obj	a1			a2			a3			a4			a5			a6			a7			a8			a9			a10			a11			a12			a13		
	I1	I2	I3	I1	I2	I3	I1	I2	I3	I1	I2	I3	I1	I2	I3	I1	I2	I3	I1	I2	I3	I1	I2	I3	I1	I2	I3	I1	I2	I3	I1	I2	I3	I1	I2	I3	I1	I2	I3
a1				0	1	1							0	1	1	1	1	0													1	0	1	0	1	1	0	1	1
a2	0	1	1										0	1	1										0	1	1				1	0	1	0	0	1	1	0	1
a3													1	0	1	1	0	1				0	1	1				1	0	0									
a4													1	0	1	1	0	1				0	1	1				0	1	1									
a5	0	1	1	0	1	1	1	0	1	1	0	1				0	1	1	1	0	1				0	1	1	1	0	1	0	1	0	0	1	1	0	1	1
a6	1	1	0				1	0	1	1	0	1	0	1	1				1	0	1				1	1	0				0	1	0	0	1	1			
a7													1	0	1	1	0	1																					
a8							0	1	1	0	1	1																											
a9				0	1	1							0	1	1	1	1	0													1	0	1	0	1	1	0	1	1
a10							1	0	0	0	1	1	1	0	1																								
a11	1	0	1	1	0	1							0	1	0	0	1	0							1	0	1							1	0	1	1	0	1
a12	0	1	1	0	0	1							0	1	1	0	1	1							0	1	1				1	0	1				1	0	1
a13	0	1	1	1	0	1							0	1	1										0	1	1				1	0	1	1	0	1			
a14	0	1	1	0	1	0							0	1	1										0	1	1				1	0	1	0	1	1			
a15							0	1	1													1	1	0															
a16													1	0	1	0	0	1							1	0	1				0	1	1	0	1	1			
a17							0	1	1				1	0	1	1	0	1				1	0	1				1	0	1									
a18	1	1	0	1	1	0	1	0	1	1	0	1	1	0	0	1	0	0	0	0	1	0	1	1	1	1	0	0	1	1	1	1	0						
a19	0	1	1	1	0	1							0	1	1										0	1	1				1	0	1	1	0	1			
a20													0	1	1																1	0	1						
a21				1	0	1							0	1	1																1	0	1	1	0	1			
a22	1	0	1	1	0	1																									1	0	1	1	0	1	1	0	1
a23							0	1	1				1	0	1	1	0	1				0	1	1				1	0	1									
a24	1	1	0	1	1	0							0	1	1	0	1	1							1	0	0				1	0	1						
a25													1	0	1	1	0	1	0	1	1																		
Totals	5			7			4			3			9			9			2			2			5			4			12			5			4		
		9			6			4			2			12			6			1			5			8			2			4			6			3	
			8			9			7			5			19			10			4			5			8			5			12			12			7

Table 5. (*Continued*)

Obj	a14			a15			a16			a17			a18			a19			a20			a21			a22			a23			a24			a25			Totals
	I1	I2	I3	I1	I2	I3	I1	I2	I3	I1	I2	I3	I1	I2	I3	I1	I2	I3	I1	I2	I3	I1	I2	I3	I1	I2	I3	I1	I2	I3	I1	I2	I3	I1	I2	I3	
a1	0	1	1										1	1	0	0	1	1							1	0	1				1	1	0				
a2	0	1	0										1	1	0	1	0	1				1	0	1	1	0	1				1	1	0				
a3				0	1	1				0	1	1	1	0	1													0	1	1							
a4													1	0	1																						
a5	0	1	1				1	0	1	1	0	1	1	0	0	0	1	1	0	1	1	0	1	1				1	0	1	0	1	1	1	0	1	
a6							0	0	1	1	0	1	1	0	0													1	0	1	0	1	1	1	0	1	
a7													0	0	1																			0	1	1	
a8				1	1	0				1	0	1	0	1	1													0	1	1							
a9	0	1	1				1	0	1				1	1	0	0	1	1													1	0	0				
a10										1	0	1	0	1	1													1	0	1							
a11	1	0	1				0	1	1				1	1	0	1	0	1	1	0	1	1	0	1	1	0	1				1	0	1				
a12	0	1	1				0	1	1							1	0	1				1	0	1	1	0	1										
a13																									1	0	1										
a14													1	1	0										1	0	1										
a15										1	0	1																0	1	1							
a16													0	0	1										1	0	1				0	1	1				
a17				1	0	1							1	0	1													0	0	1							
a18	1	1	0				0	0	1	1	0	1																			1	1	0	0	0	1	
a19																									1	0	1										
a20																						1	1	0	1	0	1										
a21																			1	1	0				1	0	1										
a22	1	0	1				1	0	1							1	0	1	1	0	1	1	0	1							1	0	1				
a23				0	1	1				0	0	1																									
a24							0	1	1				1	1	0										1	0	1										
a25													0	0	1																						
Totals	3			2			3			6			11			4			3			5			11			3			6			2			130
		6			3			3			1			8			3			2			2			0			3			6			1		106
			6			3			8			8			8			7			3			5			11			7			5			4	186

Table 6. Iterative ranking starting with intrinsic weights.

Iterated Ranks with Intrinsic Weights							
	Indicator Values			Ranks for Iteration Number			
Object	I_1	I_2	I_3	**0**	**1**	**2**	**3**
a1	4.273	5.139	4.766	17	17	17	17
a2	4.630	4.601	4.645	19	19	18	18
a3	8.226	5.983	7.500	5	5	5	5
a4	7.044	5.885	7.247	7	7	7	7
a5	3.586	6.848	6.165	12	12	12	12
a6	5.245	6.643	4.699	13	14	14	14
a7	5.928	5.524	6.926	9	9	9	9
a8	6.237	6.966	8.543	3	3	3	3
a9	4.275	5.391	5.161	16	15	15	15
a10	6.109	6.775	7.527	6	6	6	6
a11	5.756	4.151	6.216	14	11	11	11
a12	5.410	4.639	3.901	20	20	20	20
a13	4.285	4.879	3.296	22	22	22	22
a14	4.602	5.137	4.369	18	18	19	19
a15	7.498	7.789	7.654	2	2	2	2
a16	5.065	5.243	6.806	10	10	10	10
a17	7.676	6.010	9.181	1	1	1	1
a18	6.503	6.261	4.105	11	13	13	13
a19	4.430	5.090	3.657	21	21	21	21
a20	3.783	4.204	3.280	24	24	24	24
a21	3.840	4.849	3.171	23	23	23	23
a22	3.335	5.249	2.570	25	25	25	25
a23	7.801	6.488	7.544	4	4	4	4
a24	5.580	5.140	4.371	15	16	16	16
a25	5.823	5.623	7.018	8	8	8	8

Table 7. Iterative intrinsic weights.

Iteration	Rank Agreement Counts			Indicator Weights		
Number	I_1	I_2	I_3	I_1	I_2	I_3
0	364	344	412	0.325000	0.307143	0.367857
1	130	106	186	0.308057	0.251185	0.440758
2	128	104	196	0.299065	0.242991	0.457944
3	130	102	198	0.302326	0.237209	0.460465

The 112 intersecting lines divide the triangle into regions of equivalent weight vectors in the sense that composite indexes based on all weight vectors within the same region produce identical ranking for the data matrix. PMR for final ranking is given in Table 9 under Method 1.

Finally, we present rankings (Table 8) and PMR's (Table 9) for all the four methods discussed above for the 25 by 3 data matrix.

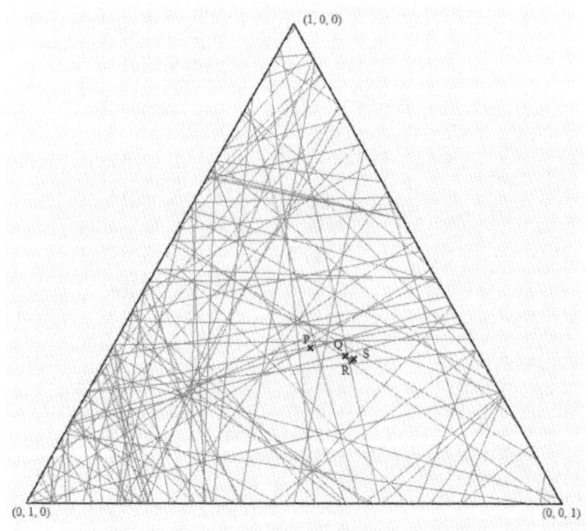

Fig. 4. Regions of equivalent weights and convergence of iterative intrinsic weights.

Table 8. Rankings by four methods.

	Ranks by various methods			
Object	**Method 1**	**Method 2**	**Method 3**	**Method 4**
$\mathbf{a}_1$	17	17	20	17
$\mathbf{a}_2$	18	18.5	18	18
$\mathbf{a}_3$	5	4.5	4	5
$\mathbf{a}_4$	7	7	7	7
$\mathbf{a}_5$	12	12	12	10
$\mathbf{a}_6$	14	10	11	11
$\mathbf{a}_7$	9	8.5	8	9
$\mathbf{a}_8$	3	2	2	3
$\mathbf{a}_9$	15	15	15	14
$\mathbf{a}_{10}$	6	6	6	6
$\mathbf{a}_{11}$	11	16	16	15
$\mathbf{a}_{12}$	20	18.5	19	20
$\mathbf{a}_{13}$	22	23	23	22
$\mathbf{a}_{14}$	19	20	17	19
$\mathbf{a}_{15}$	2	1	1	4
$\mathbf{a}_{16}$	10	13	13	13
$\mathbf{a}_{17}$	1	4.5	5	1
$\mathbf{a}_{18}$	13	11	10	12
$\mathbf{a}_{19}$	21	22	21	21
$\mathbf{a}_{20}$	24	25	25	24
$\mathbf{a}_{21}$	23	24	24	25
$\mathbf{a}_{22}$	25	21	22	23
$\mathbf{a}_{23}$	4	3	3	2
$\mathbf{a}_{24}$	16	14	14	16
$\mathbf{a}_{25}$	8	8.5	9	8

Table 9. Performance measure for rankings by four methods.

		Weights For Methods			
Indicator	**Intrinsic**	**Method 1**	**Method 2**	**Method 3**	**Method 4**
I_1	0.325000	0.302326	0.305623	0.330000	0.291866
I_2	0.307143	0.237209	0.305623	0.290000	0.272727
I_3	0.367857	0.460465	0.388753	0.380000	0.435407
Corr. Coeff.	PMR	0.99999642	0.95818697	0.98569662	0.98341741

8. Looking Forward

The illustrative example of this paper shows potential for investigating a variety of data matrices to examine patterns of performance behavior of the four ranking methods. It will be worthwhile also to investigate situations, where the features are variables, and not just indicators of common orientation. These situations are typical in applications involving variously big data. Also in multivariate nonparametric statistics involving multivariate ranking, multivariate median, image reconstruction, etc.

References

1. R. Bruggémannand, G. P. Patil, Ranking and Prioritization for Multi-indicator Systems, in *Introduction to Partial Order Applications* (Springer, New York, 2011).
2. R. Bruggémann, L. Carlson and J. Wittman (eds.), *Multi-indicator Systems and Modeling in Partial Order* (Springer, New York, 2014).
3. A. Y. Chiang, G. Li, Y. Ding and M. D. Wang, A multivariate ranking procedure to assess treatment effects, *Technical Report* (Eli Lilly and Company, Indianapolis, IN, 2008).
4. P. Diaconis and R. L. Graham, Spearman's footrule as a measure of disarray, *JRSS B* **39**, 262 (1977).
5. D. L. Donoho and M. Gasko, Breakdown properties of location estimates based on halfspace depth and projected outlyingness, *Ann. Statist.* **20**, 1803 (1992).
6. R. E. Hardie and G. R. Arce, Ranking in R^p and its use in multivariate image estimation, *SPIE Nonlinear Image Processing* **1247**, 13 (1990).
7. R. E. Hardie and G. R. Arce, Ranking in R^p and its use in multivariate image estimation, *IEEE Trans. On Circuits and Systems for Video Technology*, **1**, No. 2, June 197 (1991).
8. R. Y. Liu, R. Serfling and D. L. Souvaine (eds.), *Data Depth: Robust Multivariate Analysis, Computational Geometry and Applications* (Dimacs Series in Discrete Mathematics and Theoretical Computer Science 2006).

9. J. Mottonen, T. P. Hettmansperger, H. Oja and J. Tienari, On the efficiency affine invariant multivariate rank tests, *J. Multivariate Anal.* **66**, 118 (1998).
10. W. L. Myers and G. P. Patil, *Multivariate Methods of Representing Relations in R for Prioritization Purposes: Selective Scaling, Comparative Clustering, Collective Criteria and Sequenced Sets* (Springer, New York, 2012a). p 297.
11. W. L. Myers and G. P. Patil, *Statistical Geoinformatics for Human Environment Interface* (CRC/Chapman & Hall, New York 2012b) p 305.
12. G. P. Patil, *Inaugural Keynote Address, UNEP Panel Workshop on Sustainability Indicators* (New Delhi, India, 2011).
13. G. P. Patil, *Plenary Lecture, UNEP Panel Workshop on Green Economy Indicators* (Beijing, China, 2012).
14. G. P. Patil, *Invited Lecture on comparative knowledge discovery with partial order and composite indicators in multi-indicator information fusion systems* (DIMACS Workshop on Algorithmic Aspects of Information Fusion at Rutgers University, 2012a).
15. G. P. Patil, *Keynote Lecture on Partial Orders and Composite Indicators for Multivariate Ranking in Multivariate Nonparametric Statistics* (International Workshop on Partial Order Theory and Modeling, Berlin, Germany, 2012b).
16. G. P. Patil, *Invited Keynote Inaugural Lecture on multivariate ranking in multi-indicator systems: recent past, present, and near future.* (Annual Florida Chapter Meeting of ASA, University of South Florida, Tampa, Florida, 2015).
17. G. P. Patil, W. L. Myers and R. Bruggémann, Multivariate datasets for inference order: some considerations and explorations, in *Multi-Indicator Systems and Modeling In Partial Order*, R. Bruggémann, L. Carsen and J. Wittmann, eds., (Springer, New York, 2014).
18. G. P. Patil and S. W. Joshi, Comparative knowledge discovery with partial order and composite indicators, in *Multi-indicator Systemic Ranking, Advocacy, and Reconciliation In Multi-indicator Systems and Modeling in Partial Order*, R. Bruggémann, L. Carsen and J. Wittmann, eds., (Springer, New York, 2014).
19. G. P. Patil and C. Taillie, Multiple indicators, partially ordered sets, and linear extensions: Multi-criterion ranking and prioritization, *Environmental and Ecological Statistics* **11**, 199 (2004).

20. D. Phillips and D. Ghosh, *A two-dimensional approach to large-scale simultaneous hypothesis testing, using Voronoi tessellations* (Manuscript, 2012).
21. R. Serfling, *Depth functions in nonparametric multivariate inference*, (DIMACS Series in Discrete Mathematics and Theoretical Computer Science, 2006).
22. K. Tang, J. Astola and Y. Neuvo, Multivariate order statistic filters in color image processing, in *Proc. IEEE Internat. Conf. Syst. Eng.*, (Kobe, Japan, 1992).
23. P. Willett, *Invited Lecture on fusing database rankings in similarity-based virtual screening*, (DIMACS Workshop on Algorithmic Aspects of Information Fusion at Rutgers University, 2012).
24. K. M. Wittkowski, E. Lee *et al.*, Combining several ordinal measures in clinical studies, *Statistics in Medicine* **23**(10), 1579 (2004).
25. K. M. Wittkowski, V. Sonakya *et al.*, From single-SNP to wide-locus: genome-wide association studies identifying functionally related genes and intragenic regions in small sample studies, *Pharmacogenomics* **14**(4), 391 (2013).
26. K. M. Wittkowski and T. Song, Nonparametric methods in molecular biology in *Statistical Methods in Molecular Biology*, H. Bang, X. K. Zhou, H. L. Van Epps and M. Mazumdar, eds., (Springer,New York, 2010).
27. K. M. Wittkowski, T. Song *et al.*, U-Scores for Multivariate Data in Sports, *Journal of Quantitative Analysis in Sports* **4**(3), 7 (2008).
28. Y. Zuo, Projection-based depth functions and associated medians, *Ann Statist,* **31,** 1460 (2003).
29. Y. Zuo and R. Serfling, General notions of statistical depth function, *Ann Statist*, **28,** 461 (2000).

Statistical Aspects of SuDoKu-Based Experimental Designs

Jyotirmoy Sarkar
Department of Mathematical Sciences
Indiana University–Purdue University Indianapolis, USA

Bikas K. Sinha
Ex-Faculty, Applied Statistics Division
Indian Statistical Institute, Kolkata, India

SuDoKu is a popular combinatorial puzzle. In recent years, there has been a growing interest in applications of SuDoKu-based experimental designs. We review the applications and bring out a few finer points of data analyses, but we skip the combinatorial aspect. SuDoKu was first interpreted as a statistical experimental design by Subramani and Ponnuswamy (2009), and a correct ANOVA-based data analysis was provided by Saba and Sinha (2014), who accounted for non-orthogonality between some components of variation. Variants of SuDoKu that achieve additional orthogonality, and hence change the statistical analyses, are found in Sarkar and Sinha (2014). We extend these finer points of data analyses also to designs based on mutually orthogonal SuDoKu Latin squares.

Keywords: Confounding; Internal Blocks; Latin Squares; Mutually Orthogonal Latin Squares; Mutually Orthogonal SuDoKu Latin Squares; SuDoKu Squares.

2010 Mathematics Subject Classifications: primary 62K10; secondary 05B05.

1. The SuDoKu Square as an Experimental Design—Going One Step Beyond Latin Square Design

The (standard) *SuDoKu square* of order 9 is built on a *Latin square* of order 9 using integers 1 through 9. Aside from being a Latin square, a SuDoKu square also exhibits 9 'Internal Blocks,' each of order 3 x 3 built in it with the condition that all the integers must appear in each internal block. See Table 1 below for a SuDoKu of order 9. In this paper, we focus on the statistical aspects of SuDoKu squares in the context of their use as experimental designs. We do not dwell on combinatorial aspects of SuDoKu designs. Towards that, we refer the reader to Felgenhauer and Jarvis (2005), Russell and Jarvis (2006), Kuhl and Denley (2007), Rosenhouse and Taalman (2011) and McGuire, Tugemann and Civario (2012).

We take for granted that the readers are familiar with the scope of Latin square designs and their statistical analyses. See Rao, Mitra and Matthai (1975), for example. Recall that in the context of a Latin square design of order n, there is a factor representing row-to-row variation and another factor representing column-to-column variation. Additionally, there are n treatments whose effects are to be compared. We also know that there is an orthogonal splitting of the total variation (with $n^2 - 1$ df) into: (i) Row components with $(n-1)$ df; (ii) Column components with $(n-1)$ df; (iii) Treatment components with $(n-1)$ df and (iv) Error component with $(n-1)(n-2)$ df.

Recall that orthogonality between components of variation is a desirable feature of a design because otherwise their effects get confounded with each other. Suppose now that we want to accommodate any extra component of variation orthogonal to the three components outlined in (i) to (iii) above. Then we need what is called a *Graeco Latin square* design or an orthogonal Latin square (OLS) design. However, the existence of a Graeco Latin square design cannot be ensured for all n. For example, we know that it does not exist for $n = 6$. Furthermore, in an OLS design all pairwise combinations of the 4 components must appear exactly once. This may be prohibitive in practice, as the following scenario illustrates.

In an industrial experiment, there are 9 machines whose effects are to be compared. There are 9 suppliers of raw materials and 9 days of operation. This is an excellent and straightforward setting for the application of a Latin square design of order 9. However, if this experiment also calls for 9 operators as well, how should one proceed? Fortunately, there exists a complete set of 8 mutually orthogonal Latin squares (MOLS) of order 9. See Rao, Mitra and Matthai (1975), for example. So we can use any one pair of MOLS of order 9, and thereby accommodate the 9 operators orthogonally to each of suppliers, days and machines.

Now imagine a situation wherein the operators are not available on all 9 days of operation. Can we still have a design with (i) limited availability of the operators, and (ii) operator effect being orthogonal to at least treatment (machine) effect—retaining mutual orthogonality among the rows (suppliers), the columns (days) and the treatments (machines), as in a

Latin square design? In other words, we are asking for a combinatorial arrangement embedded within a Latin square of order 9 with this additional feature of operator x treatment orthogonality, even though operator x day is surely non-orthogonal (due to restriction on operator availability) and operator x supplier may also be non-orthogonal.

With a certain type of 'operator availability' a SuDoKu design precisely provides the required solution. Suppose that the first set of three operators are available on Days 1–3, the next set of three operators are available on Days 4–6, and the remaining three are available on Days 7–9. The reader can check that this is precisely what is achieved by the SuDoKu design of order 9, when the 9 'Internal Blocks' of the design serve as the working conditions (supplier-day-machine combinations) for the 9 operators. Table 1 illustrates this design.

Table 1. A SuDoKu of order 9 with operators available in sets of three days.

$S\backslash D$	$D1$	$D2$	$D3$	$D4$	$D5$	$D6$	$D7$	$D8$	$D9$
$S1$	$M1$	$M4$	$M7$	$M6$	$M2$	$M5$	$M9$	$M8$	$M3$
$S2$	$M9$	$M5$	$M8$	$M1$	$M7$	$M3$	$M4$	$M6$	$M2$
$S3$	$M2$	$M6$	$M3$	$M8$	$M9$	$M4$	$M5$	$M7$	$M1$
		O 1			O 4			O 7	
$S4$	$M6$	$M9$	$M5$	$M2$	$M4$	$M8$	$M3$	$M1$	$M7$
$S5$	$M7$	$M1$	$M4$	$M9$	$M3$	$M6$	$M2$	$M5$	$M8$
$S6$	$M3$	$M8$	$M2$	$M5$	$M1$	$M7$	$M6$	$M4$	$M9$
		O 2			O 5			O 8	
$S7$	$M5$	$M3$	$M1$	$M7$	$M6$	$M9$	$M8$	$M2$	$M4$
$S8$	$M4$	$M2$	$M6$	$M3$	$M8$	$M1$	$M7$	$M9$	$M5$
$S9$	$M8$	$M7$	$M9$	$M4$	$M5$	$M2$	$M1$	$M3$	$M6$
		O 3			O 6			O 9	

The reader can see that in the SuDoKu design of Table 1, with one operator assigned to each internal block, orthogonality is lost between (a) operators and rows (suppliers) and (b) operators and columns (days). That's why the usual ANOVA for a Graeco Latin square design fails to hold here. The correct analysis is provided in Saba and Sinha (2014). In particular, the total variation (with 80 df) is decomposed into: (i) Treatment (machine) classification with 8 df, (ii) Internal Block (operator) classification with 4 df (via tetra differences), (iii) Row (supplier) classification with 6 df, (iv) Column (day) classification with 6 df, (v) Row x Internal Block confounding with 2 df, (vi) Column x Internal Block confounding with 2 df, and (vii) Error with 52 df.

Remark 1. It is imperative from the operator availability constraint that a

Graeco Latin square design is not applicable in the above framework of the industrial experiment. One may wonder if there exists any SuDoKu design with the additional property of orthogonality between Internal Block (operators) and Rows (suppliers), which retains the orthogonality between Internal Block (operator) and the Treatment (machine) classifications. That such designs exist has been documented in details in Sarkar and Sinha (2014, 2015). These variants of the SuDoKu Design provide extra flexibility to the experimenters to accommodate restricted availability of operators. But they also require a different ANOVA. Specifically, if teams of three operators work on blocks of three days, then the total variation (with 80 df) is now decomposed into: (i) Treatment (machine) classification with 8 df, (ii) Internal Block (operator) classification with 6 df, (iii) Row (supplier) classification with 8 df, (iv) Column (day) classification with 6 df, (v) Column x Internal Block confounding with 2 df, and (vi) Error with 50 df. For analyzing SuDoKu designs with other patterns of 'operator availability' constraints, we refer the reader to Sarkar and Sinha (2014).

2. Mutually Orthogonal SuDoKu Latin Squares

We assume the readers are familiar with the theory of existence and construction of MOLS as laid down by Bose, Shrikhande and Parker (1960). See Raghavarao (1971, 1988) for more details. Of late, the problem of construction of Mutually Orthogonal SuDoKu Latin Squares (MOSLS) have been addressed. See, for example, Golomb (2006), Bailey, Cameron and Connelli (2008), Lorch (2009, 2010, 2013), Pedersen and Vis (2009, 2012) and Subramani (2012). Table 2 displays the complete (and maximal) set of 6 MOSLS of order 9. It is adapted from Bailey, Cameron and Connelli (2008) by randomizing the positions within each cell, the numbers/letters at each position and the blocks of rows and columns. A beautiful artistic rendition of a similar design is found in Rosenhouse and Taalman (2011, p. 51).

Let us discuss the potential application of MOSLS as experimental designs. For example, an experimenter can use the complete set of 6 (randomized) MOSLS of order 9, shown in Table 2, as an experimental design. Then on top of the treatment classification (given by the number in the top-left position of each cell), the experimenter can accommodate five more orthogonal classifications (given by the other positions within each cell) each at 9 levels with 8 df. Moreover, all these six classifications are orthogonal to the 'Internal Block' classification and also to the row and the column clas-

Table 2. A complete set of 6 mutually orthogonal SuDoKu Latin squares of order 9, with three operators available in sets of three days.

$S\backslash D$	$D1$	$D2$	$D3$	$D4$	$D5$	$D6$	$D7$	$D8$	$D9$
$S1$	$9H5$ $3c1$	$8J8$ $2a3$	$1A1$ $6g6$	$6C7$ $1j7$	$7G2$ $8e8$	$4B3$ $9h9$	$3D9$ $4f5$	$5F6$ $7b2$	$2E4$ $5d4$
$S2$	$6D6$ $9d8$	$7F4$ $1f9$	$4E9$ $8b7$	$3H8$ $5g2$	$5J1$ $4c4$	$2A5$ $7a5$	$9C2$ $6h3$	$8G3$ $3j6$	$1B7$ $2e4$
$S3$	$3C3$ $7e4$	$5G7$ $5h5$	$2B2$ $4j2$	$9D4$ $2b6$	$8F9$ $6d1$	$1E6$ $3f3$	$6H1$ $8a9$	$7J5$ $9g7$	$4A8$ $1c8$
		O 1			O 4			O 7	
$S4$	$8A2$ $5b9$	$1H3$ $4d7$	$9J7$ $7f8$	$7B6$ $6a4$	$4C4$ $3g5$	$6G9$ $2c2$	$5E8$ $9e6$	$2D1$ $1h1$	$3F5$ $8j3$
$S5$	$7E1$ $2j5$	$4D5$ $6e2$	$6F8$ $3h4$	$5A3$ $8f1$	$2H7$ $9b3$	$3J2$ $1d6$	$8B4$ $7c7$	$1C9$ $5a8$	$9G6$ $4g9$
$S6$	$5B9$ $1g3$	$2C6$ $8c6$	$3G4$ $9a1$	$8E5$ $4h8$	$1D8$ $7j9$	$9F1$ $5e7$	$7A7$ $3d2$	$4H2$ $2f4$	$6J3$ $6b5$
		O 2			O 5			O 8	
$S7$	$1J4$ $8h2$	$9A9$ $9j4$	$8H6$ $1e5$	$4G1$ $7d3$	$6B5$ $5f6$	$7C8$ $4b1$	$2F3$ $2g8$	$3E7$ $6c9$	$5D2$ $3a7$
$S8$	$4F7$ $4a6$	$6E2$ $7g1$	$7D3$ $5c3$	$2J9$ $3e9$	$3A6$ $2h7$	$5H4$ $6j8$	$1G5$ $1b4$	$9B8$ $8d5$	$8C1$ $9f2$
$S9$	$2G8$ $6f7$	$3B1$ $3b8$	$5C5$ $2d9$	$1F2$ $9c5$	$9E3$ $1a2$	$8D7$ $8g4$	$4J6$ $5j1$	$6A4$ $4e3$	$7H9$ $7h6$
		O 3			O 6			O 9	

sifications. We refer to Saba and Sinha (2014) for technical details of the statistical analyses. Suffices here to show how the total df (80) is decomposed: (i) Each of the six mutually orthogonal classifications with 8 df, (ii) Internal Blocking with 4 df (via tetra differences), (iii) Row classification with 6 df, (iv) Column classification with 6 df, (v) Row x Internal Block confounding with 2 df, (vi) Column x Internal Block confounding with 2 df, and finally (vii) Error with 12 df.

Remark 2. Although popular SuDoKus are of order 9, SuDoKus of many other orders are available in general. For order 4, we can have at most one pair of MOSLS. These are displayed in Table 3. In fact, the df breakdown shows that there cannot exist more than two MOSLS of order 4. We also exhibit a pair of MOSLS of order 8 in Table 4. There are interesting combinatorial problems involving MOSLS of various orders. We do not address these problems here. Instead, we refer the reader to the list of references.

Table 3. A pair of orthogonal SuDoKu squares of order 4, with 4 operators available in sets of two days.

$S\backslash D$	$D1$	$D2$	$D3$	$D4$
$S1$	1, 1	2, 2	3, 3	4, 4
$S2$	3, 4	4, 3	1, 2	2, 1
	O 1		O 3	
$S3$	2, 3	1, 4	4, 1	3, 2
$S4$	4, 2	3, 1	2, 4	1, 3
	O 2		O 4	

Table 4. A pair of orthogonal SuDoKu squares of order 8, with 8 operators available in sets of two days.

$S\backslash D$	$D1$	$D2$	$D3$	$D4$	$D5$	$D6$	$D7$	$D8$
$S1$	8, 8	1, 1	2, 2	4, 4	3, 3	6, 6	5, 5	7, 7
$S2$	2, 6	4, 5	8, 7	1, 3	5, 4	7, 8	3, 1	6, 2
$S3$	3, 7	7, 3	5, 6	6, 5	8, 1	4, 2	2, 4	1, 8
$S4$	5, 2	6, 4	3, 8	7, 1	2, 5	1, 7	8, 3	4, 6
	O 1		O 3		O 5		O 7	
$S5$	1, 5	8, 6	4, 3	2, 7	7, 2	5, 1	6, 8	3, 4
$S6$	4, 1	2, 8	1, 4	8, 2	6, 7	3, 5	5, 6	7, 3
$S7$	6, 3	5, 7	7, 5	3, 6	4, 8	8, 4	1, 2	2, 1
$S8$	7, 4	3, 2	6, 1	5, 8	1, 6	2, 3	4, 7	8, 5
	O 2		O 4		O 6		O 8	

3. Conclusion

We have reviewed the application of SuDoKu squares as experimental designs, and discussed some issues related to analyses of data resulting from them. SuDoKu designs and MOSLS designs are gaining attraction among experimenters because of the flexibility these designs offer. We emphasize that implementation of such designs should incorporate appropriate randomizations to avoid any potential bias, and statistical analyses of such designs must take into account the non-orthogonality between certain pairs of components of variation.

References

1. Bailey, R. A., Peter, J., Cameron, P. J. and Connelly, R. (2008). Sudoku, Gerechte designs, resolutions, affine space, spreads, reguli, and Hamming codes. *Amer. Math. Monthly* **115**(5), 383–404.
2. Bose, R. C., Shrikhande, S. S. and Parker, E. T. (1960). Further results

on the construction of mutually orthogonal Latin squares and the falsity of Euler's conjecture. *Canad. J. Math.* **12**, 189–203.

3. Felgenhauer, B. and Jarvis, F. (2005). Enumerating possible Sudoku grids. Unpublished.
4. Golomb, S. (2006). Proposed Problem 11214. *Amer. Math. Monthly* **113**(3), 268.
5. Kuhl, J. S. and Denley, T. (2007). A few remarks on avoiding partial Latin squares. *Ars Combinatoria* **106**, 313–319.
6. Lorch, J. D. (2009). Mutually orthogonal families of linear Sudoku squares. *Jour. Australas Math. Soc.* **87**, 409–420.
7. — (2010). Orthogonal combings of linear Sudoku solutions. *Australas J. Comb.* **47**, 247–264.
8. — (2013). A quick construction of mutually orthogonal Sudoku squares. Private communication.
9. McGuire, G., Tugemann, B. and Civario, G. (2012). There is no 16-Clue Sudoku: Solving the Sudoku minimum number of clues problem. Unpublished.
10. Pedersen, R. M. and Vis, T. L. (2009). Sets of mutually orthogonal Sudoku Latin squares. *College Math. Jour.* **40**, 174–180.
11. — and — (2012). The construction of a maximum set of Sudoku MOLS. Private Communication.
12. Raghavarao, D. (1988). *Constructions and Combinatorial Problems in Design of Experiments* (corrected reprint of the 1971 Wiley ed.). Dover, New York.
13. Rao, C. R., Mitra, S. K. and Matthai, A. (1975). *Formulae and Tables for Statistical Work.* Statistical Pub. Society, Calcutta.
14. Rosenhouse, J. and Taalman, L. (2011). *Taking Sudoku Seriously: The Math Behind the World's Most Popular Pencil Puzzle.* Oxford University Press, New York.
15. Russell, E., and Jarvis, F. (2006). Mathematics of Sudoku II. Unpublished.
16. Saba, F. A. and Sinha, B. K. (2014). SuDoKu as an experimental design—Beyond the traditional Latin square design. *Statistics and Applications* **12**, 15–20.
17. Sarkar, J. and Sinha, B. K. (2014). Variants of SuDoKu as useful experimental designs. *Statistics and Applications* **12**, 35–60.
18. Sarkar, J. and Sinha, B. K. (2015). SuDoKu squares as experimental designs. *Resonance* **September, 2015**, 711–725.
19. Subramani, J. and Ponnuswamy, K. N. (2009). Construction and analysis of Sudoku designs. *Model Assisted Statistics and Applications* **4**, 287–301.
20. Subramani, J. (2012). Construction of Graeco Sudoku square designs of odd order. *Bonfring International Journal of Data Mining* **2**, 37–41.

Multi Criteria Decision Making Model for Optimal Selection of Recovery Facility Location and Collection Routes for a Sustainable Reverse Logistics Network Under Fuzzy Environment

J. D. Darbari[1], V. Agarwal[2] and P. C. Jha[3]

Department of Operational Research
University of Delhi, Delhi, 110007, India
E-mail: [1]jydbr@hotmail.com; [2]vernika.agarwal@gmail.com; [3]jhapc@yahoo.com

Compliance with legislation and inclination towards sustainability have been major drivers for electronics manufacturers to implement end-of-life (EOL) and end-of-use (EOU) take back models. Designing of an efficient product recovery network entails key strategic decisions such as locating facilities with suitable capacities and determining an efficient distribution network for the reverse flow. Process of evaluation and selection of suitable location for managing recovery processes must allow the evaluation of opinions of all the stakeholders owing to its major economical, environmental as well as social impact. Multi criteria decision making (MCDM) techniques are therefore required to deal with the numerous conflicting tangible and intangible attributes. The paper aims at developing a MCDM model for an electronics manufacturing company seeking to sustainably manage its recovery processes for EOL and EOU products. It determines the optimal location of a recovery facility center (RFC) and optimal collection routes. For this purpose, the Decision Making Trial and Evaluation Laboratory (DEMATEL) is utilized to determine the interdependencies among the various conflicting criteria considered for evaluation of alternative locations. Analytic Network Process (ANP) is then applied to generate the relative importance of the locations. Locating, capacity and routing decisions are further incorporated by developing a mixed integer linear programming model under fuzzy environment which determines optimum capacity of the RFC to be opened with adoption of best technology and optimal routes of transportation with optimal selection of vehicles. The mathematical model performs trade-off between economical and environmental performance (in terms of carbon emission) of the proposed reverse logistics (RL) network.

Keywords: DEMATEL; ANP; location selection; sustainable criteria; Vehicle Routing.

1. Introduction

Due to ever increasing human population and rapid urbanization, there has been an enormous increase in the quantum of E-waste generated. In developing countries, lack of awareness among consumers and unattractive take back schemes worsen the situation as electronic products are irresponsibly discarded by consumers with no understanding of their environmental impact. The need of the hour is for companies to encourage their consumers to return the end-of-life

(EOL) and end-of-use (EOU) products by providing them with worthwhile incentives, and to look for innovative ways to generate maximum value-added recovery from the returns. On the other hand, companiessee the huge amount of returns as a nuisance and have been reluctant in enforcing effective RL practices due to their over bearing costs. Major barrier for implementation of reverse logistics (RL) is the lack of awareness of benefits of RL [1]. However, lately RL processes are being considered as potential source of revenue. Adoption of efficient RL strategy can be economically viable and can also lead to environmental benefits [2]. One of the initial difficulties though is to identify ideal locations for recovery facilities and consequently to develop an efficient collection and recovery system. Siting industrial spaces for recovery facilities is a daunting task which must take into account the following: the location must adhere to government regulations; optimum availability of resources and skilled labor force in the area; the associated economic costs and environmental hazards are minimum. Very few studies [3][4] have focused on selection of facility locations in Reverse Logistics Supply Chain (RLSC). The location selection for a Recovery Facility Centre (RFC) is a strategic decision which has socio-economic as well as environmental impact on the surrounding region. In addition to just economical benefits, environmental protection and public health considerations are of utmost importance in facility location selection. The goal of optimum selection of facility location is therefore subject to various criteria [5]. Besides the measurable variables, the decision process for a selection problem is also influenced by qualitative variables [6]. The use of multiple criteria analysis has the advantage of rendering subjective and implicit decision making more objective and analytical, with its ability to accommodate both quantitative and qualitative data. The site selection is a complex multi criteria problem which must also consider views of all the stakeholders as significant and indispensable. Qualitative techniques such as AHP and ANP are suitable to quantify the subjective judgments of the stakeholders. However, the traditional ANP can be combined with powerful technique such as the DEMATEL method developed by [7] to simplify the complex process. DEMATEL along with ANP has been successfully used in multi criteria decision making problems [8]. Once the location of RFC is identified, emphasis must be laid on choosing optimal routes for collection of returned product to attain a sustainable and efficient reverse network.

In this paper, DEMATEL based ANP (DANP) is utilized for evaluation of alternative locations for a RFC based on a number of economical and socio-environmental criteria to obtain the most suitable one. Subsequently, for the design of a economically and environmentally effective transportation network for flow of returns from CCs to the selected RFC, routes are obtained using travelling salesman problem (TSP) based on the amount of returns to be collectedand capacity of the vehicle selected. Finally, we formulate a Fuzzy Mixed Integer Linear Programming mathematical model for the planning of sustainable reverse logistics network taking into account the selected location of the RFC. The multi objective model determines the optimum capacity of the RFC, adoption of best technology and optimal routes and vehicles for collection of returned products while performing a tradeoff between cost and environmental impact of the RL model. The rest of the work is organized as follows: Section II considers case study of an electronics manufacturing company and discusses the theoretical background of the solution methodology used. Section III discusses the results generated by the application of the proposed model. Section IV summarizes the conclusion and future research scope.

2. Case Study and Proposed Methodology

The problem of E-waste management has become a perennial problem especially in Northern Capital Region of India (NCR) of India. The case-study considered here is of an electronics manufacturing company in India looking for designing a recovery network to manage their EOL and EOU returns. The company has about 19 retailers in operation in NCR, each falling into a pre-defined collection zone with one vehicle designated to each zone.

In view of legislation and environmental concerns, proper handling of electronic returns is at the forefront of the company. To begin with, the company needs to develop an efficient collection and recovery system. For this, a company owned RFC is needed to carry repair, refurbishment and dismantling of its returns and an optimum recovery network must be established. In order to design a sustainable RL network, location of the RFC also plays a major role in generating maximum potential benefits and social impact along with minimum environmental mitigation.

2.1. *Evaluation of location of RFC*

A team of company stake holders is assigned the task of choosing the best site for setting up of the RFC. Initially 5 potential location choices are identified in different zones of NCR from the list of sites authorized by thestate government as per E-waste rules issued by the Ministry of Environment and Forest, Government of India. The initial screening process is based on availability of ideal space, natural resources and work force and also proximity to retail stores which act as collection points for the returns. The most appropriate 5 locations are as follows: Maya Puri Industrial Area (A1), Badarpur (A2), Eco-Tech-1 (A3), Dwarka (A4), Bawana (A5). The initial screening helps in tremendously reducing the time and complexities involved in the final selection of the location based on many qualitative and quantitative factors such as cost decisions related to labor, land, transportation, lease term, resource availability and socio-environmental decisions such as skilled labor, community development, site use efficiency etc.

2.2. *Selection of location of RFC*

Based on an extensive literature survey [9][10][11] and due deliberation with the decision makers (DMs), following selection criteria are considered for the location of RFC: cost of labour (C1), land acquisition cost (C2), infrastructure cost (C3), transportation conditions (C4), flexibility of lease term (C5), availability of resources (C6), skilled manpower (C7), scope for community development (C8), impact on local economy (C9), possibility of inclusivity (C10), site use efficiency (C11), safe disposition options (C12) and training cost (C13).

In the study, we utilize DANP to ease the process of decision making by generating the cause and effect groups and identify the most influential criteria using DEMATEL, and subsequently generate weights of the alternatives using ANP. Because there are different degrees of influence and relationship between the criteria in the real world, assumptions of equal influence and relationship seem to be irrational. Thus, this study applies DEMATEL to establish the correlation between the mutual effects of the criteria prior to the evaluation by ANP.

The steps of the DANP methodology are as follows [12, 13]:

Step 1: Generate the group of k direct-relation matricesfrom a group of k experts, where the $(i,j)^{th}$ entry of kth matrix indicates the effect that the criteria $i$exerts on criteria j based on the perception of the kth DM (assuming the diagonal entries are zero). An integer scale of 0-4 is used to represent the effect as follows: 'No influence (0)', 'Low influence (1)', 'Medium influence (2)', 'High influence (3)', and 'Very high influence (4)' respectively.

Step 2: Calculate the direct – relation (average) matrix A whose each element a_{ij} is average of $(i,j)^{th}$ entries of the matrices obtained in *Step 1*.

Step 3: Calculate the normal direct-relation matrix D.

$$D = (d_{ij})_{nxn} = A/s\,, s = max(max\ Column\ sum, max\ Row\ sum)$$

where each element in matrix D falls between 0 and 1and atleast one row or column summation (not all) is one.

Step 4: Compute the total relation matrix T.

$$T = (t_{ij}) = lim_{m\to\infty}(D + D^2 + D^3 + \cdots . +D^m) = D(I - D)^{-1}$$

Step 5: Calculate the row-sum r(nx1) and column-sum c(1xn) of T, where r_i, the sum of the ith row in matrix T shows the total effect exerted by the ith criteria on the other criteria and c_j, the sum of jth column of matrix T shows the total effect received by criteria j from the other criteria. When $j = i$, (r_i+c_i) indicates the degree of importance that criteria i plays (see Table 1). Criteria having more values of r_i+c_i have more relationship with others and those having little values of r_i+c_i have less of a relationship with others. When $j = i$, the difference r_i-c_i indicates the net influence of criteria i on the system. If r_i-$c_i > 0$ the criteria is a net cause and is part of the dispatcher group, while criteria with r_i-$c_i < 0$ is a net receiver and part of the receiver group.

Table 1. Values of r_i, c_i, r_i+c_i and r_i-c_i.

	C1	C2	C3	C4	C5	C6	C7	C8	C9	C100	C111	C122	C133
ri	1.52	1.84	2.07	0.79	2.52	1.35	1.71	1.54	1.48	0.37	2.38	0.00	1.58
ci	1.50	1.35	1.31	1.04	0.93	1.69	2.15	1.89	1.21	1.39	1.75	1.37	1.57
ri+ci	3.02	3.19	3.38	1.83	3.45	3.04	3.85	3.44	2.70	1.76	4.13	1.37	3.16
ri-ci	0.02	0.49	0.76	-0.25	1.59	-0.34	-0.44	-0.35	0.27	-1.02	0.63	-1.37	0.01

Step 6: Draw the impact digraph-map (IDM) with r+c as the horizontal axis and r-c as the vertical axis (Fig. 1). The decision-maker must set a threshold value say α for the influence level in order to decrease the complexity of the IDM. If the influence level of an element in matrix T is higher than the threshold value, then this element is included and the others are filtered out.

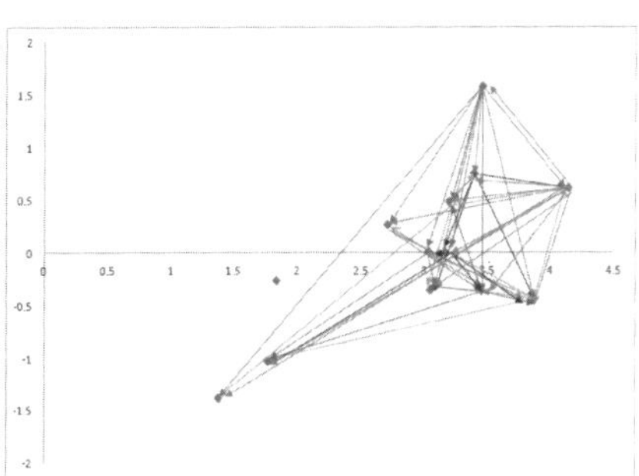

Fig. 1. IRD diagram.

Step 7: Based on the use of a ratio scale of 1-9, build pair wise comparison matrices to evaluate the effect of criteria on criteria (using interrelationships generated by IDM), criteria on alternatives and alternatives on criteria. The inverse comparison is expressed by the reciprocal ofeach number. Consistency index (CI) and Consistency Ratio (CR) for each pair wise matrix is calculated as $CI = \lambda_{max} - n/n - 1$ and $CR = \mathrm{CI}/RI$, where RI is the Random Consistency Index. The comparisons are acceptable only if CR is less than 0.10.

	C1	C2	C3	C4	C5	C6	C7	C8	C9	C10	C11	C12	C13	A1	A2	A3	A4	A5
C1	-	-	0.20	-	0.11	-	0.12	-	-	-	-	-	0.10	0.13	0.13	0.11	0.11	0.13
C2	-	-	-	-	0.09	-	-	0.13	-	-	0.09	-	-	0.06	0.09	0.08	0.09	0.06
C3	-	0.15	-	-	0.06	-	-	-	-	-	0.07	-	-	0.08	0.07	0.07	0.09	0.10
C4	-	0.12	-	-	0.11	-	-	-	-	-	0.06	-	-	0.08	0.05	0.09	0.08	0.09
C5	-	0.08	-	-	-	-	-	-	-	-	0.10	-	-	0.07	0.09	0.11	0.06	0.07
C6	-	0.17	0.12	-	0.14	-	-	0.38	-	-	0.16	-	-	0.12	0.12	0.08	0.09	0.06
C7	0.13	-	0.28	-	0.09	-	-	0.37	0.50	-	0.09	-	0.41	0.07	0.07	0.07	0.10	0.09
C8	0.17	0.25	-	-	0.08	1.00	0.20	-	0.50	-	0.08	-	0.37	0.07	0.07	0.06	0.07	0.10
C9	0.32	-	-	-	0.06	-	-	-	-	-	0.07	-	-	0.07	0.05	0.07	0.06	0.07
C10	0.10	-	-	-	-	-	0.34	0.12	-	-	0.09	-	0.13	0.07	0.07	0.07	0.05	0.06
C11	-	0.12	0.09	-	0.09	-	0.08	-	-	-	-	-	-	0.07	0.07	0.07	0.07	0.05
C12	-	0.11	0.14	-	0.08	-	-	-	-	-	0.09	-	-	0.05	0.05	0.06	0.06	0.06
C13	0.28	-	0.19	-	0.09	-	0.25	-	-	-	0.09	-	-	0.07	0.07	0.07	0.07	0.06
A1	0.09	0.07	0.42	0.33	0.05	0.33	0.32	0.10	0.08	0.33	0.08	0.33	0.42	-	-	-	-	-
A2	0.16	0.22	0.13	0.19	0.23	0.19	0.16	0.20	0.17	0.11	0.13	0.19	0.20	-	-	-	-	-
A3	0.39	0.39	0.07	0.08	0.49	0.08	0.08	0.41	0.45	0.05	0.48	0.08	0.15	-	-	-	-	-
A4	0.10	0.11	0.26	0.30	0.08	0.30	0.24	0.13	0.14	0.18	0.11	0.30	0.12	-	-	-	-	-
A5	0.26	0.22	0.13	0.11	0.15	0.11	0.21	0.15	0.15	0.33	0.20	0.11	0.11	-	-	-	-	-

Fig. 2. Unweighted super Matrix.

Step 8: The local priority vectors derived from the Eigen vectors of the pair wise comparison matrices are introduced into the unweighted super matrix (Fig. 2) which is normalized to get the weighted super-matrix. The super-matrix is converged by raising it to the power 2k+1, where k is an arbitrary large number, to get global priority vectors. The priority vector obtained is as follows: (A1, A2, A3, A4, A5) = (0.089, 0.063, 0.0795, 0.070, 0.067). Clearly Mayapuri Industrial Area (A1) with the highest weight is selected as the ideal location for RFC.

After final selection of the RFC location, the objective of the company is to have an effective RL network with optimum selection of technology, capacity and route plan for collection of the returned products from the CCS. The CCs are segregated into 4 collection zones and the collection for each zone is carried by a selected vehicle of required capacity. The RL network is configured with the help of following optimization model.

3. Mathematical Formulation of Fuzzy Multi-objective Optimization Model

Sets: Product Set indexed by$j$, Collection centres(CCs) indexed by$l$, Technology Set indexed by k, set of CCs in the rth zone denoted by I_r and $D_r = I_r \cup \{0\}$ where 0 represents RFC.

Parameters: O_l is the capacity of lth CC, O_h, C_h and E_h are the capacity, cost of setting and environmental impact of RFC of level h respectively, C_h and E_h are the cost of acquiring and per unit environmental impact of using technology level k at RFC, X_{il} are units of products in the ith category to be collected at the lth CC, O_v, C_v and E_v are capacity, hiring cost and per km environmental impact of the vth truck used for collection of returned products, d_{prqr} is distance between nodes p_r and q_r of the rth zone set D_r, h_{lr} takes value 1 if l^{th} CC is in the rth zone else takes value 0, w_i is weight of the product in the ith category.

Decision variables: A_k takes value 1 if RFC is set up with capacity level h else takes value 0, B_k takes value 1 if technology k is acquired at RFC else takes value 0, Q_r are the number of units of the product to be collected in the rth zone, S_{rv} takes value 1 if the total return load in the rth zone falls in the vth vehicle break else takes value 0, e_{pr} are real numbers used to avoid sub-touring in the routes within each rth zone, x_{prqr} takes value 1 if the truck travels from nodes p_r and q_r of the set D_r else takes value zero.
(FMOMP)

Objective functions

$$Minimize\ Z_1 \cong \sum_{r\in N}\sum_{v\in V} S_{rv}C_v + \sum_h A_h C_h + \sum_k B_k C_k$$

$$Minimize\ Z_2 \cong \sum_{r\in N}\sum_{p\in D_r}\sum_{q\in D_r} d_{pq}x_{pq}\, S_{rv}E_v + \sum_h A_h E_h + \sum_k B_k E_k$$

The symbol $\cong$ is the fuzzified version of '=' which can be read as "essentially equal to". The first objectiveminimises the cost of the RL network. The second objective minimises the carbon emission due to collection of products in and opening of facility and use of technology.

Constraints

$$X_l^{rel} = \sum_i X_{il} \qquad \forall l \tag{1}$$

$$X_l^{rel} \le O_l \qquad \forall l \tag{2}$$

$$\sum_i \sum_l w_i X_{il} h_{lr} = Q_r \qquad \forall r\epsilon N \tag{3}$$

$$Q_r \le \sum_v O_v S_{rv} \qquad \forall r\epsilon N \tag{4}$$

$$\sum_v S_{rv} = 1 \qquad \forall r\epsilon N \tag{5}$$

$$\sum_l X_l^{ret} \leq \sum_h O_h A_h \quad \forall r \epsilon N \tag{6}$$

$$\sum_h A_h = 1 \tag{7}$$

$$\sum_k B_k = 1 \tag{8}$$

$$\sum_{p_r \epsilon I_r} x_{p_r q_r} = 1 \qquad \forall\, p_r \neq q_r;\ q_r \in D_r; r = 1,2 \dots N \tag{9}$$

$$\sum_{q_r \epsilon I_r} x_{p_r q_r} = 1 \qquad \forall\, p_r \neq q_r;\ q_r \in D_r; r = 1,2 \dots N \tag{10}$$

$$\sum_{p_r \epsilon \widehat{D_r}} x_{p_r q_r} = \sum_{q_r \epsilon \widehat{D_r}} x_{p_r q_r} \ \forall p_r \neq q_r;\ q_r \in D_r; r = 1,2 \dots N \tag{11}$$

$$e_{p_r} - e_{q_r} \geq (I_r + 1) X_{p_r q_r} - I_r \ \forall\, p_r \neq q_r;\ q_r \in D_r; r = 1,2 \dots N \tag{12}$$

$$e_{p_r}, Q_r \geq 0 \ and\ integers;\ A_h, B_k, S_{rv}, x_{p_r q_r} \in \{0,1\} \tag{13}$$

Constraint (1) determines the amount of returns collected at each CC. Constraint (2) is the capacity constraint for each CC. Constraint (3) determines the number of products collected per cluster. Constraints (4) and (5) determine the choice of vehicle of least capacity required to carry the load within each zone. The capacity constraint for RFC is given jointly by equations (6) and (7). Constraint (8) determines the acquiring of a technology at RFC. Constraints (9)-(12) determine the transportation in between the zones using TSP. Constraint (13) determines non negativity and binary variables.

3.1. *Fuzzy solution algorithm*

Based on the sequential steps in [14,15], the given FMOMP is transformed to the following weighted single objective problem:
Maximise λ
subject to $\mu_{Z_g}(X) \geq w_g \lambda \ \ g = 1,2..G, x \epsilon S, w_g \geq 0, \sum_{g=1}^{G} w_g = 1, \lambda \epsilon [0,1]$
where $\lambda \epsilon [0,1]$ represents the degree of satisfaction of the DMs.

4. Result Discussion

The above proposed model is applied for strategic planning of the RL network which includes selection of appropriate capacity for the RFC, use of most suitable technology at RFC and selection of optimal vehicle and optimal routes for transportation. RFC (A1) selected has 4 choices of capacities O_h given as 500, 1000, 1500, 2000 units of products, with cost of setting up C_h as Rs500000, 5500000, 600000, 700000 and environmental impacts E_h in kg- CO_2 as 150.2, 160.3, 180.6, 20 respectively. C_k, the cost of acquiring technology levels k at RFC are Rs 200000, 300000, 400000, 500000 with E_h their per unit environmental impacts in kg CO_2 given by 17.85, 15.04, 13.57, 11.02

respectively. For each collection zone, there are 4 trucks available for transportation of the returned products from CC to RFC with variable capacities O_v given as 1500, 2000, 2000, 3000 kg, hiring cost C_v of Rs 2500, 3500, 4500, 5000 and per km environmental impacts E_v of 4.1, 3.4, 2.9, 2.1 kg CO_2 respectively. The company manufactures household electronics which are categorised into three types. w_i (i=1,2,3), weights of the product in the three categories are 6, 10, 12 kg. d_{prqr} are real distances between the nodes (RFC and 19CCs). The load collected per cluster in kg is 1478, 1562, 2050 and 878.

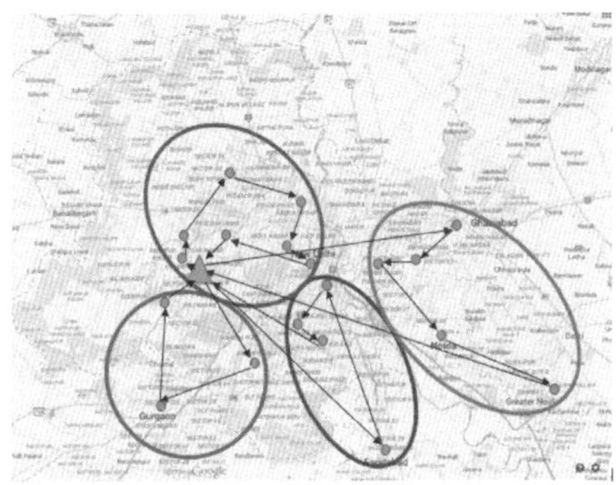

Fig. 3. Optimal routes within each zone.

At first, the multi-objective problem is solved using the above data as two separate single objective problems.The linear programming software Lingo 11 is used to solve the problems. Results confirm that the two objective functions are in conflict with each other. The first objective compels the model to select the cheapest technology 1 along with selection of cheaper vehicle with least required capacity. On the other hand, second objective function has a tendency towards choosing technology 4 as well as vehicles with least carbon emission and higher cost, therefore compromising on the cost. The price difference of for reducing the environmental impact is Rs 164150 which shows that reasonable trade-off is required. The two objectives are converted into crisp forms using the membership functions Z_1 and Z_2 constructed using Eq. (9), with upper and lower bounds as Rs 602850, Rs 910530 and 8465.3 kg CO_2, 14505.8 kg CO_2 respectively. Weights w1 and w2 of the two goals are taken as w1 = 0.6, w2 = 0.4 according to DMs' opinions.

The compromised solution is obtained after solving the equivalent crisp single objective programming problem and optimal values of cost as Rs 767000 and of environmental impact as 12599.44 kg CO_2 are generated with 77.7 % degree of satisfaction of the DMs. The optimal routes constructed are shown in

Fig. 3. The starting and ending node in each zone is the RFC: Maya Puri Industrial Area (A1). Results clearly show the trade-off between the two objectives. The model chooses the minimum required capacity level of 1000 units and technology 1 which is cheaper to give importance to cost effectiveness. On the other hand, to consider the environmental effectiveness the model chooses vehicle 4 for clusters 1 and 2 with higher cost and least emission rate even though the required capacity could be catered by cheaper vehicles. The optimal routes obtained within each zone for flow of returns however remain same in all the three cases-minimisation of only cost, minimisation of only environmental impact and simultaneously minimising both objectives. The reason is obvious, as in each case we aim to find routes within each zone with least distance of travel.

5. Conclusion

Development of a sustainable reverse supply chain network for an Indian electronics manufacturing company based in the northern capital region of the country is presented. The company already has around 19 retailers in the Delhi NCR area which act as collection points. The recovery model proposed in the study allows for the selection of the best location for a recovery facility centre taking into account sustainable criteria and an optimization model for designing an efficient routing network. DANP is utilised for selection of most suitable location of RFC for its effectiveness in developing a suitable multi criteria framework for decision making. A multi objective mathematical model is developed under fuzzy environment to determine optimal capacity of RFC to be in operation with most suitable technology and also the best routes of transportation via optimal vehicles. The main contribution of the paper lies in the development of a multi criteria decision support tool for DMs for planning a sustainable RL network model. For future work, fuzzy ANP can be combined with other multi criteria approaches in order to reduce the possibility of biasness in the subjective inputs.

References

1. S. K. Sharma, B. N. Panda, S. S Mahapatra and S. Sahu, Analysis of barriers for reverse logistics: an Indian perspective, *Int. J. of Model. and Optim.*, 1, 2, 101 (2011).
2. S. Wadhwa, J. Madaan and F. T. S. Chan, Flexible decision modeling of reverse logistics system: A value adding MCDM approach for alternative selection, *Robot. and Computer-Integr. Manuf.*, 25, 2, 460, (2009).

3. G. Kannan, A. N. Haq and P. Sasikumar, An application of the analytical hierarchy process and fuzzy analytical hierarchy process in the selection of collecting centre location for the reverse logistics multicriteria decision-making supply chain model, *Int. J. of Manag. and Decis. Mak.*, 9, no. 4, (2008), pp.350-365.
4. G. Tuzkaya and B. Gülsün, Evaluating centralized return centers in a reverse logistics network: An integrated fuzzy multi-criteria decision approach, *Int. J. of Env. Sci. & Technology*, 5, 339, (2008).
5. R. Z. Farahani, M. SteadieSeifi and N. Asgari, Multiple criteria facility location problems: A survey, Appl. Math. Model., 34, 7, 1689, (2010).
6. A. Özdağoğlu, A multi-criteria decision-making methodology on the selection of facility location: fuzzy ANP, *The Int. J. of Adv. Manuf. Technology*, 59, 5–8, 787,(2012).
7. A. Gabus, and E. Fontela, World Problems an Invitation to Further Thought within the Framework of DEMATEL. *Battelle Geneva Research Centre, Switzerland, Geneva,* (1972).
8. J. L. Shen, Y. M. Liu and Y. L. Tzeng, The Cluster-Weighted DEMATEL with ANP Method for Supplier Selection in Food Industry, JACIII, 16, 5, 567, (2012).
9. T. Burnett, Investing in inclusion, Hum. Resour. Manag. *Int. Digest*, 14, 4, 3, (2006).
10. C. L. Yang, S. P. Chuang, R. H. Huang and C. C. Tai, Location selection based on AHP/ANP approach, in Industrial Engineering and Engineering Manag.*, IEEE Int. Conf., December,* 1148, (2008).
11. Y. C. Kuo, S. T. Lu, G. H. Tzeng, Y. C. Lin and Y. S. Huang, Using Fuzzy Integral Approach to Enhance Site Selection Assessment–A Case Study of the Optoelectronics Industry., *Procedia Computer Sci.*, 17, 306, (2013).
12. G. H. Tzeng, C. H. Chiang and C. W. Li, Evaluating intertwined effects in e-learning programs: A novel hybrid MCDM model based on factor analysis and DEMATEL, *Expert Syst. with Appl.,* 32, 4, 1028, (2007).
13. T. L. Saaty, Fundamentals of decision making and priority theory with the analytic hierarchy process, *Rws Publ.* 6,(2000).
14. H. J Zimmermann, Fuzzy programming and linear programming with several objective functions. *Fuzzy sets and syst.*, 1, 1, 45, (1978).
15. Darbari, J. D., Agarwal, V., & Jha, P. C. (2015, January). Fuzzy Optimisation Approach to Supply Chain Distribution Network for Product Value Recovery. *In Proceedings of Fourth International Conference on Soft Computing for Problem Solving (pp. 487–500). Springer India.*

Optimal Allocation of SKU and Safety Stock in Supply Chain System Network

K. Gandhi

Bhavan's Usha & Lakshmi Mittal Institute of Management
K. G. Marg, New Delhi 110001, India
E-mail: gandhi.kanika@gmail.com

K. Goyal[1], A. Jha[2] and J. D. Darbari[3]

Gazelle Information Technologies, New Delhi, India
Department of Operational Research, University of Delhi, Delhi, India
E-mail: [1]kirit@gazelle.in, [2]abhinav@gazelle.in, [3]jydbr@hotmail.com

Effective inventory management and distribution has become critical to industries such as manufacturing, retailing, transportation, health, and service, as they strive to improve their customer service, cash flow and profit margin, while meeting the challenges of global competition, product proliferation, shorter life cycles and demand uncertainty. Environmental awareness throughout supply chains is also growing due to the regulatory policies legislated by governments and increasing pressure from voluntary organizations. As a result, supply chain partners need to analyze their operations such as inventory control, freight transportation, and warehousing activities for achieving an eco-efficient supply chain. At the core of inventory management is stocking control, which ensures that the right amount of stock is available to support the company's targeted fill rate in the market at minimum cost. Companies must determine and manage specific service levels so that customers across the supply chain are served in time. Otherwise, stock-outs quickly translate into lost sales. Transportation activities play a vital role in the effective management of a supply chain. In addition to being major contributor for accomplishment of service level, it has an immense impact on overall carbon footprint of the supply chain. But finding the optimal balance among these factors is not easy, especially due to the vast global market size. In this paper, an integrated inventory control and transportation model is proposed to obtain optimal stock keeping units (SKU) and Safety Stock for each product as well as each of the locations at minimum cost for the next planning horizon with environmental considerations to reduce the overall carbon footprint. Further, at the end of each period, current solution is put to test to evaluate possible deviations from prior fixed target and a modified solution is obtained, if required, for continual improvement in supply chain system design. The model has been validated through a case study.

Keywords: Stock Keeping Units; Safety Stock; Service Level; Green Supply Chain; Clustering; ABC analysis; Forecasting.

1. Introduction

The supplying and inventory strategies, currently employed throughout the supply chain, failed to satisfy the promised delivery dates of the end products. Major reasons behind this shortfall are the existence of unreliable suppliers that are unable to deliver materials on time, the inventory systems employed throughout the chain and inaccurate forecasts. These systems are established based on random ordering decisions, which often lead to a stock-out occurrence. Moreover, each member of the chain does not employ a proper safety stock policy to face the fluctuation of customer demand and suppliers' lead time.

In the consumer durable industries, supply chains typically consist of multiple stages of production facilities, warehouse and distribution centres, logistical sub-networks and end customers. When events, such as changes in the characteristics of the uncertainties, shifts in the demand to capacity ratio, introduction of a new product and retirement of a matured product, or entrance of new competitors take place, the optimal safety stock level at each stage of the supply chain needs to be re-evaluated. Thus, cost effective and agile safety stock management represents a competitive advantage for a company in a dynamic and highly competitive market. In the current market conditions, retailers are facingthe problem of increasing assortments. In shoe retailing, which is the main focus of our study, product life cycles have been decreasing. As a consequence, it is increasingly difficult to forecast sales for an individual product for tactical decisions. Moreover, retail sales are faced with extensive promotion activities and seasonality. Products are typically on promotion for a limited period of season time, e.g. two to three months, during which demand is usually substantially higher than during periods without season (see, e.g. [1]), and many of the stock-outs occur during promotions due to inaccurate forecasts. Operationally, it is clear that the accuracy of the forecast directly contributes to higher profits by reducing stock-out situations and lowering the level of safety stock. The coordination of logistics, supplier and buyer activities has been based on economic criteria for a long period regardless of the negative impacts that these activities may have on environment, basically in terms of carbon emission. Currently, the environmental concerns are becoming more relevant for firms as governments' environmental policies. Companies are highly sensitive and responsive to the carbon emissions associated with their operations. In particular, transportation is one of the major contributors to carbon emission. Clustering is one of the options to manage the connections among the entities of supply chain to avoid unnecessary multiple movements among them. The next stage shows the requirement of developing optimal route within each cluster so that the problems like total travelling and carbon emission could be managed well.

Based on this classical work, research in inventory theory has been ongoing to extend these analytical results toconditions experienced in the day-to-day operation and design of supply chains. The works of [2] and [3] usually serve as starting points for research that extends the fundamental single inventory location problem to multi-stage supply chain optimization problems. Among the extensive literature in the area, the works of [4], [5], [6] and [7] are most closely related in intent to our research, namely safety stock management in multi-stage supply chains under production capacity limits. [4] proposes a decentralized supply chain model for the determination of the base-stock level that ensures target service level or the service level that is achievable at a given base-stock level. Their analytical framework takes into account the impact of upstream service level on the downstream service level by modelling the expected replenishment lead time at each inventory location as a linear function of the service level and expected delay time of supplying upstream. [5] develops simulation based methods for estimating sensitivities of inventory costs with respect to policy parameters (base-stock levels) in multistage capacitated procurement-inventory systems. The work of [6] is similar to that of [4] in that both models pay close attention to interdependence of base-stock levels at different inventory sites and how this interdependence affects system performance. [7] models the multi-stage supply chain problem as a deterministic nonlinear optimization problem and proposed a dynamic programming algorithm to solve the problem. [8] extends the problem addressed in [7] to consider capacity limitations. They state that the safety stock amounts have to be updated by a tabulated correction factor that relates safety stock with a measure representing the degree of capacity that covers demand variations. Further relationships between demand variability, capacity, delivery lead time, and safety stocks are investigated in [8]. [9] develops a mathematical model that allows backordering at each stage. The model does not state any assumptions about the demand distribution. [10] considers an assembly system facing constant demand and discrete distributed random lead time of delivering the assembled components.

Supply chain design models have traditionally focused on minimizing fixed and operating costs without taking carbon emissions into account. Recent studies, however, started to take emissions into account. The reduction of a product carbon footprint has a stronger influence on the purchasing decisions of consumers if the most important competitors also provide this information, carbon emission turns out to be relevant competitive differentiator [11]. Since companies tend to rely on third-parties for their inbound and outbound logistics, transport emissions are rarely included in companies' carbon targets. In fact, the share of a product carbon footprint represented by transport emissions may significantly vary from one product to another depending, for instance, on the product's volume or weight, but more importantly on the employed mode of transport. In general, the trade-off in transport is that greater speed relates to greater emissions and greater costs [12]. [13] notes that carbon emissions can be effectively reduced by changing vehicle efficiency or vehicle design. In the case where a TL carrier is used, each empty truck generates a fixed amount of carbon emissions and total emissions generated by a truck increase with its load. Thus, the numbers of trucks used along with their loads determinethe retailer's carbon emissions due to transportation. [14], [15], and [16] define similar carbon emissions functions. In order to minimize carbon emissions within a supply chain network, [15] formulates a transportation problem with two modes of transportation (rail and trucks) and notes that the same structure for carbon emissions function is also observed in rail transportation.

In the current study, we want to developa coordination among all the supply chain entities discussed above. We are showing an integrated inventory control and transportation model obtain optimal SKU and Safety Stock for each product as well as each of the locations at minimum cost for the next planning horizon with environmental considerations to reduce the overall carbon footprint. Past data like past demands, their lead time for delivery and further their standard deviations helped in showing future forecasts. To forecast the demand, ABC analysis classification method is adopted. The model is explained for multi warehousing and multi demand point supply chain network. K-means clustering method has been used to create clusters among multi demand points, which have been connected with a particular warehouse using travelling salesman problem (TSP). TSP also helps in showing optimal route within each cluster and ensures one route is made in one cluster.

2. Case Study

The model developedin the paper is validated on the case problem of a shoe company. The company is showing concern for their "casual" category among many categories available. There are 32 articles in the "Casual" category and 33 stores in the Delhi NCR region. The distribution to all the 33 retail stores happens via 2 warehouses. Past 3 years demand has been provided by the company, which is used to forecast the demand for next 12 months. Total sales of past 3 years for all the stores have been averaged for each product to classify them into most demanded to least demanded categories. ABC analysis is used to show the classification. It turned out that approx 34% of the articles constitute 70% of the sales. This implies, 11 articles/products constitute 70% sales out of 32 articles. Company is more concerned for 15 retail stores out of 33.

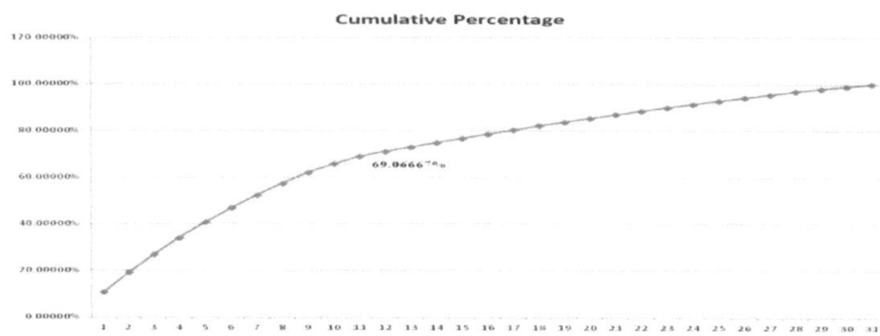

Fig. 1. ABC analysis.

In the starting, we tried to understand company's procurement and distribution process. Company transports the demanded quantity to each store individually, which costs them higher and also the carbon emission is higher because of repeat movement in the same location. We proposed use of clustering method to create small groups on the basis of distance among 15 retail stores. K-means clustering [17] has been employed to create clusters and the algorithm for the same is as follows:

The K-means clustering is the process of partitioning a set of data samples into k clusters by considering the boundary information of confusing classes. Suppose that a set of n training samples $X \in \Re^d$ and the clusters number k (0<k<n) are given. The k optimal cluster centriods $(\vec{c_1},\vec{c_2},\ldots,\vec{c_k})$ with each centroid $\vec{c_k} \in \Re^d$ can be obtained by minimizing the objective function (1).

$$\Phi_{k-mean} = \min \sum_{i=1}^{k} \sum_{\vec{x_j} \in \vec{c_i}}^{j<n} \left\| \vec{x_j} - \vec{c_i} \right\| \tag{1}$$

where, $\left\|\vec{x_j}-\vec{c_i}\right\| = \sqrt{\left(\vec{x_j}-\vec{c_i}\right)^T\left(\vec{x_j}-\vec{c_i}\right)}$ is defined as the Euclidean distance metric in $\Re^d$.

The k cluster centroids are initialized by randomly choosing k samples from the training data set. Then each sample $\vec{x_j}$ in the data set is labeled and assigned to the cluster $\vec{c_i}$ which is the nearest neighbour of the input sample according to the Euclidean distance. After processing the entire data set, each centroid is updated by calculating the mean of all data samples which are labeled to this centroid $\vec{x_j} \in \vec{c_i}$ according to (2). The iteration of finding the nearest neighbours and updating the centroids will not be terminated as long as the k cluster centroids change. This is the procedure of convergence to a local minimum in (1).

$$\vec{c}_i = \frac{1}{|\vec{c}_i|} \sum_{\vec{x}_j \in \vec{c}_i} \vec{x}_j \quad (2)$$

The K-means algorithm can be described as follows: *Step 1.* Randomly choose k cluster centroids. *Step 2.* For each training sample, find the nearest cluster centroid. *Step 3.* Assign the sample to its nearest cluster centroid. *Step 4.* Update the cluster centroids. *Step 5.* Repeat steps 2–4 until the convergence. The above method helped in creating 3 clusters as follows:

Table 1. K-means clustering.

Cluster	*Retail store index*	*X- coordinate*	*Y- coordinate*
Cluster 3	R1	27.4	42.3
Cluster 1	R4	7.8	37.1
Cluster 1	R5	11.1	36.6
Cluster 2	R6	14.6	42.1
Cluster 3	R8	26.2	47.9
Cluster 3	R9	23.9	44.5
Cluster 2	R10	14.6	43.5
Cluster 2	R12	15	47.2
Cluster 2	R13	17.9	50.3
Cluster 2	R14	16.9	45
Cluster 2	R15	16	53
Cluster 3	R2	22.7	45.8
Cluster 1	R3	21	27.8
Cluster 1	R7	12.9	23.6
Cluster 1	R11	14.7	26.7

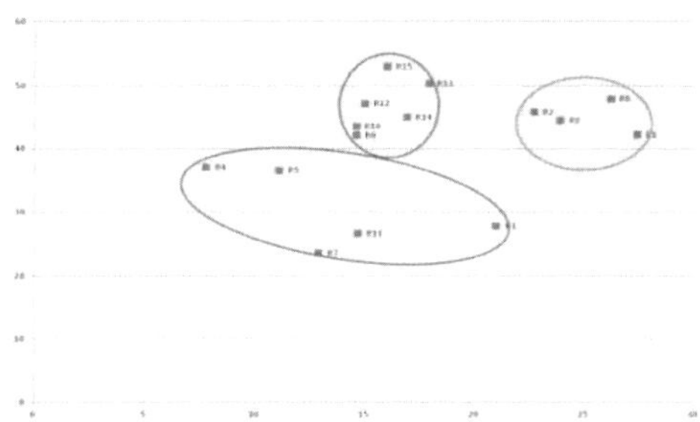

Fig. 2. Clustering as per distance.

At this stage, we have classified products and clusters as the base to formulate mathematical model for creating the coordination among many entities of supply chain.

3. Model Formulation and Solution Method

The model in the paper develops a green supply chain design that incorporates the cost of procurement, safety stock cost at 98% service level, inventory carrying cost and carbon emissions value into the objective function. The goal of the model is to simultaneously minimize procurement, inventory and logistics costs along with the environmental concerns of carbon emissions by strategically connecting warehouses with the clusters of demand points.

3.1. *Assumptions, sets, parameters and decision variables*

Initial inventory at all the warehouses and the retail stores is zero; Retail stores have no capacity restriction; Forecasted demand is used in the model; Shortages are not allowed; Safety stock has been considered at the retail stores.

Sets: i=Product i=1...P; j=warehouse j=1,..J; t=Time period t=1,...,T; r=retail store r=1,..,R; c=cluster c=1,...,C; l= vehicle load level l=1,..,L; m=carbon emission break m=1,..,M.

Parameters: h_{ijt}: per unit inventory holding cost of the ith product at the jth warehouse during the tth time period; φ_{ijt}: procurement cost of the ith product at the jth warehouse during the tth time period; h_{irt}: per unit inventory holding cost of the ith product at the rth retail store during the tth time period; d_{abjc} &T_{jc}: distance and transportation cost per km between the nodes a and b belonging to the cth cluster connected to the jth warehouse resp.; D_{irt}: forecasted demand of the ith product at the rth retail store during the tth time period and σD_{irt} is it's standard deviation; Cap_j: capacity of the jth warehouse; w_i: weight of the ith product; Pc: total number of nodes in the cth cluster; $z_{(1-\alpha)}$: service level or fill rate; IO: initial inventory at the warehouse; INI: initial inventory at the retail store; Co_{lmjct}: carbon emission value of lth vehicle load level for c^{th} cluster connected to the jth warehouse in t^{th} time period; A_{lmjct}: load threshold beyond which a particular vehicle type is used; lt_{irt}: lead time to deliver i^{th} product at retail store r in time period t and σl_{irt} is it's standard deviation.

Decision Variables: I_{ijt}: ending inventory of the ith product at j^{th} warehouse during t^{th} time period; IN_{irt}: ending inventory of ith product at r^{th} retail store during t^{th} time period; X_{ijt}: quantity procured of ith product at j^{th} warehouse during t^{th} time period; Q_{ijrt}: quantity flowing of the ith product from j^{th} warehouse to r^{th} retail store during t^{th} time period; G_{abjc}: 1, If vehicle travels from node a to node b belonging to c^{th} cluster connected to j^{th} warehouse and otherwise zero; G_{jc}: 1, if j^{th} warehouse is connected to c^{th} cluster, or 0 otherwise; Q_{jct}: weighted quantity flowing of all the products from j^{th} warehouse to c^{th} cluster; σ_{irt}: safety stock of ith product at retail store r in time period t; R_{lmjct}: 1, if l^{th} vehicle as per load gets activated for transporting weighted quantity from warehouse j to cluster c in time period t or 0 otherwise; V_{lmjct}: 1, if m^{th} carbon emission break gets activated out of lth activated vehicle or 0 otherwise; e_a and e_b are positive values that ensures no sub-routing in travelling salesman; A: number of nodes.

First objective function includes the cost of holding at warehouse and retail store, cost of procurement, cost of holding safety stock and transportation cost.

$$Min Z_1 = \sum_{i=1}^{P}\sum_{t=1}^{T}\left[\sum_{j=1}^{J}\left(h_{ijt}I_{ijt} + \phi_{ijt}X_{ijt}\right) + \sum_{r=1}^{R}h'_{irt}\left(IN_{irt} + \sigma_{irt}Z_{(1-\alpha)irt}\right)\right] \quad (3)$$

$$+ \ T^{*}\sum_{c=1}^{C}\sum_{j=1}^{J}\sum_{b=0}^{A}\sum_{a=0}^{A}d_{abjc}G_{abjc}G_{jc}T_{jc}$$

Second objective minimizes total carbon emission on the basis of vehicle load and carbon emission value.

$$Min\, Z_2 = \sum_{t=1}^{T}\sum_{c=1}^{C}\sum_{j=1}^{J}\sum_{m=1}^{M}\sum_{l=1}^{L}Co_{lmjct}R_{lmjct}V_{lmjct} \quad (4)$$

Following three equations are inventory balancing equations at warehouse which are complemented by fourth equation as a capacity constraint at warehouse:

$$I_{ijt} = I_{ijt-1} + X_{ijt} - \sum_{r=1}^{R}Q_{ijrt} \qquad \forall\, i,j,t>1 \quad (5)$$

$$I_{ijt} = IO + X_{ijt} - \sum_{r=1}^{R}Q_{ijrt} \qquad \forall\ i,j,\ t=1 \quad (6)$$

$$\sum_{t=1}^{T}I_{ijt} + \sum_{t=1}^{T}X_{ijt} \geq \sum_{t=1}^{T}\sum_{r=1}^{R}Q_{ijrt} \qquad \forall i,j \quad (7)$$

$$X_{ijt} \leq Cap_{ijt} \qquad \forall i,j,t \quad (8)$$

Next three equations explains inventory balancing equations at retail stores

$$IN_{irt} = IN_{irt-1} + \sum_{j=1}^{J}Q_{ijrt} - D_{irt} \quad \forall i,r,t>1 \quad (9)$$

$$IN_{irt} = INI + \sum_{j=1}^{J}Q_{ijrt} - D_{irt} \quad \forall i,r,t=1 \quad (10)$$

$$\sum_{t=1}^{T}IN_{irt} + \sum_{t=1}^{T}\sum_{j=1}^{J}Q_{ijrt} \geq \sum_{t=1}^{T}D_{irt} \quad \forall i,r \quad (11)$$

$$Q_{jct} = \sum_{\substack{r=1 \\ r\in c}}^{R}w_iQ_{ijrt}G_{jc} \quad \forall i,j,c,t \quad (12)$$

Equation 12 is a connector between procurement and distribution stage of the model.

Next set of equations are the travelling salesman problem equations which directly create the shortest distance path among retail stores in a particular cluster and also connects it to a particular warehouse.

$$\sum_{a=1}^{A} G_{abjc} = 1 \ \forall \ b < P_c, j, c, a \neq b \tag{13}$$

$$\sum_{b=1}^{A} G_{abjc} = 1 \ \forall \ b < P_c, j, c, a \neq b \tag{14}$$

$$\sum_{a=1}^{A} G_{abjc} = \sum_{a=1}^{A} G_{bajc} \quad \forall \ b < P_c, j, c, a \neq b \tag{15}$$

$$\sum_{j=1}^{J} G_{jc} \sum_{a=1}^{A} G_{abjc} = 1 \qquad \forall c, b, a \neq b \tag{16}$$

$$e_a - e_b + P_c G_{abjc} \leq P_c - 1 \quad \forall \, j, c, \ a \neq 1, a, b < P_c, b \neq 1, a \neq b \tag{17}$$

Following three equations check on choosing a vehicle with optimum load and minimum carbon emission.

$$Q_{jct} \geq A_{lmjct} R_{lmjct} \quad \forall \, l, m, j, c, t \tag{18}$$

$$\sum_{l=1}^{L} R_{lmjct} = 1 \quad \forall m, j, c, t \tag{19}$$

$$\sum_{m=1}^{M} \sum_{l=1}^{L} R_{lmjct} V_{lmjct} = 1 \quad \forall j, c, t \tag{20}$$

Safety stock at retail store is calculated on the basis of deviations in demand and lead time has been explained in the next equation.

$$\sigma_{irt} = \sqrt{lt_{irt} \sigma^2_{D\,irt} + D^2_{irt} \sigma^2_{l\,irt}} \qquad \forall i, r, t \tag{21}$$

$$G_{jc}, G_{abjc}, R_{lmjct}, V_{lmjct} \ are\, 0 \ or \ 1; \ X_{ijt}, \ Q_{ijrt}, \ I_{ijt}, \ IN_{irt}, \ Q_{jct}, \sigma_{irt} \geq 0$$

The above formulated model is a bi-objective formulation and very complex in structure, which cannot be solved as normal mathematical programming model. Hence goal programming method is employed with prior fixed priorities for objectives.

3.2. *Goal programming approach [18]*

Step 1: Solve the mathematical model keeping one objective at a point and the values of Z_1* and Z_2*. Step 2: Min Z_1 and Min Z_2 s.t. $Z_1 \geq Z_1$*; $Z_2 \geq Z_2$*; and all the constraints. This problem provides infeasible solution. Step 3: Min η-ρ s.t. constraints' LHS+η-ρ=RHS. Step 4: Min η-ρ s.t. constraints' LHS+η-ρ=RHS; $Z_1+\eta_1-\rho_1= Z_1^*$and $Z_2+\eta_2-\rho_2= Z_2^*$. The above process helps in reaching at a compromised solution, which creates a trade-off between objectives also.

4. Model Formulation and Solution Method

The complete data provided by the company is difficult to present in the paper due to page limitations. A small set of data and their ranges have been displayed as follows:

Table 2. Forecasted demand for P1.

P1	*T1*	*T2*	*T3*	*T4*	*T5*	*T6*	*T7*	*T8*	*T9*	*T10*	*T11*	*T12*
R1	15	17	18	18	17	10	13	18	17	15	10	18
R2	19	16	14	11	12	18	18	17	19	26	16	17
R3	3	2	3	2	3	3	4	5	2	3	2	4
R4	12	13	11	13	11	11	13	13	14	13	12	11
R5	12	11	12	11	11	11	11	11	13	12	12	12
R6	2	3	1	2	0	3	4	2	1	2	3	3
R7	15	17	14	11	11	14	14	19	16	18	16	13
R8	12	13	13	12	15	17	17	18	15	17	18	18
R9	15	24	17	13	17	17	17	21	24	15	17	12
R10	13	11	11	11	11	12	12	11	11	11	11	11
R11	12	12	12	11	11	11	11	12	12	11	11	11
R12	1	2	2	1	2	3	2	1	2	1	2	1
R13	19	16	15	16	14	15	15	19	16	12	13	11
R14	12	13	13	11	11	12	11	13	14	11	11	12
R15	17	13	12	13	12	12	12	16	13	13	15	13

Table 3. Forecasted demand for P2.

P2	*T1*	*T2*	*T3*	*T4*	*T5*	*T6*	*T7*	*T8*	*T9*	*T10*	*T11*	*T12*
R1	24	27	23	19	27	24	18	30	21	19	29	22
R2	23	20	17	20	17	24	27	22	14	23	31	10
R3	4	2	3	5	4	3	2	3	4	1	2	3
R4	12	14	11	11	14	13	13	14	11	11	12	13
R5	12	11	12	14	12	12	12	13	15	17	15	13
R6	3	2	1	2	3	2	4	2	3	2	3	1
R7	18	12	13	19	11	10	12	13	14	16	10	10
R8	20	28	24	14	23	21	24	56	14	36	24	21
R9	13	15	10	13	14	11	14	27	19	15	23	25
R10	12	6	3	2	4	3	2	3	2	2	3	0
R11	11	15	13	12	13	12	15	13	13	15	15	14
R12	13	11	11	4	11	11	12	15	12	12	13	12
R13	16	10	19	17	17	16	17	14	12	19	17	15
R14	15	14	14	16	15	15	16	15	14	17	15	14
R15	17	14	14	16	17	15	15	14	15	15	14	18

Table 4. Forecasted demand for P3.

P3	*T1*	*T2*	*T3*	*T4*	*T5*	*T6*	*T7*	*T8*	*T9*	*T10*	*T11*	*T12*
R1	11	12	16	11	19	10	15	21	13	17	26	19
R2	44	49	33	48	46	43	56	33	28	71	36	43
R3	2	3	4	1	2	3	4	2	1	3	2	2
R4	13	11	11	13	11	12	11	17	17	17	14	13
R5	19	14	12	13	12	14	15	15	16	10	10	16
R6	11	11	11	11	11	12	12	14	18	12	17	11
R7	18	19	14	18	10	14	18	27	24	18	23	13
R8	15	15	12	15	16	11	14	34	20	21	20	12
R9	48	32	18	18	13	24	37	55	42	39	39	35
R10	14	12	13	12	12	15	17	19	18	15	13	17
R11	12	14	15	16	11	12	11	18	13	17	15	17
R12	14	14	13	11	12	13	13	17	12	13	12	11
R13	13	11	18	11	14	17	15	21	11	17	16	19
R14	14	11	18	18	17	19	14	19	14	13	13	19
R15	17	6	16	4	12	12	17	19	15	15	18	12

Weight of each product is 1.5 kg. Service level or fill rate is 98%. Lead time for delivery varies from 1 day to 2 days with the standard deviation of approx 10%. On the other side demand deviation is also approx 10%. Holding cost at both the warehouses varies from Rs. 7 to Rs. 10 per item. On the other side holding cost at retail store is lower than the cost at warehouse, which does not vary product wise. However, variation takes place retail store wise. Procurement cost and transportation cost contribute major portion in the total cost.

Table 5. Procurement cost per product.

Product	P1	P2	P3	P4	P5	P6	P7	P8	P9	P10	P11
Purchase Cost	1876	2292	2372	2596	2504	2396	2156	2452	2476	2556	2676

Transportation is to take place among various retail stores of each cluster, which is considered as per the warehouse distant from the connected retail stores. From first warehouse the cost per kilometre is Rs. 12 and Rs. 13 from second warehouse to their connected retail stores. Variation in cost takes place as per the distances among all the nodes in each cluster.

Minimizing carbon emission is the second important objective for the company, for which carbon emission varies as per vehicle type. Company employs heterogeneous vehicle type in terms of load as well level of carbon emission. The model checks both the levels i.e. firstly, a truck type is chosen as per the load to be transported from warehouse to a particular cluster, and then next level option in truck chooses different level of carbon emission.

4.1. *Results and discussions*

High product cost is leading high procurement cost out of total cost. Resulting cost aspiration level as Rs.63632281 and carbon emission aspiration level as 18785 from step 1 of goal programming approach that helps in creating a trade off between the two objectives. At the end the cost and carbon emission values are bit higher than the aspiration values. The total is Rs.65932848 and carbon emission value is 20392. We are proposing a green supply chain network and reducing carbon emission with some target value becomes really essential in such case. Initially we understood the style of transportation of the company, which has given us the data on carbon emission value because of repeat movement in the same location for different retail stores. Carbon emission value provided by the company on the basis of their older transportation style is approx 22365. The model we proposed is trying to show a trade-off between the two objectives i.e. trying to minimise cost as well carbon emission values. It is observed from the results that the trade-off carbon emission value is 20392, which is approximately 10% lower than the company's exiting value.

The resulted sheets of Procured, transported quantity are as follows:

Table 6. Procured quantity at WH1.

WH 1	*T1*	*T2*	*T3*	*T4*	*T5*	*T6*	*T7*	*T8*	*T9*	*T10*	*T11*	*T12*
P1	68	87	28	64	55	83	91	68	70	63	133	93
P2	144	143	55	98	106	130	44	25	43	53	72	41
P3	187	114	55	41	83	58	158	176	94	181	189	231
P4	72	81	78	93	211	79	126	91	81	215	60	71
P5	74	80	138	71	99	132	120	85	111	71	188	97
P6	80	45	85	106	97	68	79	87	65	231	70	66
P7	97	69	59	41	98	127	120	93	118	173	176	73
P8	97	129	149	80	96	103	77	126	113	153	80	122
P9	67	86	97	91	107	88	132	114	82	97	63	138
P10	73	106	110	192	179	104	128	111	108	330	59	128
P11	79	56	56	83	79	46	50	58	69	99	90	44

Table 7. Procured quantity at WH2.

WH 2	*T1*	*T2*	*T3*	*T4*	*T5*	*T6*	*T7*	*T8*	*T9*	*T10*	*T11*	*T12*
P1	111	96	140	92	103	86	83	128	119	117	36	74
P2	69	58	133	86	96	62	159	229	140	167	154	150
P3	78	120	169	179	135	173	111	155	168	117	85	28
P4	78	72	83	88	8	77	29	65	65	17	90	70
P5	201	202	113	205	170	125	183	271	240	266	158	191
P6	67	105	59	49	62	85	75	78	96	29	120	65
P7	71	97	102	130	80	41	50	79	73	74	23	73
P8	76	51	19	95	76	91	107	86	76	43	96	53
P9	103	85	74	90	73	71	41	72	112	149	119	24
P10	106	97	102	118	133	115	103	148	160	76	291	131
P11	32	64	55	90	55	79	69	47	73	96	50	62

Transported weighted quantity from warehouse to cluster is as follows:

Table 8. Weighted quantity from WH1 to clusters.

WH1	*T1*	*T2*	*T3*	*T4*	*T5*	*T6*	*T7*	*T8*	*T9*	*T10*	*T11*	*T12*
C1	1557	1494	1365	1440	1815	1527	1687.5	1551	1431	2499	1770	1656
C2	0	0	0	0	0	0	0	0	0	0	0	0
C3	1557	1494	1365	1440	1815	1527	1687.5	1551	1431	2499	1770	1656

Table 9. Weighted quantity from WH2 to clusters.

WH 2	*T1*	*T2*	*T3*	*T4*	*T5*	*T6*	*T7*	*T8*	*T9*	*T10*	*T11*	*T12*
C1	0	0	0	0	0	0	0	0	0	0	0	0
C2	1488	1570	1573	1833	1486	1507	1515	2037	1983	1726	1833	1381
C3	0	0	0	0	0	0	0	0	0	0	0	0

Zero values in the above two tables indicate that corresponding cluster is not attached to the warehouse.

From the model results, it is observed that company shall be able to fulfil the demand and a very small quantity of safety Stock shall be required at retail stores with 98% fill rate or service level. The resulted sheet is provided safety stock level from 2 units to 5 units per article.

From Tables 8 and 9, it has been observed that cluster 1 and 3 are connected with first warehouse and cluster 2 is connected with warehouse 2. Table 1 has also already shown the connection between clusters and retails stores. Now, we are showing the route within each cluster, where all the connected retail stores have been covered with the help of TSP. The route between *warehouse 1* and *first cluster* is as; Start from *warehouse 1* then move to R_3 further in a sequence R_7, R_{11}, R_5, R_4 and returns to warehouse. Again, from *first warehouse*, the route for *third cluster* is as; Start from *warehouse 1* then move to R_9 further in a sequence R_{16}, R_8, R_1 and returns to warehouse. *Cluster 2* is connected with second warehouse and the route is as; Start from *warehouse 2* then in a sequence connect R_{12}, R_{13}, R_{14}, R_{15}, R_{10}, R_6 and finally back distance among warehouses and the retail stores.

While transporting weighted quantity from warehouse to cluster, carbon emission is the major concern that is to be kept at minimum. On one side TSP helped in creating shortest distance, which automatically helps in reducing the carbon emission. On the other side, transportation takes place using heterogeneous vehicles in terms of load and carbon emission. R_{lmjct} is a binary variable helped in find truck type with optimum load to be transported and V_{lmjct} is also a binary variable and checks on choosing the truck with lowest carbon emission out of chosen trucks as per optimum load. The model is able to provide the both. From the results, we have observed that weighted transported quantity

Q_{jct} is sufficiently large to use bigger trucks; hence generally second and third level of truck load have been used. The data provided to us is showing different vehicles with multi load and carbon emission. Here, we can interpret that, while transporting from warehouse 1 to cluster 1 and cluster 3 in first period, third truck with optimum load and minimum truck capacity has been used. Similarly the vehicle has been chosen in other time periods. The table below is only showing binary value 1 results, which indicates about a particular vehicle with best combination of vehicle load and carbon emission.

Table 10. Vehicle choosing for time period 1 and 2.

Level	*Break*	*WH*	*Cluster*	*Time*	**Rlmjct* Vlmjct**	*Level*	*Break*	*WH*	*Cluster*	*Time*	**Rlmjct* Vlmjct**
L3	M3	J1	C1	T1	1	L3	M2	J1	C1	T2	1
L1	M1	J2	C1	T1	1	L1	M1	J2	C1	T2	1
L1	M1	J1	C2	T1	1	L1	M1	J1	C2	T2	1
L3	M2	J2	C2	T1	1	L2	M2	J2	C2	T2	1
L1	M1	J1	C3	T1	1	L2	M1	J1	C3	T2	1
L1	M1	J2	C3	T1	1	L1	M1	J2	C3	T2	1

Here, we are able to show solution table only for 2 time periods. Same way we got the results for all time points and helps in calculating carbon emission.

5. Conclusion

This study analyzes an integrated procurement and transportation planning problem with carbon emissions calculations. The model minimizes the combined expense associated with the cost of procurement, holding and safety stock. Other important aspect is transportation from warehouse to retail stores. Travelling salesman problem helped in creating shortest route among connected retail stores and warehouse. In the case of many retail stores, transporting goods to each retail store individually becomes expensive, hence clustering method is proposed, that helped in creating small groups from many retail stores on the basis of distance. Heterogeneity in vehicle type is another area which can show options to transport different quantity level. In the paper, carbon emission is an important aspect for reducing it from the current level by choosing vehicle type. Therefore, heterogeneity in vehicle does not take place only for load but also for carbon emission. Each vehicle type is considered to have distinct per vehicle capacities as well as distinct emissions generations. In particular, each vehicle type is assumed to generate a specific amount of emissions due to its empty vehicle weight and the rates of emissions generated due to the loads of the vehicles depend on the vehicle type.

The proposed model is bi-objective in nature with constraints like inventory balancing, warehouse capacity, travelling salesman and load transportation thresholds for choosing vehicle. Creating coordination among many entities of supply chain shows complexity in nature. Solving it as a mathematical model is very difficult and was leading us to infeasibility. Thus, goal programming approach is used to generate a trade-off between the objectives with respect to the constraints.

One of the possible future research directions is to study the combined impact of forward supply chain with reverse supply chain network. Closed loop supply chain covers very small part of forward supply chain. Therefore, including all the possible complexities of forward supply chain in closed loop can help industry to understand and manage complexities at their end in much better way.

References

1. L. G. Cooper, P. Baron, W. Levy, M. Swisher and P. Gogos, "PromoCasttrademark: A new forecasting method for promotion planning," *Market. Sci.,* vol. 18, pp. 301–316, 1999.
2. A. Clark and H. Scarf, "Optimal policies for amulti-echelon inventory problem," *Manage. Sci.*, vol.6, pp. 475–490, 1960.
3. C. Schmidt and S. Nahmias, "Optimal policy for a two-stage assembly system under random demand," *Oper. Res., vol. 33*, pp. 1130–1145, 1985.
4. H. L. Lee and C. Billington, "Material management in decentralized supply chains," *Oper. Res.*, vol. *41*, pp. 835–847, 1993.
5. P. Glasserman and S. Tayur, "Sensitivity analysis for base-stock levels in multiechelon production-inventory systems," *Manage. Sci.*, vol. 41, pp. 263–281, 1995.
6. M. Ettl, G. E. Feigin, G. Y. Lin and D. D. Yao, "A supply network model with base-stock control and service requirements," *Oper. Res.*, vol. 48, pp. 216–232, 2000.
7. S. C. Graves and S. P. Willems, "Optimizing strategic safety stock placement in supply chains," *Manuf. & Ser. Oper. Manage.*, vol. 2, pp. 68–83, 2000.
8. C. Sitompul, E. Aghezzaf, W. Dullaert and H. Van Landeghem, "Safety stock placement problem in capacitated supply chains," *Int. J. Prod. Res.*, vol. 46, pp. 4709–4727, 2008.

9. Y. Boulaksil, J. C. Fransoo, N. G. Ernico, “Setting safety stocks in multi-stage inventory systems under rolling horizon mathematical programming models,” *OR Spectrum*, vol. 31, pp. 121–140, 2009.
10. M. A. Louly, A. Dolgui, “Calculating safety stocks for assembly systems with random component procurement lead times: a branch and bound algorithm,” *Eur. J. Oper. Res*., vol. 199, pp. 723–731, 2009.
11. B. Kalkanci, E. Ang, E. Plambeck, “Measurement and Improvement of Environmental Performance Under Voluntary Versus Mandatory Eco-labels,” *Extended Abstract, MSOM Conference*, 2012.
12. M. Joana, C. Martí, J. S. Tancrez, R. W. Seifert, “Carbon Footprint and Responsiveness Trade-offs in Supply Chain Network Design,” *Int. J. Prod. Eco*., http://dx.doi.org/10.1016/j.ijpe.2015.04.016
13. C. Rizet, M. Browne, E. Cornelis, J. Leonardi, “Assessing carbon footprint and energy efficiency in competing supply chains: review – case studies and benchmarking,” *Transport. Res. Part D: Transp. Environ*. Vol. 17, pp. 293–300, 2012.
14. K. Hoen, T. Tan, J. Fransoo, G. van Houtum, “Effect of carbon emission regulations on transport mode selection under stochastic demand,” *Flexible Ser. Manuf. J*., vol. 26, pp. 170–195, 2014.
15. S. Pan, E. Ballot, F. Fontane, “The reduction of greenhouse gas emissions from freight transport by pooling supply chains,” *Int. J. Product. Econ*., vol. 143, pp. 86–94, 2013.
16. D. Konur, “Carbon constrained integrated inventory control and truckload transportation with heterogeneous freight trucks,” *Int. J. Product. Econ*. Vol. 153, pp. 268–279, 2014.
17. S. Lloyd, “Least squares quantization in pcm,” *IEEE Trans. Inf. Theor*., vol. 28, pp. 129–137, September 2006.
18. A. Charnes and W. W. Cooper, “Goal programming and multiple objective optimizations.” *European Journal of Operational Research*, vol. 1, pp. 39-54, 1997.

Bi-Objective Optimization Model for Fault-Tolerant Embedded Systems Under Build-Or-Buy Strategy Incorporating Recovery Block Scheme

R. Kaur

Department of Computer Science, Banasthali University,
Rajasthan, India
E-mail: rrdk_07@yahoo.com

S. Arora[1], P. C. Jha[2]

Department of Operational Research, University of Delhi,
Delhi, India
E-mail: [1]stu.aro@gmail.com, [2]jhapc@yahoo.com

S. Madan

Lady Shri Ram College for Women, University of Delhi,
Delhi, India
E-mail: sushila_lsr@yahoo.com

The IT industry has enthralled the pace of development of technologies. The initial focus, which was towards the development and maintenance of hardware, has shifted towards embedded systems (a combination of hardware and software) such as routers, LCDs, automated water boilers, mobile phones, and satellites for direct telecast on televisions. With sophistication in technology, reliability issues pertaining to both software and hardware have simultaneously raised. For rendering the embedded systems fault tolerant, various redundancy techniques can be applied. Numerous models for the optimization of reliability of embedded systems have been proposed, based on different fault tolerant techniques like NVP, RB scheme but has no roots in "build-or-buy". The software components can be procured as COTS from vendor or can be built in-house. This decision is an important part of development process in which either choice can be valid depending on the situation. The proposed models are based on RB/1/1 fault tolerant technique which allows the subsystem to tolerate one hardware and one software fault. The first optimization model relinquishes to selection of hardware and software components for embedded system, incorporating build-or-buy strategy for the software components, such that overall system reliability can be maximized simultaneously minimizing the overall cost. Besides procuring the software component as COTS or developing it in-house, a pre-existing component can be made reusable after fabrication. The extended model shall incorporate reuse-build-buy decision for software components.

Keywords: Embedded system; Reliability; Optimization; Build-or-Buy; RB/1/1; Fault Tolerance.

1. Introduction

With the increasing complexity and advances in the field of technology, developing failure-free embedded systems has become indispensable. The advent of more sophisticated embedded systems supporting powerful functions, while relying on profound submicron process technologies for their fabrication, have brought reliability concerns to the forefront [20]. Higher reliability can be achieved by including redundancy in the embedded systems at hardware and/or software level in order to make them fault tolerant. Various fault tolerant architectures for embedded system were proposed based on N-Version Programming (NVP) and Recovery Block (RB) schemes [1][3-6]. The variants are NVP/0/1 architecture, NVP/1/1 architecture and RB/1/1 architecture [7-8]. Later, [9-10] proposed models for fault tolerant architectures in distributed embedded systems to optimize their overall reliability within a specified budget, with an expense of additional resources [11]. RB/1/1 is an extension of RB scheme [2], which offers a fault tolerant architecture for embedded system. The generalized architecture is *X/i/j*, where *X* is a fault tolerant technique, *i* is the number of hardware faults tolerated & *j* is the number of software faults tolerated [7]. The system underlying this architecture comprises of two hardware components, each with two independent software versions, primary (PV) & secondary (SV) as in Fig. 1. Initially, the PV is executed & its result is submitted to Acceptance Test (AT) for evaluation. If AT is passed, result is adjudged correct. Otherwise, the process rolls back to initial state & SV is executed. The system shall fail due to the occurrence of any of the following events: existence of related faults amongst both software versions; failure of AT; faults in specifications; failure of both; versions.

Studies have been conducted on series-parallel embedded systems wherein redundancy allocation problem has been formulated for a system with subsystems (SS) in series [19]. A similar architecture has been considered in this paper as in Fig. 2. Each SS may have RB/1/1, incorporating redundant hardware & software components. The intent is to increase reliability, which on the contrary, increases the cost as well. This enforces the developer to conduct a trade-off between reliability & cost.

Just as the hardware components can be procured from multiple vendors, the software components can also be procured as ready-to-use components and are known as Commercial Off-The Shelf (COTS) components [16][17]. This approach of software development is known as Component Based Development (CBD) approach [17]. It has roots in software as well as hardware and involves the purchase and integration of pre-made components [21-22]. For a module, more than one functionally equivalent COTS component may be available in market differing in cost, reliability (Rel), delivery time (DT) and the number of source lines of code (SLOC). Besides, the software components can also be developed in-house from scratch. When no COTS component is available for a functionality, developing this component in-house becomes obligatory. The decision of, whether to purchase a software component as COTS or to develop it in-house, is known as build-or-buy decision [16] [23], which depends upon various factors such as availability of expertise, specificity of requirements, etc. CBD approach promotes developing only those components in-house which can be reused to minimize the development cost and effort Some pre-existing components can be reused after fabrication [12] [18], which entails fabrication cost & time. It becomes challenging for the developer to decide optimally whether to fabricate, or develop in-house from scratch, or procure it as COTS. This decision is referred to as Reuse-or-Build-or-Buy decision [24] and depends upon several parameters.

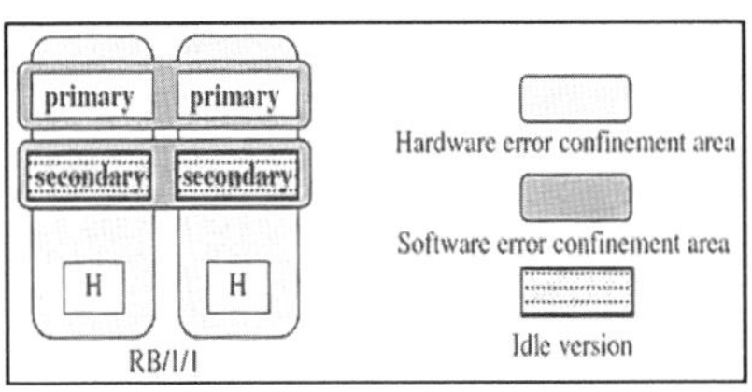

Fig. 1. RB/1/1 fault tolerant architecture [9].

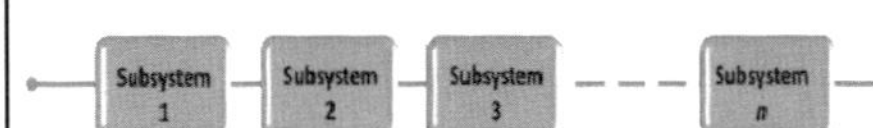

Fig. 2. Framework of subsystem based modularized embedded system.

This paper presents two optimization models, each with the objective of maximizing the reliability of fault tolerant embedded system, simultaneously minimizing the cost under constraints of DT and SLOC. The first model is based on the build-or-buy approach for software components. The extended model, in addition to the above constraints, considers reusing of some pre-existing in-house built software components after fabrication, thereby, incorporating Reuse-or-Build-or-Buy approach [12][24]. Both the models are based on RB/1/1 fault tolerant architecture and aim at optimal selection of hardware & software components as well as the redundancy levels for each SS such that system reliability is maximized while total cost is minimized.

2. Notations and Assumptions

2.1. *Notations*

n	Number of subsystems within the embedded system
m_i	Number of hardware component choices available for subsystem i; $i=1,...,n$
p_i	Number of COTS software alternatives (versions) available for subsystem i; $i=1,...,n$
C^{hw}_{ij}	Cost of using j^{th} hardware component for i^{th} subsystem; $i=1,...,n; j=1,...,m_i$
C^{sw}_{ik}	Cost of kth COTS software component available for i^{th} subsystem; $i=1,...n; k=1,...p_i$
C^{sw}_{i0}	Net cost incurred if in-house built software component is deployed in subsystem $i;\forall i$
c_i	Unitary development cost of in-house built software component for subsystem i
t_i	Estimated development time of in-house built software instance for subsystem $i;\forall i$
τ_i	Average time required to perform a test case on built component for i^{th} subsystem; $\forall i$
N^{tot}_i	Total number of tests performed on built software component for i^{th} subsystem; $\forall i$
N^{suc}_i	Number of successful tests performed on built software component for i^{th} subsystem
π_i	Testability for in-house developed software instance of subsystem i; $\forall i$
R^{sw}_{i0}	Probability that the in-house built software component for i^{th} subsystem is failure free during a single run given that N^{suc}_i test cases have been successfully performed; $\forall i$
R^{sw}_{ik}	Reliability of kth COTS software component for i^{th} subsystem; $\forall i, k$
R^{hw}_{ij}	Reliability of jth hardware component for i^{th} subsystem; $\forall i, j$
R_i	Reliability of subsystem i; $\forall i$
l_{ik}	Number of SLOC for k^{th} COTS software component available for i^{th} subsystem; $\forall i, k$
l_{io}	Number of SLOC for built software component for subsystem i; $\forall i$
L	Limit set on total SLOC of the system
d^{hw}_{ij}	Delivery time for j^{th} hardware instance of ith subsystem $\forall i, j$
d^{sw}_{ik}	Delivery time for k^{th} COTS software instance of ith subsystem; $\forall i, k$
D	Bound set on overall delivery time of the system
x_{ij}	Decision variable; $i=1,...,n; j=1,...,m_i$

$$x_{ij} = \begin{cases} 1; \text{ if } j^{th} \text{ hardware component selected for subsystem } i\text{, without redundancy} \\ 2; \text{ if } j^{th} \text{ hardware component selected for subsystem } i\text{, with redundancy} \\ 0; \text{ if } j^{th} \text{ hardware component not selected for subsystem } i \end{cases}$$

y_{ik}	Binary decision variable; $i=1,...,n; k=1,...,p_i$

$$y_{ik} = \begin{cases} 1; \text{ if } k^{th} \text{ COTS component of } i\text{th subsystem is selected} \\ 0; \text{ otherwise} \end{cases}$$

y_{io}	Binary decision variable; $i=1,...,n$

$$y_{io} = \begin{cases} 1; \text{ if in-house developed software component is selected for } i^{th} \text{ subsystem} \\ 0; \text{ otherwise} \end{cases}$$

z_i	Binary decision variable; $i=1,...,n$

$$z_i = \begin{cases} 1; \text{ if } i^{th} \text{ subsystem uses RB/1/1 redundancy technique} \\ 0; \ i^{th} \text{ subsystem does not use any redundancy technique} \end{cases}$$

P^{rv}_i	Probability of failure from the related faults between software versions in subsystem i, which has RB/1/1 redundancy; $i=1,...,n;$ $Q^{rv}_i=1-P^{rv}_i$
$P^{rv}_{i,t,t'}$	Probability of failure of subsystem i from related faults between COTS software versions t and t'; $i=1,...,n;$ $t,t'=1,...,p_i$
$P^{rv}_{i,0,t}$	Probability of failure of subsystem i from related faults between in-house developed software instance and a COTS software component t; $i=1,...,n;$ $t=1,...p_i$
P^d	Probability of failure of Acceptance Test (AT); $Q^d=1-P^d$
P^{fs}_i	Probability of failure from faults in specification, that leads to failure of all software versions in subsystem i, which has RB/1/1 redundancy; $i=1,...,n;$ $Q^{fs}_i=1-P^{fs}_i$
P^h_i	Probability of failure of a hardware in subsystem i, which has RB/1/1 redundancy

P^{fbv}_i	Probability that both software versions of subsystem *i*, which has RB/1/1 redundancy, fail independently; $\forall i$
w_i	Binary decision variable $w_i = \begin{cases} 1; \text{ if fabricated reusable software component is selected for } i^{th} \text{ subsystem} \\ 0; \text{ otherwise} \end{cases}$
C^{fab}_i	Net cost of fabricated reusable component for subsystem *i*; $\forall i$
tf_i	Estimated fabrication time for reusable software component of subsystem *i*; $\forall i$
τf_i	Average time required to perform a test case on fabricated reusable software component for subsystem *i*; $\forall i$
Nf^{tot}_i	Total number of tests performed on fabricated reusable software component for SS*i*
Nf^{suc}_i	No. of successful tests performed on fabricated reusable software component for SS*i*
πf_i	Testability forfabricated reusable software instance of subsystem-*i*; $\forall i$
R^{fab}_i	Probability that the fabricated software component for i^{th} subsystem is failure free during a single run given that Nf^{suc}_i test cases have been successfully performed; $\forall i$
lf_i	Number of SLOC for fabricated software component for SS*i*; $\forall i$
$P^{rv}_{i,fab,t}$	Probability of failure of subsystem *i* from related faults between fabricated reusable software instance & a COTS software component *t*; $i=1,\ldots,n;\ \ t=1,\ldots p_i$
$P^{rv}_{i,fab,0}$	Probability of failure of ith SS from related faults between fabricated software component & in-house developed software instance; $\forall i$

2.2. *Assumptions*

1. System is modularized into several but finite number of subsystems (SS) connected in series.
2. Each SS has finite hardware & software components with known reliability and cost.
3. Each component is non-repairable, and has 2 states, functional or failed.
4. Software & hardware failures are not identical. System failure occurs when all software versions or hardware components fail.
5. Software components for a subsystem are available as COTS or developed in-house.
6. Information for Rel, Cost, SLOC and DT of COTS components is provided by vendor.
7. Cost &Rel of built component can be specified based on the parameters of development process.
8. While COTS software versions for a subsystem may be available in abundance; atmost 1 component can be developed in-house for each SS.
9. Each subsystem (SS) either has RB/1/1 redundancy or has no redundancy.
10. SS without any redundancy has exactly one hardware and software component.
11. SS with RB/1/1 has two identical hardware & two distinct software components. Failure of individual component are s-independent [9].

12. Delivery time for hardware component(s) of a subsystem is independent of number of units being purchased. Time taken for integration of components is negligible.

3. Optimization Models

3.1. *Optimization model under build-or-buy decision*

The software components available for a subsystem (SS) can be COTS and/orin-house built. While COTS versions are available in abundance, in-house developed components for each SS can be atmost 1. In case SS incorporates RB/1/1, the two selected software versions includes at least 1 COTS component. In case of no in-house development, both components are COTS. In case of no redundancy, only 1 software component is selected which is either built or bought. Let the embedded system comprises of *n* SS, with m_i maximum number of hardware component choices & p_i number of COTS versions for i^{th} SS. The optimization model for component selection can be written as follows:

Problem (P1)

$$Max\ R = \prod_{i=1}^{n} R_i \tag{1}$$

$$Min\ C = \sum_{i=1}^{n}\left[\sum_{j=1}^{m_i} C_{ij}^{hw} x_{ij} + \sum_{k=0}^{p_i} C_{ik}^{sw} y_{ik}\right] \tag{2}$$

subject to

$$X \in S_1 = \left\{C_{i0}^{sw} = c_i\left(t_i + \tau_i N_i^{tot}\right)\right. ;\ i = 1, ..., n \tag{3}$$

$$\sum_{i=1}^{n}\sum_{k=0}^{p_i} l_{ik} y_{ik} \le L \tag{4}$$

$$\sum_{j=1}^{m_i} x_{ij} - z_i = 1\ ;\ i = 1, ..., n \tag{5}$$

$$\sum_{k=0}^{p_i} y_{ik} - z_i = 1\ ;\ i = 1, ..., n \tag{6}$$

$$x_{ij} \times x_{il} = 0\ ;\ i = 1, ..., n;\ j, l = 1, ..., m_i;\ j \ne l \tag{7}$$

$$\underset{\substack{i=1,...,n \\ j=1,...,m_i}}{\text{Max}} \frac{d_{ij}^{hw} \cdot x_{ij}}{(z_i + 1)} \le D \tag{8}$$

$$\underset{\substack{i=1,...,n \\ k=1,...,p_i}}{\text{Max}} d_{ik}^{sw} \cdot y_{ik} \le D \tag{9}$$

$$\max_{i=1,\dots,n} \left(t_i + \tau_i N_i^{tot}\right) y_{i0} \le D \tag{10}$$

$$R_{i0}^{sw} = \frac{1-\pi_i}{(1-\pi_i) + \pi_i (1-\pi_i)^{N_i^{suc}}} \quad ; i = 1,\dots,n \tag{11}$$

$$N_i^{suc} = (1-\pi_i) N_i^{tot} \quad ; i = 1,\dots,n \tag{12}$$

$$R_i = \left(1 - z_i\right)\left[\left(\sum_{j=1}^{m_i} R_{ij}^{hw} x_{ij}\right)\cdot\left(\sum_{k=0}^{p_i} R_{ik}^{sw} y_{ik}\right)\right] + z_i\left(1 - P_i\right) \quad ; \forall i \tag{13}$$

$$P_i = P_i^{rv} + Q_i^{rv} P_d + Q_i^{rv} Q_d P_i^{fs} + Q_i^{rv} Q_d Q_i^{fs} (P_i^h)^2 + Q_i^{rv} Q_d Q_i^{fs} \left(1 - (P_i^h)^2\right) P_i^{fbv} \quad ; i = 1,\dots,n \tag{14}$$

$$P_i^{rv} = \sum_{t=0}^{p_i-1} \sum_{t'=t+1}^{p_i} P_{i,t,t'}^{rv} y_{it} y_{it'} \quad ; i = 1,\dots,n \tag{15}$$

$$P_i^h = \sum_{j=1}^{m_i} \frac{\left(1 - R_{ij}^{hw}\right) x_{ij}}{\left(z_i + 1\right)} \quad ; i = 1,\dots,n \tag{16}$$

$$P_i^{fbv} = \sum_{t=0}^{p_i-1} \sum_{t'=t+1}^{p_i} \left[\left(1 - R_{it}^{sw}\right) y_{it}\right]\cdot\left[\left(1 - R_{it'}^{sw}\right) y_{it'}\right] \quad ; \forall i \tag{17}$$

$$x_{ij} \in \{0,1,2\}; \; i = 1,\dots,n; \; j = 1,\dots,m_i \tag{18}$$

$$y_{ik} \in \{0,1\}; \; i = 1,\dots,n; \; k = 0,1,\dots,p_i \tag{19}$$

$$z_i \in \{0,1\}; \; i = 1,\dots,n \quad \} \tag{20}$$

Objective function (1) maximizes system reliability, which is the product of reliabilities of its subsystems. Total cost (selected hardware & software components) is minimized in (2). Cost of built component depends upon development time & testing time, & is given in (3). Total SLOC (sum of SLOC of selected software components) should be within prescribed limit L as in (4). Constraint (5) represents the selection of exactly 1 hardware component in case of no redundancy, and 2 components if SS has redundant architecture. Similarly, selection of exactly 1 software component in case of no redundancy, & selection of 2 components if redundancy exists, is shown in (6). Eq. (7) guarantees selection of identical hardware components for a SS. Since DT of system is bounded, constraints (8), (9) & (10) ensures delivery of selected hardware & software components within D time units. Eq. (11) evaluates reliability for in-house built component, using testability & no. of successful test cases [13-14]. Number of successful test cases can be evaluated as shown in (12). Reliability of ith SS is computed in (13), which isproduct of reliabilities of selected hardware

& software component if SS has no redundancy;else it is computed on the basis of reliability for RB/1/1 architecture. If i^{th} SS has RB/1/1, then its unreliability is given by (14). Further, probability of failure of SS due to related faults between software versions is given by (15). Eq. (16) gives probability of the failure of a selected hardware component. Probability for failure of SS due to independent failures of both software versions is given by (17). Eq. (18) to (20) represent decision variables.

3.2. ***Optimization model under reuse-build-buy decision***

The proposed model is an extension of the model discussed in section 3.1. It assumes that besides COTS & in-house built software components, some pre-existing built components can be reused after fabrication. For each SS, atmost one reusable component is available.The optimization model for component selection can be written as follows:

Problem (P2)

$$Max\ R = \prod_{i=1}^{n} R_i \tag{21}$$

$$Min\ C = \sum_{i=1}^{n}\left[\sum_{j=1}^{m_i} C_{ij}^{hw} x_{ij} + \sum_{k=0}^{p_i} C_{ik}^{sw} y_{ik} + C_i^{fab} w_i\right] \tag{22}$$

subject to

$$X \in S_2 = \{\text{Constraint (3), (5), (7) to (12), (14), (16), (18) to (20)} \tag{23}$$

$$C_i^{fab} = c_i\left(tf_i + \tau f_i N f_i^{tot}\right) \ ; i = 1, ..., n$$

$$\sum_{i=1}^{n}\sum_{k=0}^{p_i} l_{ik} y_{ik} + lf_i w_i \le L \tag{24}$$

$$\sum_{k=0}^{p_i} y_{ik} + w_i - z_i = 1 \ ; i = 1, ..., n \tag{25}$$

$$\underset{i=1,...,n}{\text{Max}} \left(tf_i + \tau f_i Nf_i^{tot}\right) w_i \le D \tag{26}$$

$$R_i^{fab} = \frac{1 - \pi f_i}{(1 - \pi f_i) + \pi f_i (1 - \pi f_i)^{Nf_i^{suc}}} \ ; i = 1, ..., n \tag{27}$$

$$Nf_i^{suc} = (1 - \pi f_i) Nf_i^{tot} \ ; i = 1, ..., n \tag{28}$$

$$R_i = (1 - z_i)\left[\left(\sum_{j=1}^{m_i} R_{ij}^{hw} x_{ij}\right)\cdot\left(\sum_{k=0}^{p_i} R_{ik}^{sw} y_{ik} + R_i^{fab} w_i\right)\right] + z_i\left(1 - P_i\right); \forall\ i \tag{29}$$

$$P_i^{rv} = \sum_{t=0}^{p_i - 1}\sum_{t'=t+1}^{p_i} P_{i,t,t'}^{rv} y_{it} y_{it'} + \sum_{t=0}^{p_i} P_{i,fab,t}^{rv} y_{it} w_i \ ; i = 1, ..., n \tag{30}$$

$$P_i^{fbv} = \sum_{t=0}^{p_i-1} \sum_{t'=t+1}^{p_i} \left[\left(1 - R_{it}^{sw}\right) y_{it}\right] \cdot \left[\left(1 - R_{it'}^{sw}\right) y_{it'}\right] + \sum_{t=0}^{p_i} \left[\left(1 - R_{it}^{sw}\right) y_{it}\right] \cdot \left[\left(1 - R_i^{fab}\right) w_i\right] \; ; i = 1, \ldots, n \tag{31}$$

$$w_i \in \{0, 1\} \; ; \; i = 1, \ldots, n \quad \} \tag{32}$$

Here, objective functions (21) & (22) are for maximizing the system Rel & minimizing the overall cost respectively. Eq. (23) gives estimated fabrication cost for reusable component for i^{th} subsystem (SS). Bound on SLOC of selected software components is given by (24). Eq. (25) denotes the selection of one software component in case of no redundancy, and two software components in case of redundancy for i^{th} SS. Bound on DT for reusable software component is given by (26). Eq. (27) evaluates Rel for i^{th} SS's reusable component. No. of successful test cases performed on fabricated component is given by (28). Rel for i^{th} SS is computed by (29). If i^{th} SS has RB/1/1 redundancy, then its unreliability is computed by using (14). Some of the parameters used in (14) can be evaluated from Eq. (16), (30) & (31). Binary decision variable w_i is represented by (32).

4. Solution Procedure

To solve the bi-objective model, weighted sum method has been used [26]. Solution approach for optimization model (*P2*) has been presented, while solution approach for model (P1) shall be similar. Determine upper & lower bounds for each objective. R_{max}, upper bound of Reliability objective function in problem (P2) is determined by solving problem (*RU*). Lower bound for Reliability (R_{min}), is ascertained by solving problem (*RL*). C_{max}, upper bound for Cost objective problem (P2) is computed from problem (CU) and C_{min}, lower bound, from problem (CL). These bounds are then used to convert problem (P2) into an optimization model (P2') with an aggregated objective functions.

Problem (RU)

$$\text{Max } R = \prod_{i=1}^{n} R_i \quad \text{subject to} \quad X \in S_2$$

Problem (RL)

$$\text{Min } R = \prod_{i=1}^{n} R_i \quad \text{subject to} \quad X \in S_2$$

Problem (CU)

$$\text{Max } C = \sum_{i=1}^{n} \left[\sum_{j=1}^{m_i} C_{ij}^{hw} x_{ij} + \sum_{k=0}^{p_i} C_{ik}^{sw} y_{ik} + C_i^{fab} w_i \right] \quad \text{subject to} \quad X \in S_2$$

Problem (CL)

$$\text{Min } C=\sum_{i=1}^{n}\left[\sum_{j=1}^{m_i} C_{ij}^{hw} x_{ij}+\sum_{k=0}^{p_i} C_{ik}^{sw} y_{ik}+C_i^{fab} w_i\right]$$

subject to

$$X \in S_2$$

Assuming W_R & W_C are weights for reliability objective function & cost objective function resp., and $R' = \frac{R - R_{min}}{R_{max} - R_{min}}$ & $C' = \frac{C_{max} - C}{C_{max} - C_{min}}$, we have:

Problem (P2')

$$\begin{aligned} & Max\ W_R \cdot R' + W_C \cdot C' \\ & subject\ to \\ & \quad X \in S_2 \\ & \quad W_R + W_C = 1 \\ & \quad W_R, W_C \geq 0 \end{aligned}$$

The optimal solution for (P2') is also the efficient solution for bi-objective problem (P2).

5. Illustrative Example

An illustration of embedded system with 4 subsystems connected in series under Reuse-Build-Buy has been considered. For development of each SS, one or more hardware components can be purchased from vendor. Software components are available as COTS, in-house developed and/or reusable. Aim of the developer is selection of best-fit hardware & software components as well as redundancy levels for each SS such that the system reliability can be maximized and simultaneously overall cost can be minimized. Table 1 depicts data set for Hardware and COTS components while Table 2 presents data set for In-house-build and fabricated components.

Table 1. Dataset for hardware & COTS components.

SS	Hardware instance	Cost	Rel	D T	COTS instance	Cost	SLOC	Rel	DT
1	1	7	0.96	8	1	23	569	0.9	10
	2	4	0.945	11	2	19	582	0.93	12
2	1	11	0.83	9	1	16	410	0.91	6
	1	14	0.79	10	2	18	398	0.94	8
3	2	9	0.82	15	1	19	490	0.92	9
	3	8	0.8	14	2	21	485	0.95	7
4	1	12	0.93	11	1	27	682	0.95	15
	2	10	0.91	13	2	24	693	0.88	12
					3	22	704	0.89	11

Table 2. Dataset for in-house built and fabricated software components.

Sub-system	Unitary development/ fabrication cost	In-house built components		Fabricated reusable components	
		Estimated development time	SLOC	Estimated fabrication Time	SLOC
1	25	13	590	-	-
2	12	6	405	4	518
3	14	-	-	2	482
4	10	9	700	-	-

For simplicity, it is assumed that in RB/1/1 architecture, $P^{rv}_i = 0.002\ \forall i$, $P_d = 0.002$, $P^{fs}_i = 0.003\ \forall i$, and the values of τ_i, π_i and πf_i are 0.05, 0.002 & 0.002 resp. Limit on total SLOC & DT of the system is set as L=3500 & D=20 resp.The problem is solved by means of LINGO software [15,25] following steps in section 4. Three cases were considered, each with different set of weights W_R and W_C while keeping other parameters same. Different solutions obtained for each case have been summarized in Table 3. It can be observed that both reliability & cost increase as W_R increases & W_C decreases. The effect can also be seen on RB/1/1 redundancy which was only in subsystem-3 in case 1 but in subsystems 2 & 3 in other cases.

Table 3. Solution for component selection.

Case	W_R	W_C	SS (i)	z_i	Comp. Selected	SystemRel	SystemCost
1	0.2	0.8	1	0	$x_{11}=1, y_{12}=1$	0.614	213
			2	0	$x_{21}=1, y_{22}=1$		
			3	1	$x_{33}=2, y_{31}=1, y_{32}=1$		
			4	0	$x_{41}=1, y_{40}=1$		
2	0.5	0.5	1	0	$x_{11}=1, y_{12}=1$	0.755	240
			2	1	$x_{21}=2, y_{21}=1, y_{22}=1$		
			3	1	$x_{33}=2, y_{31}=1, y_{32}=1$		
			4	0	$x_{41}=1, y_{40}=1$		
3	0.8	0.2	1	0	$x_{11}=1, y_{10}=1$	0.817	583
			2	1	$x_{21}=2, y_{21}=1, w_2=1$		
			3	1	$x_{33}=2, y_{31}=1, w_3=1$		
			4	0	$x_{41}=1, y_{40}=1$		

6. Conclusion

Bi-objective models for fault tolerant embedded systems were formulated in order to maximize reliability & minimize cost under build-or-buy decision & reuse-or-build-or-buy decision for software components. RB/1/1 fault tolerant architecture was considered, which can tolerate one hardware fault & one software fault. With the increase in weight of reliability objective, the weight of cost objective decreases leading to subsystems which may incorporate redundant hardware & software components. This may further enhance overall system reliability as well as total cost.

References

1. L. Chen and A. Avizienis, N-Version Programming: A fault-tolerance approach to reliability of software operation. *In Proc. 8th IEEE Int. Symp. on Fault-Tolerant Computing*, 3 (1978).
2. B. Randell, System structure for software fault tolerance, *Proceedings of Int. Conference on Reliable Software, ACM,* 437 (1975).
3. R. K. Scott, J. W. Gault and D. F. McAllister, Modeling fault tolerant software reliability, *In. Proc. Third Syrup. Reliability in Distributed Software & Database Systems*, 15(1983).
4. R. K. Scott, J. W. Gault, D. F. McAllister and J. Wiggs, Experimental validation of six fault-tolerant software reliability models, *IEEE Fault Tolerant Comput. Syst.*, **14**, 102 (1984).
5. R. K. Scott, J. W. Gault and D. F. McAllister, Fault tolerant software reliability modelling, *IEEE Trans. Soft. Eng.*, **13**, 5, 582 (1987).
6. G. Levitin, Optimal structure of fault-tolerant software systems, *Reliability Engineering & System Safety*, **89**, 3, 286 (2005).
7. J.-C. Laprie, J. Arlat, C. Beounes and K. Kanoun, Definition and analysis of hardware- and software-fault-tolerant architectures, *IEEE Computer*, 39(1990).
8. M. R. Lyu (Ed.), Handbook of Software Reliability Engineering, *Mc-Graw-Hill, IEEE Computer Society Press* (1996).
9. N. Wattanapongsakorn and S. Levitan, Reliability optimization models for fault-tolerant distributed systems, *In Proc. Reliability & Maintainability Symposium, IEEE,* 193 (2001).
10. N. Wattanapongsakorn and S. P. Levitan, Reliability optimization models for embedded systems with multiple applications, *Reliability, IEEE Transactions*, **53**, 3,406 (2004).
11. O. Berman and N. Ashrafi, Optimization models for reliability of modular software systems, *Software Engineering, IEEE Trans.*, **19**, 11, 1119 (1993).
12. Z. Q. Wu, J. F. Tang, C. K. Kwong and C. Y. Chan, A model and its algorithm for software reuse optimization problem with simultaneous reliability and cost consideration, *Int. Jrnl of Innovative Computing, Information and Control,* 7, 5, 2611, (2011).

13. V. Cortellessa, F. Marinelli and P. Potena, An optimization framework for build-or-buy decisions in software architecture, *Computers & Operations Research*, **35**, 10, 3090(2008).
14. A. Bertolino and L. Strigini. "On the use of testability measures for dependability assessment." Software Engineering, IEEE Transactions on vol 22, no. 2, 1996, pp. 97.
15. L. Schrage, "Optimization modeling with LINDO, Brooks." Cole, Pacific Grove, 1997.
16. M. Bryce, T. Bryce, Make or Buy Software? *J. Syst. Munag*, **38**, 8, 6 (1987).
17. C. K. Kwong, L. Mu, J. F. Tang and X. G. Luo, Optimization of software components selection for component-based software system development, *Computers & Industrial Eng.*, **58**, 4, 618 (2010).
18. J. Sametinger, Software engineering with reusable components, *Springer Science & Business Media* (1997).
19. D. W. Coit, T. Jin and N. Wattanapongsakorn, System optimization with component reliability estimation uncertainty: a multi-criteria approach, *IEEE Transactions*, **53**, 369 (2004).
20. V. Narayanan and Y. Xie, Reliability concerns in embedded system designs, *Computer*, **39**(1), 118 (2006).
21. C. Haines, D. Carney and J. Foreman, Component-Based Software Development/ COTS Integration, *http://www.sei.cmu.edu/str/descriptions/cbsd_body.html*
22. C. Ivica, Component-based software engineering for embedded systems, *In Proc. of the 27th Int. conference on Software engineering, ACM*, 712 (2005).
23. P. C. Jha, R. Kaur, S. Narula and S. Madan, Fuzzy Multi-criteria Approach for Component based Software System under Build-or-Buy Scheme, *Proc. of the Third Int. Conference on Soft Computing for Problem Solving Advances in Intelligent Systems and Computing*, 259, 929 (2014).
24. R. Kaur, S. Arora, P. C. Jha and S. Madan, Fuzzy Multi-Criteria Approach for Modular Software System Incorporating Reuse-Build-Buy Decision Under Recovery Block Scheme, *Proc. 2nd Int. Conference on Computing for Sustainable Global Development*, 655 (2015).
25. H. Thiriez, OR software LINGO, *European J. of Operational Research*, **124**, 655 (2000).
26. M. Ehrgott, Multicriteria optimization, *Springer Science & Business Media* (2006).

Study of a Problem of Annular Cylinder Under Two-Temperature Thermoelasticity with Thermal Relaxation Parameters

Santwana Mukhopadhyay

Department of Mathematical science, IIT(BHU),
Varanasi, Uttar Pradesh, India
E-mail: mukhosant.apm@iitbhu.ac.in

Roushan Kumar

Department of Mathematics, Central university of South Bihar
Patna, State Bihar, India
E-mail: roushankumar@cub.ac.in

The present work is aimed at seeking the solution of a problem based on two-temperature thermoelaticity with one relaxation parameter as well as with two relaxation parameters in a unified way of an annular cylinder, whose both the inner and upper surfaces are assumed to be stress free and are subjected to ramp-type heating. The Eigen value method together with the Laplace transform is employed to derive the solution of the problem. Numerical values of physical quantities are computed for a suitable material and results are displayed graphically to show the distributions of the fields inside the medium and comparison are made with previous theory.

Keywords: Generalized thermoelasticity; Two temperature Theory; Vector Matrix Method; Laplace Transform.

1. Introduction

The subject generalized thermoelasticity have been developed mainly to overcome the shortcomings inherent in the classical coupled dynamical theory of thermoelastcity such as the infinite speed of thermoelastic disturbances, unsatisfactory thermoelastic response of a solid body to short laser pulses, poor description of thermoelastic behavior at low temperature. The generalized theories are characterized with the finite speed of thermal disturbance. The first two generalized themroelastic models which drew attention of researchers are Lord- Shulman model [1] and Green-Lindsay model [2]. In the model proposed by Lord and Shulman one thermal relaxation time parameter is introduced in the Fourier's law of heat conduction where as in the model developed by Green and Lindsay two relaxation times are introduced in the constitutive relations for the stress tensor and the entropy equation. Later on, another three models of thermoelasticity which have drawn the attention of researchers are the models given by Green and Naghdi [3, 4, 5]. Subsequently several studies are carried out on the basis of generalized theories of thermoelasticity. A good detail of the subject is reported in two review articles by Chandrasekharaiah [6, 7]. Different investigations relating to thermoelastic interactions in annular cylinder have been presented by Shereif and Anwar [8], Youssef and Abbas [9], Sherief and Ezzat [10] and Bagri and Eslami [11].

Theory of heat conduction on a deformable body, which depends on two different temperatures, the conductive temperature and the thermo-dynamic temperature, □, was first formulated by Chen and Gurtin [12] and Chen *et al.* [13, 14]. In 2006, Youssef [20] extended this two temperature theory in the context of the generalized theory of heat conduction introducing the thermal relaxation parameters and formulated a two-temperature theory of generalized thermoelasticity by providing the uniqueness theorem. A good detail of the subject is reported in [15–19, 21–23].

The main object of the present paper is to investigate the thermoelastic interactions in an infinitely long annular cylinder in the contexts of two-temperature thermoelasticity given by Youssef [34]. The inner and outer surfaces of the cylinder are subjected to ramp type heating. For the solution of the problem we employ eigen value approach and Laplace transform technique. Numerical inversion of Laplace transform is carried out by Bellman *et al.* [24]. Considering aluminum epoxy composite material, numerical computation is performed with the help of Mathematica software and programming language C++. The numerical results of the physical quantities are plotted in different figures and a comparative analysis of classical one-temperature generalized thermoelasic models and the two-temperature generalized thermoelastic models is presented.

It is also important to reproduce the spacing of the text and headings as shown here. Text should be slightly more than single-spaced; use a leading (which is the average distance from the base of one line of text to the base of an adjacent line) of 13 pt and 10 pt for footnotes. All headings should be separated from the text preceding it by a vertical space of about 12 pt and by 6 pt from the subsequent text. Paragraphs should have its first line indented by about 0.25 inch except where the paragraph is preceded by a heading and the abstract should be indented on both sides by 0.25 inch from the main body of the text.

2. Formulation of the Problem

The basic governing equations for thermoelastic interactions in a homogeneous isotropic elastic solid in two-temperature thermoelasticity with one relaxation parameter Youssef [20] can be written in a unified way as follows:

(i) Stress -strain temperature relation

$$\sigma_{ij} = \lambda e_{kk}\delta_{ij} + 2\mu e_{ij} - \gamma\left(1 + v\frac{\partial}{\partial t}\right)(\theta - \theta_0) \tag{1}$$

(ii) Strain displacement relation

$$e_{ij} = \frac{1}{2}\left(u_{i,j} + u_{j,i}\right) \tag{2}$$

(iii) Heat conduction equation

$$K\phi_{,ii} = \rho c_E\left(\frac{\partial\theta}{\partial t} + \tau_0\frac{\partial^2\theta}{\partial t^2}\right) + \gamma\theta_0\left(\frac{\partial}{\partial t} + n_0\tau_0\frac{\partial^2}{\partial t^2}\right)e \tag{3}$$

In these equations θ denotes the thermodynamic temperature, θ_0 is the constant reference temperature u_i is displacement component, σ_{ij} and e_{ij} are components of stress tensor and strain tensor respectively. λ and μ are Lame's elastic constants, ρ is mass density, K is thermal conductivity of the material, c_E is specific heat at constant strain. τ_0 and v are thermal relaxation parameters, $\gamma = (3\lambda+2\mu)\alpha_t$, α_t being the coefficient of linear thermal expansion. e is the dilatation and given by $e = e_{ii}$ and conductive temperature ϕ is related to thermodynamic temperature θ as

$$\phi - \theta = \alpha\phi_{,ii} \tag{4}$$

Where $\alpha > 0$, the two temperature parameter, is a scalar and called the temperature discrepancy [27].

The corresponding equations in cases of two temperature thermoelasticity with one relaxation parameter (TLS model) and two temperature thermoelasticity with two relaxation parameters (TGL model) can be extracted from the above set of equations for different sets of values of the notations α, v, n_0 and τ_0 as follows:

TLS - model : $\alpha > 0,\ \ n_0 = 1,\ \ v = 0,\ \ \tau_0 > 0$,

TGL- model: $\alpha > 0,\ \ n_0 = 0,\ \ v \geq \tau_0 > 0$.

LS -model: $\alpha = 0,\ \ n_0 = 1,\ \ v = 0,\ \ \tau_0 > 0$,

GL- model: $\alpha = 0,\ \ n_0 = 0,\ \ v \geq \tau_0 > 0$.

We consider an infinitely long isotropic elastic cylinder of inner radius '*a*' and outer radius 'b'. The center of the cavity is taken to be the origin of the cylindrical coordinate system (r, φ, z). Considering the axis-symmetric plane strain problem the equation of motion in the absence of body forces and heat sources is given by:

$$\frac{\partial \sigma_{rr}}{\partial r}+\frac{\sigma_{rr}-\sigma_{\varphi\varphi}}{r}=\rho\frac{\partial^2 u}{\partial t^2} \tag{5}$$

Where the non- zero stress components are

$$\sigma_{rr}=(\lambda+2\mu)\frac{\partial u}{\partial r}+\lambda\frac{u}{r}-\gamma\left(1+v\frac{\partial}{\partial t}\right)(\theta-\theta_0) \tag{6}$$

$$\sigma_{\varphi\varphi}=(\lambda+2\mu)\frac{u}{r}+\lambda\frac{\partial u}{\partial r}-\gamma\left(1+v\frac{\partial}{\partial t}\right)(\theta-\theta_0) \tag{7}$$

3. Boundary Conditions

Initially the medium is assumed to be at rest and undisturbed. The inner and outer surface of annular assumed to be traction free and are subjected to a ramp type heating. The boundary conditions are therefore taken as follows

$$\sigma_{rr}=0 \text{ at } r=\xi_1 \text{ and } r=\xi_2 \tag{8}$$

$$\phi\big|_{r=\xi_1}=\phi_0 h(t),\ \phi\big|_{r=\xi_1}=\phi_0' h(t) \tag{9}$$

$$\text{where } h(t)=\begin{vmatrix}0 & t\le 0\\ \dfrac{t}{t_0} & 0<t\le t_0\\ 1 & t\ge t_0\end{vmatrix}$$

where ϕ_0 and ϕ_0' are constant temperatures and $t_0\ge 0$ is fixed moment of time to rise the steady temperatures.

Applying Laplace transformation and using Eq. (3)–(9), the non dimensional form of equation of motion; heat conduction equation; stress components and boundary condition are given by

$$\left[1+w(p+\tau_0 p^2)\right]\left[\frac{\partial^2\bar{\phi}}{\partial r^2}+\frac{1}{r}\frac{\partial\bar{\phi}}{\partial r}\right]=\left(p+\tau_0p^2\right)\bar{\phi}+a_2\left(p+n_0\tau_0p^2\right)\bar{e} \tag{10}$$

$$\left(\frac{\partial^2\bar{u}}{\partial r^2}+\frac{1}{r}\frac{\partial\bar{u}}{\partial r}-\frac{\bar{u}}{r^2}\right)-a_1(1+vp)\left[\frac{\partial\bar{\phi}}{\partial r}-w\frac{\partial}{\partial r}\left(\frac{\partial^2\bar{\phi}}{\partial r^2}+\frac{1}{r}\frac{\partial\bar{\phi}}{\partial r}\right)\right]=p^2\bar{u} \tag{11}$$

$$\bar{\sigma}_{rr}=\frac{\partial\bar{u}}{\partial r}+\lambda_1\frac{\bar{u}}{r}-a_1(1+vp)\bar{T},\ \bar{\sigma}_{\varphi\varphi}=\frac{\bar{u}}{r}+\lambda_1\frac{\partial\bar{u}}{\partial r}-a_1(1+vp)\bar{T} \tag{12}$$

$$\bar{\sigma}_{rr}\big|_{r=\xi_1}=0,\ \bar{\sigma}_{rr}\big|_{r=\xi_2}=0,\ \bar{\phi}\big|_{r=\xi_1}=\phi_o\bar{h}(p),\ \bar{\phi}\big|_{r=\xi_2}=\phi_o'\bar{h}(p) \tag{13}$$

where, $\bar{h}(p)=\dfrac{\phi_0(1-e^{-t_0 p})}{t_0 p^2}$ and $\bar{e}=\dfrac{\partial\bar{u}}{\partial r}+\dfrac{\bar{u}}{r}$.

Here we have introduced the following dimensionless terms and variables:

$$r'=c_0\eta r,\ \xi_1=c_0\eta a,\ \xi_2=c_0\eta b, u'=c_0\eta u,\ \tau_0'=c_0\eta^2\tau_0,\ v'=c_0\eta^2 v,\ t'=c_0\eta^2 t,\ T=\frac{\theta-\theta_0}{\theta_0}$$

$$\varphi'=\frac{\varphi-\theta_0}{\theta_0},\ \sigma'_{rr}=\frac{\sigma_{rr}}{(\lambda+2\mu)},\quad \sigma'_{\phi\phi}=\frac{\sigma_{\phi\phi}}{(\lambda+2\mu)},\ e'=e,\ c_0{}^2=\frac{(\lambda+2\mu)}{\rho},$$

$$\eta=\frac{\rho c_E}{K},\ a_1=\frac{\gamma\theta_0}{(\lambda+2\mu)},\ a_2=\frac{\gamma}{K\eta},\ \lambda_1=\frac{\lambda}{(\lambda+2\mu)},\ w=\alpha c_0^2\eta^2.$$

4. Solution of Problem

We now take the quantities $\bar{u}$ and $\dfrac{\partial\bar{\phi}}{\partial r}$ as the state variables and introduce a differential operator $\omega=\dfrac{\partial^2}{\partial r^2}+\dfrac{1}{r}\dfrac{\partial}{\partial r}-\dfrac{1}{r^2}$.

Therefore equations (18) and (19) can be written in the matrix form as follows:

$$\omega\vec{V}(r,p)=A(p)\vec{V}(r,p) \tag{14}$$

Where, $\vec{V}(r,p)=\begin{bmatrix}\bar{u}\\ \dfrac{\partial\bar{\phi}}{\partial r}\end{bmatrix}\quad A(p)=\begin{bmatrix}N_1 & M_1\\ N_2 & M_2\end{bmatrix}$ (15)

Here, $N_1=\dfrac{p^2\left[1+w\left(p+\tau_0p^2\right)\right]}{\left[1+w\left(p+\tau_0p^2\right)+w\varepsilon\left(p+\tau_0\eta_0p^2\right)(1+vp)\right]}$

$$M_1=\frac{a_1(1+vp)}{\left[1+w\left(p+\tau_0p^2\right)+w\varepsilon\left(p+\tau_0\eta_0p^2\right)(1+vp)\right]}$$

$$N_2=\frac{p^2a_2\left(p+\tau_0\eta_0p^2\right)}{\left[1+w\left(p+\tau_0p^2\right)+w\varepsilon\left(p+\tau_0\eta_0p^2\right)(1+vp)\right]}$$

$$M_2=\frac{\left(p+\tau_0p^2\right)+a_1a_2(1+vp)\left(p+\tau_0\eta_0p^2\right)}{\left[1+w\left(p+\tau_0p^2\right)+wa_1a_2\left(p+\tau_0\eta_0p^2\right)(1+vp)\right]}$$

The characteristic equation of the matrix A(p) is given by

$$m^4-m^2\left(N_1+M_2\right)m^2+\left(N_1M_2-M_1N_2\right)=0 \tag{16}$$

Let $m_1{}^2$ and $m_2{}^2$ are eigen values of the matrix A(p), so that they satisfy

$$m_1^2 + m_2^2 = N_1 + M_2 \text{ and } m_1^2 m_2^2 = N_1 M_2 - M_1 N_2 .$$

The eigenvectors corresponding to $m_1^{\ 2}$ and $m_2^{\ 2}$ are

$$\vec{X}_i = \begin{bmatrix} M_1 \\ m_i^{\ 2} - N_1 \end{bmatrix} i{=}1{,}2 \tag{17}$$

The general solution of the system of equations (14) is therefore given by

$$\vec{V} = \sum_{i=1}^{2} \vec{X}_i(m_i)\left[A_i I_1\left(m_i r\right) + B_i K_1\left(m_i r\right)\right], \tag{18}$$

Where, $I_1\left(m_i r\right)$ and $K_1\left(m_i r\right)$ are the modified Bessel functions of order one of the first and second kind respectively.

From equation (26) and (30) the solution for $\overline{u}$ and $\dfrac{\partial \overline{\phi}}{\partial r}$ are given by

$$\overline{u} = \sum_{i=1}^{2} M_1 \left[A_i I_1(m_i r) + B_i K_1(m_i r)\right] \tag{19}$$

$$\frac{\partial \overline{\phi}}{\partial r} = \sum_{i=1}^{2} \left(m_i^{\ 2} - N_1\right)\left[A_i I_1(m_i r) + B_i K_1(m_i r)\right] \tag{20}$$

Where, A_i (p) and B_i (p) are arbitrary constants independent of *r*.

We obtain the solutions for conductive temperature and thermodynamic temperature as follows:

$$\overline{\phi} = \sum_{i=1}^{2} \frac{\left(m_i^{\ 2} - N_1\right)}{m_i}\left[-A_i I_0(m_i r) + B_i K_0(m_i r)\right] \qquad i = 1,2 \tag{21}$$

$$\overline{T} = \sum_{i=1}^{2} \left(m_i^{\ 2} - N_1\right)\left[A_i \left(\frac{1 - w m_i^{\ 2}}{m_i}\right) I_0(m_i r) - B_i \left(\frac{1 - w m_i^{\ 2}}{m_i}\right) K_0(m_i r)\right] \; i = 1,2 \tag{22}$$

Where, $I_0\left(m_i r\right)$ and $K_0\left(m_i r\right)$ are the modified Bessel functions of order zero of the first kind and second kind respectively.

The solutions for radial and circumferential stresses are derived from equation (12)

$$\overline{\sigma}_{rr} = \sum_{i=1}^{2} A_i S_{i1}^{r} + B_i S_{i1}^{\prime r} \qquad i = 1,2 \tag{23}$$

$$\overline{\sigma}_{\varphi\varphi} = \sum_{i=1}^{2} A_i S_{i1}^{\varphi} + B_i S_{i1}^{\prime \varphi} \qquad i = 1,2 \tag{24}$$

where,

$$S_{i1}^{r} = \left\{ M_1 m_i - a_1(1 + vp)(m_i^{\ 2} - N_1)\left(\frac{1 - w m_i^{\ 2}}{m_i}\right)\right\} I_0(m_i r) + M_1 \left(\frac{\lambda_1 - 1}{r}\right) I_1(m_i r) \;\; i = 1,2$$

$$S_{i1}'^{r} = \left\{ -M_1 m_i + a_1(1+vp)(m_i^2 - N_1)\left(\frac{1-wm_i^2}{m_i}\right)\right\} K_0(m_i r) + M_1\left(\frac{\lambda_1 - 1}{r}\right) K_1(m_i r) \quad i=1,2$$

$$S_{i1}^{\varphi} = \left\{ \lambda_1 M_1 m_i - a_1(1+vp)(m_i^2 - N_1)\left(\frac{1-wm_i^2}{m_i}\right)\right\} I_0(m_i r) + M_1\left(\frac{1-\lambda_1}{r}\right) I_1(m_i r) \quad i=1,2$$

$$S_{i1}'^{r} = \left\{ -\lambda_1 M_1 m_i + a_1(1+vp)(m_i^2 - N_1)\left(\frac{1-wm_i^2}{m_i}\right)\right\} K_0(m_i r) + M_1\left(\frac{1-\lambda_1}{r}\right) K_1(m_i r) \quad i=1,2$$

The boundary conditions (13) yield the constants A_i and B_i as follows:

$$\begin{bmatrix} A_1 \\ B_1 \\ A_2 \\ B_2 \end{bmatrix} = \begin{bmatrix} S_{11}^{\xi_1} & S_{11}'^{\xi_1} & S_{21}^{\xi_1} & S_{21}'^{\xi_1} \\ S_{11}^{\xi_2} & S_{11}'^{\xi_2} & S_{21}^{\xi_2} & S_{21}'^{\xi_2} \\ \phi_1^{\xi_1} & \phi_1'^{\xi_1} & \phi_2^{\xi_1} & \phi_2'^{\xi_1} \\ \phi_1^{\xi_2} & \phi_1'^{\xi_2} & \phi_2^{\xi_2} & \phi_2'^{\xi_2} \end{bmatrix}^{-1} \begin{bmatrix} 0 \\ 0 \\ \phi_0 \bar{h}(p) \\ \phi_0' \bar{h}(p) \end{bmatrix}$$

Where,

$$S_{i1}^{r}\Big|_{r=\xi_1} = S_{i1}^{\xi_1},\ S_{i1}^{r}\Big|_{r=\xi_2} = S_{i1}^{\xi_2},\ S_{i1}'^{r}\Big|_{r=\xi_1} = S_{i1}'^{\xi_1},\ S_{i1}'^{r}\Big|_{r=\xi_2} = S_{i1}'^{\xi_2}$$

$$\phi_i^{\xi_i}\Big|_{i=1,2} = \frac{(m_i^2 - N_1)}{m_i} I_0(m_i \xi_i)\Bigg|_{i=1,2},\quad \phi_i'^{\xi_i}\Big|_{i=1,2} = -\frac{(m_i^2 - N_1)}{m_i} K_0(m_i \xi_i)\Bigg|_{i=1,2}$$

5. Numerical Result and Discussion

With the help of computer programming and Bellmenmethod for the inversion of Laplace transforms the computations for the numerical solutions of the physical quantities like displacement, temperature and stresses are carried out.

The material chosen is aluminum epoxy, the physical data for which is given as follows:

$$\lambda = 7.59\times 10^{10}\ Nm^{-2},\quad \mu = 1.89\times 10^{10}\ Nm^{-2},\quad \alpha_t = 1.78\times 10^{-5}\ K^{-1}$$

$$\eta' = 8355.02\ sm^{-2},\quad c_v = 961.4\ JKg^{-1}K^{-1},\quad \rho = 2190\ Kg\, m^{-3},\quad T_0 = 296\ K$$

The other constants of the problem are taken as

$$\phi_0 = \phi_0' = 1,\ v = 0.03,\ \tau_0 = 0.02,\ w = 0.1,\ \xi = 1.$$

Taking $N = 7$ in equation (30) we get the roots (x_i) of the shifted Legendre polynomial of degree 7 and the corresponding weights (W_i) (Mathematica 5.2).

The numerical values of the physical field variables are computed for different values of r ($r \geq 1$) and dimensionless time t (t = 0.02, 0.13, 0.35, 0.69, 1.21, 2.04, 3.67) and the values are plotted in different figures (Figs. 1–4) at times t = 0.13. The numerical values of the field variables are also computed for one-temperature generalized thermoelastic models (LS-model and GL-model). The figures display the distributions of displacement, conductive temperature, thermodynamic temperature and stresses inside the annular cylinder under the interactions of mechanical and thermal fields when the inner surface and outer surface are subjected to ramp-type heating and radial stress is assumed to be zero at those surfaces. It is observed that the trend of variations of all the fields is similar for all the four models of thermoelasticity. However some significant dissimilarities among different models for different physical quantities are indicated. Figs. 1, 2, 3, 4 show the variations of fields displacement, conductive temperature, thermodynamic temperature and stressesat time t=0.13 which show the variations before the time (t=0.5) of rise of steady temperature.

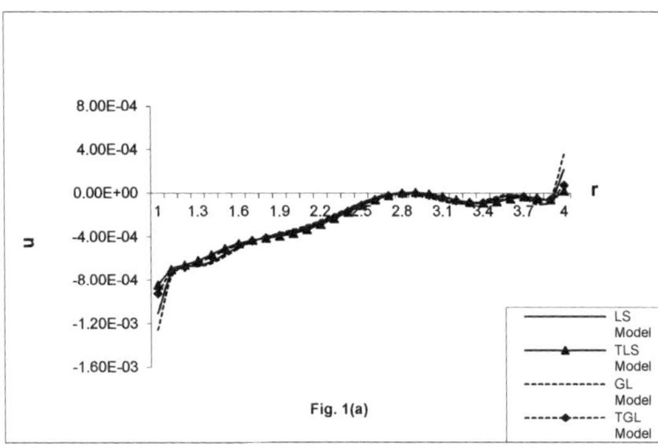

Fig. 1(a)

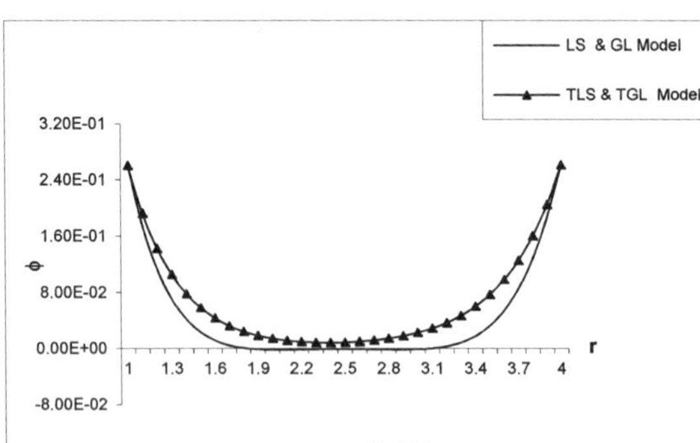

Fig. 2(a)

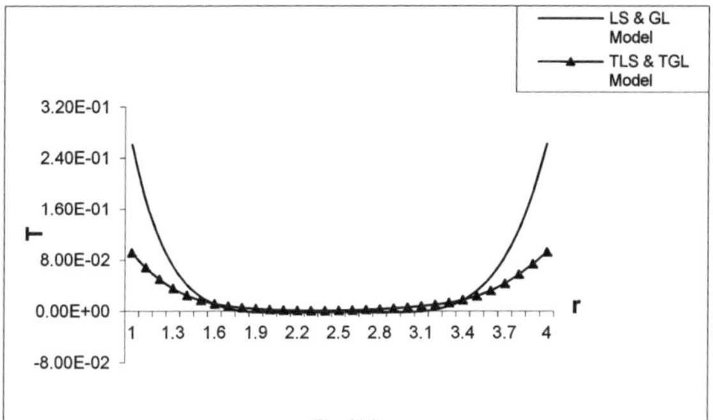

Fig. 3(a)

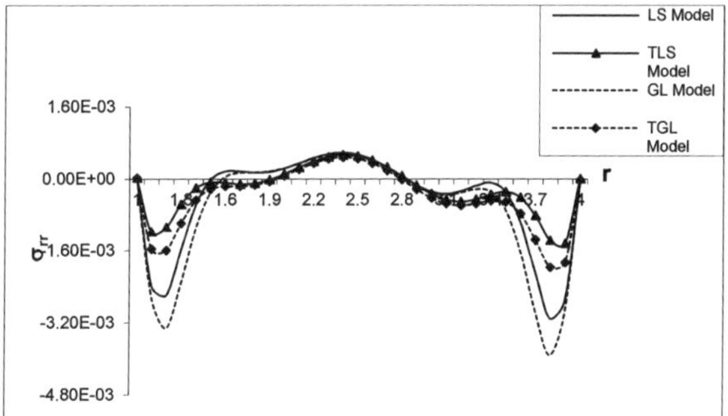

Fig.4(a)

References

1. H. W. Lord and Y. Shulman, A Generalized dynamical theory of thermoelasticity, J. Mech. Phys. Solids15 299–309 (1967).
2. A. E. Green and K. A. Lindsay, Thermoelasticity, J. Elasticity 2 1–7 (1972).
3. A. E. Green and P. M. Naghdi, A re-examination of the basic postulates of thermoemechanics, Proc. Roy. Soc. London A357253–270 (1991).
4. A. E. Green and P. M. Naghdi, On undamped heat waves in an elastic solid, J. Thermal Stresses 15253–264 (1992).
5. A. E. Green and P. M. Naghdi, Thermoelasticity without energy dissipation, J Elasticity 31 189–209 (1993).
6. D. S. Chandrasekharaiah, Thermoelasticity with second sound: A Review, Appl. Mech. Rev. 39 355–376(1986).

7. D. S. Chandrasekharaiah, Hyperbolic thermoelasticity: A Review of recent literature, Appl. Mech. Rev. 51 705–729(1998).
8. H. H. Sherief and M. N. Anwar, A problem in generalized thermoelasticity for an infinitely long annular cylinder composed of two different materials, actamechanica 80, 137–149 (1998).
9. H. M. Youssef and I. A. Abbas, Thermal shock problem of generalized thermoelasticity for an infinitely long annular cylinder with variable thermal conductivity, computational methods in science and technology 13(2), 95–100 (2007).
10. Hany H. Sherief and Magdy A. Ezzat, A problem in generalized magneto-thermoelasticity for an infinitely long annular cylinder, J. of eng. Math., 34, 387–402(1998).
11. A. Bagri and M. R. Eslami, Generalized coupled thermoelasticity of functionally graded annular disk considering the Lord–Shulman theory, composite structure, 83(2) 168–179(2007).
13. P. J. Chen and M. E. Gurtin, On a theory of heat conduction involving two-temperatures, J. Appl. Math. Phys. (ZAMP)19614–627(1968).
14. P. J. Chen, M. E. Gurtin and W. O. Williams, On the thermodynamics of non-simple elastic materials with two temperatures, J. Appl. Math. Phys. (ZAMP)20107–112(1969).
15. P. J. Chen, M. E. Gurtin and W. O. Williams, A note on non-simple heat conduction, J. Appl. Math. Phys. (ZAMP)19969–970(1968).
16. B. A. Boley and I. S. Tolins, Transient coupled thermoelastic boundary value problems in the half space, J. Appl. Mech. 29 637–646(1962).
17. D. Iesan, On thermodynamics of non-simple elastic material with two temperature, J. Appl. Math. Phys. (ZAMP) 21 583–591(1970).
18. D. E. Amos, On a half space solution of a modified heat equation, Quart. Appl. Math. 2721–33(1973).
19. W. E. Warren and P. J. Chen, wave propagation in the two temperature theory of elasticity, Acta Mech. 1621–33(1973).
20. R. Quintanilla, On existence, structural stability, convergence and spatial behavior in thermoelasticity with two temperatures, Acta Mech. 168 61–73(2004).
21. H. M. Youssef, Theory of two-temperature Generalizedthermoelasticity, IMA J. of Appl. Math. 71383–390(2006).
22. Magana, A., Quintanilla, R.: Uniqueness and growth of solution in two temperature generalized thermoelastic theories. Math. Mech. Solids. online (2008).

23. Youssef, H. M.: Problem of generalized thermoelastic infinite cylindrical cavity subjected to a ramp-type heating and loading. Arch. Appl. Mech. **75,** 553–565(2006).
24. Youssef, H. M., Al-Lehaibi, E. A.: State-space approach of two-temperature generalized thermoelasticity of one dimensional problem. Int. J Solids Struct. **44**, 1550–1562(2007).
25. R. Bellmen, R. E. Kolaba and J. A. Lockette, Numerical Inversion of the Laplace Transform, (American Elsevier Pub. Co., New York, 1966).

Multi-Criteria Advertisement Allocation Model of Multiple Advertisers on a Television Network

G. Kaur
Decision Analytics, EXL Services
Noida, Uttar Pradesh, India
E-mail: guneetsa@gmail.com

S. Aggarwal
Galgotias University
GautamBudh Nagar, Greater Noida, Uttar Pradesh, India
E-mail: sugandha.aggarwal@galgotiasuniversity.edu.in

P. C. Jha
Department of Operational Research, University of Delhi
New Delhi, India
E-mail: jhapc@yahoo.co.in

Television advertising, being one of the biggest sources of revenue for any television network, is highly practiced across the globe. The shows aired on a television network have time designated for commercial advertising in the breaks between them and also have varied sets of audience with varied TV rating points. The advertisers wanting to place their advertisements on a television network have certain non-negotiable requirements and preferences. Fulfilling the advertiser's requisitions, a TV network's problem is to maximize the revenue generated from allocating advertisements to the commercial breaks of shows appearing on their network. In this paper, we develop a multi objective mathematical programming model to determine the optimal placement of advertisements of multiple products of different advertisers, within the breaks of shows broadcasted in different day-parts on a television channel, over a planning horizon. The objectives considered are maximization of the network's revenue and minimization of the penalty cost from product conflict violation. The model also takes into consideration the advertisers' requirements for non-repetition of the same advertisement in a break, the bounds on the frequency of airings of an advertisement, preferences for locations in different day-parts and the minimum proportion specified for the appearance of advertisements' in different day-parts, as constraints in the formulated model. The model is solved through goal programming approach, to attain an optimal trade-off between the two conflicting objectives and get a compromised optimal solution. A case study of a television channel is presented to validate the model and is solved through the optimization software LINGO.

Keywords: Television Advertising; Multi Objective Optimization Model; Goal Programming; Advertisement Allocation.

1. Introduction

Commercial advertising is one of the major sources of revenue for the television networks [1]. For instance, television commercial advertising revenue in USA was 49.71 billion dollars in 2012 and is projected to grow by 3.2% by 2017 [2]. Not just the USA, the Indian TV advertising market is estimated to have grown at 14 percent in 2014 to Rs 155 billion [3]. Television advertising industry is guided by the two entities at the ends of demand and supply of commercial advertising, the advertisers and the TV networks, respectively [4]. Both these entities have their respective interests in commercial advertising and are seeking to fulfill them. There are various shows aired on a television network's channels that have time designated to commercial advertising during the breaks between them. Varied sets of audience are possessed by these shows, diversified on the basis of demographics and viewers' preferences. Television networks sell commercial time on the basis of this widespread and considerable audience, amounting to the desirable audience rating points for the shows. Advertisers demand for the schedule of their advertisements' airings to be according to certain specifications while buying the advertising time from a TV network. Some of the advertisers' requisitions encompass frequency of exposure of the same advertisement, non-violation of product conflict constraint, non-repetition of the same advertisement in the same break and getting the minimum proportion of their advertisements aired on shows of preferred day-parts. In this paper we formulate a multi-objective zero one integer programming problem considering the objectives for both, TV networks and the advertisers along with all the advertisers' requirements specified.

On any particular channel, there are different time slots within a day called as day-parts and these day-parts are characterized by different audience ratings. In general, they can be bifurcated as Early Morning (7:00 A.M.–9:00 A.M. Monday–Sunday), Morning Kids (7:00 A.M.–9:00 A.M. Monday–Sunday), Daytime (10:00 A.M.–4:00 P.M. Monday–Friday), Early Fringe (4:00 P.M.–6:00 P.M. Monday–Friday), Early News (6:00 P.M.–7:30 P.M. Monday–Saturday), Prime Time Access (7:30 P.M.–8:00 P.M. Monday–Saturday), Prime Time (8:00–11:00 P.M. Monday–Saturday & 7:00 P.M.–11:00 P.M. Sunday), Late News (11:00–11:30 P.M. Monday–Sunday), Late Fringe (11:30 P.M.–1:00 A.M. Monday–Friday), Weekend Sports (10:00 A.M.–6:00 PM Saturday–Sunday), Saturday late Fringe (11:30 P.M.–1:00 AM Saturday) [5]. The advertisers usually prefer multiple airings of their advertisements to be aired either solely in one day part, equitably in all the day-parts or in different proportions in different day-parts. Considering the case of an Indian TV channel, in this paper, we have taken majorly three segments, prime time (8:00–11:00 P.M. Monday–Sunday) with high rating points, late-afternoon to early evening (4:00 P.M.–7:00 P.M. Monday–Sunday)with medium rating points and morning to early afternoon (10:00 A.M.–4:00 P.M. Monday–Friday) with low rating points. Advertiser preferences for these day-parts are considered along with a minimum proportion of advertisements' desired to be advertised in each preferred day-part.

Advertisers are usually provided with the flexibility of specifying a range for the frequency of airing of their advertisements during the planning horizon, to create the intended effect on audience. Also, given to the history with their clients, TV networks ensure certain advertisers get some spots for their advertisements, made possible by specifying a lower bound on the frequency of airings of the advertisements. Considering the range specification with objectives and other constraints, our model allocates the most optimal frequency to each advertisement. Non-repetition of the same advertisement in the same break is also considered to ensure that the viewer's attitude is not saturated towards the reception of a product, sooner than desired.

There are multiple advertisers with multiple products competing for spots, for example, Procter & Gamble competing for Tide, Head & Shoulders, Oral B and Reckitt Benckiser for Dettol Soap, Dettol Handwash, Dettol Kitchen Gel and Oxy Life. One of the major preferences of advertisers is not to air the advertisements of competing products in the same break. Coke and Pepsi, both being carbonated beverages, are a good example of conflicting products that cannot be advertised together in the same break [6]. The preference for non-violation of product conflict constraint varies according to the product set and can be weighted accordingly. As stated by [7], products such as BMW and Mercedes cars might be more at conflict than products like Mercedes cars and Ford trucks. We have defined sets in our paper, the combinations of advertisers and their advertisements that cannot appear together in the same break. The sets vary on their level of conflict and hence, a varying penalty cost for each product type is considered. This is one of the most non-desirable violations amongst the advertisers and hence, is considered as an objective function.

In this paper we formulate a multi-objective zero one integer programming model for allocation of multiple advertisers' multiple advertisements on shows aired in different day-parts on a television channel, over a planning horizon. The objectives considered are revenue maximization of the TV network and product conflict penalty cost minimization with constraints on other advertisers' preferences and limit on break time duration. The multi-objective approach helps achieve a compromised optimal solution satisfying multiple aspects related to the decision process. The model is validated through a case and is solved using goal programming approach. Through this paper, we aim to enable the television networks to meet the advertisers' demands and maintain good relations with their clients for a longer period.The rest of the paper is organized as follows: section II gives a brief review on the television advertising models. Section III presents the model formulation and in section IV we present a case study to validate the model and the last section V concludes the paper.

2. Literature Review

Many authors have studied the problem of scheduling commercial advertisements on television broadcast network. Noticeable work has been done in the field with different objectives pertaining to television advertising such as maximization of revenue for the television network [1, 8, 9], minimizing the penalty cost generated due to deviations from the advertisers' foresaid requirements [6], maximizing viewership for each advertisement [10], evenly spacing multiple airings of an advertisement [4, 11, 12] etc. The work on scheduling of TV advertisements perceptibly started with an advertisement exchange model that re-allocates advertisements amongst different breaks for efficient utilization of break time duration [13]. The work in the field intensified after the year 2000, given to Bollapragada's work with National Broadcasting Company. Initially, a mixed integer problem was formulated to optimize the schedule of multiple airings of the same advertisement by spacing them as evenly as possible on the slots bought by the advertiser [11]. Other aspects of scheduling commercials on NBC with the objective of minimizing deviations from the product conflict constraint and equitable rotation amongst the desirable and other spots within a break were considered by [6]. The objective of minimizing the product conflicts within the same break was also considered with differential conflict weights being assigned to different pairs of similar products according to the level of conflict between them [7]. The work in the literature focusing on maximizing viewership of the advertisements and the revenue for the television networks is also extensive. Reference by [10] gave a model to simultaneously accept and place the advertisements in different time slots on a TV channel with respect to fulfillment of advertiser's requisitions of minimum audience ratings for the advertisements. Reference by [8] formulated a model which maximizes revenue by selling packages of spots to ensure required target audience ratings to the advertisers, and does not exceed the limitations of each package and time availability in general. Obtaining an optimum distribution of slots to be sold on the upfront market and the scatter market, the objective of revenue maximization is considered for the NBC by [9]. Including the concept of incompletely specified combinatorial auctions; advertisers placing their bids (A bid specifies the number of spots required, price being paid and the preferences for spots) and the TV networks accepting or rejecting them, the objective of revenue maximization is considered forprime time on a TV network [1]. Further, a problem of meeting advertiser's demands i.e. supplying promised viewer ratings for their advertisements and airing regularity requirements for multiple airings of the same advertisement with the objective of minimizing deviations from the ideal spaces between multiple airings of the same advertisement is considered by [4]. And recently, reference by [14] gave us three models for three stages of re-allocation of advertisements from the under-performing deals. The first model was to figure out the deals that needed re-allocation because of the poor performing shows, the ones they were assigned to. Second model allocates advertisements from all such deals to breaks in the upcoming week and the third schedules all the allocated spots ten minutes before the airing of the break.

This paper presents a multi-objective approach to provide the best optimal solution for scheduling problem of commercial advertisements, considering the aspects for both, TV networks and advertisers. The multi-objective zero one integer programming problem considered maximizes revenue for a television network along with minimizing the penalty cost incurred for violation of the product conflict constraint. The target audience ratings requirement is tried to be met as a constraint, that being the main stimuli motivating the advertisers to broadcast their products on a television network. Along with these, non-repetition of the same advertisement in the same break and a range for multiple airings of the same advertisement in the planning horizon is considered. To the best of our knowledge such a multi-objective approach to the problem, attaining a compromised optimal solution between the conflicting objectives and constraints has not been considered before.

3. Model Development

3.1. *Notations*

N set of advertisers; (i=1,2...N)
M set of products of an advertiser; (j=1,2...M)
W set of shows considered; (w=1,2...W)
B set of breaks in a show; (b=1,2...B)
T set of days in the planning horizon; (t=1,2...T)
R set of different product types; (r=1,2,...R)

Q_k set of shows telecasted in day part of rating k (k=1,2,3), $Q_k \in W$

S_r set consisting conflicting advertiser advertisement combinations (i,j) of product type r; (r=1,2,…,R)

c_w per unit cost of advertisement placement in the wth show

l_{ij} duration of ith advertiser's j^{th} advertisement

d_{bw} duration of bth break on wth show

u_{ij} upper bound on the airing frequency of ith advertiser's j^{th} advertisement

v_{ij} lower bound on the airing frequency of ith advertiser's j^{th} advertisement

g_{ij}^{k} preference for the day part of rating k by i^{th} advertiser for its jth advertisement

f_i^k minimum percentage of the total allocations that i^{th} advertiser aspires to have in day part of rating k

γ_{wbt}^{r} number of product conflicts in break b of show w on t^{th} day for product type r

β^r linear penalty cost per product conflict for product type r

$$y_{ijwbt} = \begin{cases} 1, \text{if } i^{th} \text{advertiser's } j^{th} \text{advertisement gets placed in } b^{th} \text{break of } t^{th} \text{day on } w^{th} \text{show} \\ 0, \text{otherwise} \end{cases}$$

3.2. *Model formulation*

The model developed is a zero one multi-objective integer-programming model, formulated to cater to the advertisements allocation problem of a television network. There are multiple advertisements of multiple advertisers to be placed within the breaks of particular shows, telecasted at different day-parts on a channel, over a planning horizon. While allocating these advertisements, the preferences of locations of the advertisers and their minimum proportion requirements from these preferences have to be met along with minimizing the penalty cost arising due to product conflict violation and maximizing the revenue for the network. Hence, the objectives are dual, maximization of the revenue for the television network and minimization of penalty costs arising from not meeting the product conflict constraint.

Problem (P1):

$$Maximize Z = \sum_{i=1}^{N} \sum_{j=1}^{M} \sum_{w=1}^{W} \sum_{b=1}^{B} \sum_{t=1}^{T} y_{ijwbt} l_{ij} c_w \, . \tag{1}$$

$$Minimize\ C = \sum_{w=1}^{W} \sum_{b=1}^{B} \sum_{t=1}^{T} \sum_{r=1}^{R} \gamma_{wbt}^{r} \beta^{r} . \tag{2}$$

subject to

$$\sum_{i=1}^{N} \sum_{j=1}^{M} y_{ijwbt} l_{ij} \leq d_{bw} . \tag{3}$$

$$\sum_{i,j\in S_r} y_{ijwbt} - \gamma_{wbt}^r \leq 1 \forall w \in W, b \in B, t \in T, r \in R. \quad (4)$$

$$\sum_{k=1}^{3}\sum_{w\in Q_k}\sum_{b=1}^{B}\sum_{t=1}^{T} y_{ijwbt} g_{ij}^k \geq v_{ij} \forall i \in N, j \in M. \quad (5)$$

$$\sum_{k=1}^{3}\sum_{w\in Q_k}\sum_{b=1}^{B}\sum_{t=1}^{T} y_{ijwbt} g_{ij}^k \leq u_{ij} \forall i \in N, j \in M. \quad (6)$$

$$\sum_{j=1}^{M}\sum_{w\in Q_k}\sum_{b=1}^{B}\sum_{t=1}^{T} y_{ijwbt} g_{ij}^1 \geq f_i^1 \sum_{k=1}^{3}\sum_{j=1}^{M}\sum_{w\in Q_k}\sum_{b=1}^{B}\sum_{t=1}^{T} y_{ijwbt} g_{ij}^k \forall i \in N. \quad (7)$$

$$\sum_{k=1}^{2}\sum_{j=1}^{M}\sum_{w\in Q_k}\sum_{b=1}^{B}\sum_{t=1}^{T} y_{ijwbt} g_{ij}^k - f_i^1 \sum_{k=1}^{3}\sum_{j=1}^{M}\sum_{w\in Q_k}\sum_{b=1}^{B}\sum_{t=1}^{T} y_{ijwbt} g_{ij}^k \geq f_i^2 \sum_{k=1}^{3}\sum_{j=1}^{M}\sum_{w\in Q_k}\sum_{b=1}^{B}\sum_{t=1}^{T} y_{ijwbt} g_{ij}^k \forall i \in N. \quad (8)$$

$$\sum_{k=1}^{3}\sum_{j=1}^{M}\sum_{w\in Q_k}\sum_{b=1}^{B}\sum_{t=1}^{T} y_{ijwbt} g_{ij}^k - (f_i^1 + f_i^2) \sum_{k=1}^{3}\sum_{j=1}^{M}\sum_{w\in Q_k}\sum_{b=1}^{B}\sum_{t=1}^{T} y_{ijwbt} g_{ij}^k \geq f_i^3 \sum_{k=1}^{3}\sum_{j=1}^{M}\sum_{w\in Q_k}\sum_{b=1}^{B}\sum_{t=1}^{T} y_{ijwbt} g_{ij}^k \forall i \in N. \quad (9)$$

$$y_{ijwbt} \in \{0,1\} \text{ and integer} \quad (10)$$

In problem (P1), the objective function (1) maximizes the total revenue obtained from placement of advertisements of multiple advertisers on shows of different day-parts on a television channel, over a planning horizon considering the per unit show cost for advertising and the objective function (2) minimizes the penalty cost arising from the product conflict violation in the breaks of the shows over the planning horizon. Constraint by (3) ensures that the total duration of the allocated advertisements in any break does not exceed the time span of that particular break. Constraint by (4) designates the total product conflicts happening in any break through allocation of advertisements, in its count, that needs to be minimized. Constraints by (5) and (6) ensure that the lower and upper bounds on the frequency specified by the ith advertiser for its jth advertisement are being met through the allocations of the jth advertisement in low (k=1), medium (k=2), and high (k=3) rating day-parts, as preferred by the advertiser. Constraints by (7), (8) and (9) ensure that of the total allocations for an advertiser's products over the planning horizon, some minimum proportion gets allocated to the preferred rating slots. Last constraint given by (10) ensures that the decision variable takes value zero or one as per the allocation.The model (P1) may lead to infeasible solution because of a huge number of advertiser preferences along with the conflicting objectives. To get a tradeoff between the above mentioned, we solve (P1) through goal programming approach.

Solution Methodology: Goal Programming

The goal programming model for (P1) can be written as follows:

Problem (P2):

$$Minimize\ g(\eta, \rho) = \lambda\eta_1 + (1-\lambda)\rho_2 \tag{11}$$

Subject to

$$\sum_{i=1}^{N}\sum_{j=1}^{M}\sum_{w=1}^{W}\sum_{b=1}^{B}\sum_{t=1}^{T} y_{ijwbt} l_{ij} c_w + \eta_1 - \rho_1 = Z^* \tag{12}$$

$$\sum_{w=1}^{W}\sum_{b=1}^{B}\sum_{t=1}^{T}\sum_{r=1}^{R} \gamma_{wbt}^{r} \beta^r + \eta_2 - \rho_2 = C^* \tag{13}$$

$$g_0(\eta, \rho, Y) = g_0(\eta^0, \rho^0, Y^0) \tag{14}$$

$$\eta_1, \rho_1, \eta_2, \rho_2 \geq 0, 1 \leq \lambda \leq 0 \tag{15}$$

and constraints (3)–(10).

The objective in (P2) given by (11) is to minimize the sum of weighted deviations from the goals where η, ρ are the negative and positive deviational variables. Constraints given by (12) and (13) are the goal constraints on the revenue and penalty cost objectives. By solving (P2) we obtain an efficient solution of model (P1) [15]. A suitable tradeoff amongst the goals is obtained at the management satisfaction level and resulting solution is one of the properly efficient solutions.

4. Numerical Illustration

To validate the model, we consider the case of a television network that wants to allocate advertisements of multiple advertisers on shows aired on one of its channels in different day-parts, within a planning horizon. A small case has been considered with the planning horizon of seven days (T=7). The shows telecasted are divided into three rating day-parts, low (k=1), medium (k=2) and high (k=3) based on the time of the day they are aired. Two shows each of the low rating day-part (W=1, 2), medium rating day-part (W=3, 4) and high rating day-part (W=5, 6) are taken and hence there are six shows in total considered for each day. The shows are of half an hour duration and each show has two breaks of 120 seconds ($\forall\, d_{bw}$) each. In total, there are two hundred advertisers (N=200) with different number of products, recognized as the prospective buyers of the advertising time for these shows in the planning horizon. Each advertiser at maximum has five products (M=5) with different time durations of advertisements (in seconds), lower and upper bounds on the frequency of appearance and preferences for day-parts (k=1, 2, 3 taken as L, M, H). Table 1 provides the data for the above components in the same order. Per unit (10s) cost of placement in a particular show is given by Table 2. There is also a minimum percentage aspired by the advertisers from the preferred day-parts. Below, we give sets consisting of conflicting advertiser advertisement combinations (i,j) of product type r; (r=1,2,…,15) and the corresponding penalty cost per conflict for that product type:

$S_1 = \{\{(1,1),(2,4),(5,1),(5,2),(7,1),(21,1),(22,1),(150,1),(200,2)\}; \beta_1 = 4000$
$S_2 = \{\{(2,3),(9,1),(10,2),(10,4),(17,1)(35,1),(50,1),(149,1)\}; \beta_2 = 6000$
$S_3 = \{\{(1,2),(1,5),(2,1),(3,5),(9,3),(21,1),(166,1),(186,1)\}; \beta_3 = 2000$
$S_4 = \{\{(1,2),(1,5),(2,1),(3,5),(9,3),(21,1),(166,1),(186,1)\}; \beta_4 = 2000$
$S_5 = \{(101,1),(111,1),(178,1),(181,1),(199,1)\}; \beta_5 = 4500$
$S_6 = \{(8,4),(9,5),(12,3),(15,2),(16,1),(21,1),(42,2)\}; \beta_6 = 4500$
$S_7 = \{(1,3),(4,2),(19,1),(39,1),(67,1),(112,1)\}; \beta_7 = 6000$
$S_8 = \{(13,2),(15,1),(23,1),(41,1),(49,1)\}; \beta_8 = 3000$
$S_9 = \{(8,1),(31,2),(52,1),(56,1),(60,1),(159,1)\}; \beta_9 = 2500$
$S_{10} = \{(66,1),(77,1),(143,1),(163,1)\}; \beta_{10} = 4500$
$S_{11} = \{(6,4),(10,1),(30,1),(31,1),(47,1),(92,1),(103,1),(121,1)\}; \beta_{11} = 500$
$S_{12} = \{(40,1),(46,1),(59,1),(124,1)\}; \beta_{12} = 2500$
$S_{13} = \{(95,1),(126,1),(184,1),(197,1)\}; \beta_{13} = 2200$
$S_{14} = \{(55,1),(64,1),(99,1),(133,1)\}; \beta_{14} = 3000$
$S_{15} = \{(29,1),(80,1),(83,1),(113,1),(135,1)\}; \beta_{15} = 2500$

The data is fed in Lingo software [16] and the problem (P1) is solved with the individual objectives to get the aspired revenue and the penalty cost. With the aspirations, the problem is solved through goal programming approach in two stages. The solution obtained is given in a compact form by Table 4 with revenue INR 13770000 and penalty cost INR 3000. The objective of product conflicts is minimized with possibly not more than one conflict in a break. None of the breaks are left vacant. A compromised optimal solution is achieved.

Table 1. Data for duration, lower, upper frequency and day part preferences.

(i,j)	**Data**	*(i,j)*	**Data**	*(i,j)*	**Data**	*(i,j)*	**Data**	*(i,j)*	**Data**
(1,1)	(20,2,3,H,M)	(15,2)	(50,0,3,M,L)	(57,1)	(50,1,3,H,M)	(110,1)	(30,1,3,H,)	(156,1)	(30,1,3,H,)
(1,2)	(40,1,3,M,L)	(15,3)	(40,2,4,H,M)	(58,1)	(50,1,3,L)	(111,1)	(50,1,3,M)	(157,1)	(20,2,4,M)
(1,3)	(20,0,3,H,L)	(16,1)	(40,1,3,H,)	(59,1)	(50,1,3,M,L)	(112,1)	(40,1,3,L)	(158,1)	(30,2,4,L)
(1,4)	(30,0,3,H,M)	(17,1)	(20,1,3,H,)	(60,1)	(30,1,3,H,)	(113,1)	(20,2,4,H,)	(159,1)	(10,3,5,H,)
(1,5)	(30,0,3,H,L)	(17,2)	(30,1,3,M)	(61,1)	(20,1,3,H,)	(114,1)	(40,1,3,M)	(160,1)	(20,2,4,M)
(2,1)	(20,1,3,H,)	(17,3)	(20,1,2,L)	(61,2)	(20,1,3,L)	(115,1)	(50,1,3,L)	(161,1)	(10,0,5,L)
(2,2)	(30,1,3,M)	(17,4)	(40,0,4,M)	(62,1)	(20,1,3,L)	(116,1)	(10,1,3,H,)	(162,1)	(20,2,4,H,)
(2,3)	(50,0,3,L)	(18,2)	(20,1,2,M)	(63,1)	(40,1,3,M)	(117,1)	(40,1,3,M)	(163,1)	(30,1,3,M)
(2,4)	(40,2,4,H,)	(19,1)	(10,2,4,H,)	(64,1)	(10,2,4,M,L)	(118,1)	(50,1,3,L)	(164,1)	(30,2,4,L)
(2,5)	(20,1,2,M,L)	(20,1)	(40,1,3,M,L)	(65,1)	(10,1,3,L)	(119,1)	(20,1,3,H,)	(165,1)	(20,2,4,H,)
(3,1)	(40,1,3,M,L)	(21,1)	(30,3,4,M,L)	(66,1)	(50,2,4,H,M)	(120,1)	(20,2,4,M)	(165,2)	(10,0,5,M)
(3,2)	(50,0,3,H,)	(22,1)	(40,3,6,L)	(67,1)	(10,2,4,H,M)	(121,1)	(50,1,3,L)	(165,3)	(10,0,5,)
(3,3)	(30,0,3,M)	(23,1)	(20,1,2,H,M,L)	(68,1)	(20,2,4,H,L)	(122,1)	(10,1,3,H,)	(166,1)	(20,1,3,L)
(3,4)	(10,2,4,H,M,L)	(23,2)	(20,1,3,M)	(69,1)	(20,1,3,M)	(123,1)	(20,1,3,M)	(167,1)	(40,0,3,H,)
(3,5)	(40,0,3,H,)	(23,3)	(10,0,4,M,L)	(70,1)	(20,2,4,M)	(124,1)	(10,2,4,L)	(167,2)	(10,2,4,M)
(4,1)	(20,4,6,M)	(24,1)	(30,1,3,M)	(71,1)	(20,1,3,L)	(124,2)	(20,2,4,H,)	(168,1)	(50,1,3,L)
(4,2)	(30,2,4,H,L)	(25,1)	(30,2,4,M,L)	(72,1)	(10,2,4,L)	(125,1)	(50,1,3,M)	(169,1)	(30,1,3,H,)
(4,3)	(40,0,3,H,)	(26,1)	(20,2,4,H,)	(73,1)	(30,1,3,H,)	(126,1)	(30,1,3,L)	(169,2)	(30,0,4,M)

Table 1. (*Continued*)

(*i,j*)	Data	(*i,j*)	Data	(*i,j*)	Data	(*i,j*)	Data	(*i,j*)	Data
(4,4)	(30,0,3,M)	(27,1)	(30,2,4,M)	(74,1)	(10,1,3,L)	(127,1)	(20,2,4,H,)	(169,3)	(20,0,4,L)
(4,5)	(40,0,3,L)	(27,2)	(40,1,3,H,)	(74,2)	(10,1,3,L)	(128,1)	(20,2,4,M)	(169,4)	(10,2,4,H,)
(5,1)	(40,2,4,H,M,L)	(27,3)	(20,0,3,H,)	(75,1)	(20,2,4,M)	(129,1)	(40,1,3,L)	(169,5)	(10,0,3,M)
(5,2)	(20,2,4,H,)	(28,1)	(10,2,4,H,L)	(76,1)	(10,1,3,H,)	(130,1)	(10,2,4,H,)	(170,1)	(50,1,3,L)
(5,3)	(10,0,3,M)	(29,1)	(30,1,3,H,M)	(77,1)	(40,2,4,M,L)	(131,1)	(40,1,3,M)	(171,1)	(30,1,3,H,)
(5,4)	(40,1,4,L)	(30,1)	(40,1,2,L)	(78,1)	(50,2,4,H,)	(132,1)	(10,2,4,L)	(172,1)	(30,2,4,M)
(5,5)	(10,0,3,H,)	(31,1)	(30,1,4H,)	(79,1)	(30,2,4,H,)	(133,1)	(30,2,4,H,)	(173,1)	(20,2,4,L)
(6,1)	(40,1,3,M)	(31,2)	(50,0,3,M,L)	(80,1)	(20,1,3,H,)	(134,1)	(10,2,4,M)	(173,2)	(30,2,4,H,)
(6,2)	(20,2,4,H,)	(32,1)	(40,2,4,H,M)	(81,1)	(30,1,3,H,M)	(135,2)	(30,1,3,L)	(173,3)	(20,1,3,M)
(6,3)	(20,0,3,H,)	(33,1)	(40,1,3,M,L)	(82,1)	(50,1,3,H,L)	(135,3)	(30,2,4,H,)	(174,1)	(30,2,4,L)
(6,4)	(30,0,4,M)	(34,1)	(30,1,3,M)	(83,1)	(10,2,4,H,)	(136,1)	(10,1,3,M)	(175,1)	(30,2,4,H,)
(6,5)	(20,0,3,M)	(35,1)	(10,1,3,L)	(84,1)	(10,2,4,H,)	(136,2)	(10,1,3,L)	(175,2)	(50,0,3,M)
(7,1)	(20,3,5,M)	(36,1)	(30,2,4,M,L)	(85,1)	(10,2,4,M)	(136,3)	(20,0,3,H,)	(176,1)	(30,1,3,L)
(7,2)	(30,1,3,L)	(36,2)	(30,1,3,M)	(86,1)	(30,1,3,L)	(137,1)	(30,1,3,M)	(177,1)	(30,1,3,H,)
(8,1)	(40,1,3,H,)	(37,1)	(40,1,3,H,)	(86,2)	(20,2,4,L)	(138,1)	(20,2,4,L)	(178,1)	(50,1,3,M)
(8,2)	(30,0,3,M)	(38,1)	(30,2,4,L)	(87,1)	(20,1,3,H,)	(139,1)	(10,2,4,H,)	(179,1)	(20,2,4,L)
(8,3)	(30,1,2,L)	(39,1)	(10,1,3,H,)	(87,2)	(20,1,3,H,)	(139,2)	(40,1,3,M)	(180,1)	(20,1,3,H,)
(8,4)	(20,0,3,H,)	(40,1)	(30,1,3,M)	(88,1)	(20,2,4,M)	(139,3)	(30,0,4,L)	(181,1)	(20,2,4,M)
(9,1)	(10,2,5,M)	(41,1)	(50,1,3,L)	(89,1)	(50,1,3,H,)	(140,1)	(30,1,3,H,)	(182,1)	(30,1,3,L)
(9,2)	(10,1,3,L)	(42,1)	(40,1,3,H,M)	(90,1)	(10,1,3,M)	(141,1)	(40,1,3,M)	(183,1)	(10,2,4,H,)
(9,3)	(40,2,3,H,)	(43,1)	(20,1,3,M,L)	(91,1)	(10,2,4,L)	(142,1)	(20,2,4,L)	(184,1)	(30,1,3,M)
(9,4)	(50,0,4,M)	(44,1)	(10,2,4,H,M)	(92,1)	(30,2,4,H,)	(142,2)	(10,1,3,H,)	(185,1)	(10,1,3,L)
(9,5)	(30,0,3,L)	(45,1)	(10,0,3,M,L)	(93,1)	(10,2,4,M)	(142,3)	(20,0,4,M)	(186,1)	(10,1,3,H,)
(10,1)	(40,1,3,H,)	(45,2)	(30,2,4,H,)	(94,1)	(40,1,3,L)	(142,4)	(10,0,4,L)	(187,1)	(30,2,4,M)
(10,2)	(20,0,3,M)	(46,1)	(50,2,4,H,M)	(95,1)	(20,2,4,H,)	(143,1)	(10,1,3,H,)	(188,1)	(30,2,4,L)
(10,3)	(10,2,4,H,)	(47,1)	(50,2,4,L)	(96,1)	(30,1,3,M)	(143,2)	(20,2,4,M)	(189,1)	(20,2,4,H,)
(10,4)	(30,0,3,L)	(48,1)	(20,2,4,H,L)	(97,1)	(20,1,3,L)	(143,3)	(30,2,4,L)	(190,1)	(10,2,4,M)
(11,1)	(30,2,4,H,)	(48,2)	(20,2,4,M)	(98,1)	(20,2,4,H,)	(144,1)	(10,1,3,H,)	(191,1)	(50,1,3,L)
(11,2)	(40,1,3,M)	(49,1)	(10,2,4,M)	(99,1)	(10,1,3,M)	(145,1)	(30,1,3,M)	(192,1)	(10,2,4,H,)
(11,3)	(10,0,3,L)	(50,1)	(40,2,4,H,)	(100,1)	(30,2,4,L)	(146,1)	(20,2,4,L)	(193,1)	(20,2,4,M)
(12,1)	(40,1,3,H,)	(51,1)	(50,1,3,L)	(101,1)	(10,1,3,H,)	(147,1)	(30,2,4,H,)	(194,1)	(30,1,3,L)
(12,2)	(10,1,3,M)	(52,1)	(20,2,4,H,M)	(102,1)	(30,2,4,M)	(148,1)	(10,1,3,M)	(194,2)	(30,0,3,H,)
(12,3)	(40,0,3,H,)	(53,1)	(50,2,4,M)	(103,1)	(50,1,3,L)	(149,1)	(20,2,4,L)	(195,1)	(20,2,4,M)
(13,1)	(40,1,3,L)	(54,1)	(50,1,3,L)	(104,1)	(10,1,3,H,)	(150,1)	(20,2,4,H,)	(196,1)	(10,2,4,L)
(13,2)	(20,1,3,L)	(54,2)	(10,2,4,H,)	(105,1)	(20,1,3,M)	(151,1)	(50,1,3,M)	(197,1)	(30,1,3,H,)
(14,1)	(20,1,3,H,)	(55,1)	(10,2,4,M)	(106,1)	(20,2,4,L)	(152,1)	(30,1,3,L)	(198,1)	(10,2,4,M)
(14,2)	(10,2,4,M)	(55,2)	(50,1,3,H,)	(107,1)	(40,2,4,H,)	(153,1)	(30,1,3,H,)	(199,1)	(30,1,3,L)

Table 2. Advertisement cost (per 10 second).

W1	W2	W3	W4	W5	W6
5000	6000	12000	11000	23000	25000

Table 3. Solution table.

(i,j)		(i,j)		(i,j)		(i,j)		(i,j)		(i,j)		(i,j)	
(1,1)	(0,2,1)	(10,1)	(0,0,1)	(31,2)	(0,0,0)	(65,1)	(1,0,0)	(103,1)	(1,0,0)	(139,2)	(0,1,0)	(169,3)	(0,0,0)
(1,2)	(1,0,0)	(10,2)	(0,0,0)	(32,1)	(0,2,0)	(66,1)	(0,1,1)	(104,1)	(0,0,1)	(139,3)	(0,0,0)	(169,4)	(0,0,2)
(1,3)	(0,0,0)	(10,3)	(0,0,2)	(33,1)	(1,0,0)	(67,1)	(0,2,0)	(105,1)	(0,1,0)	(140,1)	(0,0,1)	(169,5)	(0,0,0)
(1,4)	(0,0,0)	(10,4)	(0,0,0)	(34,1)	(0,1,0)	(68,1)	(0,0,2)	(106,1)	(2,0,0)	(141,1)	(0,2,0)	(170,1)	(1,0,0)
(1,5)	(0,0,0)	(11,1)	(0,0,2)	(35,1)	(1,0,0)	(69,1)	(0,1,0)	(107,1)	(0,0,2)	(142,1)	(2,0,0)	(171,1)	(0,0,1)
(2,1)	(0,0,1)	(11,2)	(0,1,0)	(36,1)	(0,2,0)	(70,1)	(0,2,0)	(108,1)	(0,2,0)	(142,2)	(0,0,1)	(172,1)	(0,2,0)
(2,2)	(0,1,0)	(11,3)	(0,0,0)	(36,2)	(0,1,0)	(71,1)	(1,0,0)	(109,1)	(1,0,0)	(142,3)	(0,0,0)	(173,1)	(2,0,0)
(2,3)	(0,0,0)	(12,1)	(0,0,1)	(37,1)	(0,0,1)	(72,1)	(3,0,0)	(110,1)	(0,0,1)	(142,4)	(0,0,0)	(173,2)	(0,0,2)
(2,4)	(0,0,2)	(12,2)	(0,1,0)	(38,1)	(2,0,0)	(73,1)	(0,0,1)	(111,1)	(0,1,0)	(143,1)	(0,0,1)	(173,3)	(0,1,0)
(2,5)	(0,1,0)	(12,3)	(0,0,0)	(39,1)	(0,0,1)	(74,1)	(1,0,0)	(112,1)	(1,0,0)	(143,2)	(0,2,0)	(174,1)	(3,0,0)
(3,1)	(1,0,0)	(13,1)	(1,0,0)	(40,1)	(0,1,0)	(74,2)	(1,0,0)	(113,1)	(0,0,2)	(143,3)	(2,0,0)	(175,1)	(0,0,2)
(3,2)	(0,0,0)	(13,2)	(2,0,0)	(41,1)	(1,0,0)	(75,1)	(0,2,0)	(114,1)	(0,1,0)	(144,1)	(0,0,2)	(175,2)	(0,0,0)
(3,3)	(0,0,0)	(14,1)	(0,0,1)	(42,1)	(0,0,1)	(76,1)	(0,0,1)	(115,1)	(1,0,0)	(145,1)	(0,1,0)	(176,1)	(1,0,0)
(3,4)	(2,0,0)	(14,2)	(0,3,0)	(43,1)	(1,0,0)	(77,1)	(1,1,0)	(116,1)	(0,0,1)	(146,1)	(2,0,0)	(177,1)	(0,0,1)
(3,5)	(0,0,0)	(14,3)	(1,1,0)	(44,1)	(0,1,1)	(78,1)	(0,0,2)	(117,1)	(0,1,0)	(147,1)	(0,0,2)	(178,1)	(0,1,0)
(4,1)	(0,4,0)	(15,1)	(1,1,0)	(45,1)	(0,0,0)	(79,1)	(0,0,2)	(118,1)	(1,0,0)	(148,1)	(0,1,0)	(179,1)	(2,0,0)
(4,2)	(1,0,1)	(15,2)	(0,0,0)	(45,2)	(0,0,2)	(80,1)	(0,0,1)	(119,1)	(0,0,1)	(149,1)	(2,0,0)	(180,1)	(0,0,1)
(4,3)	(0,0,0)	(15,3)	(0,2,0)	(46,1)	(0,2,0)	(81,1)	(0,0,1)	(120,1)	(0,2,0)	(150,1)	(0,0,4)	(181,1)	(0,2,0)
(4,4)	(0,0,0)	(16,1)	(0,0,2)	(47,1)	(2,0,0)	(82,1)	(0,0,1)	(121,1)	(1,0,0)	(151,1)	(0,1,0)	(182,1)	(1,0,0)
(4,5)	(0,0,0)	(17,1)	(0,0,1)	(48,1)	(1,0,0)	(83,1)	(0,0,2)	(122,1)	(0,0,1)	(152,1)	(1,0,0)	(183,1)	(0,0,2)
(5,1)	(1,0,1)	(17,2)	(0,0,0)	(48,2)	(0,3,0)	(84,1)	(0,0,2)	(123,1)	(0,1,0)	(153,1)	(0,0,1)	(184,1)	(0,1,0)
(5,2)	(0,0,2)	(17,3)	(1,0,0)	(49,1)	(0,2,0)	(85,1)	(0,2,0)	(124,1)	(3,0,0)	(154,1)	(0,1,0)	(185,1)	(1,0,0)
(5,3)	(0,0,0)	(17,4)	(0,0,0)	(50,1)	(0,0,2)	(86,1)	(1,0,0)	(124,2)	(0,0,2)	(155,1)	(1,0,0)	(186,1)	(0,0,1)
(5,4)	(1,0,0)	(18,2)	(0,1,0)	(51,1)	(1,0,0)	(86,2)	(2,0,0)	(125,1)	(0,1,0)	(156,1)	(0,0,1)	(187,1)	(0,2,0)
(5,5)	(0,0,0)	(19,1)	(0,0,2)	(52,1)	(0,1,1)	(87,1)	(0,0,2)	(126,1)	(1,0,0)	(157,1)	(0,2,0)	(188,1)	(2,0,0)
(6,1)	(0,1,0)	(20,1)	(1,1,0)	(53,1)	(0,2,0)	(87,2)	(0,0,1)	(127,1)	(0,0,2)	(158,1)	(2,0,0)	(189,1)	(0,0,4)
(6,2)	(1,0,2)	(21,1)	(2,1,0)	(54,1)	(2,0,0)	(88,1)	(0,2,0)	(128,1)	(0,2,0)	(159,1)	(0,0,3)	(190,1)	(0,4,0)
(6,3)	(0,0,0)	(22,1)	(3,0,0)	(54,2)	(0,0,2)	(89,1)	(0,0,1)	(129,1)	(1,0,0)	(160,1)	(0,2,0)	(191,1)	(2,0,0)
(6,4)	(0,0,0)	(23,1)	(1,0,1)	(55,1)	(0,2,0)	(90,1)	(0,1,0)	(130,1)	(0,0,2)	(161,1)	(0,0,0)	(192,1)	(0,0,2)
(6,5)	(0,0,0)	(23,2)	(0,1,0)	(55,2)	(0,0,1)	(91,1)	(2,0,0)	(131,1)	(0,1,0)	(162,1)	(0,0,2)	(193,1)	(0,2,0)
(7,1)	(0,3,0)	(23,3)	(0,0,0)	(55,3)	(2,1,0)	(92,1)	(0,0,2)	(132,1)	(2,0,0)	(163,1)	(0,1,0)	(194,1)	(1,0,0)
(7,2)	(1,0,0)	(24,1)	(0,1,0)	(56,1)	(0,0,1)	(93,1)	(0,2,0)	(133,1)	(0,0,2)	(164,1)	(2,0,0)	(194,2)	(0,0,1)
(8,1)	(0,0,1)	(25,1)	(2,0,0)	(57,1)	(0,0,1)	(94,1)	(1,0,0)	(134,1)	(0,2,0)	(165,1)	(0,0,2)	(195,1)	(0,3,0)
(8,2)	(0,0,0)	(26,1)	(0,0,1)	(58,1)	(1,0,0)	(95,1)	(0,0,2)	(135,2)	(2,0,0)	(165,2)	(0,0,0)	(196,1)	(4,0,0)
(8,3)	(1,0,0)	(27,1)	(0,2,0)	(59,1)	(1,0,0)	(96,1)	(0,1,0)	(135,3)	(0,0,2)	(165,3)	(0,0,0)	(197,1)	(0,0,1)
(8,4)	(0,0,0)	(27,2)	(0,0,1)	(60,1)	(0,0,1)	(97,1)	(1,0,0)	(136,1)	(0,1,0)	(166,1)	(1,0,0)	(198,1)	(0,4,0)
(9,1)	(0,2,0)	(27,3)	(0,0,0)	(61,1)	(0,0,1)	(98,1)	(0,0,2)	(136,2)	(1,0,0)	(167,1)	(0,0,1)	(199,1)	(2,0,2)
(9,2)	(1,0,0)	(28,1)	(0,0,2)	(61,2)	(1,0,0)	(99,1)	(0,1,0)	(136,3)	(0,0,0)	(167,2)	(0,2,0)	(200,1)	(0,0,1)
(9,3)	(0,0,2)	(29,1)	(0,1,0)	(62,1)	(1,0,0)	(100,1)	(2,0,0)	(137,1)	(0,2,0)	(168,1)	(1,0,0)	(200,2)	(0,3,0)
(9,4)	(0,0,0)	(30,1)	(1,0,0)	(63,1)	(0,1,0)	(101,1)	(0,0,1)	(138,1)	(2,0,0)	(169,1)	(0,0,1)		
(9,5)	(0,0,0)	(31,1)	(0,0,1)	(64,1)	(0,4,0)	(102,1)	(0,2,0)	(139,1)	(0,0,2)	(169,2)	(0,0,0)		

5. Conclusion

This paper caters to the interests of both, the television network and the advertiser, with the formulated optimization model that allocates a set of advertisements from multiple advertisers in breaks appearing within different shows of a channel, over a planning horizon. The objectives considered are dual, minimizing the penalty cost incurred on product conflict occurrence in a break and maximizing the television network's revenue. The model formulated is a multi-objective zero one integer programming model with the constraints on the time inventory available for advertising in each break, the bounds on the frequency specified by the advertiser for each of its advertisements and the minimum proportion an advertiser wants to allocate to the preferred rating slots. With the nature of the objectives considered and the presence of multiple constraints, it is not possible to meet all the criteria with the aspired objective values. Hence, goal-programming approach is used to achieve an optimal trade-off between the conflicting objectives with equal weightage assigned to both the objectives. The criteria specified for rating requirements is being met to a satisfactory level for most of the advertisers, with the limited time available and limited preferred spots.

References

1. F. GhassemiTari and R. Alaei, Scheduling TV commercials using genetic algorithms, *Int. J. of Prod. Res.*, vol. 51, pp. 4921–4929, 2013.
2. Broadcast TV advertising revenue in the United States from 2004 to 2013 (in billion U.S. $) http://www.statista.com/statistics/183366/ad-revenue-in-us-broadcast-television-since-2004.
3. Indian Media & Entertainment Industry Report 2015 – KPMG.
4. A. García-Villoria and S. Salhi, Scheduling commercial advertisements for television, *Int. J. of Prod. Res.*, vol. 53, pp. 1198–1215, 2015.
5. J. Sissors and R. Baron, *Advertising media planning*, (New York: McGraw-Hill, 2010).
6. S. Bollapragada and M. Garbiras, Scheduling commercials on broadcast television, *Oper. Res.*, vol. 52, pp. 337–345, 2004.
7. D. R. Gaur, R. Krishnamurti, R. Kohli, Conflict Resolution in the scheduling of television commercials,*Op. Res.*, vol. 57, pp. 1098–1105, 2009.
8. T. Benoist, E. Bourreau and B. Rottembourg, The TV-break packing problem, *Eur. J. of Op. Res.*, vol. 176, pp. 1371–1386, 2007.
9. S. Bollapragada and S. Mallik, Managing on-air ad inventory in broadcast television, *IIE transactions*, vol. 40, pp. 1107–1123, 2008.
10. R. Bai and J. Xie, Heuristic Algorithms for Simultaneously Accepting and Scheduling Advertisements on Broadcast Television, *Journal of Information and Computer Science*, Vol. 1(4), pp. 245–251, 2006.

11. S. Bollapragada, M. R. Bussieck and S. Mallik, Scheduling commercial videotapes in broadcast television, *Op. Res*., vol. 52, pp. 679–689, 2004.
12. M. J. Brusco, Scheduling advertising slots for television, *J. of the Op. Res. Society*, vol. 59, pp. 1363–1372, 2008.
13. A. R. Brown, Selling television time: an optimization problem, *The Comput. J*., vol. 12, pp. 201–207, 1969.
14. Seshadri, S., Subramanian, S. and Souyris, S. (2015). Scheduling Spots on Television. Working Paper. http://www.ssouyris.com/. (In Press)
15. M. A. Geoffrion, Proper efficiency and the theory of vector maximization. *J. of Math. Anal. and Appl*., vol no. 22, pp. 618, 1968.
16. LINDOSYSTEM (2015) http://home.anadolu.edu.tr/~nila/PROGRAM/KILAVUZ/LINGO%20users%20ma nual.pdf (accessed on 2nd February, 2015).

Computation of Maximum Likelihood Estimates in Three Parameter Weibull for Censored Data

Sanjeeva Kumar Jha
Department of Statistics
North-Eastern Hill University
Shillong-793022, India
E-mail: skjha@nehu.ac.in

Three parameter Weibull distributions do not satisfy the usual regularity conditions for maximum likelihood estimation of parameters. Again, as the threshold parameter becomes very close to the smallest observation the log-likelihood function becomes unbounded. Further, a shape parameter with a value lass than unity makes the density J-shaped which excludes the availability of a consistent local maximum. The presence of censoring increases the complexity further. The present paper performs maximum likelihood estimation through Differential Evolution, an evolutionary computation method which does not require the differentiability of the likelihood function, for censored Weibull data. It successfully obtains maximum likelihood estimates and their precision for simulated data sets.

Keywords: Three parameter Weibull; right censoring; maximum likelihood estimation; differential evolution.

1. Introduction

"Inference for the three parameter Weibull model is not so well developed as the two parameter case. In particular, estimating the parameters by the method of maximum likelihood (ML) is complicated by the fact that the likelihood equations may have no solutions; that is, the likelihood function may have no stationary point. In other cases the solutions of the likelihood equations may converge to a saddle point and not to the maximum of the likelihood function. Exact interval estimation based on ML estimates has not been addressed in the literature." The preceding lines, as reported by McCool on page 299 of his book, published in the year 2012, [5], are sufficient enough to indicate upon the complexity of ML estimation in three parameter Weibull distributions. The complexity in ML estimation increases further if apart from all the three parameters being unknown, the observations are censored.

The present paper successfully obtains the ML estimates together with their precision measures in three parameter right censored Weibull distribution by applying the technique of Differential Evolution (DE). The Author has not come across any such application of DE as presented in the paper. The paper is divided into 5 sections. Section 1 introduces the Problem while Section 2 discusses the problem experienced in maximum likelihood (ML) estimation of parameters in three parameter Weibull distribution, Section 3 describes the Differential Evolution method in brief; Section 4 presents the application of DE on simulated data sets and Section 5 presents the results and discussions.

2. The Problem

The classical Weibull distribution introduced by W. Weibull [10], is defined by the following probability density function,

$$f(x;\alpha,\beta,\theta)=\frac{\alpha}{\beta}\left(\frac{x-\theta}{\beta}\right)^{\alpha-1}\exp\left\{-\left(\frac{x-\theta}{\beta}\right)^{\alpha}\right\};\ \theta\leq x<\infty\ ;\ 0<\alpha,\beta. \quad (1)$$

where, x is usually the elapsed time, α is the shape parameter, β is the scale parameter and θ is the threshold parameter. The notation $W(\alpha, \beta, \theta)$ will be used throughout the paper to represent a Weibull distribution with shape parameter α, a scale parameter β and a threshold parameter θ.

Among the various methods available for parameter estimation the ML is the most used for its efficiency and desirable asymptotic properties. Given an uncensored (complete) random sample of size n the likelihood function for a realization of sample values $x_1, x_2, \ldots, x_n$, is

$$L=\prod_{i=1}^{n} f(x_i). \quad (2)$$

Where, L depends only upon the unknown parameters of the distribution. The value of a parameter that maximizes L is called the maximum likelihood estimate of the parameter, which can be obtained in several ways. In case of right censored samples, with r denoting the number of failures and n-r denoting the number of censored observations, the likelihood function becomes,

$$L=K\prod_{i=1}^{r} f(x_i)\prod_{i=r+1}^{n}\left[1-F(x_i)\right]. \quad (3)$$

Where, F is the corresponding distribution function. The constant K varies with the type of censoring but is independent of parameters of the distribution and is not relevant in finding the maximizing value of the parameter. A failure contributes to a term f(x) evaluated at the failure life. A censored observation contributes a term [1-F(x)] evaluated at the censoring time. Out of the several types of censoring schemes only the two basic types of planed censoring, namely, Type I and Type II censoring are considered in the paper. Type I censoring refers to a testing up to a pre determined time and the Type II refers to testing up to a predetermined number of failures. Under both the above types of censoring only samples having right censored observations will be considered. An observation is called right censored if it is known that the exact value of the random variable is greater than the bounding value of a test. For a right censored observation it is only known that the exact time to failure exceeds the time at which the testing is terminated.

In cases of Weibull distribution with unknown threshold θ, an unknown α or a value of $\alpha < 1$ can make the log likelihood infinite, Smith [8]. When all the three parameters are unknown it has been observed that for the shape parameter $\alpha < 1$, the ML estimate do not exist as likelihood function is not bounded Rockette *et al.*, [7]. Unbounded likelihood problem is experienced in 3-parameter Weibull for when the threshold parameter is close to the smallest observation; the distribution becomes highly skewed with the bulk of probability located at or near the smallest observation. The value of log likelihood for this observation tends to infinity as the threshold parameter approaches the smallest observation, Cheng *et al.* [1]. Again for $1 < \alpha \leq 2$, ML estimates exist but do not satisfy the usual regularity conditions and lead to inefficient estimates of the parameters. For ML the likelihood equations need to be solved numerically Dodson [3], these methods provide biased estimates and the bias is not known Causineau [2].

It is important to note that the above problems pertaining to ML estimation for Weibull, is primarily due to the gradient based techniques applied for optimizing the log likelihood function. The problems may subside in case the likelihood function can be optimized directly without using any gradient based technique. Differential evolution is an optimization technique which does not require the differentiability of the function to be optimized.

3. Differential Evolution (DE)

Differential Evolution (DE), proposed by Storn and Price [9], is a simple real parameter optimization algorithm having a cycle of four stages — initialization, mutation, crossover and selection. It has three main control parameters-population size NP, mutation scale factor F and crossover probability Cr. The performance of the program depends on the choice of these three control parameters. Population of size NP is generated with vectors $x \in R^d$, ad-dimensional Euclidean space. Each of the next generations involve creation of a new population of size NP from the current population using differential mutation of the current population members by creating a trial mutant vector

$v_i = x_0 + F^*(x_1 - x_2)$ for three distinct vectors of the currentgeneration x_o, x_1 and x_2, and F being the mutation scale factor. Mutation is continued to generate d mutants with a crossover probability Cr, that controls the fraction of the trial mutantvector which are to be considered in light of the boundary conditions. Next the objective function is evaluated for the next generation values, and if found better, in the optimization sense, replaces the old value otherwise the old value remains. The process gives rise to an improved new generation ofcandidate individuals than the previous generation. The search process is repeated till convergence.

4. Application

Four data sets were generated from Weibull distribution, each of size 100, using Minitab 16 for different shape parameter values from known inconsistent solution region of $\alpha \leq 2$. The data sets generated were from W(0.5, 100, 10), W(1, 100,10), W(1.5, 100, 10) and W(2, 100, 10). The data sets are presented in the appendix. For each of the data sets ML estimation is performed for 20%, 30% and 50% right censoring intensities at sample sizes 50 and 100, using DE. The source code for DE available in Mishra [6], with suitable modifications, is used to obtain the simultaneous ML estimates of the parameters together with their 95% Bootstrap confidence interval and presented in Tables 1–3. (The source code is also available atthe URL address: http://www.nehu-economics.info/). Table 1 presents the results for Shape parameter while Table 2 and Table 3 present results for the threshold parameters. Further, for each of the data sets Bootstrap mean and standard deviation are also computed together with Root Mean Square Error (RMSE) and presented in Tables 4–5, for shape and scale parameters respectively, (RMSE = $\sqrt{(\text{bias}^2 + \text{variance})}$).

Table 1. ML estimates for shape parameter with 95% CI at different censoring intensity.

Weibull and right Censoring intensity		50 Observations	100 observations
W(0.5,100,10)	20%	0.3741 (0.1875, 0.4900)	0.3344 (0.2622, 0.4552)
	30%	0.3403 (0.1808, 0.4077)	0.3028 (0.2260,0.3852)
	50%	0.2737 (0.1402, 0.3469)	0.2409 (0.1709, 0.3119)
W(1,100,10)	20%	0.4915 (0.2344, 1.0025)	0.6495 (0.4078, 0.9720)
	30%	0.4359 (0.1662, 0.7786)	0.5828 (0.3732, 0.8995)
	50%	0.32891 (0.1684, 0.6853)	0.4562 (0.2498, 0.7148)
W(1.5,100,10)	20%	1.0706 (0.2883, 1.6979)	1.1847 (0.7052, 1.5012)
	30%	0.5596 (0.2628, 1.4772)	1.0608 (0.4358, 1.3481)
	50%	0.4106 (0.1704, 1.0662)	0.5993 (0.3012, 1.1056)

Table 1. (*Continued*)

Weibull and right Censoring intensity		50 Observations	100 observations
W(2,100,10)	20%	1.4186 (0.2875, 2.0985)	1.4892 (0.9551, 1.8971)
	30%	1.2297 (0.2427, 1.6617)	1.3083 (0.4501, 1.6679)
	50%	0.4456 (0.1865, 1.3068)	0.6464 (0.2922, 1.2611)

Table 2. ML estimates for scale parameter with 95% CI at different censoring intensity.

Weibull and right Censoring intensity		50 Observations	100 observations
W(0.5,100,10)	20%	205.0142 (50.86, 407.56)	115.3488 (62.55, 265.75)
	30%	305.4198 (88.69, 122.559)	178.0495 (71.90, 498.97)
	50%	1210.1274 (27.02, 31.0)	828.6898 (197.56, 11056.0)
W(1,100,10)	20%	113.1296 (64.40, 215.91)	110.5674 (79.97, 159.18)
	30%	156.3186 (67.018, 660.44)	139.5906 (90.79, 229.46)
	50%	519.8471 (121.56, 7342.27)	321.9737 (138.99, 1459.02)
W(1.5,100,10)	20%	112.0810 (83.2460, 198.711)	111.0145 (91.38, 134.86)
	30%	148.6290 (99.31, 319.75)	124.916 (95.90, 212.79)
	50%	401.4274 (139.78, 5748.13)	252.7100 (131.38, 937.12)
W(2,100,10)	20%	103.3632 (88.30, 166.56)	99.4831 (87.1575, 114.96)
	30%	113.7477 (83.05, 290.080)	109.1107 (88.39, 215.74)
	50%	340.5069 (103.97, 3402.29)	215.6769 (123.20, 1040.35)

Table 3. ML estimates for threshold parameter with 95% CI at different censoring intensity.

Weibull and right Censoring intensity		50 Observations	100 observations
W(0.5,100,10)	20%	10.0083[a]	10.0083[a]
	30%	10.0083[a]	10.0083[a]
	50%	10.0083[a]	10.0083[a]
W(1,100,10)	20%	11.2595[a]	11.2595[a]
	30%	11.2595[a]	11.2595[a]
	50%	11.2595[a]	11.2595[a]
W(1.5,100,10)	20%	22.8650 (17.2191, 23.0533)	11.9497 (10.7986, 12.2382)
	30%	23.0533 (21.5880, 23.0533)	12.1528 (11.7299, 12.2386
	50%	23.0533[a]	12.1528 (12.1670, 12.2338)
W(2,100,10)	20%	12.3480 (6.0700, 13.9868)	12.8094 (7.5875, 13.9868)
	30%	13.2201 (9.5194, 13.9868)	13.3819 (11.3534, 13.9868)
	50%	13.9868 (13.8020, 13.9868)	13.9868 (13.7398, 13.9868

Note: [a]Length of Confidence interval is zero.

For each of the computations the control parameters for DE set as, NP=70, F= 0.5, Cr=0.9. These control parameter values are arrived at on the basis of the recommendations provided by Storn *et al.* [9], Kukkonen *et al.* [4] and Mishra [6]. The Bootstrap results are based on a sample of size 100 as well as a sample of size 50. The sample of size 50 consists of the first 50 observations for each of the sample of size 100.

Table 4. Bootsrap sample mean, standard deviation and RMSE for shape parameter.

Weibull and right Censoring intensity		50 Observations Mean	SD	RMSE	100 Observations Mean	SD	RMSE
W(0.5,100,10)	20%	0.3209	0.0756	0.194402	0.3464	0.0469	0.160601
	30%	0.2867	0.0603	0.22166	0.3055	0.0433	0.199261
	50%	0.228	0.0565	0.277806	0.2443	0.0367	0.25832
W(01,100,10)	20%	0.5725	0.2111	0.47678	0.7195	0.134	0.310864
	30%	0.474	0.178	0.555302	0.6466	0.1363	0.378773
	50%	0.3814	0.1617	0.639385	0.4809	0.1349	0.536342
W(1.5,100,10)	20%	0.9937	0.4162	0.65541	1.2145	0.1593	0.326935
	30%	0.8105	0.3564	0.776164	0.9428	0.2598	0.614791
	50%	0.542	0.27	0.995321	0.7192	0.2356	0.815571
W(2,100,10)	20%	1.3372	0.4679	0.811316	1.4838	0.195	0.551804
	30%	1.1234	0.44628	0.983663	1.2858	0.2584	0.759508
	50%	0.6789	0.4039	1.381463	0.8206	0.2934	1.215347

Table 5. Bootsrap sample mean, standard deviation and RMSE for scale parameter.

Weibull and right Censoring intensity		50 Observations Mean	SD	RMSE	100 Observations Mean	SD	RMSE
W(0.5,100,10)	20%	174.55	90.46	117.2208	126.17	50.32	56.71835
	30%	349.34	296.57	387.4586	199.04	91.61	134.9122
	50%	3661.77	741.63	3638.162	1630.42	2326.47	2784.717
W(1,100,10)	20%	122.66	38.48	44.65631	113.79	20.24	24.49126
	30%	215.75	151.63	190.7609	144.46	35.42	56.84424
	50%	1425.99	4373.44	4570.036	379.4	321.09	425.6326
W(1.5,100,10)	20%	119.54	23.29	30.40125	111.94	12.03	16.94947
	30%	153.01	51.12	73.64316	134.61	29.15	45.25013
	50%	877.9	2593.99	2708.12	290.95	203.59	279.125
W(2,100,10)	20%	111.44	19.03	22.20393	100.518	7.38	7.398157
	30%	136.61	78.65	86.75318	114.47	24.59	28.53154
	50%	567.75	761.39	893.591	246.26	228.76	271.52

5. Results and Discussions

The Differential evolution technique could provide maximum likelihood estimates for parameters simultaneously, in three parameter right censored Weibull distributions for simulated datasets considered in the article. It could also provide the exact confidence interval together with root mean square error (RMSE).

Table 1 presents the ML estimates of shape parameter and the corresponding 95% Bootstrap confidence interval for each of the four datasets with 20%, 30% and 50% right censoring as well as for the sample sizes of 50 and 100. The confidence interval shortens with increasing sample size for different values of the true shape parameters and also for different censoring intensities. The confidence interval extends with increasing values of the true shape parameter for different sample sizes and different censoring intensities. In case of a sample of size 100 the confidence interval shortens with increasing censoring intensity in cases where true shape value is not greater than unity and the same happens for across the shape parameters when the sample size is of 50.

The Bootstrap mean, standard deviation and the corresponding RMSE for the shape parameter for different sample sizes are presented in Table 4. In case of sample of size 100 the bias decreases with increasing censoring intensity while RMSE increases with increasing censoring intensity for all the four datasets. In case of sample of size 50 the bias as well as the RMSE increases with increasing censoring intensity for all the four datasets. For both the sample sizes the bias increases with increasing true shape parameter value.

Table 2 presents the ML estimates of scale parameter and the corresponding 95% Bootstrap confidence interval for each of the four datasets with 20%, 30% and 50% right censoring as well as for the sample sizes of 50 and 100. The confidence interval for scale parameter shortens, in general, with increasing true shape parameter value for both of the sample sizes considered here. In a Weibull distribution with a fixed shape parameter value the confidence interval elongates with increasing censoring intensity.

Table 3 presents the ML estimates of threshold parameter and the corresponding 95% Bootstrap confidence interval for each of the four datasets with 20%, 30% and 50% right censoring as well as for the sample sizes of 50 and 100. The confidence interval for threshold parameter for the true shape value of up to unity is zero. That is, the method is super efficient for estimating the threshold parameter when the true shape parameter is not more than unity, as the variance is zero in this case. For a shape parameter up to unity the distribution is of inverted J shaped and the estimated value for threshold, in this case, is the minimum sample observation at which the global maxima occurs. However, here the confidence interval shortens with increasing censoring intensity in cases where it existed.

The Bootstrap mean, standard deviation and the corresponding RMSE for the shape and scale parameters for different sample sizes are presented in Tables 4 and 5, respectively. The bias increases with the increase in censoring intensity for all the datasets at both the sample sizes. The bias reduces with increasing value of the true shape parameter.

In case of the scale parameter the bias as well as the RMSE increased with the increasing censoring intensity for all the four datasets considered. The

RMSE reduces for increasing sample sizes, the results obtained at two sample sizes of 50 and 100 are presented in Table 5.

As has been reported in literature, the simultaneous ML estimation of Weibull parameters are not available, McCool [5]. Reference on the precision measures for the parameters such as exact confidence interval, root mean square errors etc., are not available in literature too. The problems that are reported are with respect the complete (uncensored) random samples from a three parameter Weibull. In case of censored random samples too, the simultaneous estimation of parameters and the estimation of their precision measures are absent in literature. 'Differential Evolution' has the power to resolve both the issues in respect to ML estimation in 3-parameter Weibull distribution with complete as well as censored random samples as shown in the paper.

Acknowledgement: The author is thankful to Prof. S. K. Mishra, Department of Economics, North-Eastern Hill University, Shillong, India, for his advice and support and for putting the source code of 'Differential Evolution' in public domain at the URL address: http://www.nehu-economics.info/, without which the present paper would not have been possible at this time.

Appendix: Simulated datasets (row wise):

1. W (0.5, 100, 10)

10.88589	63.6305	576.3075	107.9199	286.8068	116.1406	13.6834	10.04348
146.992	1142.299	25.23058	34.94055	1331.772	13.31093	1311.162	363.5392
158.7856	1403.68	26.7229	183.3061	56.08411	3364.586	119.9661	151.1806
56.53749	58.6249	25.51703	15.86971	82.96416	602.3135	58.53645	11.39896
42.82177	104.8166	33.93667	10.05499	79.95167	121.0585	266.1623	10.01004
18.29679	13.36657	23.77903	20.93802	10.53781	553.7906	120.7856	11.42792
10.00835	28.08812	116.6467	15.92383	258.8084	36.44171	40.03888	21.07629
147.7905	10.6598	10.07553	10.29192	1645.454	138.0974	18.42445	132.0623
44.85352	43.21325	216.8453	874.575	29.0993	76.97382	10.08527	30.72493
23.00528	50.47386	159.4781	73.4306	23.4946	15.62508	10.22917	46.12065
134.588	12.38457	224.0975	36.81452	501.1157	10.33112	31.11667	10.18694
16.19082	80.95329	22.42679	115.884	13.60368	10.0144	157.3871	11.91613
256.5949	23.23809	16.3467	371.6559				

2. W (1, 100, 10):

47.49365	295.1655	134.1384	58.32444	434.0472	116.933	11.2595	387.8843
34.40515	395.0495	27.23795	133.9897	55.87218	135.3203	88.29591	43.77226
128.4002	104.2123	72.25289	115.0368	26.53809	33.28041	149.6989	89.09069
14.99806	87.36182	48.06647	165.4061	229.2109	17.11071	201.9444	18.67324
220.517	22.70854	41.3113	52.02028	55.89891	276.9058	45.11195	84.78072
218.8556	20.10533	17.74928	66.08979	24.40962	118.9987	11.55825	59.64108
48.63375	105.4963	52.41224	73.76266	157.4282	63.05239	72.7209	45.99832

28.63289	215.9302	85.00178	27.3089	48.05678	69.12445	19.0777	86.72892
63.34687	104.3303	57.56328	22.30018	135.6429	102.6268	57.38987	46.6116
81.30443	139.0088	153.4838	140.0552	93.14742	28.26073	47.98886	16.85729
21.08593	25.99897	47.83381	81.79443	49.26247	129.82	58.99666	35.33675
270.0333	131.1984	250.9283	17.94736	148.535	107.098	93.13852	462.0037
149.961	114.2861	72.46065	399.0561				

3. W (1.5, 100, 10):

167.9125	97.72681	62.68049	94.17963	157.6733	182.2022	169.7751	130.162
135.1891	38.87289	136.5477	53.55421	62.99484	91.02476	343.7928	163.5941
29.44855	68.67554	73.38809	72.42	70.068	152.717	123.5195	29.24991
90.86742	168.324	172.6514	84.87927	74.52873	104.9567	71.34661	74.88485
126.4875	33.39032	23.05331	69.00029	151.356	132.9488	57.0583	239.5405
152.1866	197.2168	141.8735	76.44568	238.0119	157.8169	59.35942	93.83019
87.81452	28.72347	222.151	29.69787	69.25049	109.3646	155.3238	50.46603
89.26636	115.1868	72.29168	73.00929	33.03837	31.75766	29.14215	104.9976
42.25148	133.8173	381.9531	95.56237	168.1405	95.6525	80.21905	31.98132
70.81795	132.7257	99.36178	38.01508	60.1404	36.44472	120.2334	39.79363
100.5378	123.844	56.2011	120.8139	12.23876	28.52478	91.0206	61.12111
105.3082	50.34851	145.9647	38.30483	74.96395	71.23637	134.0334	50.66302
79.89302	20.11814	136.6417	86.21994				

4. W (2, 100, 10):

89.93747	42.75763	110.2538	185.1665	13.98683	63.75666	66.44992	169.771
198.6103	34.74639	70.15021	38.7088	79.75191	68.08543	122.8832	49.1645
150.1009	137.044	151.5859	125.0042	103.3946	57.70559	71.02444	71.23749
28.48026	149.3145	50.16073	107.8491	104.0207	127.2183	28.65072	37.93944
100.492	70.94973	119.1672	98.2574	85.7604	158.1271	72.01637	110.3442
95.69552	45.75431	52.55324	129.8862	164.9224	176.8113	136.711	123.1034
29.18118	85.51219	101.104	106.6789	88.83583	179.4847	154.7584	115.9942
93.3887	26.80991	134.3632	76.32264	49.74649	23.95366	117.5087	31.10771
147.7742	56.76707	92.25778	28.44296	130.1985	86.88968	79.57844	69.88057
137.4948	118.1158	48.18291	51.21879	105.7173	26.71556	138.7066	32.77757
69.20591	88.53731	56.06408	178.4145	58.61763	76.04012	126.6377	127.5203
100.8514	94.50437	108.6658	128.8352	79.02293	59.02912	69.05671	34.90643
95.87717	67.19489	102.7858	138.5175				

References

1. R. C. H Cheng and L. Traylor, Non-Regular Maximum Likelihood Problems, *J. Royal Stat Society. Series B (Methodologica)*. **57**, No. 1, pp. 3–44, 1995.
2. D. Cousineau, Fitting the Three-Parameter Weibull Distribution: Reviewand Evaluation of Existing and New Methods, *IEEE Trans. Dielectrics and Electrical Insulation*, **16**, No. 1, pp. 281–288, 2009.
3. B. Dodson, *The Weibull Analysis Handbook*, 2nd ed. ASQ QualityPress, Milwaukee.

4. S. Kukkonen and J. Lampinen, An Empirical of Control Parametersfor Generalized Differential Evolution, Online at ieeexplore.ieee.org/stamp/stamp.jsp?tp=arnumber=1688553, pp. 2002–2009, 2006.
5. J. I. McCool, *Using the Weibull Distribution*, John Wiley Sons, Hoboken, New Jersy, 2012.
6. S. K. Mishra Performance of Differential Evolution and Particle Swarm Methods on Some Relatively Harder Multi-modal Benchmark Functions, *Munich Personal RePEc Archive (MPRA)* Paper No. **1743**, Online at http://mpra.ub.uni-muenchen.de/1743/, 2006.
7. H. Rockette, C. Antle and L. A. Klimko Maximum likelihood estimation with the Weibull model, *J. American Statistical Association*. **69**, pp. 246–249, 1974.
8. R. L. Smith, Maximum Likelihood estimation in a class of non-regularcases, *Biometrika*, **72**, pp. 67–90, 1985.
9. R. Storn and K. Price, Differential Evolution - A Simple and efficient heuristic for Global Optimization over continuous spaces, *J. of Global Optimization*. **11**, pp. 341–359, 1997.
10. W. A. Weibull, Statistical distribution function of wide applicability, *J. Appl. Mech. Trans. ASME* **18** (3), pp. 293–297, 1951.

On Statistical Quality Control Techniques Based on Ranked Set Sampling

Md Sarwar Alamand

Department of Preventive and Social Medicine
A. N. M. Medical College, Gaya - 823 001, Bihar, India
E-mail: sarwar_stat@yahoo.co.in

Arun Kumar Sinha

Department of Statistics
Central University of South Bihar,
Patna - 800 014, India
E-mail: arunkrsinha@yahoo.com

Rahbar Ali

Department of Statistics, Patna University,
Patna - 800 005, India
E-mail: rahbar.007@gmail.com

Ranked set sampling is a cost-effective sampling technique, which has been used in many real life situations effectively. In this paper, we have made an attempt to explore its applications for improving statistical quality control techniques, which are usually based on simple random sampling; An illustration is given to show its applications in statistical quality control techniques.

Keywords: Ranked set sampling; Simple random sampling; Statistical quality control; Average run length (ARL); $\bar{X}$ control chart.

1. Introduction

Ranked set sampling (RSS) was first introduced by McIntyre (1952) to increase the efficiency of the estimation of the population mean. The method is useful when the variable of interest is very expensive or difficult to measure, but it could be easily ranked at a negligible cost. It is a cost effective sampling procedure as compared with the commonly used simple random sampling (SRS) in the situations where ordering of units with respect to the characteristics of interest can be done easily on the basis of outside information but the exact measurement of the units is difficult and expensive. Many modifications to RSS have been made since McIntyre (1952).

This method is more efficient while estimating the population mean. However, the gain of RSS over SRS is little when estimating the population variance. Perron and Sinha (2004) showed that for more than one cycle, it is possible to construct a class of quadratic unbiased estimates of σ^2 in both balanced and unbalanced cases. They obtained a minimum variance unbiased quadratic nonnegative estimate of σ^2 within a certain class of quadratic estimates. See Sinha (2005 and 2014) for more detailed discussion on the estimation of the population variance by RSS.

2. Ranked Set Sampling Methods

In this method we partitioned the selected first phase sample into sets of equal size. In order to plan an RSS design, we must, therefore, choose a set size that is typically small, around three or four, to minimize the ranking error. Let us refer to this as the set size *m*, where *m* is the number of sample units allocated to each set. Next, we proceed as follows:

Step 1: We randomly select m^2 sample units from the population of interest.

Step 2: Then we allocate the m^2 selected units as randomly as possible into *m* sets, each of size *m*. Without using the exact measurements of the characteristic of interest, the units are ranked within each set based on a perception of relative values for this characteristic. This could be based on personal judgment or done with measurements of a covariate that is correlated with the variable of interest.

Step 3: Next, we select a sample for actual analysis by including the smallest ranked unit from the first set, then the second smallest ranked unit from the second set, continuing in this manner until the largest ranked unit is selected from the last set.

Step 4: The steps 1 through 4 are repeated for *r* cycles until the desired sample size, $n = mr$, is obtained for analysis.

This procedure is referred to as MRSS with balanced (equal) allocation because each rank order consists of *r* observations for each set. Further, one could utilize the prior information, if available, about the population for obtaining a more efficient estimator of the population parameter of interest. For a skewed population the numbers of quantifications of the ith rank, r_i is taken as proportional to the standard deviation of the rank order, $\sigma_{(i,m)}$. Because of unequal number of quantifications this allocation could be termed as MRSS with unbalanced allocation (MRSSUA). The data stricture is given in Table 1.

Table 1. Data structure of a ranked set sample of size *mr*.

Cycles	Data					
1						
$X_{(1:m)1}$	$X_{(2:m)1}$	...	$X_{(i:m)1}$	...	$X_{(m:m)1}$	
2						
$X_{(1:m)2}$	$X_{(2:m)2}$	...	$X_{(i:m)2}$	...	$X_{(m:m)2}$	
	.	.	.	.	.	
	.	.	.	.	.	
j						
$X_{(1:m)j}$	$X_{(2:m)1}$	...	$X_{(i:m)j}$	...	$X_{(m:m)j}$	
	.	.	.	.	.	
	.	.	.	.	.	
r						
$X_{(1:m)r}$	$X_{(2:m)r}$	...	$X_{(i:m)r}$	...	$X_{(m:m)r}$	
Population Mean	$\mu_{(1:m)}$	$\mu_{(2:m)}$	...	$\mu_{(i:m)}$	...	$\mu_{(m:m)}$
Sample Mean	$\overline{X}_{(1:m)}$	$\overline{X}_{(2:m)}$	...	$\overline{X}_{(i:m)}$	...	$\overline{X}_{(m:m)}$
Population Variance						
$\sigma^2_{(1:m)}$	$\sigma^2_{(2:m)}$	...	$\sigma^2_{(i:m)}$	...	$\sigma^2_{(m:m)}$	
Sample Variance	$s^2_{(1:m)}$	$s^2_{(2:m)}$	...	$s^2_{(i:m)}$	...	$s^2_{(m:m)}$

Salazar and Sinha have (1997) used RSS in statistical quality control first time, and some other authors, who contributed to this area later on, include Muttlak and Al-Sabah (2003), Al-Nasser and Al-Rawwash (2007), Al-Omari and Al-Nasser (2011) and Pongpullponsak and Sontisamran (2010).

3. McIntyre's Estimators

3.1. *Equal allocation*

Suppose that $X_{(i:m)j}$ denotes the ith order statistic under perfect ranking in *j*th cycle, for $i = 1, 2, \ldots, m$ and $j = 1, 2, \ldots, r$. Also, $E(X_{(i:m)j}) = \mu_{(i:m)}$ and $var(X_{(i:m)j}) = \sigma^2_{(i:m)}$. The McIntyre's estimator, $\hat{\mu}_{(MRSS)}$, of the population mean, μ, is defined as follows:

$$\hat{\mu}_{(MRSS)} = \frac{1}{mr}\sum_{i=1}^{m}\sum_{j=1}^{r} X_{(i:m)j} \tag{1}$$

Also,

If $\hat{\mu}_{(i:m)} = \frac{1}{r}\sum_{j=1}^{r} X_{(i:m)j}$ then $\hat{\mu}_{(MRSS)} = \frac{1}{m}\sum_{i=1}^{m}\hat{\mu}_{(i:m)}$.

Here $E(\hat{\mu}_{(i:m)}) = \mu_{(i:m)}$; $E(\hat{\mu}_{MRSS}) = \frac{1}{m}\sum_{i=1}^{m}\hat{\mu}_{(i:m)} = \mu$

$$\text{a } var(\hat{\mu}_{MRSS}) = \frac{1}{m^2 r}\sum_{i=1}^{m}\sigma^2_{(i:m)} \tag{2}$$

This is also expressed as

$$var(\hat{\mu}_{MRSS}) = \frac{1}{mr}\left[\sigma^2 - \frac{1}{m}\sum_{i=1}^{m}\left(\mu_{(i:m)} - \mu\right)^2\right]$$

The alternative expression is useful when the variance of the ith order statistics is not available. If we denote the SRS estimator of the population mean with the same sample size $n = mr$ by $\hat{\mu}_{SRS}$ and then $var(\hat{\mu}_{SRS}) = \frac{\sigma^2}{n}$.

The unbiased estimator of the population variance σ^2 based on a ranked set sample when the number of cycle is more than one is given by

$$\hat{\sigma}^2_{MRSS} = \frac{mr - m + 1}{m^2 r(r-1)} \sum_{i=1}^{m} \sum_{j=1}^{r} \left(X_{(i:m)j} - \hat{\mu}_{(i:m)}\right)^2 + \frac{1}{m} \sum_{i=1}^{m} \left(\hat{\mu}_{(i:m)} - \mu_{MRSS}\right)^2$$

See Norris, Patil and Sinha (1995), and Sinha (2005) for the result. The relative precision (RP) of the MRSS estimator $\hat{\mu}_{MRSS}$, as compared with simple random sample (SRS) estimator $\hat{\mu}_{SRS}$, with the same sample size, $n = mr$ is computed as follows:

$$RP = \frac{var(\hat{\mu}_{SRS})}{var(\hat{\mu}_{MRSS})} = \frac{\sigma^2/mr}{\frac{1}{m^2 r}\sum_{i=1}^{m} \sigma^2_{(i:m)}} = \frac{\sigma^2}{\overline{\sigma^2}} \tag{3}$$

where, $\overline{\sigma^2} = \frac{1}{m}\sum_{i=1}^{m} \sigma^2_{(i:m)}$

3.2. *Unequal allocation*

McIntyre (1952) suggested taking the sample size of each rank order proportional to its standard deviation while sampling asymmetrical populations. It means that if r_i denotes the number of sets having quantified units with rank i, then $r_i \alpha \sigma_{(i:m)}$ for $i = 1, \ldots, m$. This leads to

$$r_i = \frac{n\sigma_{(i:m)}}{\sum_{i}^{m} \sigma_{(i:m)}} I = 1, \ldots, m. \tag{4}$$

Let r_i ($\neq 0$) measurements be made corresponding to the ith rank for $i = 1, 2, \ldots, m$. The resulting sample size is $n = \sum_{i=1}^{m} r_i$. The unbiased estimator of the population mean is thus given by

$$\hat{\mu}_{(MRSSUA)} = \frac{1}{m}\sum_{i=1}^{m} \frac{T_i}{r_i} \tag{5}$$

And

$$var(\hat{\mu}_{MRSSUA}) = \frac{1}{m^2}\sum_{i=1}^{m} \frac{\sigma^2_{(i:m)}}{r_i} \tag{6}$$

where $T_i = \sum_{j=1}^{r_i} X_{(i:m)j}$

The relative precision (RP_{ua}) of $\hat{\mu}_{MRSSUA}$, as compared with simple random sample (SRS) estimator, $\hat{\mu}_{SRS}$, with the same sample size, n, is computed as follows:

$$RP_{ua} = \frac{var(\hat{\mu}_{SRS})}{var(\hat{\mu}_{MRSSUA})} = \frac{\sigma^2/n}{\left(\frac{1}{m}\right)^2 \sum_{i=1}^{m} \frac{\sigma^2_{(i:m)}}{r_i}} \tag{7}$$

Where $\bar{\sigma} = \frac{1}{m}\sum_{i}^{m} \sigma_{(i:m)}$

On comparing Neyman allocation with equal allocation, the relative precision is obtained as

$$RP_{eq:ney} = \frac{var(\bar{X}_{(m)eq})}{var(\bar{X}_{(m)ney})} = \frac{\overline{\sigma^2}}{\bar{\sigma}^2} \qquad (8)$$

The estimator of the population variance and that of the population variance of the ith rank order based on unequal sample sizes below given.

$$\hat{\sigma}^2_{MRSSUA} = \sum_{i=1}^{m} \frac{mr_i - m + 1}{m^2 r_i (r_i - 1)} \sum_{j=1}^{r_i} \left(X_{(i:m)j} - \bar{X}_{(i:m)}\right)^2 + \frac{1}{m} \sum_{i=1}^{m} \left(X_{(i:m)} - \bar{X}_{(m)r}\right)^2$$

See Norris, Patil and Sinha (1995) for this expression.

4. Comparing the Control Charts Using Average Run Length

Muttlak and Al-Sabah (2001) compared the newly developed charts using Average Run Length (ARL) to the Shewhart control chart. They showed that the newly charts are more efficient than the Shewhart charts i.e. they have smaller average run length. Finally they collected a real life data set and used to construct quality control charts for the newly suggested control charts.

The ARL which is defined as

$$\text{ARL} = \frac{1}{\alpha}$$

where α is type I error. If the process gets out of control, then the ARL is written in terms of type II error β as follows

$$\text{ARL} = \frac{1}{1-\beta}$$

Based on the ARL criterion, the process remains in control with mean μ_0 and standard deviation σ_0 and sometimes it may get out of control in terms of a mean shift of the amount $\delta\sigma_0/\sqrt{m}$ where δ is nonnegative and selected to dominate the shift in the mean μ. We conduct a simulation study under the normality assumption with mean $\mu_0$and variance σ_0^2 assuming the ranking is perfect. When the process is under control, the ARL is around 370 implying that the out-of-control signal will flash once every 370 observed samples.The results obtained by AL-Nasser and M. Al-Rawwash (2007) and AL- Nasser *et al.* (2013) are given in Table 2.

Table 2. Average run length using different methods when $m = 4$.

δ	SRS	RSS
0.0	369.412	349.040
0.1	341.712	312.304
0.2	312.982	229.410
0.3	256.012	166.750

Table 2. (*Continued*)

δ	SRS	RSS
0.4	200.798	115.942
0.5	156.121	76.704
1.0	45.324	14.149
1.8	8.881	2.480
2.6	2.952	1.193
3.4	1.518	1.013

4.1. *Statistical quality control charts*

A control chart is a graphical technique used for continuously monitoring whether the manufacturing process is in a state of statistical control or not. Its primary objective is to detect quickly the presence of assignable causes of process shifts so that investigation of the process and corrective measures may be taken before many non-conforming units are manufactured. Generally, a control chart is an effective tool for identifying the assignable (special) causes of variation in process variability as well as estimating the parameters of the production process. Since the first work of Shewhart (1924), the construction of quality control charts has undergone series of modifications with new methods being suggested. All methods reported in the literature are based on SRS, which to a certain extent is considerably less effective in estimating the population mean compared with a new sampling method, such as RSS and its modifications. Only a very few cases have been reported on the use of RSS and its modifications to develop control charts for monitoring the process mean and these are shown to dominate the classical control charts based on SRS, see Salazar and Sinha (1997) followed by Muttlak and Al-Sabah (2003) and others. Salazar and Sinha (1997) developed an average control chart based on RSS when the population has normal distribution and the process has various shift values. They calculated various average run length values using computer simulations and concluded that every average control chart based on their established methods is more useful than that of Shewhart. The control chart used in this study is given in the following subsections.

4.2. *Quality control chart using SRS*

Let X_{ij} when i=1,2,...n and j =1,2...r, be the ith unit in the jth order or sample size n, and $X_{ij} \sim N(\mu, \sigma^2)$. When the population mean and variance are μ and σ^2, respectively. The Shewhart control chart for

$$\bar{X}_j = \frac{1}{n}\sum_{i=1}^{n} X_{ij} \quad ; j = 1, 2, \dots n \tag{9}$$

is defined by

$$UCL = \mu + 3\frac{\sigma}{\sqrt{n}}, CL = \mu, LCL = \mu - 3\frac{\sigma}{\sqrt{n}} \quad (10)$$

When UCL, CL, and LCL represent upper control limit, central line, and lower control limit respectively. After obtaining the values, the population mean $\bar{X}_j$,j=1,2,…,r can be plotted. However, in actual measurement, the mean μ and variance σ^2 are unknown so both μ and σ are estimated from the collected data, where the unbiased estimator for μ is

$$\bar{\bar{X}}=\frac{1}{r}\sum_{j=1}^{r}\bar{X}_j \quad (11)$$

but

$$\bar{s} = \frac{1}{r}\sum_{j=1}^{r} S_j \quad (12)$$

where

$$s_j = \left[\frac{1}{n-1}\sum_{i=1}^{n}\left(X_{ij} - \hat{X}_j\right)^2\right]^{1/2} \quad (13)$$

is a biased estimate for σ. We can use $\bar{s}/c_4$ as unbiased estimator where c_4 is calculated from

$$c_4 = \left(\frac{2}{n} - 1\right)^2 \frac{(n-2)!}{[(n-1/2)!]} \quad (14)$$

and the control chart for sample mean can be expressed as

$$\text{UCL}=\bar{\bar{X}}+3\frac{\bar{S}}{C_4\sqrt{n}}\ , \quad \text{CL}=\bar{\bar{X}}, \quad \text{LCL}=\bar{\bar{X}}-3\frac{\bar{S}}{C_4\sqrt{n}} \quad (15)$$

4.3. *Quality control chart using RSS*

The RSS mean $\bar{X}_{rss.j}$ at the jth cycle can be plotted in the control chart based on RSS that was first proposed by Salazar and Sinha (1997). These are given below.

$$UCL = \bar{X}_{rss} + 3\sigma_{\bar{X}_{rss}}, CL = \bar{X}_{rss}, LCL = \bar{X}_{rss} - 3\sigma_{\bar{X}_{rss}}$$

When

$$\sigma_{\bar{X}_{rss}} = \sqrt{\frac{1}{n^2}\sum_{i=1}^{n}\sigma_{(i;n)}^2} \quad (16)$$

where an unbiased estimate for The population mean under RSS is

$$\bar{X}_{rss} = \frac{1}{r}\sum_{j=1}^{r}\bar{X}_{rss.j} \quad (17)$$

An estimate for the standard deviation was suggested by Muttlak and Al-Sabah (2003) and it is given below

$$\hat{\sigma}_{\bar{X}_{rss}} = \left[\frac{1}{n}\hat{\sigma}_{\bar{X}_{rss}}{}^2 - \frac{1}{n^2}\sum_{i=1}^{n}\left(\bar{X}_{(i)} - \bar{X}_{rss}\right)^2\right]^{1/2} \quad (18)$$

$$\hat{\sigma}_{rss}^2 = \frac{1}{nr-1}\sum_{i=1}^{n}\sum_{j=1}^{r}\left(X_{(i;n)j} - \bar{X}_{rss}\right)^2 \tag{19}$$

and

$$\bar{X}_{(i)} = \frac{1}{r}\sum_{j=1}^{r} X_{(i;n)} \tag{20}$$

is the estimate for population mean of the ith order statistic. Subsequently, the control chart can be constructed by using $\bar{X}_{rss}$ and $\hat{\sigma}_{\bar{X}_{rss}}$ as the equation below,

$$UCL = \bar{X}_{rss} + 3\hat{\sigma}_{\bar{X}_{rss}} CL = \bar{X}_{rss}, LCL = \bar{X}_{rss} - 3\hat{\sigma}_{\bar{X}_{rss}} \tag{21}$$

5. Illustration

For the illustration we have taken the data for 20 days from the original data of 28 days from Levine, *et al.* (2006), page 767. It is given in Table 3.

A hotel claimed that the customers' luggage would arrive in their rooms within 14 minutes of their check in. The required data were recorded over a 4-week period and subgroups of five deliveries were selected from the evening shift on each day for analysis.

Table 3. Luggage delivery times in minutes.

Days	**Luggage Delivery Times in Minutes**			**Average**	
	1	**2**	**3**	$\bar{X}$	s_i
1	6.70	11.70	9.70	9.37	2.52
2	7.60	11.40	9.00	9.33	1.92
3	9.50	8.90	9.90	9.43	0.50
4	9.80	13.20	6.90	9.97	3.15
5	11.00	9.90	11.30	10.73	0.74
6	8.30	8.40	9.70	8.80	0.78
7	9.40	9.30	8.20	8.97	0.67
8	11.20	9.80	10.50	10.50	0.70
9	10.00	10.70	9.00	9.90	0.85
10	8.60	5.80	8.70	7.70	1.65
11	10.70	8.60	9.10	9.47	1.10
12	10.80	8.30	10.60	9.90	1.39
13	9.50	10.50	7.00	9.00	1.80
14	12.90	8.90	8.10	9.97	2.57
15	7.80	9.00	12.20	9.67	2.27
16	11.10	9.90	8.80	9.93	1.15
17	9.20	9.70	12.30	10.40	1.66
18	9.00	8.10	10.20	9.10	1.05
19	9.90	10.10	8.90	9.63	0.64
20	10.70	9.80	10.20	10.23	0.45
		Total		192.00	27.58

For construction a mean $\bar{X}$ chart for first 20 days for n=3, we consider the Table 3.

The center line of the control data is the average of the samples:

$$\bar{\bar{X}} = \frac{192.00}{20} = 9.60,\ \bar{s} = \frac{27.58}{20} = 1.3788,\ c_4 = 0.8862,\ \hat{\sigma} = \frac{1.3788}{0.8862} = 1.556$$

The control limits are under SRS are obtained as shown below

$$\text{UCL} = 9.60 + 3 * \frac{1.556}{\sqrt{3}} = 12.295,\ \text{CL} = 9.60,\ \text{LCL} = 9.60 - 3 * \frac{1.556}{\sqrt{3}} = 6.905$$

Table 4. Data for RSS $m = 3$.

S.No	Lowest	Middle	Highest	RSS Sample Mean
1	8.20	10.20	7.80	8.73
2	7.60	8.60	8.40	8.20
3	7.00	9.00	10.20	8.73
4	8.20	8.80	10.20	9.07
5	6.90	10.20	10.20	9.10
6	8.70	9.00	9.90	9.20
7	7.00	8.30	10.00	8.43
8	9.10	8.40	9.80	9.10
9	8.80	10.20	8.70	9.23
10	9.00	7.50	9.40	8.63
11	8.80	10.00	11.00	9.93
12	7.10	9.00	9.90	8.67
13	6.90	8.90	9.60	8.47
14	6.70	7.50	7.80	7.33
15	8.00	8.40	9.80	8.73
16	8.70	9.50	9.60	9.27
17	8.80	9.50	10.80	9.70
18	8.70	8.60	9.00	8.77
19	9.00	8.60	9.40	9.00
20	6.70	8.60	9.60	8.30
Total	159.90	178.80	191.10	176.59
Mean	8.00	8.94	9.56	8.83

From equation (17) and (20)

$$\bar{X}_{rss} = 8.83, \bar{X}_1 = 8.00, \qquad \bar{X}_2 = 8.94, \qquad and\ \bar{X}_3 = 9.56$$

Using the equation (18) and (19) for an estimate standard deviation

$$\hat{\sigma}^2_{rss} = \frac{60.33}{59} = 1.124\ and\ \hat{\sigma}_{\bar{X}_{rss}} = \frac{1.124}{3} - \frac{1.235}{9} = 0.4873$$

we get,

$$UCL_{RSS} = \bar{X}_{rss} + 3\hat{\sigma}_{\bar{X}_{rss}} = 8.83 + 3 * 0.4873 = 10.292$$

$$CL_{RSS} = \bar{X}_{rss} = 8.83$$

$$LCL_{RSS} = \bar{X}_{rss} - 3\hat{\sigma}_{\bar{X}_{rss}} = 8.83 - 3 * 0.4873 = 7.367$$

Table 5. Control limits of SRS and RSS.

Quality Control Chart for sample size 3			
Limit	SRS	RSS	Difference
UCL	12.295	10.292	2.003
CL	9.600	8.830	0.770
LCL	6.905	7.367	0.462

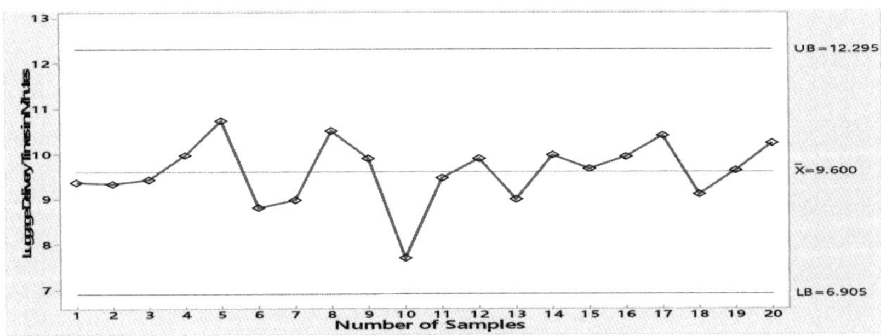

Figure 1. Chart based on SRS for size $m = 3$.

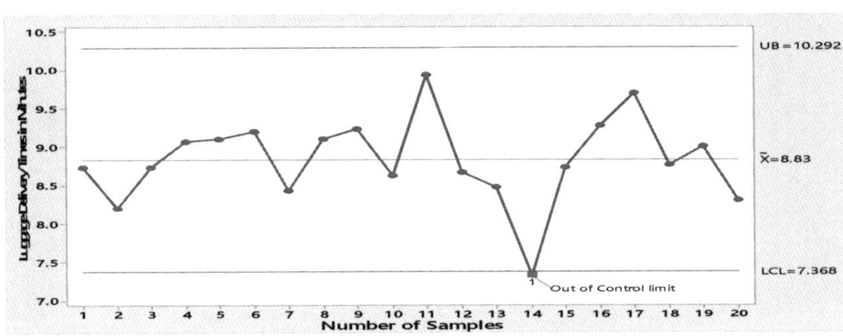

Figure 2. Chart based on RSS for size $m = 3$.

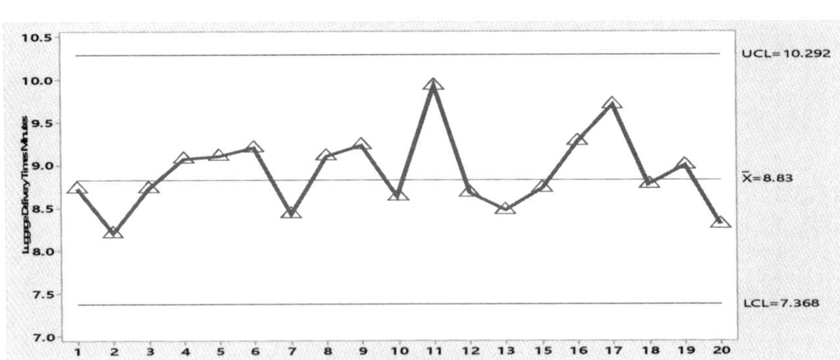

Figure 3. Chart based on RSS for size $m = 3$ (after removing out of control sample).

We have selected nine values with replacement from each of the first twenty rows to obtain the ranked set sample of size $m = 3$. These are given in Table 3 and Table 4 which provides the control limits for SRS and RSS Figures 1 and 2 depict the control charts based on SRS and RSS respectively. Note that the value of the UCL under RSS is smaller than that of RSS but the LCL is largerfor RSS than that for SRS. This happens because of smaller variance of mean under RSS than SRS.

References

1. A. K. Sinha, On some recent developments in ranked set sampling. *The Bulletin of Informatics and Cybernatics*, **37**, 137–160 (2005).
2. A. K. Sinha, Ranked set sampling methods, for vegetation research. *Int. J. Mendel*, **31(1-2)**, 13–22 (2014).
3. A. D. Al-Nasser and M. Al-Rawwash, A control chart based on ranked data, *Journal of Applied Sciences,* 7(14), pp. 1936–1941 (2007).
4. A. I.Al-Omari, K. Ibrahim & M. I. Syam, Investigating the use of Stratified Percentile Ranked Set Sampling method for estimating the population mean. *Proyecciones Journal of Mathematics*, **30(3)**, 351–368 (2011).
5. A. Pongpullponsak and P Sontisamran, Statistical quality control based on ranked set sampling for multiple characteristics, *Proc. Int. Conf. Sustainable Greater Mekong Sub region*; 26–27, August, Bangkok, Thailand (2010).
6. D. C. Montgomery, *Introduction to Statistical Quality Control, 6th edition.* John Wiley & Sons (2009).
7. F. Perron, and B. K. Sinha, Estimation of Variance based on a ranked set sample, *Journal of statistical planning and inference*, **120**, 21–28 (2004).
8. G. A. McIntyre, A method of unbiased selective sampling using ranked sets. *Australian Journal of Agricultural Research*, **59**, 385–390 (1952).
9. H. A. Muttlak and Sabah Al, Statistical Quality Control Using Ranked Set Sampling. *Technical Report no. FT/2000-15*, Department of Mathematical Sciences, KFUPM (2001).
10. H. A. Muttlak and W. S. Al-Sabah, Statistical quality control based on ranked set sampling, *J. Appl. Stat.*; 30: 1055–1078 (2003).
11. G. P. Patil, A. K. Sinha and C. Taillie, Finite population corrections for ranked set sampling, *Annals of the Institute of Statistical Mathematics*, **47** (4), 621–636 (1995).
12. R. D. Salazar and A. K. Sinha, Control chart *x* based on ranked set sampling. ComunicacionTecica No. 1-97-09 (PE/CIMAT)(1997).

13. R. C. Norris, G. P. Patil and A. K. Sinha, Relative precision of ranked set sampling: Comparison of the regression estimator, *Environmetrics*, **4**(4), 399–412 (1995, 1993a).
14. W. A. Shewhart, Economic Control of Quality of Manufactured Product. Van Nostrand (1931).
15. Levine, Stephen, Krehbiel and Brenson, Statistics for Managers using Microsoft Excel, Prentice Hall of India, New Delhi (2006).

Approximate Solution for Nonlinear Oscillator with Cubic and Quintic Nonlinearities

Jitendra Singh

Central University of South Bihar
Patna, Bihar 800014, India
E-mail: jitendrasingh@cub.ac.in
www.cub.ac.in

In this article, an approximate solution for Duffing equations with cubic and quantic nonlinearities is obtained using the Differential Transform Method and Pade approximation technique. The concept of Differential Transform Method is briefly introduced and applied it to given problem to derive solution of nonlinear equation. The major concern of this paper is successfully use of Adomian polynomial to assess the nonlinearities. The results are compared with the numerical solution by fourth-order Runge–Kutta method and found with good agreement. Results are shown by graphs for whole range of time domains accurately.

Keywords: Duffing Equation; Differential Transform Method; Pade Approximation; Fourth order Runge-Kutta Method; Adomian Polynomial.

1. Introduction

Nonlinear oscillation in engineering and applied mathematics has been a topic to intensive research for many years[1-3]. In the past, there have been analytical and numerical methods for solving these complicated nonlinear systems. The perturbation method involving expansion over a small parameter has been the most common analytical techniques for nonlinear oscillation systems. In many cases, the nonlinear equation has been replaced by a related linear equation that approximates the original nonlinear equation closely enough to provide useful solution. However, such linearization is not often feasible and the original nonlinear differential equation itself must be considered[4]. There are some most common methods for constructing approximate analytical solutions to the nonlinear oscillator equation. These methods include the harmonic balance method (HB)[5], the elliptic Lindstedt-Poincare method (LP)[6] and multiple scales method (MSM)[7] are widely used to obtain approximate solution of nonlinear oscillators.

The Duffing equation is a well-known nonlinear differential equation which is related to many practical engineering systems such as the classical nonlinear spring system with odd nonlinear restoring characteristics and more recently in different physical phenomena. There have been many variations of Duffing equation, for instance, the Duffing-harmonic equation[8] and the cubic-quintic Duffing equation. Due to the presence of fifth power nonlinearity, the cubic-quintic Duffing equation inherits strong nonlinearity and thus accuracy of approximate analytical methods becomes extremely demanding.

The differential transform method (DTM) was first applied in the engineering domain by Momani[9]. DTM provides an efficient explicit and numerical solution with high accuracy, minimal calculation, avoidance of physically unrealistic assumptions. By using DTM, we obtain a series solution, in practice a truncated series solution. This series solution does not exhibit the periodic behavior which is characteristic of oscillator equations and gives a good approximation to the true solution in a very small region. In order to improve the accuracy of DTM, we use an alternative technique which modifies the series solution for non-linear oscillatory system as follows: we first apply the Laplace transformation to the truncated series obtained by DTM, then convert the transformed series into a meromorphic function by forming its Pade approximants, and finally adopt an inverse Laplace transform to obtain an analytical solution, which may be periodic or a better approximation solution than the DTM truncated series solution. The first connection between series solution methods such as an Adomian decomposition method and Pade approximants was established by Jiao[10] *et al.* S. Nourazar and A. Mirzabeigy[11] have solved the nonlinear doffing oscillator with damping effect using the modified differential transform method.

The nonlinearities are computed in terms of Adomian polynomials. Finally, a numerical comparison has been made between the present method and the fourth-order Runge-Kutta method.

2. Differential Transform Method (DTM)

The differential transform method is a semi-numerical-analytic-technique that formulizes the Taylor series in a totally different manner. With this technique, the given differential equation and related initial conditions are transformed into a recurrence equation that finally leads to the solution of a system of algebraic equations as coefficients of a power series solution. This method is useful for obtaining exact and approximate solutions of linear and nonlinear differential equations. There is no need for linearization or perturbation. The basic definitions of differential transformation are introduced as follows:

Definition 2.1. If $f(t)$ is analytic in the time domain T, then it will be differentiated continuously with respect to time t

$$\phi(t,k) = \frac{d^k f(t)}{dt^k}, \quad \forall t \in T. \tag{1}$$

For, $t = t_i$, then $\phi(t,k)$=$\phi(t_i,k)$, where k belong to the set of non-negative integers, denoted as the K -domain. Therefore, Eq. (1) can be rewritten as

$$F(k) = \frac{1}{k!}\left[\frac{d^k f(t)}{dt^k}\right]_{t=t_i}, \quad \forall k \in K, \tag{2}$$

Where, $F(k)$ is called the differential transform of $f(t)$ in the K-domain.

Definition 2.2. If $f(t)$ can be represented by the Taylor series, then it can be represented as

$$f(t) = \sum_{k=0}^{\infty} F(k)(t-t_i)^k, \tag{3}$$

Eq. (3) is called the inverse transform of $F(k)$. With the symbol DT denoting the differential transform process. We obtain

$$DT^{-1}(F(k)) = f(t), \tag{4}$$

Table1. The fundamental operations of the differential transform method.

Time function	Transformed function
$w(t) = \alpha u(t) \pm \beta v(t)$	$W(k) = \alpha U(k) \pm \beta V(k)$
$w(t) = \frac{d^m u(t)}{dt^m}$	$W(k) = \frac{(k+m)!}{k!} U(k+m)$
$w(t) = u(t)v(t)$	$W(k) = \sum_{l=o}^{k} U(l)V(k-l)$
$w(t) = t^m$	$W(k) = \delta(k-m) = \begin{cases} 1, & if\ k = m \\ 0, & if\ k \neq m \end{cases}$
$w(t) = \exp(t)$	$W(k) = \frac{1}{k!}$
$w(t) = \sin(\omega t + \alpha)$	$W(k) = \left(\frac{\omega^k}{k!}\right)\sin\left(\frac{k\pi}{2} + \alpha\right)$
$w(t) = \cos(\omega t + \alpha)$	$W(k) = \left(\frac{\omega^k}{k!}\right)\cos\left(\frac{k\pi}{2} + \alpha\right)$

3. The Adomian Polynomial and Their Evaluation

The Adomian polynomials are indispensable in nonlinear analyses by the ADM. Let N be a nonlinear operator acting upon an unknown function u. For treating functional equations including such nonlinear terms like $N(u)$, ADM entails decomposition of $N(u)$ into an infinite summation of the Adomian polynomials, $A_n s$, corresponding to N as

$$N(u) = \sum_{n=0}^{\infty} A_n, \tag{5}$$

Where, $A_n s$ are classically suggested to be obtained by

$$A_n = A_n(u_0, u_1, ...u_n) = \frac{1}{n!}\left[\frac{d^n}{d\lambda^n} N\left(\sum_{i=0}^{\infty} \lambda^i u_i\right)\right]_{\lambda=0}, \tag{6}$$

and $u_i s$ are components to which the function u is decomposed to, as in

$$u = \sum_{i=0}^{\infty} u_i, \tag{7}$$

Recently, HoomanFatoorehchi and Hossein Abolghasemi[12] proposed a novel technique to obtain the differential transform of nonlinearities by the Adomian polynomials

$$DT\{N(u)\} = A_k(U(0), U(1), ...U(k)). \tag{8}$$

4. Pade Approximants

A pade approximants is the ratio of two polynomials constructed from the coefficients of the Taylor series expansion of a function $y(x)$. The $[L/M]$ Pade approximants to a function $y(x)$ are given by

$$\left[\frac{L}{M}\right] = \frac{P_L(x)}{Q_M(x)}, \tag{9}$$

Where, $P_L(x)$ is a polynomial of degree at most L and $Q_M(x)$ is a polynomial of degree at most M. The formal power series

$$y(x) = \sum_{i=0}^{\infty} a_i x^i, \tag{10}$$

$$y(x) - \frac{P_L(x)}{Q_M(x)} = O(x^{L+M+1}), \tag{11}$$

Determine the coefficients of $P_L(x)$ and $Q_M(x)$ by the equation. Since we can clearly multiply the numerator and denominator by a constant and leave $[L/M]$ unchanged, we impose the normalization condition

$$Q_M(0)=1, \tag{12}$$

Finally we require that $P_L(x)$ and $Q_M(x)$ have no common factors.

If we write the coefficient of $P_L(x)$ and $Q_M(x)$ as

$$\left.\begin{aligned} P_L(x) &= p_0+p_1x+p_2x^2+\ldots\ldots\ldots\ldots+p_Lx^L, \\ Q_M(x) &= q_0+q_1x+q_2x^2+\ldots\ldots\ldots\ldots+q_Mx^M \end{aligned}\right\}. \tag{13}$$

then by Eq. (12) and Eq. (13), we multiply Eq. (11) by $Q_M(x)$, which linearizes the coefficient equations. We can write out Eq. (11) in more detail as

$$\left.\begin{aligned} &a_{L+1}+a_Lq_1+\ldots\ldots\ldots\ldots+a_{L-M+1}q_M=0, \\ &a_{L+2}+a_{L+1}q_1+\ldots\ldots\ldots\ldots+a_{L-M+2}q_M=0, \\ &\vdots \\ &a_{L+M}+a_{L+M-1}q_1+\ldots\ldots\ldots\ldots+a_Lq_M=0, \end{aligned}\right\}, \tag{14}$$

$$\left.\begin{aligned} &a_0=p_0, \\ &a_1+a_0q_1=p_1, \\ &a_2+a_1q_1+a_0q_2=p_2, \\ &\vdots \\ &a_L+a_{L-1}q_1+\ldots\ldots+a_0q_L=p_L \end{aligned}\right\}, \tag{15}$$

To solve these equations, we start with Eq. (14), which is a set of linear equations for all the unknown q 's. Once the q 's are known, then Eq. (15) gives an explicit formula for the unknown p 's, which complete the solution. If Eq. (14) and Eq. (15) are non-singular, then we can solve them directly and obtain Eq. (16)

$$\left[\frac{L}{M}\right]=\frac{\det\begin{bmatrix} a_{L-M+1} & a_{L-M+2} & \cdots & a_{L+1} \\ \vdots & \vdots & \ddots & \vdots \\ a_L & a_{L+1} & \cdots & a_{L+M} \\ \sum_{j=M}^{L} a_{j-M}x^j & \sum_{j=M-1}^{L} a_{j-M+1}x^j & \cdots & \sum_{j=0}^{L} a_j x^j \end{bmatrix}}{\det\begin{bmatrix} a_{L-M+1} & a_{L-M+2} & \cdots & a_{L+1} \\ \vdots & \vdots & \ddots & \vdots \\ a_L & a_{L+1} & \cdots & a_{L+M} \\ x^M & x^{M-1} & \cdots & 1 \end{bmatrix}}. \tag{16}$$

To obtain diagonal Pade approximants of different order such as [2/2], [4/4] or [6/6], we have used calculus software, Mathematica for computation.

5. Solution of the Problem by DTM

Duffing equations describe many kinds of nonlinear oscillatory systems in physics, mechanics and engineering[1-3]. They are famous in nonlinear dynamics and have been presented considering various types of nonlinearity.

The unforced, undamped cubic-quintic Duffing equation is given by

$$\frac{d^2u}{dt^2}+\alpha u+\beta u^3+\gamma u^5=0, \tag{17}$$

where u is displacement and α, β and γ are arbitrary constants. Initial condition are as follows:

$$u(0)=a, \tag{18}$$

$$\frac{du(0)}{dt}=0, \tag{19}$$

where a is constant. We are taking the differential transform of both sides of Eq. (17), we obtain the following recurrence relation:

$$U(k+2)=\frac{1}{(k+1)(k+2)}\left\{\begin{matrix}-\alpha U(k)-\beta A_k\left(U(0),U(1),......U(k)\right)\\ -\gamma B_k\left(U(0),U(1),......U(k)\right)\end{matrix}\right\}, \tag{20}$$

the initial condition given in Eq. (18) and Eq. (19) can be transformed at $t=0$ as

$$U(0)=a, \tag{21}$$

$$U(1)=0, \tag{22}$$

and A_k's and B_k's are given by the Adomian polynomials

$$DT\left\{u^3\right\} = A_k\left(U(0), U(1), U(2), \ldots\ldots. U(k)\right), \tag{23}$$

$$DT\left\{u^5\right\} = B_k\left(U(0), U(1), U(2), \ldots\ldots. U(k)\right), \tag{24}$$

for the value of $k = 0,1,2,3.....$ we obtain the following expression

$$U(2) = -\frac{1}{2}\left(\alpha a + \beta a^3 + \gamma a^5\right), \tag{25}$$

$$U(3) = 0, \tag{26}$$

$$U(4) = \frac{\alpha^2 a}{24} + \frac{1}{6}\alpha\beta a^3 + \frac{1}{8}\beta^2 a^5 + \frac{1}{4}\alpha\gamma a^5 + \frac{1}{3}\beta\gamma a^7 + \frac{5}{24}\gamma^2 a^9, \tag{27}$$

$$U(5) = 0, \tag{28}$$

$$U(6) = -\left(\frac{\alpha^3}{720}a + \frac{5}{144}\alpha^2\beta a^3 + \frac{17}{240}\alpha\beta^2 a^5 + \frac{3}{80}\beta^3 a^7 + \frac{71}{720}\alpha^2\gamma a^5 + \frac{101}{360}\alpha\beta\gamma a^7 \right.$$
$$\left. + \frac{3}{16}\beta^2\gamma a^9 + \frac{31}{144}\alpha\gamma^2 a^9 + \frac{193}{720}\beta\gamma^2 a^{11} + \frac{17}{144}\gamma^3 a^{13}\right), \tag{29}$$

$$U(7) = 0, \tag{30}$$

$$U(8) = \frac{\alpha^4}{40320}a + \frac{13\alpha^3\beta}{2520}a^3 + \frac{47\alpha^2\beta^2}{2240}a^5 + \frac{3\alpha\beta^3}{112}a^7 + \frac{7\beta^4}{640}a^9$$
$$+ \frac{319\alpha^3\gamma}{10080}a^5 + \frac{47\alpha^2\beta\gamma}{336}a^7 + \frac{221\alpha\beta^2\gamma}{1120}a^9 + \frac{5}{56}\beta^2\gamma a^{11} + \frac{59}{448}\alpha^2\gamma^2 a^9 \tag{31}$$
$$+ \frac{143}{420}\alpha\beta\gamma^2 a^{11} + \frac{69}{320}\beta^2\gamma^2 a^{13} + \frac{49}{228}\alpha\gamma^3 a^{13} + \frac{208}{1008}\beta\gamma^3 a^{15} + \frac{565}{8064}\gamma^4 a^{17},$$

now the following approximate series solution is obtained

$$u(t) = \sum_{k=0}^{8} U(k)t^k, \tag{32}$$

since this series solution does not exhibit the periodic behavior that is characteristic of the oscillatory system Eq. (17) and Eq. (18). In order to improve the accuracy of the differential transform solution Eq. (32), we implement the Pade approximant of order [4/4] and got the desired accuracy.

For the sake of generalities, If we take $\alpha = 1, \beta = -1, \gamma = 10$ and $a = 1$, we got the desired accurate solution

$$u(t) = 0.988807\cos(2.81534t) + 0.0111916\cos(13.9008t). \tag{33}$$

We also find the numerical results using Runge-Kutta forth order method and both the results are compared with graphs Figs. (1)–(3) for small as well as large time. We are found our results are best approximation of the given nonlinear cubic-quintic Duffing equation.

The exact frequency of the given nonlinear oscillation are given by following expression

$$\omega_e = \frac{\pi k_1}{2\int_0^{\frac{\pi}{2}} \left(1 + k_2 \sin^2 t + k_3 \sin^4 t\right)^{-\frac{1}{2}} dt}, \tag{34}$$

Where

$$k_1 = \sqrt{\alpha + \frac{\beta a^2}{2} + \frac{\gamma a^4}{3}}, \tag{35}$$

$$k_2 = \frac{3\beta a^2 + 2\gamma a^4}{6\alpha + 3\beta a^2 + 2\gamma a^4}, \tag{36}$$

$$k_3 = \frac{2\gamma a^4}{6\alpha + 3\beta a^2 + 2\gamma a^4}. \tag{37}$$

For given oscillation the exact frequency is obtained $\omega_e = 2.85598$ and approximate frequency $\omega_a = 2.85588$ which is correct up to three decimal place.

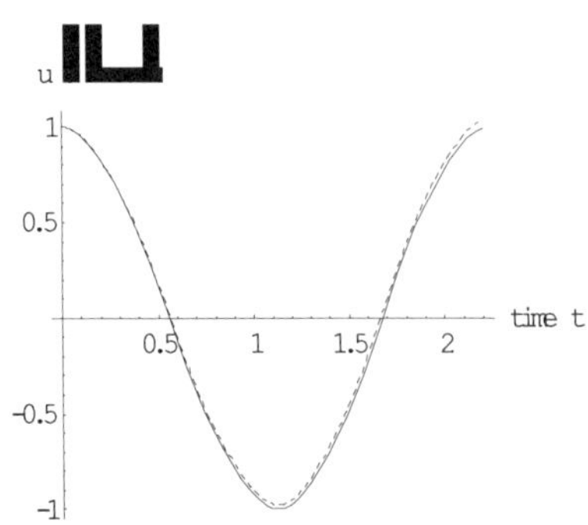

Fig. 1 $u(t)$ versus t $\alpha = 1, \beta = -1, \gamma = 10, a = 1,\ t = 2.2,\ \cdots\cdots DTM, -Runge\,Kutta$.

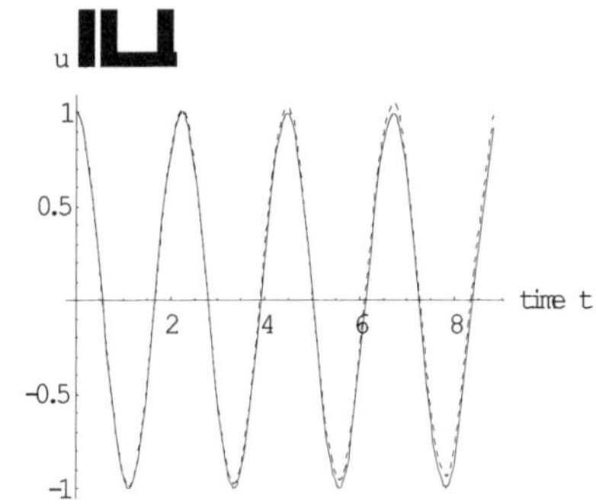

Fig. 2. $u(t)$ versus t $\alpha = 1, \beta = -1, \gamma = 10, a = 1, \; t = 8.8, \cdots\cdots DTM, -Runge\, Kutta$.

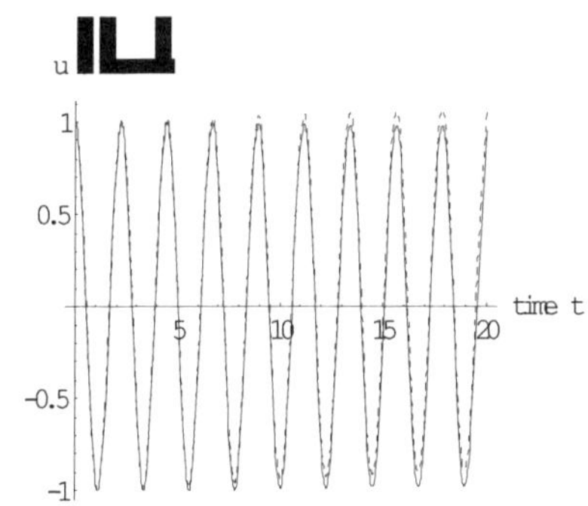

Fig. 3. $u(t)$ versus t $\alpha = 1, \beta = -1, \gamma = 10, a = 1, \; t = 20, \cdots\cdots DTM, -Runge\, Kutta$.

6. Conclusion

The Differential Transform Method (DTM) is an efficient method for calculating periodic solutions for non-linear oscillatory systems. All the results and graphs show that the results of the present method are in excellent agreement with those obtained by fourth-order Runge-Kutta method and exact method even for small as well as large value of time. Pade approximant improve DTM's truncated series solution in the convergence rate. The results also predict that the method is a very promising one and might find wide applications.

Acknowledgments

Author is very thankful to U.G.C. for the scheme "Start-Up Grant for Newly Recruited Faculty" and to grant the fund UGC-BSR Research Start-Up Grant project No. F.20-6(20)/2012(BSR).

References

1. A. H. Nayfeh, D. T. Mook, *Nonlinear Oscillations,* New York, 1979.
2. A. H. Nayfeh, *Problems in perturbation*, Wiley, New York, 1985.
3. P. Hagedorn, *Non-linear Oscillations*, Clarendon, Oxford, 1998, translated by Wolfram Stadler.
4. R. E. Mickens, Oscillations in Planar Dynamic Systems, *Word Scientific*, Singapore, 1996.
5. A. H. Nayfeh, *Perturbation Methods*, John Wiley, New York, 1973.
6. A. H. Nayfeh and D. T. Mook, *Nonlinear Oscillations*, John Wiley, New York, 1979.
7. N. Krylov and N. Bogolioubov, *Introduction to Nonlinear Mechanics*, Princeton University Press, Princeton, NJ, 1943.
8. N. Bildik, A. Konuralp, F. Bek, S. K¨uc¸ ¨ukarslan, Solution of different type of the partial differential equation by differential transform method and Adomian decomposition method, *Appl. Math. Comput.* **172**, 551, (2006).
9. S. Momani and V. S. Erturk, Solution of non-linear oscillators by the modified differential transform method, *Comp. Math. App.*, **55**, 833, (2008).
10. Y. C. Jiao, Y. Yamamoto, C. Dang and Y. Hao, An aftertreatment technique for improving the accuracy of Adomian's decomposition method, *Comput. Math. Appl.* **43**, 783, (2002).
11. S. Nourazar and A. Mirzabeigy, Approximate solution for nonlinear doffing oscillator with damping effect using the modified differential transform method, *ScientiaIranica B*, 364, (2013).
12. H. Fatoorehchi and H. Abolghasemi, Improving the differential transform method: A novel technique to obtain the differential transforms of nonlinearities by the Adomian polynomials, **37**, 6008, (2013).

Fuzzy DEA Cross-Efficiency Model for Ranking and Performance Evaluation Using Ideal and Anti-Ideal Decision Making Units

Seema Gupta[1] and K. N. Rajeshwari
School of Mathematics
Devi Ahilya University, Indore, M.P., India
[1]E-mail: seemadavv@gmail.com
www.dauniv.ac.in

P. C. Jha
Department of Operations Research
University of Delhi, Delhi

Cross efficiency evaluation, regarded as a Data Envelopment Analysis (DEA) extension tool, is an effective approach for ranking Decision Making Units (DMUs) which consumes multiple inputs and produces multiple outputs. In addition to self-evaluation assessment, as done in traditional DEA models, peer-evaluation assessment is carried out, wherein; the efficiency of the target DMU is evaluated using the weights determined by other peer DMUs. The classical cross-efficiency DEA models, using crisp data, cannot effectively deal with real-world problems having vagueness and/or uncertainty. This paper introduces a novel Fuzzy DEA Cross-Efficiency model which evaluates performance by constructing two virtual DMUs viz. the ideal DMU and the anti-ideal DMU, having fuzzy parameters. An ideal DMU is constructed using minimum inputs to yield maximum outputs whereas an anti-ideal DMU is constructed using maximum inputs to produce minimum outputs. These virtual DMUs serve as reference points for performance evaluation and ranking of considered DMUs. A numerical illustration to demonstrate the applicability of the proposed model is also presented.

Keywords: Data Envelopment Analysis; cross-efficiency; Ideal DMU; Anti-ideal DMU; fuzzy set theory.

1. Introduction

Data Envelopment Analysis (DEA) is one of the multi-criteria techniques for performance evaluation of decision making units (DMUs) with multiple inputs and outputs. It has been widely accepted as an effective technique for measuring efficiency to decision making units since it was first introduced by Charnes, Cooper and Rhodes, further defined by [1]. The efficiency is defined as weighted sum of outputs to weighted sum of inputs in DEA. Thus we determine the weights which maximize efficiency of target DMU relative to all considered

DMUs and hence it differentiates between efficient and non-efficient units. However in many cases we require more than self-efficiency evaluation as to how does others perceives us. Hence cross efficiency scores are calculated using weights of others in efficiency evaluation. There may be different approaches to cross efficiency evaluation [2][3][4]. Some noteworthy models are proposed in [5][6][7][8]. One approach could be to maximize own efficiency at the same time minimize the efficiency of the rest all known as aggressive approach. Other approach could be to maximize the efficiency scores of all the units including self known as benevolent approach [3]. Yet another approach could be neutral [7]. There are other cross efficiency evaluation model existing in the prior art. Ref. [5] modeled cross efficiency method by concept of game theory finding the resultant game cross efficiency. The proposed extension maintains the peer evaluation concept which converges to Nash equilibrium. Ref. [9] used a cross efficiency concept in DEA, as an approach towards resource and cost allocation problem. The concept of virtual ideal and virtual anti ideal DMU to determine the input and output weights in the cross efficiency model are also used in a number of studies [10][11][12]. Many applications are also found in literature of these models [13][14][15].

The intensity of global competition and ever-increasing economic uncertainties made itessential to device suitable tools to evaluate the efficiency in uncertain environment. There are different programming methods used to model these uncertainties, for example stochastic programming and stochastic dynamic programming methods [16][17][18]. A further and well-known alternative method to successfully address uncertainty is the fuzzy set theory, where fuzzy numbers are applied to represent the imprecise quantity [19]. This paper extends the concept of cross efficiency evaluation using ideal and anti-ideal DMU given in [10] and extended by [11] by incorporating fuzzy set in the study.

The rest of the paper is organized as follows. Section II gives basic models for DEA self and cross efficiency evaluation. Section III gives mathematical formulation to model the concept. Section IV proposes methodological approach towards its solution. The next sections illustrate the applicability of the proposed models via a numerical example. The last section concludes the research work.

2. Preliminaries

2.1. *Basic self efficiency model*

Suppose there are n numbers of DMUs, each having m number of inputs and s number of outputs. Let index for DMU is j, for input isi, and for output is r. The self efficiency of target DMU k is determined by the following formulation.

$$\max \theta_{kk} = \sum_{r=1}^{s} u_{rk} y_{rk}$$

$$s.t. \sum_{i=}^{m} v_{ik} x_{ik} = 1,\ \sum_{r=1}^{s} u_{rk} y_{rj} - \sum_{i=1}^{m} v_{ik} x_{ij} \leq 1 \qquad for\ j = 1,2,...,n \quad (1)$$

$$u_{rk} \geq 0 \qquad r = 1,2,...,s,\ v_{ik} \geq 0 \qquad i = 1,2,...,m$$

2.2. *Basic cross efficiency model*

There are n efficiency scores θ_{jk} out of which one score is self efficiency score as θ_{kk} and rest (n-1) are cross efficiency scoresusing weights of all other DMUs $j = 1,2,..,n\ j \neq 0$

$$\theta_{kk} = \frac{\sum_{r=1}^{s} u_{rk} y_{rk}}{\sum_{i=1}^{m} v_{ik} x_{ik}},\ \theta_{kj} = \frac{\sum_{r=1}^{s} u_{rj} y_{rk}}{\sum_{i=1}^{m} v_{ij} x_{ik}} \qquad j = 1,2,...,n \quad j \neq k \qquad (2)$$

2.3. *Triangular fuzzy number*

Among various fuzzy numbers, triangular fuzzy numbers are commonly used because of its generic and simplistic nature. Hence triangular fuzzy numbers are considered in this paper.

The inputs and outputs of the considered DMU is taken as fuzzy hence for a DMU k inputs and outputs are defined as $\tilde{x}_{ik} = (x_{ik}^{l}, x_{ik}^{m}, x_{ik}^{u})$ i=1,2,..s, and $\tilde{y}_{rk} = (y_{rk}^{l}, y_{rk}^{m}, y_{rk}^{u})$.

2.4. *Fuzzy ideal DMU*

An fuzzy ideal DMU is a hypothetical DMU consuming least inputs and producing maximum possible outputs wherein the inputs and outputs are fuzzy numbers.

2.5. *Fuzzy anti-ideal DMU*

An fuzzy ideal DMU is a hypothetical DMU consuming least inputs and producing maximum possible outputs wherein the inputs and outputs are fuzzy numbers.

2.6. *Mean de-fuzzification*

It is one of the ranking method in fuzzy set theory. For a triangular fuzzy number $\tilde{a} = (a^l, a^m, a^u)$ the formula is given as

$$val(a) = \frac{a^l + 2a^m + a^u)}{4} \tag{3}$$

3. The Model

The virtual fuzzy ideal DMU and fuzzy anti ideal DMU are created using the inputs and outputs of the considered set as

3.1. *Fuzzy ideal DMU*

$$\text{Fuzzy ideal DMU } = \{\tilde{y}_r^{\max}, \tilde{x}_i^{\min}\} \quad r = 1,2,...,s \quad i = 1,2,...,m \tag{4}$$

wherein

$$\tilde{y}_r^{\max} = (y_r^{\max l}, y_r^{\max m}, y_r^{\max u}), y_r^{\max l} = \max_j(y_{rj}^l), y_r^{\max m} = \max_j(y_{rj}^m), y_r^{\max u} = \max_j(y_{rj}^u),$$

$$\tilde{x}_i^{\min} = (x_i^{\min l}, x_i^{\min m}, x_i^{\min u}), x_i^{\min l} = \min_j(x_{ij}^l), x_i^{\min m} = \min_j(x_{ij}^m), x_i^{\min u} = \min_j(x_{ij}^u).$$

3.2. *Fuzzy anti-ideal DMU*

$$\text{Fuzzy anti-ideal DMU} = \{\tilde{y}_r^{\min}, \tilde{x}_i^{\max}\} \quad r = 1,2,...,s \quad i = 1,2,...,m \tag{5}$$

wherein

$$\tilde{y}_r^{\min} = (y_r^{\min l}, y_r^{\min m}, y_r^{\min u}), y_r^{\min l} = \min_j(y_{rj}^l), y_r^{\min m} = \min_j(y_{rj}^m), y_r^{\min u} = \min_j(y_{rj}^u),$$

$$\tilde{x}_i^{\max} = (x_i^{\max l}, x_i^{\max m}, x_i^{\max u}), x_i^{\max l} = \max_j(x_{ij}^l), x_i^{\max m} = \max_j(x_{ij}^m), x_i^{\max u} = \max_j(x_{ij}^u).$$

The first model is to determine the input output weights which will minimize the distance from ideal DMU. Second model is to determine the weights which will maximizing the distance from fuzzy anti-ideal DMU. Third model is used to determine weights when the distance between fuzzy ideal DMU and fuzzy anti-ideal DMU is being maximized and the forth model determines the weights when the relative closeness index is being maximized.
Mathematically

3.3. *Model I: Minimizing the distance form fuzzy ideal DMU*

$$
\begin{aligned}
\min & \sum_{r=1}^{s} u_{rk}(\tilde{y}_r^{\max} - \tilde{y}_{rk}) + \sum_{i=1}^{m} v_{ik}(\tilde{x}_{ik} - \tilde{x}_i^{\min}) \\
s.t. \quad & \sum_{i=1}^{m} v_{ik}\tilde{x}_{ik} = \tilde{1}, \\
& \sum_{r=1}^{s} u_{rk}\tilde{y}_{rk} = \theta_{kk}^{*}, \\
& \sum_{r=1}^{s} u_{rk}\tilde{y}_{rj} - \sum_{i=1}^{m} v_{ik}\tilde{x}_{ij} \le \tilde{0}, \qquad j = 1, 2, \ldots, n \\
& u_{rk} \ge 0, \qquad r = 1, 2, \ldots, s, \\
& v_{ik} \ge 0, \qquad i = 1, 2, \ldots, m.
\end{aligned}
\tag{6}
$$

3.4. *Model II: Maximizing the distance from fuzzy anti-ideal DMU*

$$
\begin{aligned}
\max & \sum_{r=1}^{s} u_{rk}(\tilde{y}_{rk} - \tilde{y}_r^{\min}) + \sum_{i=1}^{m} v_{ik}(\tilde{x}_i^{\max} - \tilde{x}_{ik}) \\
s.t. \quad & \sum_{i=1}^{m} v_{ik}\tilde{x}_{ik} = \tilde{1}, \\
& \sum_{r=1}^{s} u_{rk}\tilde{y}_{rk} = \theta_{kk}^{*}, \\
& \sum_{r=1}^{s} u_{rk}\tilde{y}_{rj} - \sum_{i=1}^{m} v_{ik}\tilde{x}_{ij} \le \tilde{0}, \qquad j = 1, 2, \ldots, n \\
& u_{rk} \ge 0, \qquad r = 1, 2, \ldots, s, \\
& v_{ik} \ge 0, \qquad i = 1, 2, \ldots, m.
\end{aligned}
\tag{7}
$$

3.5. *Model III: Maximizing the distance between fuzzy ideal DMU and fuzzy anti-ideal DMU*

$$\max \sum_{r=1}^{s} u_{rk}(\tilde{y}_r^{\max} - \tilde{y}_r^{\min}) + \sum_{i=1}^{m} v_{ik}(\tilde{x}_i^{\max} - \tilde{x}_i^{\min})$$

$$\begin{aligned} s.t. \quad & \sum_{i=1}^{m} v_{ik}\tilde{x}_{ik} = \tilde{1}, \\ & \sum_{r=1}^{s} u_{rk}\tilde{y}_{rk} = \theta_{kk}^{*}, \\ & \sum_{r=1}^{s} u_{rk}\tilde{y}_{rj} - \sum_{i=1}^{m} v_{ik}\tilde{x}_{ij} \le \tilde{0}, \qquad j = 1,2,\ldots,n \\ & u_{rk} \ge 0, \qquad r = 1,2,\ldots,s, \\ & v_{ik} \ge 0, \qquad i = 1,2,\ldots,m. \end{aligned} \tag{8}$$

3.6. *Model IV: Maximizing the relative closeness index*

$$\max \sum_{r=1}^{s} u_{rk}(\tilde{y}_{rk} - \tilde{y}_r^{\min}) + \sum_{i=1}^{m} v_{ik}(\tilde{x}_i^{\max} - \tilde{x}_{ik})$$

$$\begin{aligned} s.t. \quad & \sum_{r=1}^{s} u_{rk}(\tilde{y}_r^{\max} - \tilde{y}_r^{\min}) + \sum_{i=1}^{m} v_{ik}(\tilde{x}_i^{\max} - \tilde{x}_i^{\min}) = \tilde{1}, \\ & \sum_{r=1}^{s} u_{rk}\tilde{y}_{rk} - \theta_{kk}^{*} \sum_{i=1}^{m} v_{ik}\tilde{x}_{ik} = \tilde{0}, \\ & \sum_{r=1}^{s} u_{rk}\tilde{y}_{rj} - \sum_{i=1}^{m} v_{ik}\tilde{x}_{ij} \le \tilde{0}, \qquad j = 1,2,\ldots,n \\ & u_{rk} \ge 0, \qquad r = 1,2,\ldots,s, \\ & v_{ik} \ge 0, \qquad i = 1,2,\ldots,m. \end{aligned} \tag{9}$$

4. Proposed Methodology

The following steps are proposed to solve the fuzzy DEA cross efficiency models formulated in section III.

1. In the first step the fuzzy ideal DMU and fuzzy anti-ideal DMU are determined from the considered set of DMU using the formula as given in (4) and (5).

2. Then using the mean defuzzification method (3) the fuzzy inputs and fuzzy outputs of each DMU is given a value.
3. The self efficiency for each of the DMU is calculated using the basic CCR model (1).
4. The self efficiency of each DMUs and two virtual DMUs are used to calculate optimal weights for it so as to achieve different objectives as detailed in four different models (6), (7), (8) and (9).
5. These weights of target DMU are used to calculate the cross efficiency scores for other DMUs in each case using formula as given in (2).
6. The average of cross efficiency scores for a DMU give it a final ranking.

5. Numerical Example

To illustrate the potential applicability of the proposed methodology in cross efficiency evaluation of Decision Making Units having fuzzy data, we take numerical example of 15 hospitals and wards performance under the uncertainty developing from ongoing reforms. The data is taken from [13] of a health care system. In this healthcare system assessment,seven performance indices are considered, which are categorized into inputs and outputs. The inputs are taken as

(Input1) The number of doctors;
(Input2) The number of nurses;
(Input3) The number of other employees and administrative staff;
(Input4) The total number of available beds.

Out of these four inputs (1), (2) and (4) are triangular fuzzy numbers and (3) input is crisp. The outputs are taken as

(Output1) The total number of discharges in the year;
(Output2) The total yearly days of hospitalization of all patients;
(Output3) The total number of surgeries.

All these outputs are fuzzy triangular numbers. The unit of all outputs is in thousands. All the maximal outputs and minimal inputs form the fuzzy IDMU and all the minimal outputs and maximal inputs from the fuzzy ADMU as determined in Table 1.

Table 1. Ideal and anti-ideal DMU.

	Inputs				outputs		
	I1	I2	I3	I4	O1	O2	O3
Ideal DMU	(34, 34, 35)	(88, 90, 89)	120	(101, 101, 128)	(48.9, 54.4, 59.8)	(366.5, 407.2, 447.9)	(94.6, 105.1, 115.6)
Anti-ideal DMU	(933, 933, 1033)	(1782, 1782, 1955)	1762	(1392, 1392, 1532)	(3.7, 4.1, 4.5)	(26.8, 29.8, 32.7)	(0.0, 0.0, 0.0)

The fuzzy triangular numbers are defuzzified using mean defuzzification method and then self efficiency evaluation is performed for each of the 15 DMUS. Table 2 gives these self efficiency scores. These self efficiency scores are used in the proposed models to get the ranking from different perspectives.

Table 2. Self efficiency scores.

DMU	D1	D2	D3	D4	D5	D6	D7	D8
Self eff. score	0.949	1	0.923	1	0.876	1	1	0.801
DMU	D9	D10	D11	D12	D13	D14	D15	
Self eff. score	1	0.819	1	1	0.699	1	1	

Table 3 gives average efficiency score and raking of DMUs when the objective is taken as minimizing the distance form ideal DMU. Efficiency scores and ranking of DMUs when other three models are used is also calculated. Table 4 gives average efficiency score and raking of DMUs from different approaches so to enable a comparative analysis when different objectives are taken into consideration.

Table 3. Model I Average cross efficiency and ranking by minimizing distance from fuzzy ideal DMU.

	Target DMU							
DMU	D1	D2	D3	D13	D14	D15	Avg.	rank
D1	0.95	0.74	0.93	0.90	0.41	0.75	0.74	6
D2	0.98	1.00	0.98	0.81	0.36	1.00	0.76	5
D3	0.91	0.75	0.92	0.81	0.36	0.79	0.72	7
D4	0.94	0.67	1.00	1.00	0.43	0.65	0.72	8
D5	0.66	0.43	0.68	0.73	0.33	0.50	0.63	11
D6	1.00	0.95	1.00	1.00	0.52	0.99	0.88	2
D7	1.00	0.55	0.92	0.96	0.45	0.63	0.85	3
D8	0.40	0.19	0.39	0.52	0.19	0.21	0.48	15
D9	1.00	0.66	1.00	1.00	0.44	0.74	0.88	1
D10	0.63	0.37	0.63	0.64	0.22	0.38	0.54	13
D11	0.66	0.47	0.66	0.63	0.27	0.59	0.68	9
D12	0.54	0.61	0.59	0.53	0.18	0.76	0.68	10
D13	0.68	0.51	0.67	0.70	0.34	0.52	0.55	12
D14	1.00	0.32	1.00	1.00	1.00	0.28	0.52	14
D15	0.94	0.82	1.00	0.77	0.31	1.00	0.85	4

Table 4. Ranking from different approaches.

DMU	Method 1	Method 2	Method 3	Method 4
D1	6	10	10	6
D2	5	11	11	7
D3	7	8	8	8
D4	8	9	9	5
D5	11	7	7	11
D6	2	4	4	3
D7	3	3	3	4
D8	15	14	13	15
D9	1	2	2	1
D10	13	15	15	14
D11	9	6	6	9
D12	10	5	5	10
D13	12	13	14	13
D14	14	12	12	12
D15	4	1	1	2

6. Conclusion

This paper proposes a new approach to fuzzy DEA problems wherein cross efficiency evaluation is performed to give ranking to different DMU using the concept of virtual fuzzy DMUs. Two virtual DMUs viz. fuzzy ideal DMU and fuzzy anti-ideal DMU are formed for evaluating the cross efficiencies of the considered DMUs. Four different models are formulated based on the distance measurement from these fuzzy virtual DMUs. The ranking of DMUs from different models are given at the end. The proposed method can be applied on any fuzzy data set DEA Problem as it is generic and structured. Different other fuzzy ranking methods can be used in the future and results can be compared. Other possibilistic and stochastic approaches can also be applied.

References

1. R. D. Banker, A. Charnes and W. W. Cooper, "Some models for estimating technical and scale inefficiencies in data envelopment analysis", Management science, vol. 30, no. 9, 1078–1092, 1984.
2. T. R. Sexton, "The methodology of data envelopment analysis", New Directions for Program Evaluation, vol. 32, pp. 7–29, 1986.
3. J. Doyle and R. Green "Efficiency and cross-efficiency in DEA: Derivations, meanings and uses", J. of the operational research society, pp. 567–578, 1994.
4. L. Liang, J. Wu, W. D. Cook and J. Zhu, "Alternative secondary goals in DEA cross-efficiency evaluation", Int. J. of Production Economics, vol. 113, no. 2, pp. 1025–1030, 2008.

5. L. Liang, J.Wu, W. D. Cook and J. Zhu, "The DEA game cross-efficiency model and its Nash equilibrium", Operations Research, vol. 56, no. 5, pp. 1278–1288, 2008.
6. Y. M. Wang and K. S. Chin, "Some alternative models for DEA cross efficiency evaluation", Int. J. of Production Economics, vol. 128, no. 1, pp. 332–338, 2010a.
7. Y. M. Wang and K. S. Chin "A neutral DEA model for cross efficiency evaluation and its extension", Expert Systems with Applications vol. 37, pp. 3666–3675, 2010b.
8. M. Toloo and T. Tomáš, "Two alternative approaches for selecting performance measures in data envelopment analysis", Measurement, 2015J.
9. Du, W. D. Cook, L. Liang and J. Zhu, "Fixed cost and resource allocation based on DEA cross-efficiency", Euro. J. of Operational Research, vol. 235, no. 1, pp. 206–214, 2014.
10. Y. M. Wang, and L. Ying "DEA efficiency assessment using ideal and anti-ideal decision making units", Applied Mathematics and Computation, vol. 173, no. 2, pp. 902–915, 2006.
11. Y. M. Wang, K. S. Chin and L. Ying, "Cross-efficiency evaluation based on ideal and anti-ideal decision making units", Expert Systems with Applications vol. 38.8, pp. 10312–10319, 2011.
12. M. Zeydan and C. Cüneyt, "A new decision support system for performance measurement using combined fuzzy TOPSIS/DEA approach", Int. J. of Production Research, vol. 47, no. 15, pp. 4327–4349, 2009.
13. M. Dotoli, N. Epicoco, M. Falagario and F. Sciancalepore, "A cross-efficiency fuzzy Data Envelopment Analysis technique for performance evaluation of Decision Making Units under uncertainty", Computers & Industrial Engineering, vol. 79, pp. 103–114, 2015.
14. E. C. Wang and W. Huang, "Relative efficiency of R&D activities: A cross-country study accounting for environmental factors in the DEA approach", Research Policy vol. 36, no. 2 , pp. 260–273, 2007.
15. R. H. Green, J. R. Doyle and W. D. Cook, "Preference voting and project ranking using DEA and cross-evaluation", Eur. J. of Operational Research, vol. 90, no. 3, pp. 461–472, 1996.
16. L. Li and Z. B. Zabinsky, "Incorporating uncertainty into a supplier selection problem", Int. J. of Production Economics, vol. 134.2, pp. 344–356, 2011.
17. R. U. Bilsel and A. Ravindran, "A multiobjective chance constrained programming model for supplier selection under uncertainty",

Transportation Research Part B: Methodological, vol. 45, no. 8, 1284–1300, 2011.

18. J. L. Zhang and M. Y. Zhang, "Supplier selection and purchase problem with fixed cost and constrained order quantities under stochastic demand", Int. J. of Production Economics, vol. 129, no. 1, pp. 1–7, 2011.
19. O. Kabak and F. Ülengin, "Possibilistic linear-programming approach for supply chain networking decisions", Euro. J. of Operational Research, vol. 209.3, pp. 253–264, 2011.

Poverty Analysis Using Scan Statistic Methods

Arun Kumar Sinha[1] and Mukesh Kumar[2]

Department of Statistics
Central University of South Bihar
Patna 800014, India
E-mail: [1]arunkrsinha@cub.ac.in, [2]mukesh.kumar@cub.ac.in
www.cub.ac.in

The paper discusses the changing concept of poverty in general and points out the limitations of the poverty line generally used for identifying the poor. Instead of traditional rule based on "income" and "expenditure" approach the new approach is discussed, which appears more relevant in the present scenarios. Using scan statistic methods the most likely and secondary clusters of districts are identified by the SaTScan Software.

Keywords: Poisson Distribution; SaTScan Software; Spatial Impact.

1. Introduction

India faces two major challenges, which include poverty and environmental degradation. The poverty alleviation programme has been on the national agenda for almost sixty-nine years. The Indian National Congress constituted a National Planning Committee headed by Jawaharlal Nehru in 1938 to get rid of the awful poverty of the people. The importance of reduction of poverty and provision of other basic needs has been emphasized in all Five-Year Plans particularly since the Fifth Five-Year Plan. The government has two approaches:promoting economic growth anddirecting action towards poverty alleviation. To get the complete idea of poverty it should be measured in terms of deprivations and not merely in terms of income as the poverty reveals itself in the deprivation of the lives. It is thus a denial of choices and opportunities of living a tolerable life. Opportunities and choices most basic to human development are denied to lead a long, healthy, creative life and to enjoy a decent standard of living, freedom, dignity, self-respect and the respect of others. The traditional index is popularly known as head-count-ratio but it is also called "poverty incidence ratio" as the incidence is one of the important dimensions of the phenomenon called poverty. For more details related to the definition and the measurement of poverty, see Chaubey (1995), Singh (1997) and Sen (2003). In this paper we have used scan statistic to locate the poorest areas employing SatScan software.

2. Poverty and United Nations Millennium Development Goals

The UN Research Institute for Social Development (UNRISD) has worked on the measurement of levels of living rather than poverty at both national and sub-national (local) levels. It has investigated a number of commonly used concepts including income distribution and poverty lines.

The Millennium Development Goals come from the Millennium Declaration signed by 189 countries, including 147 heads of State and Government in September 2000. There are eight Millennium Development Goals (MDGs). These are to eradicate extreme hunger and poverty (Goal 1), to achieve universal primary education (Goal 2), to promote gender equality and empower women (Goal 3), to reduce child mortality (Goal 4), to improve maternal health (Goal 5), to combat HIV/AIDS, malaria and other diseases (Goal 6), to ensure environmental sustainability (Goal 7) and to develop a global partnership for development (Goal 8). The goals and targets are interrelated and should be seen as a whole. There are three targets of the Goal 1. Target A aims to halve, between 1990 and 2015, the proportion of people whose income is less than one dollar a day while target B directs to achieve full and productive employment and decent work for all, including women and young people and target C focuses tom halve, between 1990 and 2015, the proportion of people who suffer from hunger. The approach of hotspot detection and prioritization would help achieve the targets. They show a partnership between the developed countries and the developing countries to generate an environment at the national and global levels, for further discussion, see HDR (2004), which is favorable to development and elimination of poverty.

3. Poverty Related Work in India

In India a rich and varied literature has appeared on the incidence of poverty and the concept of poverty line during the decade of the seventies. But the most important of them included the landmark work of Dandekar and Rath (1971), Rao and Sukhatme (1977). In fact, Rao and Sukhatme criticized the methods and results of Dandekar and Rath who defined poverty in terms of household consumer expenditure, making use of the estimates of minimum calorie requirement for determining the point at which expenditure was insufficient. Sukhatme found the Dandekar and Rath's estimates inappropriate because of inter-personal and intra-personal variations in calorie requirements. But Rao proposed that poverty should be identified with deficiency in total living, which included not only energy requirements but also balanced diet needed for health and other components of basic needs essential for human existence at a tolerable level. Dandekar expressed difficulty in defining a tolerable level of human existence and identifying its basic needs. See Dandekar (1981). Further Scott and Mathew (1983) suggested that poverty should be expressed in terms of a small number of indicators, most of them non-monetory covering the main components of levels of living. The indicators should be expressed in terms of a distribution in the beginning and not a cut-off point, which is not absolute. They have suggested nine main indicators, which include type of house, electricity, drinking water and sanitation, education, regular use of footwear, household and individual possessions and level of furnishing, average household income per adult, size of land held and food consumption.

3.1. *Changing scenarios of the BPL*

Though Scott and Mathew (1983) developed the new approach for determining the levels of living and poverty for Kerala, probably for the first time, the Ministry of Rural Development, Government of India adopted the method for all States and Union Territories (UT) of India from the tenth five-year plan (2002–2007). The Ministry conducts a Below Poverty Line (BPL) Census at he beginning of Five-Year Plan. The basic objective of this Census was to identify the households (not an individual) living below the poverty line who could be assisted under various anti-poverty programmes initiated by the Ministry of Rural Development of the Government of India. The reason for adopting the new approach is that the data obtained through the BPL survey conducted for the previous plans had revealed discrepancies between the survey results and the official State wise estimates of poverty made by the Planning Commission based on the NSSO consumer Expenditure Surveys. For improving the methodology for identifying the BPL households in the BPL census for the tenth Five-Year Plan the Ministry had constituted an Expert Group administrators, academicians, planners and representatives of the States and UT. Using the recommendations of the Expert Committee it was proposed to follow a "normative approach" for identification of the rural poor by using a "score based ranking" based on relative deprivations discovered by certain socio-economic indicators in contrast to the "income" and "expenditure" approach in the BPL Censuses 1992 and 1997 respectively. A schedule consisting of questions for thirteen main indicators was constructed. The indicators include (1) size group of operational holding of land, (2) type of house, (3) average availability of normal wear clothing (per person in pieces), (4) food security, (5) sanitation, (6) ownership of consumer durables, (7) literacy status of the highest literate adult, (8) status of the household labour force, (9) means of livelihood (10) status of children (5–14 years, any child), (11) status of type of loan, (12) reason for migration from household and (13) preference for assistance. As each indicator has five options, from zero to four, the total score comes as 52, and for determining the BPL the households getting upto 13 were considered BPL. The State Governments / UT Administrations have the flexibility to decide the cut-off scores for identifying and sub-categorization of households into various groups.

Though the estimation of poverty and the identification of poor are closely related, the estimation work is given to the Planning Commission and is carried out through sample survey consumer expenditure conducted by the NSSO. The identification of the poor is done through door to door survey with hundred percent coverage. It is worth to point out that the estimation of poverty helps in assessing the magnitude of poverty and the identification of the specific households living below the poverty line is necessary for targeting them for various poverty alleviation programmes.

4. Scan Statistics and Methodology

The standard purely spatial scan statistic imposes a circular window on map. The window is, in turn, centered on each of several possible grid points positioned throughout the study region. For each grid point, the radius of the window varies continuously in sizes from zero to some upper limit specified by the user. In similar manner the space- time scan statistics is defined by a cylindrical window with a circular (or elliptic) geographic base and with height corresponding to time. The temporal scan statistic uses a window that moves in one dimension, time, defined in the same way as height of the cylindrical used by the space-time scan statistics. For each such district and scanning window, the alternative hypothesis is that there is an elevated risk within the window as compared to outside. Under the Poisson assumption, the likelihood function for a specific window is proportional to:

$$\left(\frac{c}{E[c]}\right)^{c}\left(\frac{C-c}{C-E[c]}\right)^{C-c} I(\,)$$

where C is the total number of WPV cases in Bihar in a year, c is the observed number of cases within the window, and E[c] is the expected number of cases within the window under the null-hypothesis. I(.) is an indicator function. Since the analysis is conditioned on the total number of cases observed, C-E[c] is the expected number of cases outside the window. SaTScan has been set to scan only for clusters with high rates, and thus I(.) is equal to 1 when the window has more cases than expected under the null-hypothesis, and 0 otherwise.

For the Bernoulli model the likelihood function is

$$\left(\frac{c}{n}\right)^{c}\left(\frac{n-c}{n}\right)^{n-c}\left(\frac{C-c}{N-n}\right)^{C-c}\left(\frac{(N-n)-(C-c)}{N-n}\right)^{(N-n)-(C-c)} I(\,)$$

where c and C are defined as above, n is the total number of cases and controls within the window, while N is the combined total number of cases and controls in the data set. See Kuldroff (2006) for more detail.

4.1. *Data set*

Table 1 provides the number of households above the poverty line (APL) and below the poverty line. This also shows the proportion of the BPL households as well as the districts with their poverty ranks. Kishanganj gets the first poverty rank one because it has the highest proportion of the BPL households while Siwan gets the lowest poverty rank with the lowest proportion of the BPL households. The above poverty line (APL) households and the BPL households together constitute total households in an area

Table 1. The proportions of BPL families and the poverty ranks of the districts of Bihar in 2007.

Districts	APL Households	BPL Households	Proportion of BPL Households	Ranked Districts	Poverty Ranks
Araria	246662	304793	0.552707	Kishanganj	1
Araval	65045	80274	0.552399	Purnia	2
Aurangabad	268766	226526	0.457358	Khagaria	3
Banka	219231	206944	0.485585	Madhubani	4
Begusarai	333222	319388	0.489401	East Champaran	5
Bhagalpur	222821	279170	0.556126	Supaul	6
Bhojpur	211880	256221	0.547363	WestChamparan	7
Buxar	175441	170332	0.492612	Vaishali	8
Darbhanga	324612	371134	0.533433	Sitamadhi	9
Gaya	292000	360848	0.552729	Sheohar	10
Gopalganj	259953	265491	0.505270	Muzaffarpur	11
Jahanabad	104016	96258	0.480632	Madhepura	12
Jamui	134443	195234	0.592198	Munger	13
Kaimur	164695	187599	0.532507	Saharsa	14
Katihar	271529	295576	0.521202	Jamui	15
Khagaria	113190	233715	0.673715	Nalanda	16
Kishanganj	71078	262574	0.786970	Bhagalpur	17
Lakhisarai	95934	76706	0.444312	Gaya	18
Madhepura	144168	235255	0.620034	Araria	19
Madhubani	325723	662909	0.670532	Araval	20

Table 1. (*Continued*)

Districts	APL Households	BPL Households	Proportion of BPL Households	Ranked Districts	Poverty Ranks
Munger	91814	136619	0.598070	Bhojpur	21
Muzaffarpur	353393	596244	0.627865	Darbhanga	22
Nalanda	249526	330943	0.570130	Kaimur	23
Nawada	200021	193116	0.491218	Katihar	24
West Champaran	249904	480386	0.657802	Patna	25
Patna	387022	403781	0.510596	Gopalganj	26
Purnia	143029	482114	0.771206	Shekhpura	27
East Champaran	296306	594443	0.667352	Samastipur	28
Rohtas	284513	282017	0.497797	Rohtas	29
Saharsa	143897	212465	0.596206	Buxar	30
Samastipur	459912	459922	0.500005	Nawada	31
Saran	426958	405443	0.487077	Begusarai	32
Shekhpura	59465	59492	0.500113	Saran	33
Shivhar	46215	80569	0.635482	Banka	34
Sitamadhi	253881	444998	0.636731	Jahanabad	35
Siwan	423950	315461	0.426638	Aurangabad	36
Supaul	154130	305815	0.664895	Lakhisarai	37
Vaishali	265448	470215	0.639172	Siwan	38

5. SaTScan Software

SaTScan is a free software that analyzes spatial, temporal and space-time data using the spatial, temporal, or space-time scan statistics. It helps for the following

- To do geographical surveillance of disease, to detect spatial or space-time disease clusters, and to identify if there are statistically significant.
- Apply to check whether a disease is randomly distributed over space, over time or over space and time etc.
- To test statistical significance of disease clusters.
- To perform repeated time – periodic disease surveillance for early detection possibility the disease.

The SaTScan V7.0.3 was run to identify the most likely clusters and secondary clusters based on Bernoulli and Poisson models. The proportion of the BPL families in the seven districts forming the most likely clusters is computed as 0.63587.

Table 2. The Bernoulli and Poisson based most likely and secondary clusters.

Bernoulli Model	Poisson Model
Most Lilkely Clusters: Figure (a) Madhubani, Sheohar, Sitamarhi, Muzaffarpur, Darbhanga, East Champaran, and West Champaran Secondary Clusters I: Figure (b) Purnea Secondary Clusters II: Figure (d) Saharsa, Madhepura, Khagaria and Munger Secondary Clusters III: Figure (e) Vaishali Secondary Clusters IV: Figure (f) Jamui	Most LilkelyClusters: Figure (a) Madhubani, Sheohar, Sitamarhi, Muzaffarpur, Darbhanga, East Champaran, and West Champaran Secondary Clusters I: Figure (c) Araria, Kishanganj and Purnea, Secondary Clusters II: Figure (d) Saharsa, Madhepura, Khagaria and Munger Secondary Clusters III: Figure (e) Vaishali Secondary Clusters IV: Figure (f) Jamui

6. Conclusions

We observed that both Bernoulli and Poisson models indicate the same most likely clusters. Also, the secondary clusters II, III and IV have the same districtsunder the two models. District Purnea appears as a secondary cluster I under the two models whileAraria and Kishanganj are the two additional districts under the Poisson model.

7. Figures

Figure (a).

Secondary Clusters I: Purnea(Bernoulli Model)

Figure (b).

Secondary Clusters I :Araria, Kishanganj, Purnea (Poisson Model)

Figure (c).

Secondary Clusters II: Saharsa, Madhepura, Khagaria, Munger (Bernoulli and Poisson Models)

Figure (d).

Secondary Clusters III : Vaishali (Bernoulli and Poisson Models)

Figure (e).

Secondary Clusters IV: Jamui (Bernoulli and Poisson Models)

Figure (f).

References

1. P. K. Chaubey, *Poverty Measurement*: *Issue, Approaches and Indices*, Wiley Eastern Ltd., New Delhi. (1995).
2. V. M. Dandekar and N. Rath, Poverty in India. *Economic and Political Weekly*, Vol. XVI, No. 31, Nos. 1 and 2. (1971).
3. V. M. Dandekar, Reply to Rao and Sukhatme, *Economic and Political Weekly*, Vol. XVI, Nos. 22, 23, 30, 31, 32. (1981).
4. Human Development Report, *Beyond scarcity: Power, Poverty and the global water crisis.* United Nations Development Programme (UNDP), Palgrave Macmillan, New Yark, USA. (2004).
5. V. K. R. V. Raoand P. V. Sukhatme, Dandekar and Rath's Methods and Results. *Bulletin of the International Statistical Institute's Proceedings of the 41st Session*, Vol. XLVII. (1977).
6. W. Scottand, N. T. Mathew, *Levels of living and poverty in Kerala*, A Development Monitoring Service at the Loacal Level, Volume 11, United Nations Research Institute for Social Development (UNRISD), 1211 Geneva 10, Switzerland. (1983).
7. AmartyaSen, *Poverty and Famines.*Oxford University Press, New Delhi 110001. (2003).
8. P. Singh., Poverty Measurement: Rational and Relevance. *Journal of the Indian Society of Agricultural Statistics*, Vol. L, No. 3, 211–239. (1997).
9. M. Kulldorff, *SatScan user guide* (2006), http//www.satscan.org.

Joint Performance Evaluation Data Envelopment Analysis Problem: An Interactive Approach

Riju Chaudhary

Department of Mathematics, University of Delhi
Delhi, India
riju.chaudhary@gmail.com

Pankaj Kumar Garg

Rajdhani College, University of Delhi
Delhi, India
gargpk08@gmail.com

P. C. Jha

Department of Operations Research, University of Delhi
Delhi, India
jhapc@yahoo.com

Data Envelopment Analysis (DEA) has been recognized, over the recent years, as a valuable analytical research tool for performance evaluation of several, similar entities engaged in different activities. Characterized by single (multiple) input(s) and output(s), this technique distinguishes between efficient and inefficient units, thereby forming an efficient frontier. It measures the level of efficiency of non-frontier units and identifies benchmarks against which such inefficient units can be compared. In the classical DEA approach, an optimization model is formulated and solved to evaluate the efficiency score of each Decision Making Unit (DMU) separately. The Joint Optimization DEA model presentedin this paper extends the performance measurement DEA technique by evaluating the performance of all DMUs simultaneously. An Interactive method is designed which considers the gap between the target and achieved values of inputs and outputs of the DMUs and provides the decision maker with an appropriate framework to choose the most preferred solution.

Keywords: Data Envelopment Analysis (DEA); Slack Analysis; Target Setting; Interactive Multi-Objective Optimization; Joint Optimization; Performance Evaluation.

1. Introduction

Survival and growth of any business entity in today's competitive world is possible not only if it is an efficient entity in itself but is also relatively efficient among all peer entities. This inspired researchers and scholars to develop reliable and rational analytical tools to measure efficiency scores of entities characterized by single (multiple) input(s) and output(s), independently and relatively. Several parametric and non-parametric techniques have been developed over time for evaluation of efficiency and estimation of best-practice relationship between inputs and outputs. Data Envelopment Analysis (DEA) is one such non-parametric, data-oriented, quantitative technique, which by evaluating efficiency scores of business entities, referred to as Decision Making Units (DMUs), distinguishes between efficient and inefficient ones. The efficient units form what is known as the production (efficient) frontier, thereby providing reference units for inefficient DMUs to benchmark against. Not only this, DEA helps the decision maker to identify the sources of inefficiency and correspondingly determine the improvement to be made in each of the considered inputs and output inspired by the seminal work on measurement of productive efficiency by [1], the concept of Data Envelopment Analysis was introduced by [2] in 1978. Since then, several improved and more reliable DEA models and their extensions have been developed. The construction and analysis of Pareto-efficient frontier production functions by a Data Envelopment Analysis method was presented by [3] in 1985. A framework for incorporating ordinal data factors into the standard ratio DEA model was developed by [4] in 1996. Reference [5] proposed a practical approach to the cost allocation of fixed resources or costs across a set of competing entities in an equitable manner. In 2014, [6] presented in his paper, formulations and properties of Weighted Additive Data Envelopment Analysis models associated with dataset standardization techniques.

The task of efficiency measurement and decision making is often very cumbersome due to prevalence of complex economic and behavioral attributes and conflicting objectives. In the classical DEA models, as many optimization problems are formulated and solved for evaluating efficiency scores, as the number of considered DMUs. The Joint Optimization DEA model presented in this paper reduces the cumbersome and time consuming job of solving multiple optimization problems by the development of a single optimization problem which simultaneously evaluates the performance of all DMUs.

Unlike the conventional DEA models which do not consider decision makers' preference information, an interactive approach is presented in this paper to incorporate the same. The solution obtained using this approach is both feasible as well as satisfying to the decision maker. Based on empirical production functions, [7] presented an Interactive Multi-Objective Linear Programming procedure for setting goals to achieve desired outputs. In the paper by [8], feasible, potential performance is computed and presented to the decision maker and is worked upon for improvements until either the DMU becomes efficient or the decision maker is satisfied. Reference [9] also proposed an Interactive Multi-Objective method as a target setting DEA approach and used Analytic Hierarchy Process (AHP) to compute relative priorities of input-output changes.

In this paper, a Joint Performance Evaluation Data Envelopment Analysis Model is presented. An Interactive approach, which allows for a progressive articulation of decision makers' preferences, is developed to compute improved inputs and outputs of all considered DMUs together and arrive at a feasible and desirable solution.

Rest of the paper is structured as follows: Section II states the necessary assumptions made in the DEA models under consideration. The conventional DEA models have been discussed in Section III and the Joint Performance Evaluation Model using Interactive approach has been presented in Section IV.

2. Assumptions

DEA can only be applied to a set of homogenous DMUs; homogenous in the sense that each DMU must consume the same bunch of inputs to yield the same bunch of outputs. Thus, we consider a set of n comparable DMUs, each consuming m common inputs and generating s common outputs. Let be the input-output combination for the jth DMU,

Characterized by m inputs and s outputs, the envelopment set, generally referred to as the production possibility set (PPS), can be defined as

An input-output combination is feasible if, and only if, In DEA; the specification of the production possibility set is determined uniquely by the *minimum extrapolation principle* (Banker, Charnes, and Cooper, 1984). According to this principle, production possibility set should be *theminimal set* that satisfies the following imposed production assumptions [10]:

(1) All observed input-output combinations are feasible.

(2) The production possibility set, PPS, is convex.

Hence, if, and then

(3) Inputs are freely disposable.

Hence, if any, then

(4) Outputs are freely disposable.

Hence, if any, then

Arranging the input-output data in matrices $X = (x^j)$ and $Y = (y^j)$, the production possibility set, PPS, satisfying all four assumptions can mathematically be defined as: {

Since DEA requires very few assumptions, it has become an increasingly popular technique for measuring efficiency, even in cases where other approaches have not been successful to measure efficiency due to complex (often unknown) nature of the relations between multiple inputs and multiple outputs involved in DMUs.

3. The DEA Models

3.1. *Basic DEA model*

3.1.1. *Notations*

Consider a set of n homogenous. Let DMU_j denote the jth and DMU_0 denote the target *DMU*. Let $x_{ij} \in R$ be the column vector of inputs consumed by DMU_j, $x_{i0} \in R$ be the column vector of inputs consumed by the target and $X \in R$ be the matrix of inputs of all. Similarly, let be the column vector of outputs produced by, be the column vector of outputs produced by the target and be the matrix of outputs of all. Also, let denote the column vector of a linear combination of n. Let the value of the objective function (efficiency score of the target) in the classical DEA models be denoted by. Let the column vector of input weights be represented by and the column vector of output weights be represented by

Classical DEA models have been mathematically expressed using the above mentioned assumptions and notations.

3.1.2. *The model*

The objective of this DEA model is to determine weight vectors and, corresponding to the input vector and output vector respectively, which would maximize the efficiency of the target. The Basic DEA model is a fractional programming problem. Its equivalent linear programming problem, using the Charnes-Cooper method of conversion.

3.2. *Joint optimization DEA model*

The Joint Optimization DEA model [8] aims at evaluating efficiency of all considered DMUs simultaneously.

3.2.1. *Notation*

Let h_k, θ_k represent the efficiency scores of the kth DMU, x_{ij} represent the amount of input i consumed by the jth DMU and y_{rj} represent the amount of output r produced by the jth DMU. Let v_{ij} denote the weight of the ith input for the jth DMU and u_{rj} denote the rth output for the jth DMU. Also let λ_{kj} refer to the proportion of jth DMU in reference unit for the kth DMU.

3.2.2. *The model*

The dual of (JP), is mathematically stated as:

4. Joint Performance Evaluation DEA Models Using Interactive Approach

4.1. *Interactive DEA approach*

The Dual of the Joint Optimization DEA model, incorporating slacks is presented below.

Slacks indicate the gap between the actual values and targeted values of each input and output of all considered DMUs. The gross input improvements and the gross outputs improvements are then computed as follows:

Before proceeding with the Interactive approach for Joint Performance Evaluation, efficiency score for each DMU is calculated using the Joint Optimization DEA model to categorize the DMUs into efficient and inefficient ones. It is followed by the Interactive DEA procedure which begins by taking the current performance profile (input-output level) of all DMUs together as the initial performance profile. Adjustments are made to the input-output levels, iteration by iteration, until a satisfactory, feasible performance profile is identified for each DMU. This procedure involves the following steps:

Step 1. Compute the efficiency score of all DMUs simultaneously, using the Joint Optimization DEA model discussed above (either using (JP) or (JD))

Step 2. Distinguish between efficient and inefficient DMUs (efficient units being the ones having efficiency score, and inefficient units having efficiency score,)

Step 3. Make no change to the inputs and outputs of the efficient DMUs and perform Slack Analysis for inefficient DMUs

Step 4. Taking into consideration the results obtained from Slack Analysis, use the Joint Performance Evaluation Model incorporating Interactive approach to set targets for input-output parameters for inefficient DMUs. The same is discussed in the following steps

Step 5. Set the current performance profile of each DMU as the initial feasible performance profile

Step 6. At each iteration, t, interact with the decision maker of each DMU to acquire information on all inputs consumed and outputs produced and categorize them into discretionary and non-discretionary sets. Denote the set of non-discretionary inputs by and the set of non-discretionary outputs by

Step 7. Categorize the discretionary inputs and outputs into the following sets:

(i) Set of inputs and outputs for which only improvement (decrease in inputs, increase in outputs) is desired. Denote the set of such inputs and outputs by and, respectively

(ii) Set of inputs and outputs for which no change is allowed (inputs and outputs remain at the same level). Denote the set of such inputs and outputs by and, respectively

(iii) Set of inputs and outputs for which worsening (increase in inputs, decrease in outputs) may be allowed. Denote the set of such inputs and outputs by and, respectively

Step 8. Seek priorities for all inputs and outputs from the decision makers of all DMUs and correspondingly assign weights to them

Step 9. Set lower and upper bounds on changes of current input and output levels as desired by the decision maker

Step 10. Formulate and solve the Joint Performance Evaluation DEA model (discussed later in this section) and present the obtained solution (obtained performance profile) to all the decision makers

Step 11. Stop the iterative process for only those DMUs whose decision makers find the offered solution optimal (satisfying and feasible). For the remaining, proceed to the next step

Step 12. For those DMUs whose decision makers are not satisfied with the offered solution, set the obtained performance profile as their current performance profile and repeat the above steps for such DMUs only (Refer to Fig. 1.)

This process is carried out until an optimal solution is obtained for all the DMUs.

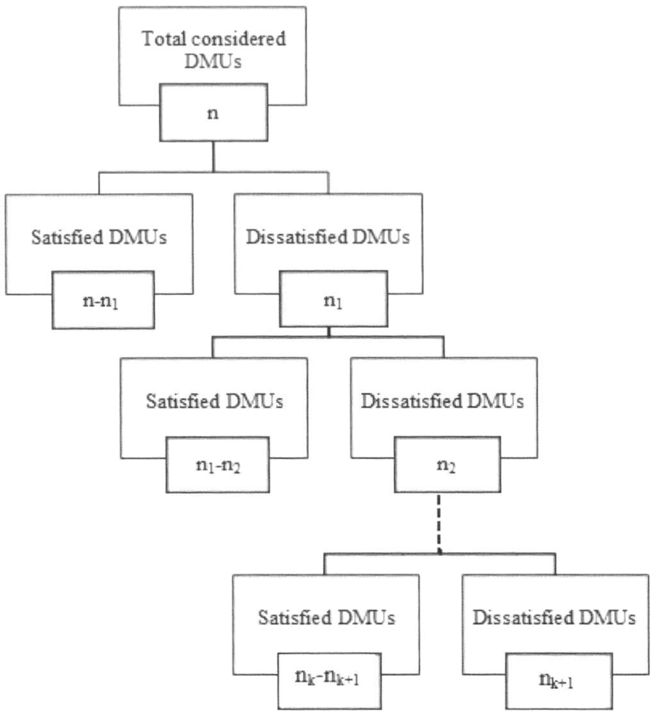

Fig. 1. Process for interactive DEA approach.

4.2. *Joint performance evaluation DEA model*

4.2.1. *Notations*

Let be the amount of input i consumed by DMU_j, be the amount of output r produced by DMU_j, 't' be the current iteration index of the interactive process, be the amount of input i consumed by the DMU at the current iteration level, 't' (for the first iteration, t=1 and initial input is), be the amount of output r produced by the DMU at the current iteration level, 't' (for the first iteration, t=1 and initial output is). Let be the maximum improvement amount of input i consumed by the DMU from the current performance profile and be the maximum worsening amount of input i consumed by the DMU from the current performance profile. Also, let be the maximum improvement amount of output r produced by the DMU from the current performance profile and be the maximum worsening amount of output r produced by the DMU from the current performance profile. Let refer to the proportion of DMU in reference unit for the DMU at iteration ‘t’.

Let be the amount of input i consumed by the DMU by the new current performance profile and bethe amount of output r produced by the DMU by the new current performance profile at the iteration. Also, let be the proportional reduction of input i consumed by the DMUfrom the current performance profile and be the proportionalincrease of output r produced by the DMUfrom current operation point.

4.2.2. *The model*

The DEA model to be solved at each iteration *t*, assuming variable returns to scale (VRS), can be formulated as:

In the above model, the following assumptions are made:

1.

2.

3.

4. For non-discretionary inputs and outputs, the constraints ensure that the linear combination projection is not worse than the corresponding initial values.

5. Conclusion

In this paper, a Joint Performance Evaluation Data Envelopment Analysis Model has been presented using an Interactive approach, which allows for a progressive articulation of decision makers’ priorities. The approach involves an iterative process wherein the model is formulated in accordance with the observations resulting from Slack Analysis and preferences of the decision makers. The decision maker projects the current performance profile, by improving some selected inputs and outputs, maintaining some of them intact and allowing others to worsen.

In the first iteration, the model is solved jointly for n DMUs, to compute improved inputs and outputs of all considered DMUs together to arrive at an improved, feasible solution. Assuming the resulting solution as desired, hence optimal, for DMUs, the second iteration is carried out for the remaining DMUs. Proceeding in this manner, assuming the resulting solution as optimal for DMUs, the next iteration is carried out for the remaining DMUs. The process is continued until an optimal solution is achieved for all the DMUs.

References

1. M. J. Farrell, "The Measurement of Productive Efficiency," J. of Royal Statistical Soc. Series A General, vol. 120, no. 3, pp. 253–290, 1957.
2. A. Charnes, W. Cooper and E. Rhodes, "Measuring the efficiency of decision making units," Eur. J. of Oper. Res., vol. 2, no. 6, pp. 429–444, 1978.
3. A. Charnes, W. Cooper, B. Golany, L. Seiford and J. Stutz, "Foundations of data envelopment analysis for Pareto-Koopmans efficient empirical production functions," J. of Econom., vol. 30, no. 1–2, pp. 91–107, 1985.
4. W. Cook, M. Kress and L. Seiford, "Data envelopment analysis in the presence of both quantitative and qualitative factors," J. of the Oper. Res. Soc., vol. 47, no. 7, pp. 945–953, 1996.
5. Cook, Wade D. and Joe Zhu, "Allocation of shared costs among decision making units: a DEA approach," *Computers & Operations Research* 32.8 (2005): 2171–2178.
6. K. Chen and M. Kou. (2014). "Weighted Additive DEA Models Associated with Dataset Standardization Techniques." [Online]. Available at http://mpra.ub.uni-muenchen.de/55072/.
7. B. Golany, "An interactive MOLP procedure for the extension of DEA to effectiveness analysis", J. of the Oper. Res. Soc., vol. 39, no. 8, p. 725, 1988.
8. T. Post and J. Spronk, "Performance benchmarking using interactive data envelopment analysis," Eur. J. of Oper. Res., vol. 115, no. 3, pp. 472–487, 1999.
9. S. Lozano and G. Villa, "Multiobjective target setting in data envelopment analysis using AHP," Computer. and Oper. Res., vol. 36, no. 2, pp. 549–564, 2009.
10. W. Cooper, L. Seiford and K. Tone, Data envelopment analysis. Boston: Kluwer Academic, 2007.

Stochastic Modeling of a Repairable System Under Different Weather Conditions

S. C. Malik

Department of Statistics,
M.D. University, Rohtak-124001 (Haryana)
E-mail: sc_malik@rediffmail.com

An effort has been made to develop stochastic model for a repairable system of two non-identical units working in different weather conditions. The system starts operation with the original unit (called main unit) and the other substandard unit (called duplicate unit) is kept a spare in cold standby. The repair activities have been tackled by a single server immediately on need basis. The operation and repair of the units are not allowed in abnormal weather. The system has been analyzed at different epochs by adopting semi-Markov process and regenerative point technique. The distributions of failure time of the units and change of weather conditions follow negative exponential while that of repair times of the units are taken as arbitrary with different probability density functions. The expressions for some important measures of system effectiveness are derived in steady state. The behavior of some reliability measures have been observed for arbitrary values of various parameters and costs.

Keywords: Repairable System, Non-identical Units, Weather Conditions, Reliability Measures and Stochastic Modeling.

1. Introduction

Over the past few decades, the stochastic modeling of repairable systems of identical units have been done extensively by the researchers keeping in view of their frequent use in modern technology and management sectors. Osaki and Asakura (1970), Goel and Sharma (1989) and Kadyan *et al.* (2012) have also investigated reliability models of such systems with different repair policies. But, sometimes it is very difficult to afford a high cost identical unit in spare and in such a situation, the non-identical unit (may be a substandard unit) might be kept as spare in cold standby not only to improve the reliability of the system but also to minimize overall cost of the system. Mokkadis *et al.* (1989) and Chander *et al.* (2007) discussed standby systems of non-identical units with different sets of assumptions on failure and repair. Further, it is a matter of fact that it is not possible to keep the environmental conditions under control which may fluctuate due changing climate and other natural catastrophic. In view of this fact in mind, Goel *et al.* (1985), Gupta and Goel (1991), Gupta *et al.* (2010) and Malik and Deswal (2012) have analyzed reliability measures of repairable systems operating under different weather conditions.

Hence, in the present paper, a repairable system of two non-identical units – one is original (called main unit) and other is a substandard unit (called duplicate unit) has been analyzed stochastically in detail under two weather conditions – normal and abnormal. The environmental conditions when satisfied to the system correspond to normal weather; otherwise, it is supposed that the system is in abnormal weather. Initially, the system is operative with main unit and duplicate unit is kept a spare in cold standby. Both units have direct complete failure from normal mode. Each unit is capable of performing the same set of functions with different degree of reliability and desirability. There is a single server who visits the system immediately whenever needed to do repair of the failed unit in normal weather only. The operation and repair of the units are not allowed in abnormal weather as a precautionary measure to avoid excessive damage to the system. However, operation and repair of the units are as usual in normal weather. The units work as new after repair.

The distributions of failure time of the units and change of weather conditions follow negative exponential while that of repair times of the units are taken as arbitrary. All random variables are statistically independent. The switch devices and repairs are perfect. The expressions for various measures of system effectiveness such as transition probabilities, mean sojourn times, mean time to system failure (MTSF), availability, busy period of the server and profit function in steady state are derived using semi-Markov process and regenerative point technique. The behavior of some important reliability measures has been observed for arbitrary values of various parameters and costs. The application of the present work can be visualized in a software industry where application software is run through two different databases-one is initially operative and other is kept in cold standby.

2. Notations

λ/λ_1	:	Constant failure rate of Main /Duplicate unit
β/β_1	:	Constant rate of change of weather from normal to abnormal/abnormal to normal weather
$g(t)/G(t)$	:	pdf/cdf of repair time of Main unit
$g_1(t)/G_1(t)$	:	pdf/cdf of repair time of Duplicate unit

3. System Description

In this section, a stochastic model for the system is developed as per the assumptions. The following are the possible transition states of the system:

State S_0: The main unit is operative and the duplicate unit is in cold standby.

State S_1: Main unit is under repair after failure and duplicate unit is operative.

State S_2: Both units are waiting for operation due to abnormal weather.

State S_3: Main unit is failed but waiting for repair due to abnormal weather and duplicate unit is waiting for operation due to abnormal weather.

State S_4: Main unit is continuously under repair from previous state and duplicate unit is failed but waiting for repair.

State S_5: Main unit is in cold standby and duplicate unit is operative.

State S_6: Main unit is operative and duplicate unit is under repair.

State S_7: Main unit is waiting for repair due to abnormal weather and duplicate unit is continuously waiting for repair due to abnormal weather.

State S_8: Main unit is under repair and duplicate unit is waiting for repair continuously from previous state.

State S_9: Main unit is in cold standby and duplicate unit is waiting for operation due to abnormal weather.

State S_{10}: Main unit is waiting for operation due to abnormal weather and duplicate unit is waiting for repair due to abnormal weather.

State S_{11}: Main unit is waiting for repair and duplicate unit is under repair continuously from previous state.

State S_{12}: Main unit is waiting for repair continuously from previous state due to abnormal weather and duplicate unit is waiting for repair due to abnormal weather.

State S_{13}: Main unit is waiting for repair continuously from previous state and duplicate unit is under repair.

The states S_0, S_1, S_2, S_3, S_5, S_6, S_9 and S_{10} are regenerative while the states S_4, S_7, S_8, S_{11}, S_{12} and S_{13} are non-regenerative.

4. Transition Probabilities and Mean Sojourn Times

Simple probabilistic considerations yield the following expressions for the non-zero elements

$$p_{ij} = Q_{ij}(\infty) = \int_0^\infty q_{ij}(t)\,dt \tag{1}$$

we have

$$p_{01} = \frac{\lambda}{\beta+\lambda}, \qquad p_{02} = \frac{\beta}{\beta+\lambda}, \qquad p_{13} = \frac{\beta}{\beta+\lambda_1}(1-g^*(\beta+\lambda_1)),$$

$p_{14} = \frac{\lambda_1}{\beta+\lambda_1}(1\text{-}g^*(\beta+\lambda_1))$, $\quad p_{15} = g^*(\beta+\lambda_1)$,

$p_{20}=1$, $p_{31}=1$, $p_{46}=g^*(\beta)$, $p_{47} = 1\text{-}g^*(\beta)$, $p_{56} = \frac{\lambda_1}{\beta+\lambda_1}$ $p_{59}=\frac{\beta}{\beta+\lambda_1}$, $p_{60}=g^*_1(\beta+\lambda)$,

$p_{6,10} = \frac{\beta}{\beta+\lambda}(1\text{-}g_1^*(\beta+\lambda))$, $p_{6,11} = \frac{\lambda}{\beta+\lambda}(1\text{-}g_1^*(\beta+\lambda))$, $p_{78} = 1$ $p_{86} = g^*(\beta)$,

$p_{87} = 1\text{-}g^*(\beta)$, $p_{95} = 1$, $p_{10,6} = 1$, $p_{11,1} = g_1^*(\beta)$, $p_{11,12} = g_1^*(\beta)$ $p_{12,13} = 1$, $p_{13,1} = g^*_1(\beta)$, $p_{13,12} = 1\text{-}g^*_1(\beta)$

$p_{16.4} = \frac{\lambda_1}{\beta+\lambda_1}(1\text{-}g^*(\beta+\lambda_1))\, g^*(\beta)$, $p_{16.4(7,8)^n} = \frac{\lambda_1}{\beta+\lambda_1}(1\text{-}g^*(\beta+\lambda_1))\,(1\text{-}g^*(\beta))$,

$p_{61.11} = \frac{\lambda}{\beta+\lambda}(1\text{-}g_1^*(\beta+\lambda))\, g_1^*(\beta)$, $p_{61.11,(12,13)^n} = \frac{\lambda}{\beta+\lambda}(1\text{-}g_1^*(\beta+\lambda))(1\text{-}\, g_1^*(\beta))$

It can be easily verified that sum of all possible transition probabilities from a particular state is one. The mean sojourn times (μ_i) in the state S_i are

$\mu_0=\frac{1}{\beta+\lambda}$, $\mu_1 = \frac{1}{\beta+\lambda_1}(1\text{-}g^*(\beta+\lambda_1))$, $\mu_2=\frac{1}{\beta_1}$, $\mu_3 = \frac{1}{\beta_1}$, $\mu_4=\frac{1}{\beta}(1\text{-}g^*(\beta)$,

$\mu_5=\frac{1}{\beta+\lambda_1}$, $\mu_6=\frac{1}{\beta+\lambda}(1\text{-}g_1^*(\beta+\lambda))$, $\mu_7 = \frac{1}{\beta_1}$ $\mu_8 = \frac{1}{\beta}(1\text{-}g^*(\beta))$, $\mu_9 = \frac{1}{\beta_1}$,

$\mu_{10}=\frac{1}{\beta_1}$, $\mu_{11}=\frac{1}{\beta}(1\text{-}g_1^*(\beta))$, $\mu_{12}=\frac{1}{\beta_1}$, $\mu_{13}=\frac{1}{\beta}(1\text{-}g^*_1(\beta))$,

$$\mu'_1= \frac{(1-g^*(\beta+\lambda_1))(\beta\beta_1 g^*(\beta)+\lambda_1(\beta+\beta_1)(1-g^*(\beta)))}{\beta\beta_1(\beta+\lambda_1)g^*(\beta)},$$

$$\mu'_6= \frac{(1-g_1^*(\beta+\lambda))(\beta\beta_1 g_1^*(\beta)+\lambda(\beta+\beta_1)(1-g_1^*(\beta)))}{\beta\beta_1(\beta+\lambda)g_1^*(\beta)}$$

5. Reliability And Mean Time To System Failure (MTSF)

Let $\phi_i(t)$ be the c.d.f. of first passage time from regenerative state S_i to a failed state. Regarding the failed state as absorbing state, we have the following recursive relations for $\phi_i(t)$:

$$\phi_i(t) = \sum_j Q_{i,j}(t)(S)\phi_j(t) + \sum_k Q_{i,k}(t) \tag{2}$$

where S_j is an un-failed regenerative state to which the given regenerative state S_i can transit and S_k is a failed state to which the state S_i can transit directly. Taking LST of above relations (2) and solving for $\phi_0^{**}(s)$, we get

$$R^*(s) = \frac{1-\phi_0^{**}(s)}{s} \tag{3}$$

The reliability of the system can be obtained by taking inverse Laplace transform of (3).

The mean time to system failure (MTSF) is given by

$$MTSF = \lim_{s\to 0}\frac{1-\phi_0^{**}(s)}{s} = \frac{N_1}{D_1} \tag{4}$$

where

$$N_1=p_{56}(1-p_{6,10})((1-p_{13})(\mu_0+p_{02}\mu_2)+p_{01}(\mu_1+p_{13}\mu_3))+p_{01}p_{15}((1-p_{6,10})(\mu_5+p_{59}\mu_9)+p_{56}(\mu_6+p_{6,10}\mu_{10}))$$

$$D_1= p_{01}p_{56}((1-p_{6,10})(1-p_{13})-p_{15}p_{60})$$

6. Steady State Availability

Let $A_i(t)$ be the probability that the system is in up-state at instant 't' given that the system entered regenerative state S_i at t = 0. The recursive relations for $A_i(t)$ are given as

$$A_i(t) = M_i(t) + \sum_j q_{i,j}^{(n)}(t) \copyright A_j(t) \tag{5}$$

Where, S_j is any successive regenerative state to which the regenerative state S_i can transit through n transitions and $M_i(t)$ is the probability that the system is up initially in state S_i is up at time t without visiting to any other regenerative state, we have $M_0(t)=e^{-(\beta+\lambda)t}$, $M_1(t)=e^{-(\beta+\lambda_1)t}\overline{G(t)}$, $M_5(t)= e^{-(\beta+\lambda_1)t}$,

$$M_6(t)= e^{-(\beta+\lambda)t}\overline{G_1(t)} \tag{6}$$

Taking LT of above relations (5) and (6) and solving for $A_0^*(s)$. The steady state availability is given by $A_0(\infty) = \lim_{s\to 0} sA_0^*(s) = \frac{N_2}{D_2}$

$$N_2= p_{56}(1-p_{13})(p_{60}\mu_0+p_{01}\mu_6)+p_{01}(1-p_{6,10})(p_{56}\mu_1+p_{15}\mu_5)$$

$$D_2= p_{56}(1-p_{13})(p_{60}\mu_0+p_{01}\mu_6'))+p_{01}(1-p_{6,10})(p_{56}\mu_1'+p_{15}\mu_5)$$

7. Busy Period Analysis For Server

Let $B_i(t)$ be the probability that the server is busy in repairing the unit at an instant 't' given that the system entered state S_i at t = 0. The recursive relations $B_i\,(t)$ for are as follows:

$$B_i\,(t) = W_i\,(t) + \sum_j q_{i,j}^{(n)}(t) © B_j\,(t) \tag{7}$$

where S_j is any successive regenerative state to which the regenerative state S_i can transit through n transitions and W_i (t) be the probability that the server is busy in state S_i due to hardware failure up to time t without making any transition to any other regenerative state or returning to the same via one or more non-regenerative states and so

$$W_1(t) = e^{-(\beta+\lambda_1)t}\,\overline{G(t)} + (\lambda_1 e^{-(\beta+\lambda_1)t} © 1)\,\overline{G(t)}\,,$$

$$W_6(t) = e^{-(\beta+\lambda)t}\,\overline{G_1(t)} + (\lambda e^{-(\beta+\lambda)t} © 1) \tag{8}$$

Taking LT of above relations (7) and (8) and solving for B_0^* (s). We can obtain the time for which server is busy due to repair of the failed unit as

$$B_0^*(\infty) = \lim_{s\to 0} sB_0^*(s) = \frac{N_3}{D_2}$$

where $N_3 = p_{01}\, p_{56}\, (W_1^*(0)\,(1-p_{6,10}) + W_6^*(0)\,(1-p_{13}))$ and D_2 is already defined.

8. Expected Number Of Visits By The Server

Let N_i (t) be the expected number of visits by the server in (0, t] given that the system entered the regenerative state S_i at t = 0. The recursive relations for N_i (t) are given as

$$N_i\,(t) = \sum_j Q_{i,j}^{(n)}(t)(S)\left[\delta_j + N_j\,(t)\right] \tag{9}$$

where S_j is any regenerative state to which the given regenerative state S_i transits and δj=1, if S_j is the regenerative state where the server does job afresh, otherwise δj=0. Taking LST of relations (9) and solving for $\tilde{N}_0(s)$. The expected number of visits per unit time by the server is given by

$$N_0(\infty) = \lim_{s\to 0} sN_0^{**}(s) = \frac{N_4}{D_2}$$

where $N_4 = p_{01}\, p_{56}\, ((1-p_{6,10})\,(1 + p_{15} - p_{14}\, p_{47}) - (1 - p_{13})\, p_{6,11}\, p_{11,1})$ and D_2 is already specified.

9. Profit Analysis

The profit incurred to the system model in steady state can be obtained as

$$P=K_0A_0-K_1B_0-K_2N_0 \quad (10)$$

where

K_0=Revenue per unit up-time of the system

K_1=Cost per unit for which server is busy

K_2= Cost per unit visit by the server and A_0, B_0, N_0 are already defined.

10. Numerical Results For Reliability Measures

Table 1. MTSF vs. normal weather rate (β_1).

Normal Weather Rate(β_1) ↓	α=2,α_1=2.5,β=0.01, λ=0.5,λ_1=0.6	α=1.5	α_1=2	β=0.05	λ=0.3	λ_1=0.4
1.1	11.02792208	9.734759	10.42727	11.42532	16.99522	15.77851
1.2	11.01964286	9.727451	10.41944	11.38393	16.98246	15.76667
1.3	11.01263736	9.721267	10.41282	11.3489	16.97166	15.75664
1.4	11.00663265	9.715966	10.40714	11.31888	16.96241	15.74805
1.5	11.00142857	9.711373	10.40222	11.29286	16.95439	15.74061
1.6	10.996875	9.707353	10.39792	11.27009	16.94737	15.73409
1.7	10.99285714	9.703806	10.39412	11.25	16.94118	15.72834
1.8	10.98928571	9.700654	10.39074	11.23214	16.93567	15.72323
1.9	10.98609023	9.697833	10.38772	11.21617	16.93075	15.71866
2.0	10.98321429	9.695294	10.385	11.20179	16.92632	15.71455

Table 2. Availability vs. normal weather rate (β_1).

Normal Weather Rate (β_1) ↓	α=2,α_1=2.5,β=0.01, λ=0.5,λ_1=0.6	α=1.5	α_1=2	β=0.05	λ=0.3	λ_1=0.4
1.1	0.952288701	0.93391	0.944042	0.950654	0.969052	0.967458
1.2	0.952322813	0.933956	0.944081	0.950824	0.969075	0.967482
1.3	0.952351679	0.933995	0.944115	0.950968	0.969094	0.967502
1.4	0.952376422	0.934029	0.944144	0.951091	0.96911	0.967519
1.5	0.952397867	0.934058	0.944169	0.951198	0.969124	0.967534
1.6	0.952416632	0.934083	0.94419	0.951292	0.969137	0.967547
1.7	0.952433191	0.934106	0.94421	0.951375	0.969148	0.967558
1.8	0.95244791	0.934126	0.944227	0.951448	0.969157	0.967569
1.9	0.95246108	0.934144	0.944242	0.951514	0.969166	0.967578
2	0.952472933	0.93416	0.944256	0.951573	0.969174	0.967586

Table 3. Profit vs. normal weather rate (β_1).

Normal Weather Rate (β_1)↓	K_0=5000, K_1=350, K_2=300 α=2,α_1=2.5,β=0.01, λ=0.5,λ_1=0.6	α=1.5	α_1=2	β=0.05	λ=0.3	λ_1=0.4
1.1	4577.52164	4476.754344	4531.372	4564.752	4704.575	4693.31
1.2	4577.68561	4476.97648	4531.562	4565.568	4704.684	4693.425
1.3	4577.824362	4477.164458	4531.724	4566.259	4704.776	4693.522
1.4	4577.9433	4477.325595	4531.862	4566.851	4704.856	4693.605
1.5	4578.046384	4477.465256	4531.981	4567.365	4704.924	4693.677
1.6	4578.136586	4477.587467	4532.086	4567.814	4704.985	4693.74
1.7	4578.21618	4477.695306	4532.178	4568.211	4705.038	4693.796
1.8	4578.286932	4477.791167	4532.261	4568.563	4705.085	4693.845
1.9	4578.350238	4477.876941	4532.334	4568.879	4705.127	4693.89
2	4578.407215	4477.95414	4532.4	4569.163	4705.165	4693.929

11. Conclusion

For the particular case $g(t)=\alpha e^{-\alpha t}$ and $g_1(t)=\alpha_1 e^{-\alpha_1 t}$, the behavior of Mean Time to System Failure (MTSF), availability and profit functions has been observed for arbitrary values of various parameters and costs as shown respectively in Tables 1, 2 and 3. It is revealed that MTSF keeps on decreasing with the increase of normal weather rate (β_1) and failure rates of the units. And, it increases with increase of abnormal weather rate (β) and repair rates of the units. The results show that availability and profit of the system model go on increasing as normal weather rate (β_1) and repair rates increase while their values decline with the increase of abnormal weather rate and failure rates of the units. Thus, a system of non-identical units which operate in different weather conditions can be made more reliable and profitable to use by increasing repair rate of the unit if operation and repair activities are not allowed in abnormal weather.

References

1. S. Osaki and T. Asakura, A two-unit standby redundant systems with repair and preventive maintenance, *Journal of Applied Probability* **7**, 3(1970).
2. L. R. Goel, G. C. Sharma and R. Gupta, Cost analysis of a two unit cold standby system under different weather conditions, *Microelectronics Reliability*, **25**, 4(1985).
3. L. R. Goel and G. C. Sharma Stochastic analysis of a 2-unit standby system with two failure modes and slow switch, *Microelectronics Reliability*, **29**, 4(1989).

4. G. S. Mokkadis, S. S. Elias and S. W. Labib, On a Two Dissimilar Unit Standby System with Three Modes and Administrative Delay in Repair, *Microelectronics Reliability*, **29**, 4(1989).
5. R. Gupta and R. Goel, Profit analysis of a two-unit cold standby system with abnormal weather condition. *Microelectronics Reliability*, **31**, 1(1991).
6. S. Chander, M. Singh and M. Kumari, Cost-benefit analysis of a stochastic model of an electric transformer and generator, *Journal of Pure and Applied Mathematika Sciences*, **LXVI**, 1-2(2007).
7. R. Gupta, P. Sharma and V. Sharma, V. Cost-benefit analysis of a two duplicate-unit parallel system with repair/replacement and correlated lifetimes of the units, *Rajasthan Academy of Physical Sciences*, **9**, 4(2010).
8. S. C. Malik and S. Deswal, Stochastic Analysis of a Repairable System of Non-Identical Units with Priority for Operation and Repair Subject to Weather Conditions. *International Journal of Computer Applications*, **49**, 14(2012).

Estimation of Risk Surfaces and Identification of District Boundaries for Tuberculosis in North-Eastern Indian States

Sanjeeva Kumar Jha[1] and Ningthoukhongjam Vikimchandra Singh[2]

Department of Statistics
North-Eastern Hill University
Shillong, India-793022
E-mail: [1]skjha@nehu.ac.in, [2]n.vikim12@gmail.com

The recent surge of Tuberculosis (TB) cases in India and in its seven North-Eastern States in particular, has introduced a renewed interest in the estimation of TB risk surfaces in the neighborhood and identification of units having elevated risk. The present paper maps the TB risk surfaces on the basis of district level incidence rates for the seven neighboring North-eastern states of India. The risk surface is represented with a set of random effects through Bayesian Hierarchical approach. In the present case, the random effects are modeled through conditional autoregressive (CAR) prior exhibiting a single level of spatial smoothness. Also, attempt has been made to identify the risk boundaries having elevated risk through localized spatial structure by modeling the weighted contiguity matrix for geographically adjacent areas as binary random quantities.

Keywords: Disease incidence; spatial regression; Bayesian hierarchical modeling; disease risk surface; boundary detection.

1. Introduction

India accounts for one-fourth of the global annual incidence of Tuberculosis (TB) cases [1]. In the year 2013, out of the estimated global annual incidence of 9 million TB cases, 2.1 million were estimated to have occurred in India. The mortality rate due to TB in the same year was 240 per hundred thousand populations [2], [3]. One of the biggest issues faced in tackling the disease is the number of undiagnosed cases. According to the World Health Organization (WHO) out of 9 million estimated TB cases, only 5.7 million were both detected and notified to national TB programmes (NTPs) and also that 24% of world's total missed TB cases are shared by India alone [4]. A recent review suggests that the average cost of the treatment of TB patients and affected families is more than half of their yearly income and the TBis four times more common to people in the lowest socio-economic quintile as compared to the highest [5].

TB control programmes also aim at identifying the causal factors responsible for its elevated incidence rates. It has been reported that its burden follows a strong socio-economic gradient between and within areas with the poorest having the highest risk [6], [7]. As per Census 2011, in India 65.8% of households use firewood, crop residue and cow-dung cake as cooking fuels and 87.32% of the households are cooking inside the house. Smoke and indoor air pollution from these fuels have been recognized as independent risk factors for TB [8], [9], [10]. Delay in diagnosis and treatment of TB increases the duration in which active cases are infectious, thereby sustaining TB transmission and have a positive association between the duration of delay to TB treatment and household infection rates [11].

The uneven spread of TB incidents in India calls for the identification of TB risks in smaller administrative areas to facilitate the effective management for its prevention and control. The seven North-Eastern States, namely, Arunachal Pradesh, Assam, Manipur, Meghalaya, Mizoram, Nagaland and Tripura experience elevated TB burden as compared to other states in India [12]. The present paper aims to estimate the risk surfaces on the basis of district wise TB incidence data for the seven North-eastern states. It also aims to identify the district boundaries which are geographically adjacent but have different TB risk.

The paper is divided into four sections. Section 1 is introductory in nature and Section 2 presence the methodology. Spatial ecological regression is performed using Conditional autoregressive (CAR) based Leroux model and boundary detection is done using Lee and Mitchel model [17], [19]. Section 3 covers the spatial analysis of data together with the obtained results. Section 4 present discussions.

2. Methodology

2.1. *Ecological regression with covariate*

Given n non-overlapping contiguous regions with observed and expected number of TB cases as, $Y = (Y_1, Y_2, Y_3, \dots, Y_n)$ and $E = (E_1, E_2, E_3, \dots, E_n)$ respectively. The expected number of cases in an area, district in the present case, depends upon the population size and the cases of TB incidence. The standardized incidence ratio (SIR) is the simplest measure of disease risk. The SIR for area k will be represented by $\widehat{R}_k$ and is given by $SIR_k = \widehat{R}_k = \frac{Y_k}{E_k}$. Sub-region having values of disease risk more than unity are labelled as having elevated risk in relation to the disease incidence. In case the population at risk in a sub-region is small region, the $\widehat{R}_k$ is biased. A Bayesian Hierarchical model approach may be used to overcome the problem, as discussed [13], [14], [15]

$$Y_k | E_k, R_k \sim Poisson(E_k R_k) \tag{1}$$

$$ln(R_k) = \mu + x'_k\beta + \varphi_k$$

In the above model, the disease risk R_k of sub-region k is modeled through a combination of an intercept μ, a set of p covariates information $x'_k = (x_{k1}, x_{k2}, x_{k3}, \dots, x_{kp})$ and a random effects φ_k. The vector of regression parameters are denoted by $\beta = (\beta_1, \beta_2, \beta_3, \dots, \beta_p)$.

The above model becomes unsuitable in presence of over-dispersion or/and spatial correlation. In this situation conditional auto-regressive (CAR) prior is used for modeling the random effect $\varphi = (\varphi_1, \varphi_2, \varphi_3, \dots, \varphi_n)$, which induces spatial correlation into random effects via a binary $n \times n$ neighbourhood matrix **W** [16]. The model is implemented using Markov chain Monte Carlo simulation (MCMC) method.

2.1.1. *Leroux model*

A number of priors have been proposed with general class of CAR models, Leroux *et al.* (1999) prior model is one of the most accepted one [17], [18]. It is given as n univariate full conditional distribution as below

$$\varphi_k|\varphi_{-k}, W, \tau^2, \rho, \mu \sim N\left(\frac{\rho\sum_{j=1}^{n} w_{kj}\varphi_j + (1-\rho)\mu}{\rho\sum_{j=1}^{n} w_{kj} + 1 - \rho}, \frac{\tau^2}{\rho\sum_{j=1}^{n} w_{kj} + 1 - \rho}\right) \quad (2)$$

The conditional expectation of a φ_k is a weighted mixture of spatial smoothing to the neighbouring random effects with weight ρ and non-spatial smoothing to a global mean μ with weighted $(1-\rho)$. The conditional variance is a weighted average of spatially dependent $(\rho \neq 0)$ and being constant if random effects are independent $(\rho = 0)$, and inversely proportional to the numbers of neighbouring areas. When $(\rho = 1)$, the intrinsic model proposed by Besag *et al.* (1991) is obtained [16].

2.2. *Boundary detection*

The boundaries in the risk surfaces are likely to occur where neighbouring populations are very different. Lee and Mitchel (2012) proposed a model to capture the presence or absence of boundary in the risk surfaces between geographically adjacent areas (k, j), as binary random quantities using a small number of regression parameters $\alpha = (\alpha_1, \alpha_2, \alpha_3, \dots, \alpha_q)$ for q non-negative dissimilarity matrices $z_{kj} = (z_{kj1}, z_{kj2}, z_{kj3}, \dots, z_{kjq})$, where $z_{kji} = |z_{ki} - z_{ki}|/\sigma_i$ for $i = 1,2,3, \dots, q$, taking these fixed at one [23]. However, if areas (k, j) are not share a common border, then w_{kj} is fixed at zero. The proposed neighbourhood model is

$$w_{kj}(\alpha) = \begin{cases} 1, \ if \ exp\left(-\sum_{i=1}^{q} z_{kji}\alpha_i\right) \geq 0.5 \ and \ j \sim k \\ 0, \ if \ otherwise \end{cases} \quad (3)$$

Intercept term is not present in this model, so that two areas with homogeneous population $\left(z_{kji} = 0\right)$ cannot have a boundary between the mandif $\alpha_i > \frac{-ln(0.5)}{max(z_{kji})}$, then the ith dissimilarity metric has not solely identified any boundaries because $exp\left(-\alpha_i z_{kji}\right) > 0.5$ for all $k \sim j$. If $exp\left(-\sum_{i=1}^{q} z_{kji}\alpha_i\right)$ is less than 0.5, the model detects a boundary in the risk surface of contiguous areas.

3. A Spatial Analysis of Tuberculosis

The study region comprised of the seven North-eastern states of India divided into 82 contiguous districts. The software ArcGIS was used to create the shape file for the study region.

3.1. *Data*

District level number of TB smear positive diagnosed registered patients at Revised National TB Control Programme (RNTCP) for the year 2013 was used to demonstrate the applications of Leroux CAR model and Boundary detection model [1]. The expected number of TB cases is compute by internal standardization method. The population proportion is the ratio of kth district population to the total population of the entire North-eastern states. The percentage of households cooking inside house using wood, cow dung cake and crop residue as cooking fuels, is taken as a covariate [17], [19].

3.2. *Modelling*

The global Moran's I for the study area was obtained as 0.2174 with a p-value of 0.003, indicating the presence of significant spatial correlation. The significant spatial correlation makes the generalized linear model (GLM) unsuitable in analysing the present data. Therefore, the disease risk surface is constructed using ecological regression model, as presented in the Sub-section 2.1, and the boundary detection is performed using the model presented in Sub-section 2.2. Inference for ecological regression and boundary detection models are performed on the basis of 1,00,000 MCMC samples generated from 6 Markov chains, with a burn-in period of 20,000 and a thinning period of 10. In both the cases the residuals exhibited insignificant spatial correlation.

3.3. *Result*

The results for both the analyses are presented in Table 1. In both the analyses only one covariate, namely, cooking fuel which includes the percentage of households cooking inside the house using either firewood or crop residue or cow-dung and their combinations as cooking fuel, is considered. The table presents the exponentiated estimated parameter values together with their 95% credible intervals. The results obtained by using ecological regression model suggest that one percent increase in the number of households using the said cooking fuel is related to reduction of 172 cases per ten thousand populations. The result is difficult to explain unless the information on missed TB cases is considered. The possible explanation is presented into the discussions section. The 95% credible interval for the exponentiated estimated value is 0.9769 to 0.9898. The estimated values of the covariates were used to construct the district level estimated TB risk surfaces for the study region and presented in Figure 2.

Table 1. Estimated coefficients with associated 95% credible intervals.

Ecological regression model			
Covariate	$Exp(\beta_i)$	95% credible interval	
Intercept	2.5943	(1.6153, 3.8409)	
Cooking fuel [a]	0.9828	(0.9769, 0.9898)	
Boundary detection model			
Covariate	$Exp(\beta_i)$	95% credible interval	α_i^{min}
Intercept	0.8262	(0.8058, 0.8463)	
Cooking fuel [a]	1.0265	(1.0127, 1.0293)	1.0095

[a]Percentage of household which cook inside the house using firewood, crop residue and cow-dung cake as a cooking fuel.

For boundary detection, the estimated exponentiated parameter for cooking fuel obtained was as 1.0265. The threshold value 1.0095 is smaller than the lower limit of the 95% credible interval of 1.0127 to 1.0293, therefore the cooking fuel has the ability to identify risk boundaries. This covariate has identified 36 boundaries in the risk surfaces accounting for 8.91% of the total boundaries in the study region. The risk surfaces with boundaries are presented in Figure 3.

The advantage of boundary detection model over the ecological regression model is to identify the location of any boundaries in disease risk surface, so the available covariates were used to construct dissimilarity metrics, rather than being incorporated into the linear predictor. In the bottom part of Table 1, the estimated parameter for cooking fuel is 1.0265 and its credible interval is more than the threshold value. Hence, the dissimilarity metric has effect in identifying risk boundaries (white dotted line) in the risk surface. The cooking fuel covariate has identified 36 boundaries in the risk surface, which is 8.91% of the total number of boundaries in the study region.

3.4. *Risk surface*

The risk surfaces for the TB in 82 districts of seven North-eastern Indian states are presented in Figures 1–3. Figure 1 presents the map for SIR for TB in which 10 out of 16 district of Arunachal Pradesh exhibit elevated TB risk. East Kameng, Papum Pare and Upper Siang are the worst affected districts. In Assam, out of 27 districts, 13 districts have elevated risk and Tinsukia and Dibrugarh are most affected districts among them. Dimapur and Kohima are the districts having high elevated risk in the state of Nagaland. In Meghalaya, the East Khashi hills district is the worst affected. Saiha, Lawngtlai and Aizawl are the districts having elevated risk in the state of Mizoram. Among the 7 states, Manipur and Tripura have in general, lower risk than the average.

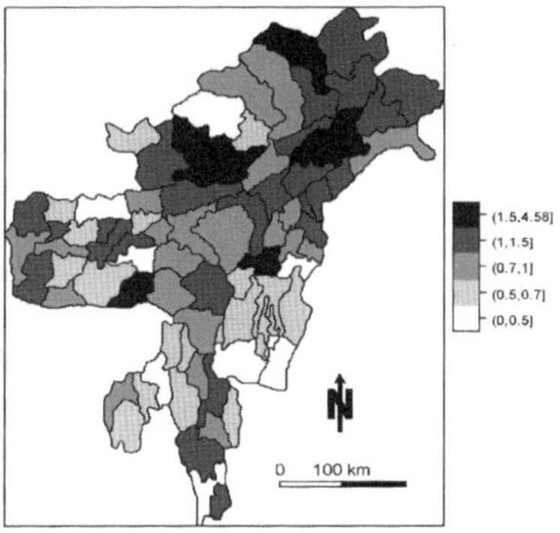

Fig. 1. Tuberculosis risk surface using SIR.

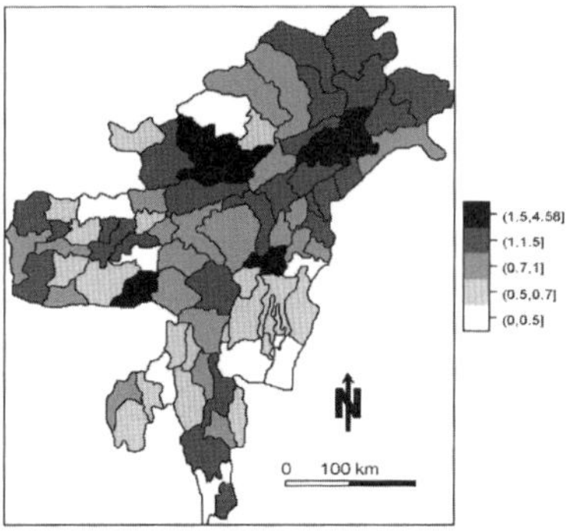

Fig. 2. Tuberculosis risk surface using Ecological regression model.

Figure 2 presents the estimated disease risk surfaces obtained through ecological regression. It is similar to Figure 1 except the district West Siang of Arunachal Pradesh has now come down to having a lower than the average risk. Finally, the spatial risk pattern from the boundary detection model obtained by taking cooking fuel as dissimilarity metricis presented in Figure 3.

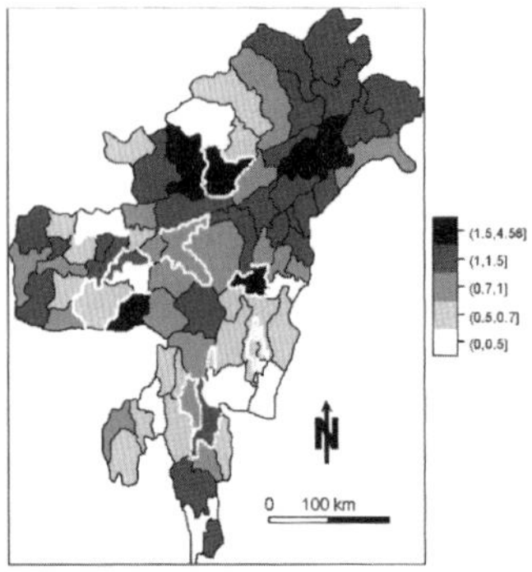

Fig. 3. Tuberculosis risk surface using boundary detection model.

4. Discussion

The paper performs ecological regression and boundary detection based on district level TB incidence data for the North-eastern states of India using Leroux CAR and Boundary detection models and maps the risk surfaces. Distict level incidence data for TB smear positive diagnosed within registered patients at RNTCP were used as response for the analysis.

The ecological regression model was able to capture the spatial correlation effectively. Both of the spatial regression and boundary detection models provided similar spatial risk pattern for the study area. Even, though they have different global spatial correlation of 0.41 and 0.99 respectively.

The present findings are surprising in the sense that increasing the percentage of households using firewood, crop residue and cow-dung inside the house, reduces TB incidence. Such households are expected to belong to the lower socio-economic strata of the society. It is in contradiction to the findings that TB is more common to people in lowest socio-economic quintile as compared to the highest quintile and that the poorest have the highest risk. Given the information that a huge percentage of global missed TB cases are from India, the present results indicate that the detection and reporting of TB cases is not taking place from within the lower socio-economic section of the population.

A better relationship could have been obtained with the availability of data at household level. We could access the data at the district level only and hence were used for the present analysis.

Acknowledgments

We would like to thank the R Core Team (2014). R: Alanguage and environment for statistical computing. R Foundation for Statistical Computing, Vienna, Austria. The URL address is, http://www.R-project.org/. We extend our thanks to Duncan Lee (2013) for developing the useful R package,'CARBayes'.

References

1. Central TB Division, TB India 2014, March 2014, http:// www.tbcindia.nic.in/pdfs/TB%20INDIA%202014.pdf.
2. World Health Organization, Tuberculosis: WHO Global Tuberculosis Report 2014, http://www.who.Int/tb/publications/factsheet_global.pdf ?ua=1.
3. World Health Organization, Global Tuberculosis Report 2014: Key indicators for the WHO South-East Asia Report, http://www.who.int/tb/publications/global_report/indicators_south_east_as ia_region.pdf?ua=1.
4. K. Sinha, WHO, 2014, http://timesofindia.indiatimes.com/india/24-of-global-missed-TB-cases-in-India-WHO/articleshow/44921039.cms.
5. S. Swaminathan and Ch. Mehra, 2014, http://www.thehindu.com /opinion/op-ed/tackling-tb-fighting-indias-silent-epidemic/article6493937.ece.
6. K. Lonnroth, E. Jaramillo, B. G. Willioms, C. Dye and M. Raviglione, Drivers of tuberculosis epidemics: the role of risk factors and social determinants, Social Science and Medicine, 64,12, pp. 2240–2246, 2009.
7. M. Muniyandi, R. Ramachandran, P. G. Gopietal, The prevalence of tuberculosis in different economic strata: a community survey from South India, International Journal of Tuberculosis and Lung Disease, 11, 9, pp. 1042–1045, 2007.
8. V. K. Mishra, R. D. Retherford and K. R. Smith, Biomass cooking fuels and prevalence of tuberculosis in India, International Journal of Infectious Diseases, 5, 5, pp. 119–129, 1999.
9. C. Kolappan and R. Subramani, Association between biomass fuel and pulmonary tuberculosis: a nested case-control study, Thorax, 64, 8, pp. 705–708, 2009.
10. N. Bruce, R. Peter-Padilla and R. Albalak, Indoor air pollution in developing countries: a major environmental and public health challenge, Bulletin of the World health Organization, 78, 9, pp. 1078–1092, 2000.

11. J. Golub, S. Bur, W. A. Croninetal. Delayed tuberculosis diagnosis and tuberculosis transmission, International Journal of Tuberculosis and Lung Disease, 10, 1, pp. 24–30, 2006.
12. P. Venkatesan, R. Srinivasan and C. Dharuman, Bayesian Conditional Auto regressive model for mapping Tuberculosis prevalence in India, International Journal of Pharmacentical Studies and Research, 3, 2012.
13. P. Elliott, J. Wakefield, N. Best and D. Briggs, Spatial: epidemiology: methods and applications, 1st ed. Oxford University Press, 2000.
14. S. Banerjee, B. Carlin and A. Gelfand, 1st ed. Hierarchical modeling and analysis for spatial data, Chapman and Hall, 2004.
15. A. Lawson, 1st ed. Bayesian disease mapping: hierarchical modeling in spatial epidemiology, Chapman and Hall, 2008.
16. J. Besag, J. York and A. Mollie, Bayesian image restoration with two applications in spatial statistics, Ann Inst Stat Math, 43, pp. 1–59, 1991.
17. B. Leroux, X. Lei and N. Breslow, Estimation of disease rates in small areas: a new mixed model for spatial dependence, In: M. Halloran, D. Berry editors. Statistical models in epidemiology, the environment and clinical trials. New York: Springer-Verlag, pp. 135–178, 1999.
18. D. Lee, A comparison of conditional auto regressive models used in Bayesian disease mapping, Spatial and Spatio-temporal Epidemiology, 2, pp. 79–89, 2011.
19. D. Lee and R. Mitchell, Boundary detection in disease mapping studies, Biostatistics, 0, 0, pp. 1–12, Oct., 2011. doi:10.1093/biostatistics/kxr036.

Optimal Advertisement Allocation for Product Promotion on Television Channels

A. Kaul, S. Aggarwal and P. C. Jha
Department of Operational Research, University of Delhi,
Delhi, 110007, India
E-mail: arshia.kaul@gmail.com, sugandha_or@yahoo.com, jhapc@yahoo.com

A. Gupta
School of Business, Public Policy and Social Entrepreneurship, Ambedkar University
Delhi, 110006, India
E-mail: anshu@aud.ac.in

Advertising for a product is imperative for a firm for sustenance in a competitive market. It is performed through various media vehicles of which television media with a large reach to the audience has always attracted the firms to advertise their product. In this media, an advertiser has multiple options available for placement of its commercial in terms of different categories of channels, time slots of telecasting different programs and number of breaks within a program. The problem faced by an advertiser is to select the appropriate combination among these multiple options for placement of its commercial. In this paper, we formulate a multiple objective programming problem for an advertiser carrying out an advertising campaign through different categories of television channels. The model optimally places an advertiser's product in breaks within multiple programs scheduled on different day parts of television channels over a planning horizon, minimizing the cost incurred and maximizing the reach of the advertisement for its product. With these two objectives, the model is constrained on diversification of budget among the channels, bounds on the frequency of commercial in time slot of each channel and preferences of the advertiser for allocation of advertisement. The trade-off between the two objectives is achieved by solving it through goal programming technique. A case study for a firm advertising a product on multiple television channels is presented to validate the model and solved using optimization software LINGO.

Keywords: television channels; multi-objective programming; advertisement allocation; planning horizon.

1. Introduction

As the market has evolved over the years the approach firms use to advertise for their product is changing with addition of new media vehicles to the composition of the media mix. However, television as a traditional media has maintained its' importance till date. Along with a large reach to potential purchasers, the visual impact of television ensures that a higher audience remains tuned in over time [1]. It is estimated that television advertising in India will maintain the largest market share of 45.8%, which has increased from 44.5% in 2014 [2]. With such a large share of investment in this medium, television advertising forms an important asset and a lot of research on scheduling advertisements in this medium is being carried out.

Television as a media has grown manifold. There has been an exponential rise in the number of channels available for viewing and advertising. Different channels provide a variety of options (such as different time slots, programs and breaks) to advertisers for buying positions for advertisements while advertising their product. The advertisers could either sponsor a program, buy air-time to place advertisements within programs of a channel or could buy air-time in between two programs to place their advertisements [1]. In this paper we develop a mathematical model for an advertiser to determine the optimal allocation of their advertisements for a product across a number of television channels where advertisements are to be placed within the programs.

In television, the day-parts (slots) in which the advertisement is allocated plays a significant role. Advertisements placed in different slots are able to impact different audiences. A particular day for a television schedule is divided into time slots dependent on the kind of audience that watch programs in the slot. In general, it can be bifurcated into time dependent day-parts defined as Early Morning (7:00 A.M.–9:00 A.M.; Monday–Sunday), Daytime (10:00 A.M.–4:00 P.M.; Monday–Friday), Prime Time Access (7:30 P.M.–8:00 P.M.; Monday–Saturday), Prime Time (8:00–11:00 P.M.; Monday–Saturday & 7:00 P.M.–11:00 P.M.; Sunday), Weekend Sports (10:00 A.M.–6:00 PM; Saturday–Sunday), etc. [3]. In each slot there are multiple programs that can be placed of different durations. Within a program certain stipulated duration is allocated for advertisements by regulatory authorities of the respective country. This stipulated duration is used to telecast a number of commercials within the program breaks. Here we have considered the combinations of each of these channels, day parts and programs for possible placement of commercial for an advertiser. The choice of these combinations is made by the advertiser on the basis of product type. For example, an advertiser who wants to advertise a product suited for children would target a channel apt for them, more together with the programs viewed by children in the potential day parts in other channels.

The aim of a firm advertising for the product through television is to ensure that its commercial reaches the maximum number of potential purchasers of the product. The higher the impact of advertisement in the minds of potential purchasers the higher is the chance that the potential purchaser would be converted to an actual purchaser for a product. Thus an important aspect to be considered from an advertiser's point of view while scheduling their advertisement is maximizing the reach of the product through optimal placement of its commercials. In recent years it has been observed that Television Viewers in Thousands (TVT) is a more effective method to understand the audience watching programs on television [4]. TVT data is available in general for each week and on a daily basis for different programs. We measure the reach of the commercial through the product of average expected viewership data and the percentage of the target audience of the product possibly watching the program and aim at maximizing the advertiser's reach.

In this paper, we propose a multi-objective mathematical programming model for an advertiser with objectives of maximizing the advertising reach to the target audience and minimizing the cost incurred in television advertisement campaign over a planning horizon. The campaign is carried out through multiple channels and multiple programs which are aired in different slots of a day. The model is subject to the constraints of diversification of budget in each channel, limit on the frequency of advertisements in the slots and preferences of the advertiser for allocation of the advertisement. Goal programming technique is used to achieve a trade-off between the conflicting objectives under the imposed constraints.

The rest of the paper is divided into the following sections; Section 2 discusses the literature in research focused on optimization in television advertising. Section 3 discusses in detail the proposed multi-objective mathematical model for a firm advertising through television. In Section 4 a case study is presented to validate the mathematical model. Section 5 concludes the paper and gives future directions.

2. Literature Review

There are a number of researchers who have discussed the optimization models related to advertising on television. Each research gives a different dimension to the multifaceted problem of placement of advertisements on television. In the early stages of the research done in this field, a television scheduling model was given by [5] in which they proposed a binary integer programming model with the objective of maximization of weekly ratings of a network. The model is solved under the constraints on scheduling based on show parts and time slots. A mathematical programming problem developed by [6] focused on placement of advertisement of the product at a particular location in multiple programs of a channel with the objective to maximize the viewership of the potential customers of the product under the budget constraint. An algorithm is proposed for solving the model. A mixed integer programming problem was proposed by [7] with an objective to minimize the error between the actual and ideal distance of multiple airings of advertisements of an advertiser on television. The objective is achieved under the constraints on the number of advertisements to be placed in each break and the order of advertisements. Also [8] proposed a mathematical model to minimise penalty incurred by the television channel when advertiser objectives are not met while placing advertisements in multiple breaks of shows aired on a television channel. The model is developed with the constraints on advertising time, the product conflict constraint and limit on the number of advertisements that can be placed in a particular break of a program. Further [9] discussed a model with an objective to maximize the revenue of the television channel by placing advertisements in multiple breaks of a program telecast. The model is established on the concept of selling advertising air-time in bundles instead of single slots to the advertisers. Constraint programming method is proposed to analyze the experiment for the TV-break packing problem. The model given by [10] is discussed from the point of view of the advertiser. They proposed a multi objective model to determine the total duration of a company's advertisement in different broadcast over the advertising campaign span. The objective is to maximize the impact of advertisements and minimize the cost of advertising with advertising break constraint, mature product constraint and diversification constraint. A mathematical model was discussed by [11] to maximize the opportunity to see of the advertisement placed on multiple channels of television subject to the budgetary limitation and limit on duration of advertising. The authors proposed a step-wise project-advancement technique to solve the mathematical model. Advertising strategy of a theme-park in Taiwan which is advertising for two facilities is given as an application. Recently [12] proposed a model to maximise the revenue of the television channel. The concept of combinatorial bids was used in which multiple advertisers bid for the placing advertisements in the multiple breaks that are aired in the programs during prime time on television. The model is solved under constraints on the limit of air-time for breaks, limit on number of advertisements that can be placed in a break and product conflict constraints. A model given by [13] is a multiple objective model to maximise the ratings and minimise the cost of a television channel when programs are scheduled on different slots available on different days of the week. The model is validated for the case of an Italian Television channel.

In the extant literature researchers have discussed the two perspectives —

the channel perspective and the advertiser perspective. Although there is an extensive literature available in this field, it is observed that limited number of researchers have delved upon the complete multi-dimensional television advertising problem from the perspective of advertiser. The motivation of this research paper is to give a near-complete optimization model from an advertiser perspective for the placement of advertisements on television. The objective of the mathematical model is to maximise the reach of advertisements and simultaneously minimise the cost of advertising. The model is developed for a firm which desires to advertise a product by placing advertisements in breaks of the programs aired in different slots on multiple channels over a planning horizon. The trade-off between the two conflicting objectives is obtained using goal programming.

3. Model Development

3.1. *Notations*

d Index for day; $d = 1,2,\ldots,T$

m Index for television channels; $m= 1,2,\ldots,M$

s Index for day-parts/slots in a television channel; $s = 1,2,\ldots,S$

p Index for programs/shows in a slot; $p = 1,2\ldots,P_s$

b Index for advertisement breaks in a program; $b = 1,2,\ldots,B_p$

T Time span of advertisement campaign (in days)

R_{dmsp} Expected television viewership of program p in slot s of channel m telecast on day d

β_{dmsp} Percentage of target audience viewing program p in slot s telecast of channel m on day d

C_{dms} Per unit advertising cost of a commercial telecast in slot s on channel m telecast on day d

L_{ms} Lower bound on frequency of the commercial specified by an advertiser in slot s of channel m over the planning horizon

U_{ms} Upper bound on frequency of the commercial specified by an advertiser in slot s of channel m over the planning horizon

q_m Diversification level of commercials among the M channels

V Set of all possible combinations of commercials in slot s of channel m telecast on day d: $V=\{(d,m,s):d\in\{1,2,\ldots,D\},m\in\{1,2,\ldots,M\},s\in\{1,2,\ldots,S\}\}$

H Set of advertiser preferences in slot s of channel m telecast on day; d ($H \subseteq V$)

F Fixed duration of the commercial for the product

$$x_{dmspb} = \begin{cases} 1 & \text{if advertisement is placed in break } b \text{ of program } p \text{ in slot } s \text{ of channel } m \text{ on day } d \\ 0 & \text{otherwise} \end{cases}$$

3.2. *Model formulation*

Here we formulate an advertisement planning model for a firm that plans to promote a consumable product with advertisement of fixed duration through multiple television channels over a planning horizon. On each television channel, multiple slots/day parts are considered on which a number of shows are telecast. Each show has multiple number of breaks of differing duration available for advertisement placements of various products. Based on the product that needs to be advertised the preferences are given by the advertiser for the possible placements of their commercial. The aim of an advertiser is to achieve maximum reach through the advertisement placements with minimum cost for advertising. The model proposed here is for an advertiser who wants to determine the optimal placement of the product over a planning horizon considering the two objectives and the model constraints. The multi objective programming problem is formulated as follows:

(P1):

$$Maximise\ Z = \sum_{d=1}^{T}\sum_{m=1}^{M}\sum_{s=1}^{S}\sum_{p=1}^{P_s}\sum_{b=1}^{B_p} R_{dmsp}\beta_{dmsp}x_{dmspb} \quad (1)$$

$$Minimise\ C = \sum_{d=1}^{T}\sum_{m=1}^{M}\sum_{s=1}^{S}\sum_{p=1}^{P_s}\sum_{b=1}^{B_p} C_{dms}x_{dmspb} \quad (2)$$

subject to

$$\sum_{d=1}^{T}\sum_{s=1}^{S}\sum_{p=1}^{P_s}\sum_{b=1}^{B_p} x_{dmspb} \geq q_m \sum_{d=1}^{T}\sum_{m=1}^{M}\sum_{s=1}^{S}\sum_{p=1}^{P_s}\sum_{b=1}^{B_p} x_{dmspb}\ \forall m = 1,2,\dots,M \quad (3)$$

$$\sum_{d=1}^{T}\sum_{p=1}^{P_s}\sum_{b=1}^{B_p} x_{dmspb} \geq L_{ms}\ \forall m = 1,2,\dots,M, s = 1,2,\dots,S \quad (4)$$

$$\sum_{d=1}^{T}\sum_{p=1}^{P_s}\sum_{b=1}^{B_p} x_{dmspb} \leq U_{ms}\ \forall m = 1,2,\dots,M, s = 1,2,\dots,S \quad (5)$$

$$x_{dmspb} = 0\ \forall p = 1,2,\dots,P_s, b = 1,2,\dots,B_p, (d,m,s) \notin H \quad (6)$$

$$x_{dmspb} \in \{0,1\}\ and\ integer \quad (7)$$

The first objective Z given in (1) is corresponding to the reach maximization through the placement of the product's commercial in television channels over a planning horizon computed as a product of the expected number of viewers of the channel and the percentage of target potential viewers of the channels for the advertisements placed. The second objective C given in (2) is to minimize the advertising cost incurred through the placement of the product's commercial in television channels over a planning horizon. The model objectives are achieved under the constraints given by (3)–(7). First constraint given by (3) ensures that total number of commercials placed over a planning horizon are diversified over the channels with diversification level q_m in each of the M channels. Thus a minimum proportion of the total budget is allotted to each of the channels rather than allocating it to a limited number of channels.

Constraints through (4) and (5) impose lower and upper bounds on the number of commercials to be placed in each slot of a television channel over a planning horizon specified by an advertiser. Constraint given by (6) is to cater to the advertiser preferences for placement of commercial in slots of television channels in each time period of a planning horizon. The constraint (7) is to ensure that the decision variable takes binary values.

The model given by (P1) is first solved taking reach objective subject to the constraints (3)–(7). The budget (A) is decided by the management of the firm. This gives reach and the cost aspirations. Owing to the conflicting nature of the objectives it is not possible for the firm to achieve the aspirations set on both objectives simultaneously, thus the model is solved through the goal programming approach to obtain a trade-off between the objectives.

The goal programming model for (P1) can be written as follows:

(P2):

$$Minimise\ g(\eta,\rho) = \gamma\eta_1 + (1-\gamma)\rho_2 \tag{8}$$

subject to

$$R + \eta_1 + \rho_1 = 1 \tag{9}$$

$$B + \eta_2 + \rho_2 = 1 \tag{10}$$

$$\sum_{d=1}^{T}\sum_{s=1}^{S}\sum_{p=1}^{P_s}\sum_{b=1}^{B_p} x_{dmspb} \geq q_m \sum_{d=1}^{T}\sum_{m=1}^{M}\sum_{s=1}^{S}\sum_{p=1}^{P_s}\sum_{b=1}^{B_p} x_{dmspb}\ \forall m = 1,2,\dots,M \tag{11}$$

$$\sum_{d=1}^{T}\sum_{p=1}^{P_s}\sum_{b=1}^{B_p} x_{dmspb} \geq L_{ms}\ \forall m = 1,2,\dots,M, s = 1,2,\dots,S \tag{12}$$

$$\sum_{d=1}^{T}\sum_{p=1}^{P_s}\sum_{b=1}^{B_p} x_{dmspb} \leq U_{ms}\ \forall m = 1,2,\dots,M, s = 1,2,\dots,S \tag{13}$$

$$x_{dmspb} = 0\ \forall p = 1,2,\dots,P_s, b = 1,2,\dots,B_p, (d,m,s) \notin H \tag{14}$$

$$R = Z/Z^* \tag{15}$$

$$B = C/A \tag{16}$$

$$x_{dmspb} \in \{0,1\}\ and\ integer \tag{17}$$

$$\eta_1, \eta_2, \rho_1, \rho_2 \geq 0 \tag{18}$$

The objective in (P2) given by (8) is to minimize the sum of weighted deviations from the goals and as the goals have different units of measurement, they are normalized by dividing the reach and budget goals by their respective aspirations as given in (15) and (16). Constraints given by (9) and (10) are the goal constraints on the reach and advertisement cost objectives. Keeping in view the weights that the management may want to assign to the two objectives, on solving the goal programming we are able to establish the allocations of advertisements in different breaks of the programs in different slots in channels over a planning horizon. The solution to the model (P2) is obtained through

LINGO [14] software. The optimal solution obtained from (P2) is an efficient solution for (P1) [15].

4. Numerical Illustration

To validate the model developed in the last section a case study for a firm that needs to advertise for a product best suited for men through multiple television channels (CH) over a planning horizon is considered. The management has shortlisted two channels each in the news (CH1, CH2), general entertainment (CH3, CH4) and sports (CH5, CH6) category. Therefore, in total there are six television channels chosen as per the decision of the management of the firm. Depending on the channel type there are different time slots (*S*) available as discussed in the introduction section. Here two slots early morning and prime time in news channels, a single slot of prime time in general entertainment channels and a single slot of long duration (i.e. weekend sports slot) in the sports channels is considered. The time span of the advertisement planning is of one week consisting of five weekdays (D1 D5) and two days (D6, D7) of weekend. Length of a typical commercial of the firm is taken to be of 30 seconds.

There are multiple programs which are aired during a particular time slot (day-part). For both the news channels (CH1 and CH2), two programs (P1, P2) are telecast of duration one hour each in first slot. And four programs (P1–P4), with two programs of duration one hour and two of half an hour are telecast in the second slot. The same pattern of programs is followed for D1–D7 in news channels. In CH3, six programs (P1–P6) of half an hour each is considered for D1–D6 and a single program (P1) of three hours is considered on D7. In CH4, six programs (P1–P6) of half an hour each is considered for D1–D5 and three programs (P1–P3) of one hour each on D6 and D7. For both the sports channels, two programs (P1, P2) of one and a half hours and one program (P3) of five hours is considered on D6 and one program (P1) of eight hours is considered on D7. In each program multiple breaks are available for the placement of commercial and the breaks considered are in accordance to the time stipulation of the regulatory authorities. In news and general entertainment channels a half an hour program consists of three breaks and one hour program consists of six breaks. And in sports channel, one and a half hour program consists of nine breaks, a five hour program consists of twenty breaks and an eight hour of thirty two breaks.

The aim of the advertiser is to determine the optimal placement of commercial in the specified channels with the objectives of maximizing reach and concurrently minimizing the cost of advertising. The advertiser prefers the channels, slots and days according to the target audience of the product. Set of advertiser preferences in slot s of channel m telecast on day d is given as (in the form of (d, m, s)):$H = \{(1,1,1), (1,1,2), (1,2,2), (1,4,1), (2,1,1), (2,1,2), (2,4,1), (3,1,2), (3,2,1), (3,2,2), (3,3,1), (3,4,1), (4,2,1), (4,2,2), (4,3,1), (5,3,1), (6,5,1), (6,6,1), (7,1,1), (7,1,2), (7,2,1), (7,2,2), (7,4,1), (7,5,1), (7,6,1)\}$

The expected viewership of different programs on slots of television channels and the percentage of target audience watching the program (computed through a sample survey) on slots of television channels over the planning period is given by Table 1. The cost for placement of commercial (per 30 seconds) varies according to the time slots of television channels and its type for different days as given in Table 2. The lower (min) and upper (max) bounds on the number of commercial to be placed in each slot of a television channel over a planning horizon specified by an advertiser is given in Table 3.

Table 1. Expected viewership in '000 (% target audience).

	CH1		CH2		CH3		CH4		CH5		CH6	
	D1–6	D7	D1–6	D7	D1–6	D7	D1–5	D6–7	D6	D7	D6	D7
S1-P1	7545 (0.55)	7980 (0.6)	8020 (0.7)	8234 (0.72)	6890 (0.68)	4800 (0.38)	3450 (0.52)	2560 (0.4)	9700 (0.72)	13800 (0.67)	12800 (0.75)	14500 (0.6)
S1-P2	8960 (0.58)	9180 (0.65)	9100 (0.75)	9345 (0.8)	8760 (0.78)	–	4780 (0.44)	3420 (0.48)	12500 (0.83)	–	8700 (0.58)	–
S1-P3	–	–	–	–	12000 (0.84)	–	5460 (0.58)	5600 (0.52)	11200 (0.78)	–	9500 (0.68)	–
S1-P4	–	–	–	–	7890 (0.72)	–	2940 (0.38)	–	–	–	–	–
S1-P5	–	–	–	–	5430 (0.54)	–	4230 (0.56)	–	–	–	–	–
S1-P6	–	–	–	–	6250 (0.59)	–	2800 (0.3)	–	–	–	–	–
S2-P1	6780 (0.38)	7050 (0.45)	7345 (0.52)	7656 (0.55)	–	–	–	–	–	–	–	–
S2-P2	9234 (0.45)	9567 (0.52)	9750 (0.6)	1080 (0.64)	–	–	–	–	–	–	–	–
S2-P3	5678 (0.67)	6234 (0.72)	6200 (0.58)	6456 (0.64)	–	–	–	–	–	–	–	–
S2-P4	5890 (0.56)	6125 (0.62)	5890 (0.6)	6235 (0.63)	–	–	–	–	–	–	–	–

Table 2. Advertising cost (per 30 seconds).

	CH1		CH2		CH3		CH4		CH5		CH6	
	D1-6	D7	D1-6	D7	D1-6	D7	D1-5	D6-7	D6	D7	D6	D7
S1	15300	12940	18770	13670	98580	120400	94250	87650	67560	73250	58560	62430
S2	35890	30890	42340	38940	-	-	-	-	-	-	-	-

Table 3. Upper and lower bounds.

	CH1		CH2		CH3		CH4		CH5		CH6	
	MIN	MAX	MIN	MAX	MIN	MAX	MIN	MAX	MIN	MAX	MIN	MAX
S1	12	36	12	36	18	48	24	64	12	44	12	44
S2	24	64	24	64	-	-	-	-	-	-	-	-

The data is used for the mathematical model coded in LINGO and the problem is solved by taking the reach objective subject to all constraints to get the aspiration on reach. The aspired reach obtained is 2,44,47,17,000 (Z^*). A budget of Rs.1,36,00,000 is set for advertising by the management of the firm. With these values the problem (P2) is solved through goal programming approach. The reach obtained is 1,82,32,17,719 and the budget utilized is Rs.1,36,00,440. The solution for the placement of commercials is given in Table 4 for each channel on each day of the planning week.

As per discussion with the management of the firm, sensitivity analysis on the budget was performed to analyse the variation that can be achieved in the value of reach with different values of budget. It was observed that with an increase in budget there was an increase in achieved reach. By increasing the budget by 10% to Rs.1,49,60,000 reach obtained was 1,96,94,33,000 and with a decrease of budget by 10% to Rs.1,22,40,000 the reach achieved was 1,66,19,77,000. Thus it can be observed that the firm can increase its budget to achieve a higher value for reach. Keeping in view the conditions in the market as well as the profits earned by the firm, the value for budget can be decided upon. For a firm to increase its market presence, it must be able to reach the maximum number of potential purchasers for the product and in turn increase its profit.

5. Conclusion

The proposed model is for a firm which is advertising for its consumable product. The firm wants to advertise for its product by placing commercials in multiple breaks aired in different slots in programs telecast on multiple channels over a planning horizon. The objectives of the firm are to maximise the reach for advertisements of the product as well as minimise the cost accrued for advertising. The model determines the optimal allocation of the advertisements of the firm's product in different channels, day parts, programs, and breaks over a planning horizon. The goal programming technique is used to establish the trade-off between the two objectives given the constraints on diversification of budget, limits on frequency of advertisements placed in slots and the preference set of advertiser for placements. A case study is presented for a firm which is advertising its single product. In the current mathematical model we have considered the advertisement campaign for a single product, the future scope of the model can be to consider a mathematical model which considers multiple products of a firm while scheduling.

Table 4. Solution table.

D1			D2		D3				D4		D5	D6		D7				
CH1	CH2	CH4	CH1	CH4	CH1	CH2	CH3	CH4	CH2	CH3	CH3	CH5	CH6	CH1	CH2	CH4	CH5	CH6
x_{11112}	x_{12221}	x_{14131}	x_{21111}	x_{24131}	x_{31221}	x_{32111}	x_{33121}	x_{34131}	x_{42111}	x_{43121}	x_{53121}	x_{65121}	x_{66111}	x_{71111}	x_{72111}	x_{74121}	x_{75111}	x_{76111}
x_{11113}	x_{12222}	x_{14132}	x_{21112}	x_{24132}	x_{31222} x_{31223}	x_{32112}	x_{33122}	x_{34132}	x_{42112}	x_{43122}	x_{53122}	x_{65122}	x_{66112}	x_{71112}	x_{72112}	x_{74126}	x_{75112}	x_{76112}
x_{11121}	x_{12223}	x_{14133}	x_{21113}	x_{24133}	x_{31224} x_{31225}	x_{32113}	x_{33123}	x_{34133}	x_{42113}	x_{43123}	x_{53123}	x_{65123}	x_{66113}	x_{71113}	x_{72113}	x_{74131}	x_{75113}	x_{76113}
x_{11122}	x_{12224}		x_{21114}	x_{24153}	x_{31226}	x_{32114}	x_{33131}		x_{42114}	x_{43131}	x_{53131}	x_{65124}	x_{66114}	x_{71114}	x_{72114}	x_{74132}	x_{75114}	x_{76114}
x_{11123}	x_{12225}		x_{21115}			x_{32115}	x_{33132}		x_{42115}	x_{43132}	x_{53132}	x_{65125}	x_{66115}	x_{71115}	x_{72115}	x_{74133}	x_{75115}	x_{76115}
x_{11124} x_{11125}	x_{12226}		x_{21116}			x_{32116}	x_{33133}		x_{42116}	x_{43133}	x_{53133}	x_{65126}	x_{66116}	x_{71116}	x_{72116}	x_{74134}	x_{75116}	x_{76116}
x_{11126}			x_{21121}			x_{32121}			x_{42121}			x_{65127}	x_{66117}	x_{71121}	x_{72211}	x_{74135}	x_{75117}	x_{76117}
x_{11221}			x_{21122}			x_{32122}			x_{42122}			x_{65128}	x_{66118}	x_{71122}	x_{72212}	x_{74136}	x_{75118}	x_{76118}
x_{11222} x_{11223}			x_{21123}			x_{32123}			x_{42123}			x_{65129}	x_{66119}	x_{71123}	x_{72213}		x_{75119}	x_{76119}
x_{11224}			x_{21124}			x_{32124}			x_{42124}			x_{65131}	x_{66131}	x_{71124}	x_{72214}		x_{751110}	x_{761110}
x_{11225}			x_{21125}			x_{32125}			x_{42125}			x_{65132}	x_{66132}	x_{71125}	x_{72215}		x_{751111}	x_{761111}
x_{11226}			x_{21126}			x_{32126}			x_{42126}			x_{65133}	x_{661315}	x_{71126}	x_{72216}		x_{751112}	x_{761112}
			x_{21221}						x_{42221}			x_{65134}	x_{661316}	x_{71221}	x_{72121}		x_{751113}	x_{761113}
			x_{21222}						x_{42222}			x_{65135}	x_{661317}	x_{71222}	x_{72122}		x_{751114}	x_{761114}
			x_{21223}						x_{42223}			x_{65136}	x_{661318}	x_{71223}	x_{72123}		x_{751115}	x_{761115}
			x_{21224}						x_{42224}			x_{65137}	x_{661319}	x_{71224}	x_{72124}		x_{751116}	x_{761116}
			x_{21225}						x_{42225}			x_{65138}	x_{661320}	x_{71225}	x_{72125}		x_{751117}	x_{761117}
			x_{21226}						x_{42226}			x_{65139}		x_{71226}	x_{72126}		x_{751118}	x_{761118}
												x_{651310}		x_{71231}	x_{72231}		x_{751119}	x_{761119}
												x_{651311}		x_{71232}	x_{72232}		x_{751120}	x_{761120}
												x_{651312}		x_{71233}	x_{72233}		x_{751121}	x_{761121}
												x_{651313}		x_{71241}	x_{72241}		x_{751122}	x_{761122}
												x_{651314}		x_{71242}	x_{72242}		x_{751123}	x_{761123}
												x_{651315}		x_{71243}	x_{72243}		x_{751124}	x_{761124}
												x_{651316}					x_{751125}	x_{761125}
												x_{651317}					x_{751126}	x_{761126}
												x_{651318}					x_{751127}	x_{761127}
												x_{651319}					x_{751128}	x_{761128}
												x_{651320}					x_{751129}	x_{761129}
																	x_{751130}	x_{761130}
																	x_{751131}	x_{761131}
																	x_{751132}	x_{761132}

References

1. G. Belch, M. Belch and K. Purani, *Advertising and Promotion: An Integrated Marketing Communications Perspective*, 7th edn. (Tata McGraw Hill, India, 2010).
2. A. Dina, GroupM says ad spends to grow 12.6% in 2015 (2015), http://www.livemint.com/Consumer/nAhthSk6371gEVIQiZ929J/Indian-advertising-expenditure-to-touch-Rs49000-crore-in-20.html.
3. J. Sissors and R. Baron, *Advertising media planning*, (McGraw-Hill, New York, 2010).
4. A. L. Chougule, Is TVT a game changer? (2013), http://archive.indianexpress.com/news/is-tvt-a-game-changer-/1164459/.
5. C. Kelton and L. Schneider Stone, Optimal television schedules in alternative competitive environments, *Eur. J. Oper. Res.* 104, 3, 451 (1998).
6. A. Mihiotis and I. Tsakiris, A mathematical programming study of advertising allocation problem, *Appl. Math. Comp.* 148, 2, 373 (2004).
7. S. Bollapragada, M. Bussieck and S. Mallik, Scheduling Commercial Videotapes in Broadcast Television,*Oper. Res.* 52, 5, 679 (2004).
8. S. Bollapragada and M. Garbiras, Scheduling Commercials on Broadcast Television, *Oper. Res.*, 52, 3, 337 (2004).
9. T. Benoist, E. Bourreau and B. Rottembourg, The TV-Break Packing Problem, *Eur. J. Oper. Res.* 176, 3, 1371 (2007).
10. Pérez-Gladish, I. Gonzalez, A. Bilbao-Terol and M. Arenas-Parra, Planning a TV advertising campaign: A crisp multiobjective programming model from fuzzy basic data, *Omega* 38, 1-2, 84 (2010).
11. Y. Jeng and F. Chiu, Allocation model for theme park advertising budget, *Qual. Quant.* 44, 2, 333 (2010).
12. F. Ghassemi Tari and R. Alaei, Scheduling TV commercials using genetic algorithms, *Int. J. Prod. Res.*, 51, 16, 4921, (2013).
13. T. Lupo, Non-dominated "trade-off" solutions in television scheduling optimization, *Int. Trans. Oper. Res.* (2014).
14. LINDO SYSTEM (2015), http://home.anadolu.edu.tr/~nila/PROGRAM/KILAVUZ/LINGO%20users%20manual.pdf.
15. A. Geoffrion, Proper efficiency and the theory of vector maximization, J. Math. Anal. Appl. 22, 3, 618 (1968).

Fitting Linear Regressions: Development and Scope*

Pranesh Kumar†
Dept. of Mathematics and Statistics, University of Northern British Columbia
Prince George, BC V2N 4Z9, Canada
E-mail: pranesh.kumar@unbc.ca
www.unbc.ca/kumar-dr-pranesh

J. N. Singh
Dept. of Mathematics and Computer Science, Barry University 11300 NE
Second Avenue, Miami Shores Florida 33161, USA
E-mail: jsingh@barry.edu

Some pioneering works on the methods of fitting functional relationship which are often used in estimating linear regression models are discussed. This fundamental problem in the theory of errors has drawn special attention of the leading mathematicians and scientists, in particular, since eighteenth century. We discuss some current investigations based on analysis of the Lp-norm regression which raise questions about the appropriateness and applications of the least squares based regressions in making predictions.

Keywords: Function fitting, Linear regression, Least squares estimation, Lp-norm.

1. Function-Fitting Problem and Pioneering Works

A fundamental problem in the theory of errors was of fitting functions which had drawn attention of many leading mathematicians and scientists over past few centuries. This problem is well known now as the problem of fitting a cause and effect relationship, i.e., a standard linear regression model expressing relationship between a response variable (y) and the p-covariates (x) defined by n linear equations:

$$y_i = \beta_0 + x_{i1}\beta_1 + \dots + x_{ip}\beta_p + \varepsilon_i, \quad i = 1,2,\dots,n, \tag{1.1}$$

or, in matrix notation,

$$Y = X\beta + \varepsilon, \tag{1.2}$$

where ε is the vector of random errors. Methods of fitting functions attempt to obtain the best estimates of the unknown model constants ($\beta's$). The pioneering

* This work is supported by UNBC Research Seed Grant, 2014.
† Work partially supported by author's NSERC Discovery Grant.

approaches to solve this problem considered applications of algebraic and geometric methods. Attention was restricted to the special cases when there were either the same number of observations as there were unknowns, or, n discordant observations on a single unknown quantity. Mayer [18] published a paper on problem of modelling observed celestial positions of certain lunar features and established the linear relationship

$$\beta - (90° - h) = \alpha \sin(g - k) - \alpha \sin\theta \cos(g - k), \qquad (1.3)$$

where α = angle between pole of the moon and pole of celestial sphere, β = lunar (selenography) latitude of the chosen feature, k = celestial longitude of the ascending node of the moon's orbit, k+θ = celestial longitude of the rising equinoctial point, g = celestial longitude of the chosen lunar feature, $90° - h$ = celestial latitude, g and h are observed; k is from the Euler's lunar tables. Boscovich [1, 2] outlined an objective procedure for determining suitable values for the parameters of the fitted model from a greater number of observations. An unconstrained variant of the fitting procedure is now known as the L_1 -norm or least absolute deviations procedure. Boscovich suggested to determine values of slope coefficient y and intercept z which best fit n equations: $a_i = z + yp_i, i = 0,1, \ldots, n - 1$, where a_i = length of a measured one degree arc of the meridian, θ_i = the latitude of its midpoint, $p_i = {}^1/_2\{versed\ sine(2\theta_i)\} = sin^2(\theta_i)$. In 1757, Boscovich formulated the principle that the values of y and z be chosen such that the corrections $v_i = z + yp_i - a_i, i = 0,1, \ldots, n - 1$, sum to zero and have minimum absolute $\sum|v_i|$. Laplace's first variant of Boscovich's algorithm from its geometrical framework is found in a paper published in 1793 [11]. In 1799, Laplace [12] made a small but significant change in the Boscovich's method that instead of choosing y and z to minimize the unweighted sum of the absolute errors $\sum|v_i|$ subject to the unweighted adding-up constraint $\sum v_i = 0$, he chose y and z to minimize the weighted absolute sum $\sum|l_i v_i|$ subject to the weighted constraint $\sum l_i v_i = 0$, where l_i are given weights. Laplace [13] developed the minimax procedure by considering the problem of correcting the elements of a formula describing the celestial longitude of Saturn. Selecting 24 observations from a 200-year period corresponding to instances when the Sun and Saturn appeared in diametrically opposite parts of the sky, he obtained 24 equations. Laplace provided results which were comparable with those later obtained from the method of least squares. The algebraic reformulation of the problem and introduction of weighted observations lead to Laplace's development of an asymptotic distribution theory for this procedure. Legendre's method of least squares is discussed in the first four pages (pp. 72–75) of a nine page appendix attached to his 1805 work on the determination of the orbits of comets. The procedure discussed by Laplace [12] is restricted to two unknowns but can be applied to any number of equations of the form $e_i = a_i + b_i x + c_i y + d_i z + \text{etc}, i = 0,1, \dots, n-1$, in any lesser number of unknowns. Legendre [17] chose values of unknowns $x, y, z,$ etc to minimize the sum of squared errors $E = \sum(a_i + b_i x + c_i y + f_i z + \text{etc})^2$. Legendre noted that these *équations duminimum* may also be obtained by multiplying all the terms in each of the equations by the coefficients of one of the unknowns and summing the result. There is, therefore, the same number of equations as there are unknowns. These *équations duminimum* may be solved for the unknowns by the ordinary method (which he did not describe). He asserted that the solution found by this procedure corresponds to a minimum sum of squares function $\sum e_i^2$. However, he did not prove this result. Similarly, to determine the position of a point (x, y, z) in three dimensional space from n observations on this point, $(a_i, b_i, c_i), i = 1,2, \dots, n,$ Legendre suggested that the sum of squared Euclidean distances $\sum[(a_i - x)^2 + (b_i - y)^2 + (c_i - z)^2]$ is minimized when $x, y,$ and z are set to their respective arithmetic means: $x = \sum \frac{a_i}{n}, y = \sum \frac{b_i}{n}, z = \sum \frac{c_i}{n}$. Fitted point (x, y, z) represents center of gravity of n given points. Gauss's least sum of even powers criterion [7]: "Legendre's general object was to choose the values of the unknowns x, y, z etc. to make the errors e_0, e_1, etc as small as possible. Without any alternative, he asserted that the sum of the squares of the errors was the simplest criterion." Gauss [7] offered that the sum of the fourth powers, $e_0^4 + e_1^4 + e_2^4 + etc$, the sum of the sixth powers, $e_0^6 + e_1^6 + e_2^4 + etc$, or, the sum of even powers as alternatives to the sum of squared errors. Of these criteria, simplest is the sum of squared errors which Gauss used in practical calculations. Gauss continued with the observation that in the limit as the even power selected becomes infinite, we have the criterion which chooses unknowns to minimize the largest absolute error. Laplace [14] considered the characterization of the minimax procedure: "I have found from numerous examples that when this power does not exceed the square, the results differ little from those which give the systems that minimize largest errors; this is the new advantage for the method of least squares of errors of observations," (Translation by Farebrother [6]). In 1809 and 1815, Bowditch [3, 4] offered an alternative generalization of the Boscovich's procedure: Instead of restricting himself to a single adding-up constraint as Gauss had done, he suggested that one should impose $q-1$ such constraints to eliminate all but the q unknowns and to reduce the problem to one which could be solved by the Laplace's algebraic procedure [Stigler (1966, p. 55)]. It may thus be noted that the algebraic phase of the calculus of observations reached its climax in 1805 with the publication of

Legendre's description of the method of least squares. Despite the significant contributions which Gauss made to the theory of the method of least absolute errors in 1809, he probably abandoned this method in favor of more accessible method of least squares. Laplace [15] also adopted a Bayesian (inverse probabilistic) approach to the fitting problem. He suggested six optimality criteria which may be used to derive a mean from the posterior density function.

2. The L_p-norm Regression Estimation Problem

The methods of estimation, based on the criterion of least squares or minimum absolute deviation, belong to a general class of L_p-norm based estimation methods.

Definition: The L_p-norm of a residual vector ε in (1.2) is

$$\| \varepsilon \|_p = \begin{cases} (\Sigma |\varepsilon_i|^p)^{1/p}, \text{ for } p \in [1; \infty), \\ \max |\varepsilon|, \text{ for } p \to \infty. \end{cases} \tag{2.1}$$

The L_p-norm regression estimation problem is to obtain estimates of model parameters ($\beta's$) by minimizing the L_p-norm of the residual vector ε, that is, to determine βwhich attains the minimum L_p-norm for the difference between $y$and $X\beta$ for $1 \leq p < \infty$, or maximizing L_p-norm of residual vector ε for $p = \infty$.

2.1. *The L_1-norm regression estimation*

Letting $p = 1$ in (2.1), the L_1-norm regression problem requires minimizing the norm $\| Y - X\beta \|_1$ with respect to β. In terms of linear programming problem, this is equivalent to

$$\min \Sigma\, t_i: \; -t_i \leq x_i^T \beta - y_i \leq t_i, \;\; i = 1, \ldots, n. \tag{2.2}$$

Boscovich [1] first proposed this criterion to fit the functional relationship.

2.2. *The L_2-norm regression estimation*

The L_2-norm based regression problem corresponds to letting $p = 2$ in (2.1) which requires minimizing the norm $\| X\beta - Y \|_2$ with respect to β. This is equivalent to determining β which results in $\min \Sigma(Y - X\beta)^2$. Estimators of β using the L_2-norm are known as the least squares estimators (LSE). Legendre [17] and Gauss [7] proposed the criterion to minimize the sum of the squares of the errors to determine β and, thereafter, the method of least square became the most popular estimating technique. The main reasons for the least squares method's popularity are easy computations and due to the fact that when residuals are independent and identically normally distributed, the least squares estimators are the best linear unbiased estimators as well as they are equivalent to the maximum likelihood estimator. It is however noted that the least squares estimates are sensitive to the departures from the assumptions, in particular, from the assumption of normally distributed errors.

2.3. *The L_∞-norm regression estimation*

Letting $p \rightarrow \infty$ in (2.1), the L_∞-norm regression problem requires maximizing the norm $\| X\beta - y \|_\infty$ with respect to β. In terms of linear programming problem, this problem is equivalent to [21] minimizing a variable t such that:

$$\min\, t\colon\ -t \leq x_i^T \beta - y_i \leq t,\quad i = 1, \dots, n. \tag{2.3}$$

Laplace [15] and Edgeworth [5] have shown that the L_∞-norm estimator is preferable to the least squares estimators when estimating a linear regression model with fat-tailed residuals. This minimization problem is often referred to as the Chebyshev approximation.

3. The L_p-norm Regression Studies

The Nyquist [19] prepared a monograph on the results of L_p-norm estimation. In his monograph, relations between L_p-norm estimators and the class of maximum likelihood type estimators (M-estimators) and the class of estimators that are linear combinations of order statistics (L-estimators) are given. The concept of the optimal L_p-norm estimator that possesses the smallest asymptotic variance is discussed and also, the L_p-norm based methods for estimating multicollinear regression models, with serially dependent residuals and interdependent systems. He discussed theorems on existence, uniqueness and asymptotic distributions of the L_p-norm estimators and also the geometrical interpretations of the L_p-norm estimators. Monte Carlo simulations indicated that an L_p-normestimator with p < 2 may be superior to the least squares estimator, when the distribution of residuals possess an infinite variance.

Nyquist [19] has proved that there always exists an L_p-norm estimator and that estimator is uniquefor $1 < p < \infty$. For $p = 1$, a necessary and sufficient condition on the X matrix is proved for the unique estimation. Also a condition on the residual distribution is proved that $\lim_{n\rightarrow\infty} E(\| Y - X\beta \|_1)/n$ has a uniqueminimum. For fixed regressors X and independent and identically distributed residuals ε, it is observed that for sufficiently small p, $\sqrt{n}\big(\hat{\beta}_p - \beta\big)$, where $\hat{\beta}_p$ is the L_p-norm estimator, is asymptotically normally distributed with mean zero and variance $w_p^2\ /Q$, where w_p^2 is

$$=\begin{cases} \{2F_\varepsilon'(0)\}^{-2}, \text{ for } p\ = 1, \\ E\ [\ |\varepsilon_1|^{2p-2}\ ]/\{(p-1)E[|\varepsilon_1|^{p-2}]\quad ^2\}\ \ , \text{ for } \ 1 < p < \infty, \end{cases} \tag{3.1}$$

F_ε is the distribution function of the residuals and $Q = \lim_{n\to\infty} X'X/n$. Special cases of stochastic regressors, heteroscedastic and linear dependent residuals are discussed.

The optimal L_p -norm estimator was defined which possesses the smallest asymptotic variance. The concept of the optimal L_p-norm estimator and the problem of chosing L_p-norm estimator hasbeen discussed and illustrated with some examples.

Further, possibilities to construct L_p-norm based estimators for estimating multicollinearity in regression models and regression models with serially dependent residuals, are discussed. Nyquist also discussed L_p-norm based estimators of interdependent systems.

Nyquist [17] finally concluded that there are a large number of models with residuals distributed such that there is an L_p-norm estimator superior to the least squares estimators. In particular, when residuals are fat-tailed or have contaminated normal distribution, alternatives with p < 2 are recommended. When actual residual distribution is known, it is possible to derive the optimal L_p-norm estimator (i.e. the L_p-norm estimator with the smallest asymptotic variance). When the residual distribution is only known and belong to a class of distributions, a minimax procedure may be used to choose estimator. With this procedure, the precision of the estimators will often considerably increase and, in a few cases, slightly decreased, compared to the least squares estimator. Therefore, the members of the class of L_p-norm estimators with $p \neq 2$ are viable alternatives to the least squares estimator and may be used when the residual distribution is different from the normal distribution.

4. Model Adequacy Measures

For assessing model adequacy, most commonly used measure based on estimated residuals is the coefficient of determination R^2 which is defined as the proportion of the total variance of y that is explained by the model

$$R^2 = 100\left[1 - \frac{\Sigma\ \varepsilon^2}{\Sigma(y-\bar{y}\ )^2}\right]. \tag{4.1}$$

Kumar and Kashanchi [8, 9] suggested alternative model adequacy measure based on the estimated residuals as

$$\|R^2\|_1 = 100\left[1 - \frac{\Sigma\,|\varepsilon|}{\Sigma|y-\bar{y}\,|}\right]. \tag{4.2}$$

It may be noted that the numerator $\Sigma|\varepsilon|$ and denominator $\Sigma|y - \bar{y}\,|$ terms in $\|R^2\|_1$ are the L_1-norm while in case of R^2, these are L_2-norm. Either measure compares how well model fits. A higher $\|R^2\|_1$ or R^2 indicates a better fit.

5. Comparative Studies of L_p-norm Regressions

We discuss in what follows the main findings of some recent studies to examine the appropriateness of the linear regression fits using the criteria of the L_1-, L_2- and L_∞-norms. The appropriateness of fitted models was compared with respect to making predictions for the entire range of data, middle-half and lower- and upper- five or ten percent data.

5.1. *Value-at-risk: Iran's crude oil prices and exports*

Kumar and Kashanchi [9] have studied risk analysis for the Iran's crude oil prices and crude oil exports from January, 1997 to December, 2008. The distribution of crude oil prices were skewed to the right with suspected outliers. Distribution of crude oil export was approximately symmetric with few suspected outliers on both ends (Figure 5.1.1).

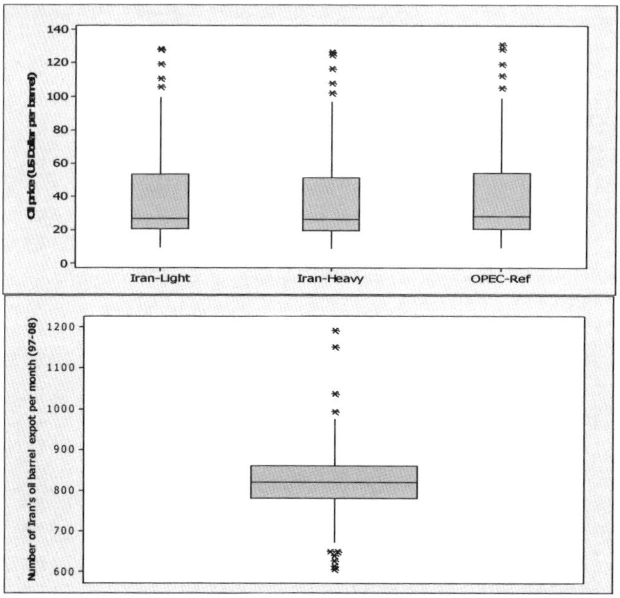

Fig. 5.1.1. Box plots of the crude oil prices and crude oil exports, January, 1997 – December, 2008.

The Value-at-Risk (VaR) was calculated in terms of the relative changes of the monthly crude oil prices. Risk was the unexpected rare event of observing either a very low relative change (drop in oil price) or a very high relative change (rise in oil price).

Following indicative conclusions were drawn:

- Models in order of goodness of fit: L_1 , L_2 , L_∞-norm regression models.
- Model adequacy: R^2 values were small for fitted models because of lower degree of linear association and assumptions of optimality of linear models not met. Values of $\|R^2\|_1$ measure were high (> 90%) for each of three models.
- Predicted values of export for lower and upper percentiles of crude oil prices (Figure 5.1.2) are close enough. Absolute relative error was minimum for the L_1-regression model. The L_2-(least square) regression model is not the best to predict tail values.
- For the crude oil price and export data series, the L_1-regression model is recommended over the L_2-(least square) regression model and the L_∞-regression model.

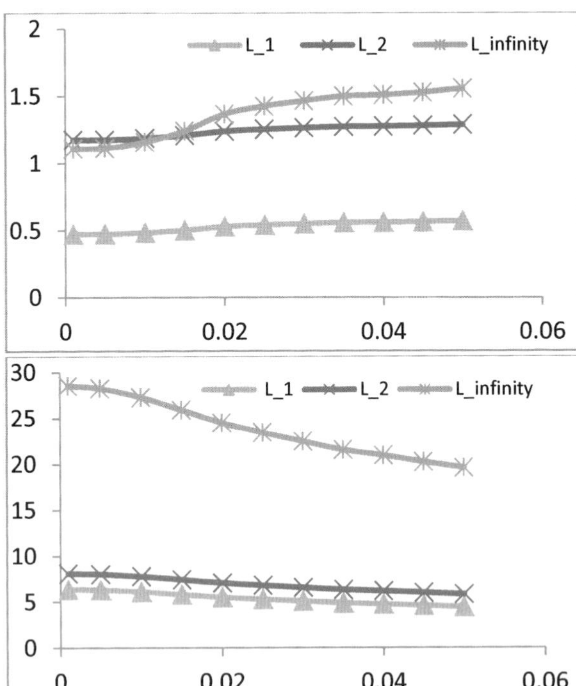

Fig. 5.1.2. Predicted absolute relative errors (%) for lower and upper tails.

5.2. *Comparison of L_p regressions: normal and non-normal populations*

Kumar and Kashanchi [8] investigated the performance of L_p-norm regressions by analyzing six data sets which ranged in size from small ($n = 16$) to large ($n = 64$) and possessed different types of (normal and non-normal) probability distributions. Residual analysis of fitted models indicated:

- Linear relationship: In all cases, residuals are more orless spread randomly about the zero line. This suggests that the assumption that the relationship is linear is reasonable.
- Error Variance: The residuals showed no increasing or decreasing trend and roughly formed a horizontal band around the zero line. This suggests that the variances of residuals were constant.
- Independence of Residuals: The residual lag plot by plotting residual (i) against lagged residual $(i-1)$ indicates the dependency of the residual terms. A randompattern in a lag plot suggests that the residuals are independent. This assumption appears tohold good for all models.
- Normality and Outlier detection: The Shapiro-Wilk statistic and probability values indicated that residuals have normal distributions.

Table 5.2. L_1, L_2 and L_∞-norm regressions and model adequacy measures.

Data Set	L_p Norm	Regression Coefficient	Intercept	$\|R^2\|_1$	R^2
A	L_1	12.2518	15201	**95.72**	60.98
	L_2 (LSE)	10.9820	16310	95.50	**62.62**
	L_∞	8.8329	17423	94.04	38.59
B	L_1	0.6154	7.6923	**86.52**	51.79
	L_2 (LSE)	0.5866	8.9565	86.41	**53.72**
	L_∞	0.6522	8.0495	86.03	52.23
C	L_1	-0.0003	0.3714	**80.11**	5.93
	L_2 (LSE)	-0.0003	0.3749	80.03	2.95
	L_∞	-0.0002	0.3767	78.47	**9.68**
D	L_1	0.0463	0.0475	**83.96**	55.95
	L_2 (LSE)	0.0434	0.1329	83.94	**56.31**
	L_∞	0.0667	-0.8135	79.53	36.03
E	L_1	20.8393	43.8264	**93.76**	27.80
	L_2 (LSE)	16.9229	49.7796	93.67	**29.37**
	L_∞	29.8039	29.6033	93.15	11.55
F	L_1	0.8254	1.0495	**92.47**	35.57
	L_2 (LSE)	0.8337	0.9867	92.42	**35.69**
	L_∞	1.2097	-0.5095	89.24	13.32

In conclusion, it was noted (Table 5.2) that for three data sets, L_2-norm models performed better than L_1 and L_∞-norm regression models. In fourth case, the L_1-norm resulted in the best model, however, $L_2 \approx L_1$ using measures $\| R^2 \|_1$ and

L_∞-norm is the best model according to R^2 criterion. In the last two data sets, L_1-norm is the best model, however, $L_2 \approx L_1$ based on measure $\| R^2 \|_1$. L_2-norm is best according to R^2 criterion, however, $L_1 \approx L_2$. Thus, both estimated models L_1- and L_2 -norm linear regressions are good competitors for these two populations.

5.3. *Kinesiology experiment: physiological processes and anatomy of human body*

Kumar and Singh [10] analyzed the interrelationship of physiological processes and anatomy of the human body with respect to movement for 53 subjects enrolled in the study. Each subject performed a standard exercise at gradually increasing levels and oxygen uptake and expired ventilation were recorded. Both variables showed high correlation and were approximately normal with large variances. It was noted that the predicted expired ventilation values except the tails were close enough to the actual values of the expired ventilation (Tablc 5.3.1).

Table 5.3.1. Absolute relative errors lower- and upper-tails.

Lower-tail (10%)		
L_1	L_2	L_∞
1.1158	1.0262	1.0532
1.1073	1.0007	1.0315
0.9912	0.8802	0.9068
0.9867	0.8784	0.9042
0.9071	0.7963	0.8192
Upper-tail (10%)		
L_1	L_2	L_∞
1.1158	1.0262	1.0532
1.1073	1.0007	1.0315
0.9912	0.8802	0.9068
0.9867	0.8784	0.9042
0.9071	0.7963	0.8192

Absolute relative error analysis of the lower tail (10%) data indicated the model performance (in decreasing order): $L_2 \approx L_1 \approx L_\infty$ and in the upper tail (10%) data: $L_2 \approx L_1 \approx L_\infty \approx L_2$, L_1 (Figure 5.3.2).

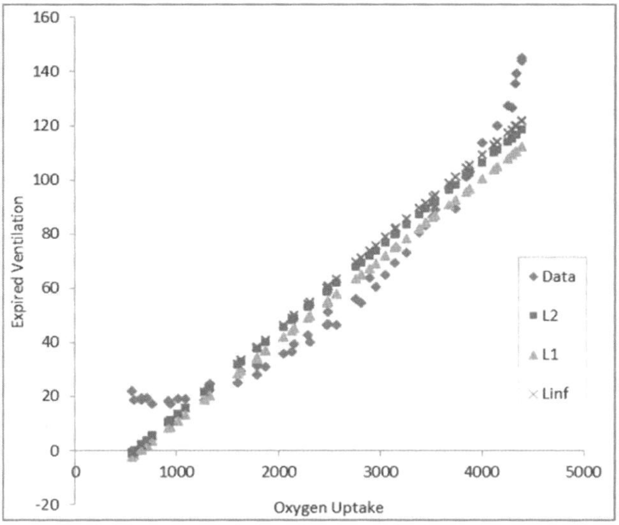

Fig. 5.3.2. L_1, L_2andL_∞-norm regressions.

6. Concluding Remarks

The method of least-squares proposed by Legendre [17] and Gauss (7) is the most popular technique in data analysis for analyzing the linear relationship between response and factors influencing the response. The least squares estimation (LSE) is simple, algebraically highly developed and is implemented using efficient and easy-to-use computer programs. However, studies have shown that the LSE based linear regression may not be the optimal model when one or others of its assumptions fail. We discussed investigations on the comparative performance of the L_1 , L_2 , L_∞-based regressions by considering several bivariate populations representing small to large sizes and normal and non-normal distributions. We have noted that our findings were in agreement with those in some other studies that the L_2-norm (LSE) based linear regression models were not always the best models. Our exploratory studies have indicated need for further investigations on problems like: distributional properties of the L_p-norm regression models; effects of deviatingfrom the assumptions of LSE on the L_p-norm regressions; statistical inference issues, like interval estimation, hypothesis testing and prediction bands. For given application data, howto determine the optimal choice of the L_p-norm.

References

1. R. J. Boscovich, De literaria expeditione per ponti_ciam ditionem et synopsis amplioris operis, *Scientiarum et Artum Instituto atque Academia Commentarii*, **4**, pp. 353–396 (1757).
2. R. J. Boscovich, *De Solis ac Lunae Defectibus* (Paris, 1779).
3. N. Bowditch, Elements of the comet of 1811, *Mme. Amer. Acad. Arts Sci.*, **3**, pp. 313–325 (1815).
4. N. Bowditch, Observations of the comet of 1807, *Mme. Amer. Acad. Arts Sci.*, 3, part 1, pp. 1–17 (1809).
5. F. Y. Edgeworth, On observations relating to several quantities, *Phil. Mag.*, 5th Series, **24**, pp. 222–223 (1887).
6. R. W. Farebrother, *Fitting Linear Relationships: A History of the Calculus of Observations 1750–900* (Springler-Verlag, New York Inc, 1999).
7. C. F. Gauss, *Theoria motus corporum coellestium in sectionibus conicis solem ambientium* (Hamburg, 1809).
8. P. Kumar and F. Kashanchi, Linear Regression Models Using L_1, L_2, L_∞-Norms, *Proceedings of COMPSTAT-2014*, 97–102 (2014).
9. P. Kumar and F. Kashanchi, On the L_p-norm regression models for estimating value-at-risk, *Serdica J. Computing*, **8**, 3, 101–114 (2014).
10. P. Kumar and J. N. Singh, Regression Model Estimation Using Least Absolute Deviations, Least Squares Deviations and Minimax Absolute Deviations Criteria, *Int. Jour. of Comp. Sci. and Electronics Eng. (IJCSEE)*, **3**, 4 (2015).
11. P. S. Laplace, *Sur quelques points du système du monde* (Mme. De l'Academie Royale des Sciences de Paris, 1793).
12. P. S. Laplace, *Théorie de Analytique des Probabilités* (Courcier, Paris, 1812).
13. P. S. Laplace, *Treatié des Méchanique Céleste* (Tome II, J. B. M. Dubrat, Paris, 1799).
14. P. S. Laplace, *Deuxieme Supplement to Laplace-1812* (1818).
15. P. S. Laplace, *Mémoir sur la probabilitiés des causes par les événements* (Mme. De l'Academie Royale des Sciences de Paris 6, 1788).
16. P. S. Laplace, *Theorie de Jupiter et de Saturne* (Mme. De l'Academie Royale des Sciences de Paris, 1788).
17. A. M. Legendre, *Nouvelles methodes pour la determination des orbites des cometes* (Mme. Courcier, Paris, 1805).
18. J. T. Mayer, Abhandlung über die Umwälzung des Monds um seine Axe, *Kosmographische Nachrichten und Sammlungen*, **1**, 52–183, (1748).

19. H. Nyquist, *Recent Studies on L_p-norm Estimation* (Doctoral thesis, University of Umea, 1980)
20. S. M. Stigler, *The History of Statistics: The Measurement of Uncertainty Before 1900* (Harvard University Press, 1990).
21. L. N. Vaserstein, *Introduction to Linear Programming* (Prentice Hall, 2003).

The Impact of Family Planning on Fertility in Jharkhand State

Dilip Kumar
Population Research Centre
Department of Statistics, Patna University
Patna-800005, Bihar (India)
E-mail: dilip360@gmail.com

In the present study an attempt has been made to evaluate the impact of family planning on fertility in Jharkhand State through the Prevalence Model. If prevalence levels of both programme and non-programme contraception are known, this technique permits the estimation of gross natural and potential fertility for assessing births averted. With the emergence of the National Family Health Survey (NFHS: 2005-06) to monitor family planning and health activities, this method becomes a useful tool. Of special, interest is the ability of the procedure to yield estimates by age group as well as by type of contraceptive methods used. In the study, the standard method-specific use-effectiveness levels weight observed use and prevalence level by method have been applied. Of the total births averted in Jharkhand State by programme contraception, 80 per cent of births were averted by Sterilization users in 2005–06 while the spacing methods users contributed to about 20 per cent of the birth prevention. The spacing methods need to be strengthening for the greater use. With regard to the births averted by non-programme contraception/natural methods, the main contribution was made by the users of Rhythm of about 45 per cent, which was followed by the users of Withdrawal of 43 per cent and by the other methods of 13 per cent. Of the total birth averted in Jharkhand State, the contribution of programme contraception and non-programme contraception/natural methods is about 85 per cent and 15 per cent respectively in 2005–06. The programme contraception has the dominance role to control fertility however the non-programme contraception/natural methods use should also be enhanced at the places where accessibility of programme contraception is poor.

Keywords: Births Averted; Prevalence Model; Potential Fertility Rate; Natural Fertility Rate; Programme Contraception; Non-Programme Contraception/Natural Method.

1. Background

With the rapid growth of national family planning programme in 1990 onward, family planning administrators and policy makers felt an increasing need to evaluate the fertility impact of the programme yearly. For such, probably the most widely used measure of fertility impact is the number of births averted by a programme in a given year. An estimation of the number of births averted is typically obtained by subtracting observed fertility (i.e. fertility with programme) from potential fertility (i.e. fertility without the programme) and multiplying the differences by the appropriate base of the population. Since potential fertility is an unobservable quantity, it has to be estimated indirectly. Gross and net potential fertility are two major types of potential fertility used in the study. The gross potential fertility is the fertility that would prevail if all use of programme contraception were eliminated, without switching to the non-programme contraception. The net potential fertility is the fertility that actually would be observed if there had never been a programme. In that case, many who would have been programme users would have obtained supplies from non-programme sources. This substitution would tend to make net potential fertility lower than the gross potential fertility.

The use of contraception greatly reduces the chances of conceiving, but except in the case of sterilization, the chance is not zero. To take this contraceptive failure into account, some methods introduce a penalty for accidental failure. The simplest way to do this consists of multiplying potential fertility by 'e', the contraceptive use-effectiveness. In the present study Prevalence Model has been used in the State of Jharkhand to evaluate the number of births averted by programme and non-programme contraception efforts.

2. Prevalence Model

The prevalence model is based on age-specific and method specific prevalence rates of both programme and non-programme contraceptives. This information helps in getting the gross potential fertility and subsequently the number of births averted by programme and non-programme contraceptive efforts.

Age-specific model: The age-specific model requires information on prevalence of programme and non-programme contraception by age, age-specific fertility rates which have been taken from the NFHS data of Jharkhand State. The Census figures have been used for the female population of reproductive age. Bongaarts, J (1993) provides new estimates of gross and net impact on fertility reductions from family planning (FP) programs for 31 developing countries in Africa, Latin America, and Asia. A comparison is made of net and gross measures, and the interaction with the level of development is identified. The conclusion is reached that FP has been crucial in reducing fertility in many countries. Without FP, the total annual number of births in the late 1980s would have been 164 million instead of 120 million. There is no agreed upon measure for determining the impact of FP on fertility and estimates has ranged from 3–40%. Discrepancies in results are due to the use of multiple methodologies, of which some are unsuitable or unreliable for normal evaluation due to difference research objectives and due to conceptual differences in measurement of gross versus net impact. Gross impact refers to the reduction due to the use of contraception available from program sources. Net impact measures the reduction achieved by the presence of the program. Net and gross impact varies within each country, with net impact the smaller of the two. Gross impact usually was measured with statistics on acceptors. The measures of gross potential fertility, gross natural fertility and births averted are obtained as follows:

$$NAF_a = AF_a / \{1 - C_a (u_a' + u_a'')\} \tag{1}$$

$$PAF_a = AF_a \{1 - C_a (u_a'')\} / \{1 - C_a (u_a' + u_a'')\} \tag{2}$$

$$BA_{a=}(PAF_a - AF_a) POP_a \tag{3}$$

$$BAN_a = (NAF_a - PAF_a) POP_a \tag{4}$$

Where;

a = age group of women, a = 15–19.
u_a' = prevalence of programme contraception, by age.
u_a'' = prevalence of non-programme contraception, by age.
AF_a = age-specific fertility rate.
PAF_a = potential age-specific fertility rate.
NAF_a = natural age-specific fertility rate.
BA_a = birth averted by programme contraception, by age.
BAN_a = birth averted by non-programme contraception, by age.
POP_a = number of women in age group a.
C_a = elasticity coefficient by age.

In order to estimate the gross potential fertility and natural fertility, information on elasticity coefficient of sterility and use-effectiveness by age of women is utilized.

C (15 – 19) = 0.620
C (20 – 24) = 0.620
C (25 – 29) = 0823
C (30 – 34) = 0.940
C (35 – 39) = 1.022
C (40 – 44) = 1.309
C (45 – 49) = 1.898

The method-specific model: It drives the number of births averted by each programme and non-programme method through the data on prevalence and use-effectiveness of contraception for both sectors. Estimates of births averted are obtained by the following equations:

$$BA_m = BA.\ u_m'\ .e_m' / (u'\ .e') \tag{5}$$

$$BAN_m = BAN.\ u_m''\ .e_m'' / (u''.e_m'') \tag{6}$$

Where;

u_m' = prevalence of programme method 'm'.
u_m'' = prevalence of non-programme method 'm'.
e_m' = use-effectiveness of programme method 'm'.
e_m'' = use-effectiveness of non-programme method 'm'.

$$u' = \sum_m u_m'$$

$$u'' = \sum_m u_m''$$

$$e' = \sum_m u_m'\ .e_m' / u'$$

$$e_m'' = \sum_m u_m''.e_m'' / u''$$

3. Data Estimates

National Family Health Survey (NFHS-3) is the third in the NFHS series of surveys in 2005–06. The first NFHS was conducted in 1992–93, and the second (NFHS-2) was conducted in 1998–99. All three NFHS surveys were conducted under the stewardship of the Ministry of Health and Family Welfare (MoHFW), Government of India, New Delhi.

In Jharkhand, NFHS-3 was based on a sample of 2,483 households that is representative at the State level and within the State at the urban and rural levels. Women age 15–49 were eligible for interview in all NFHS-3 sample households, while men age 15–54 were eligible for interview in a subsample of households. NFHS-3 interviewed 2,983 women age 15–49 and 996 men age 15–54 in Jharkhand State to obtain information on population, health, and nutrition.

The data of the National Family Health Survey (NFHS) in Jharkhand State during 2005–06 were utilized to estimate the number of births averted by programme and non-programme sources separately. The prevalence information by age and method was obtained for currently married females who are currently using and contraceptive method. The female population of reproductive ages 15–49 years was estimated from the 2001 population census figures.

The data problem arose when programme and non-programme methods were to be sorted out. Some of the programme methods and contraceptive services, such as condoms and pills, are available outside the programme at the private clinic. However, the data of such services are not available. It was thus assumed that all modern methods were offered by the programme and were termed programme methods. All traditional methods were considered non-programme methods. This assumption is fairly reasonable because modern contraceptives, such as condoms and pills, are widely distributed through the programme.

4. Results

The application of the pertinent formulae yielded the estimates of gross natural fertility and gross potential fertility of Jharkhand State (Table 1). The difference between gross potential and gross natural fertility, on the one hand, and observed fertility, on the other hand, provided the basis for estimating births averted by programme and non-programme contraception for the State (Table 2). The results show that young fertile women avert the majority of births by non-programme methods compared to the programme methods. The births averted by programme contraception are concentrated among women aged 20–29 years in the State. The findings confirm an earlier study where mean age of use was found to be high in the early thirties (Kumar, D; 1990). It is noted that, in general, the effectiveness or impact of the program resembles a bell-shaped curve, i.e., in the initial phases pregnancy reduction increased to reach a plateau and then declined in the remaining phases. This may represent a cyclical occurrence and pregnancy reduction may again increase. Continual follow-up is necessary for an extended time period to analyze any additional trends in fertility reduction.

5. Summary

The results presented in Table 3 summarize the outcome of the study. The silent feature of the study is that it is not based on the service statistics and most of the data are obtained from the State level survey. In the study, the standard method-specific use-effectiveness levels weight observed use and prevalence level by method. Of the 179076 births averted in Jharkhand State by programme contraception 143598 (80.2 per cent) births were averted by sterilization users in 2005–06 while the spacing methods users contributed to only 19.8 per cent of the birth prevention. The spacing methods need to be strengthen for the greater use. With regard to the 31015 births averted by non-programme contraception, the main contribution was made by the users of Rhythm (44.7 per cent) that was followed by the users of Withdrawal (42.6 per cent) and by the other methods (12.8 per cent). Of the total birth averted in Jharkhand State, the contribution of programme contraception and non-programme contraception is about 85 per cent and 15 per cent in 2005–06. The programme contraception has the dominance role to control fertility however the non-programme contraception use should also be enhanced at the places where accessibility of programme contraception is poor.

Table 1. Prevalence of Programme and non-programme contraception, observed fertility rates and estimated natural and gross fertility rates by age group in Jharkhand State, 2005–06.

Age group	Prevalence of		Observed fertility, 2005	Elasticity coefficient			Natural fertility rate	Gross potential fertility rate
	Programme contraception	Non-programme contraception			$1 - C_a$ $(u_a' + u_a'')$	$1 - C_a$ (u_a'')		
a	u_a'	u_a''	AF_a	C_a			NAF_a	PAF_a
15-19	4.3	2.4	53.7	0.620	0.95846	0.98512	56.0	55.2
20-24	17.4	3.6	249.3	0.620	0.8698	0.97768	286.6	280.2
25-29	31.8	6	181.2	0.823	0.688906	0.95062	263.0	250.0
30-34	19.1	2.6	116.6	0.940	0.79602	0.97556	146.5	142.9
35-39	24.8	3.4	53.2	1.022	0.711796	0.96525	74.7	72.1
40-44	22.9	1.1	35.2	1.309	0.68584	0.98560	51.3	50.6
45-49	20.9	3.3	18.9	1.898	0.540684	0.93737	35.0	32.8

Table 2. Gross fertility and gross birth averted by the programme and non-programme contraception by age group in Jharkhand State, 2005–06.

Age group	Female Population, 2006	Gross fertility effect of:		Birth averted by	
		Programme Contraception	Non-programme contraception	Programme methods	Non-programme methods
a	POP_a	PAF_a- AF_a	NAF_a -PAF_a	BA_a	BAN_a
15-19	1586000	0.00149	0.00083	2369(1.3%)	1322(4.3%)
20-24	1238000	0.03092	0.00640	38279(21.4)	7920(25.5%)
25-29	1055000	0.06884	0.01299	72624(40.6%)	13703(44.2%)
30-34	997000	0.02630	0.00358	26220(14.6%)	3569(11.5%)
35-39	929000	0.01894	0.00260	17598(9.8%)	2413(7.8%)
40-44	818000	0.01538	0.00074	12585(7.0%)	605(1.9%)
45-49	678000	0.01387	0.00219	9401(5.2%)	1484(4.8%)
Total	7301000			179076(100.0%)	31015(100.0%)

Table 3. Estimated birth averted by programme and non-programme contraception in Jharkhand State, 2005–06.

Methods	Prevalence of		Use effectiveness of		Estimated birth averted by	
	Programme contraception	Non-programme contraception	Programme contraception	Non-programme contraception	Programme contraception	Non-programme contraception
	u_m'	u_m''	e_m'	e_m''	BA_m	BAN_m
Oral pills	3.8		0.9		20635 (11.5%)	
IUD	0.6		0.95		3439 (1.9%)	
Condom	2.7		0.7		11403 (6.4%)	
Tubectomy	23.4		1.0		141185 (78.8%)	
Vasectomy	0.4		1.0		2413 (1.3%)	
Rhythm		2.1		0.5		13858 (44.7%)
Withdrawal		2		0.5		13198 (42.6%)
Other methods		0.6		0.5		3959 (12.8%)
Total	30.9	4.7			179076 (100.0%)	31015 (100.0%)

Acknowledgements

The author is thankful to the Ministry of Health and Family Welfare, Government of India, New Delhi for given the permission and regular fund for completion of this study.

References

1. Bongaarts, John, A Prevalence Model for Evaluating the Fertility Effect of Family Planning Programmes: Age-specific and Method specific Results, Studies to Enhance the Evaluation of Family Planning Programmes, United Nations, New York, ST/ESA/SER.A/87, pp. 246 (1985).
2. Bongaarts, John, The fertility impact of family planning programs', New York, Population Council, Research Division Working Papers No. 47, pp. 35 (1993).
3. Census of India, 2001, Age Level Data of the States of India, India Level (In CD).

4. International Institute for Population Sciences (IIPS) and ORC Macro. 2008. *National Family Health Survey (NFHS-3), India, 2005–06: Jharkhand.* Mumbai: IIPS.
5. Kumar, Dilip, Evaluation of Family welfare and MCH Programmes in Some PHCs of Patna district, Population Research Centre, Patna University, Patna, PRC Mimeograph Series No. 118, pp. 91 (1990).
6. Ministry of Health and Family Welfare, Government of India, New Delhi 'Family Welfare Statistics in India, 2009.
7. Potter, Robert G, Estimating births averted in a family planning program, in S. J. Behrman, Leslie Corsa, and Ronald Freedman (eds.), Fertility and Family Planning: A World View. Ann Arbor: University of Michigan Press, pp. 413–434(1969).

Spatial Analysis of AFP Surveillance Strategy for Polio Eradication in India

Pankaj Srivastava

Ministry of Corporate Affairs
Govt. of India, New Delhi-110001, India
E-mail: Pankaj.srivastava@gov.in

Arun Kumar Sinha

Department of Statistics,
Central University of South Bihar, Patna-800014, India
E-mail: arunkrsinha@cub.ac.in

This paper demonstrates the application of spatial statistical tools for effective governance by way of designing an epidemiologically guided public healthcare system for eradication of poliomyelitis from India. Spatial scan statistic tools detect high priority areas or 'hotspots' for focused management response to interrupt the transmission of wild poliovirus in Bihar, India. This paper highlights probable lacunae in the existing acute flaccid paralysis surveillance as well as suggests a modified surveillance mechanism while detecting 'hotspot' districts in Bihar for effective management intervention for interrupting the transmission of wild poliovirus in the state.

Keywords: Spatial scan statistic, poliomyelitis, wild poliovirus, acute flaccid paralysis surveillance, geographical hotspot, SaTScan.

1. Introduction

Childhood immunization is one of the most cost-effective health interventions. Government of India is committed to the reduction of morbidity and mortality due to vaccine preventable diseases and to the establishment of reliable surveillance for vaccine preventable diseases. The 41st World Health Assembly in May 1988, passed a resolution calling for the global eradication of poliomyelitis (or, polio) by the year 2000. However, the status could not be achieved till 2010 and wild polioviruses are still found in approximately 10 countries in the world including India. The last polio case was detected in January 2011. India was removed from the list of polio endemic countries by WHO in February 2012.

1.1. *Epidemiology of polio*

Polio is a highly infectious disease caused by a virus that invades the nervous system, and can cause total paralysis in a matter of hours. The polioviruses belong to the genus *Enterovirus* in the family *Picornaviridae* and comprise three closely related serotypes that can cause paralysis. These three serotypes of poliovirus, viz., wild poliovirus type-1 (WPV1), type-2 (WPV2) and type-3 (WPV3) have slightly different protein composition. In tropical countries like India, seasonality of poliovirus is less clearly defined and it occurs more commonly in summer and early autumn in temperate climates. Transmission of poliovirus is primarily person-to-person via the faecal-oral route. Poliovirus multiplies in the intestines and is spread through the feces. The virus is intermittently excreted for up to 2 months or more after infection, with maximum excretion occurring just before paralysis and during the first 14 days after onset of paralysis. On the average, the incubation period from exposure to the virus to the onset of first symptoms is 7–10 days. Poliovirus is highly communicable and all un-immunized persons are susceptible to poliomyelitis, see Park (2005).

1.2. *Acute flaccid paralysis surveillance*

Surveillance for polio is conducted through investigation of Acute Flaccid Paralysis (AFP) cases. AFP is defined as sudden onset of weakness and floppiness in any part of the body in a child less than 15 years of age. AFP Surveillance indicates a criterion used to decide whether or not a suspected case of poliomyelitis should undergo epidemiological and laboratory investigation. All cases of AFP are reported and their stools specimen collected within 14 days of onset of paralysis for laboratory investigations. Within 90 days of paralysis onset, all cases should undergo final classification, either as confirmed polio, non-polio AFP (NPAFP) or compatible with poliomyelitis. If stool specimens are inadequate, final classification of the AFP case as either "NPAFP" or "compatible with polio" will depend on the results of 60 days follow-up examination. If the 60 days follow-up examination shows no residual weakness, the case is classified as "NPAFP". The cases are discarded as NPAFP only if it is confident that the case is not "compatible with polio". NPAFP rate is actually the incidence of AFP caused by diseases other than polio. If any doubt exists with respect to classification of the case, then it is classified as "compatible with polio".

The rate at which NPAFP cases are detected over time in each geographic area also helps to assess the sensitivity of the surveillance system. The rate at which NPAFP cases occur in a community is called the NPAFP rate. If NPAFP rate is "less than two per one lakh children in India", then the surveillance system is probably missing cases of AFP. (WHO, 1997; Field Guide, Govt of India, 2005).

$$\text{NPAFP Rate} = \frac{\text{Reported NPAFP cases of children} < 15 \text{ Years age}}{\text{Total Children} < 15 \text{ Years Age}} \times 100000 \qquad (1)$$

For all reported AFP cases, "2 stool specimen" are to be collected by the 14th day after the onset of paralysis to maximize the chances of isolating poliovirus. The adequate stool specimen collection should be at least 80 per cent, i.e, at least for 80 per cent of all reported AFP cases, '2 stool specimens' should be collected before 14th day of the onset of paralysis and send to laboratory for investigation/confirmation in a good condition. Adequate Stool Rate has been defined as

$$\textit{Adequate Stool Rate} = \frac{\#\ \textit{Adequate stool specimen}}{\textit{Total AFP cases}} \times 100 \qquad (2)$$

2. Prevalence of Polio in India

Till 2009, India was the most populous country where transmission of WPV has never been interrupted. Chart 1 below shows the prevalence of polio due to strains of WPV1 and WPV3.

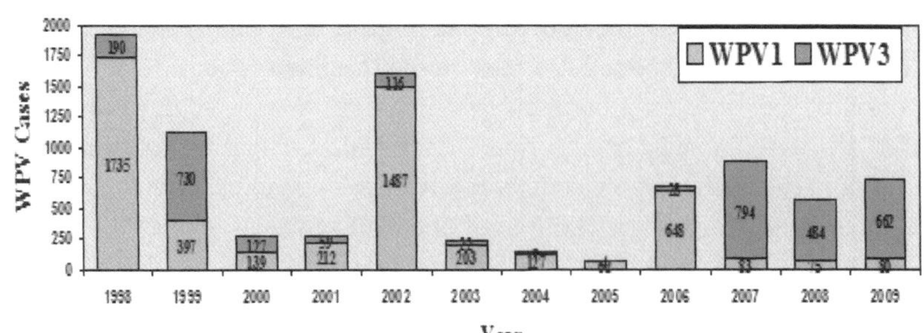

Chart 1. Incidences of WPV cases in India

WPV transmission persists in two of India's northern states of Bihar and Uttar Pradesh, which account for approximately 94 per cent of all reported WPV cases in the country during the past 5 years (WHO Reports, 2009). The numbers of polio cases in Bihar and Uttar Pradesh vis-à-vis India have been given in Table 1 below.

Table 1. WPV cases in Bihar and Uttar Pradesh vis-à-vis all India.

Year	Bihar	Uttar Pradesh	All India
1999	115	757	1126
2000	49	178	265
2001	27	216	268
2002	121	1242	1600
2003	18	88	225
2004	39	82	134
2005	30	29	66
2006	61	548	676
2007	503	341	874
2008	233	305	559
2009	117	602	741

3. Present Study and Data Sources

The present study aims to assess and review whether AFP surveillance is functioning adequately at district level and in all geographical areas of Bihar, especially in areas considered at high-risk of continuing virus transmission. We also aim to detect spatial 'hot-spots' or high priority districts in the state of Bihar for polio eradication program in India. District wise AFP surveillance related data in Bihar for the year 2009 as below in Table 2 have been used to demonstrate the tools and methodologies developed in the present study (NPSP, 2009).

Table 2. District wise AFP surveillance parameters in Bihar, 2009.

District	AFP	WPV	Adequate Stool Rate	NPAFP Rate	Polio Compatible Cases
Araria	342	2	86.55	32.35	66.67
Arwal	134	0	90.3	46.84	100
Aurangaabad	277	0	81.95	27.23	100
Banka	190	0	93.16	22.58	100
Begusarai	427	13	81.5	38.71	45.83
Bhagalpur	374	2	85.29	31.99	60
Bhojpur	275	0	91.27	26.13	-
Buxar	183	0	93.44	26.26	100
Champaran(e)	358	0	83.24	17.42	100
Champaran(w)	355	0	82.82	20.86	100
Darbhanga	683	14	81.41	42.51	51.72
Gaya	532	0	89.66	30.62	100
Gopalganj	312	0	86.22	25.04	100
Jamui	278	1	95.32	42.96	50
Jehanabad	244	0	90.98	48.63	100
Kaimur	166	0	93.37	23.73	-
Katihar	278	2	76.98	24.53	75
Khagaria	371	16	80.05	54.14	30.43
Kishanganj	141	0	83.69	22.4	100

Table 2. (*Continued*)

District	AFP	WPV	Adequate Stool Rate	NPAFP Rate	Polio Compatible Cases
Lakhisarai	267	1	92.51	72	0
Madhepura	416	7	91.35	50.78	53.33
Madhubani	597	0	81.91	29.7	100
Munger	305	0	84.26	57.97	100
Muzaffarpur	378	1	85.45	20.89	83.33
Nalanda	486	1	90.33	41.69	85.71
Nawada	300	0	87.33	33.16	100
Patna	952	14	84.66	47.66	17.65
Purnia	237	0	86.08	16.85	100
Rohtas	319	0	84.64	25.2	100
Saharsa	414	23	81.4	46.35	41.03
Samastipur	806	16	81.64	47.42	44.83
Saran	432	0	81.71	25.76	100
Sheikhpura	143	1	95.8	52.63	50
Sheohar	68	0	88.24	23.22	100
Sitamarhi	429	0	79.95	32.57	100
Siwan	343	0	90.67	24.13	100
Supaul	333	3	90.09	33.02	62.5
Vaishali	343	0	88.92	24.31	100

Source: National Polio Surveillance Project.

4. Tools and Methodology

The high priority districts as per polio incidences reported in a year were identified spatial scan statistics (Kulldorff and Nagarwala 1995) using the SaTScan. The high priority districts with high WPV rates vis-à-vis total AFP cases have identified with purely spatial discrete Poisson distributional assumption. Under the null hypothesis, the expected number of WPV cases in each district is proportional to the number of AFP cases in the district. The likelihood ratio test statistic for the null hypothesis has been derived for each district and for varying size of scanning window. For each such district and scanning window, the alternative hypothesis is that there is an elevated risk within the window as compared to outside. Under the Poisson assumption, the likelihood function for a specific window is proportional to:

$$\left(\frac{c}{E[c]}\right)^{c}\left(\frac{C-c}{C-E[c]}\right)^{C-c} I(\,) \tag{3}$$

where C is the total number of WPV cases in Bihar in a year, c is the observed number of cases within the window, and E[c] is the expected number of cases within the window under the null-hypothesis. I(.) is an indicator function. Since the analysis is conditioned on the total number of cases observed, C-E[c] is the expected number of cases outside the window. SaTScan has been set to scan only for clusters with high rates, and thus I(.) is equal to 1 when the window has more cases than expected under the null-hypothesis.

Discrete Poisson model requires confirmed wild poliovirus cases and the total AFP cases in the particular district during year 2009 along with geographical coordinates (latitude and longitude) for each district as given in Table 3. These basic inputs are provided to SaTScan software in a text format as coordinates, case and population files.

Table 3. District wise WPV cases and total AFP cases in Year 2009.

District	Latitude	Longitude	Year	WPV Cases	AFP Cases
Araria	26.09	87.31	2009	2	342
Arwal	25.14	84.40	2009	0	134
Aurangaabad	24.46	84.23	2009	0	277
Banka	24.53	86.56	2009	0	190
Begusarai	25.25	86.08	2009	13	427
Bhagalpur	25.14	86.59	2009	2	374
Bhojpur	25.34	84.40	2009	0	275
Buxar	25.35	84.00	2009	0	183
Champaran (e)	26.40	84.55	2009	0	358
Champaran (w)	26.49	84.30	2009	0	355
Darbhanga	26.10	85.54	2009	14	683
Gaya	24.48	85.00	2009	0	532
Gopalganj	26.28	84.26	2009	0	312
Jamui	24.55	86.13	2009	1	278
Jehanabad	25.13	84.59	2009	0	244
Kaimur	25.04	83.39	2009	0	166
Katihar	25.33	87.34	2009	2	278
Khagaria	25.31	86.27	2009	16	371
Kishanganj	26.06	87.57	2009	0	141
Lakhisarai	25.12	86.06	2009	1	267
Madhepura	25.55	86.47	2009	7	416
Madhubani	26.27	85.08	2009	0	597
Munger	25.24	86.29	2009	0	305
Muzaffarpur	26.07	85.23	2009	1	378
Nalanda	25.07	85.25	2009	1	486
Nawada	24.54	85.33	2009	0	300
Patna	25.37	85.13	2009	14	952
Purnia	25.47	87.31	2009	0	237
Rohtas	24.36	83.59	2009	0	319
Saharsa	25.54	86.36	2009	23	414
Samastipur	25.52	85.47	2009	16	806
Saran	24.29	86.19	2009	0	432
Sheikhpura	25.09	85.50	2009	1	143
Sheohar	26.31	85.18	2009	0	68
Sitamarhi	26.36	85.30	2009	0	429
Siwan	26.14	84.21	2009	0	343
Supaul	26.07	86.36	2009	3	333
Vaishali	25.41	85.13	2009	0	343

5. Steps for Executing SaTScan Software

There are four main steps. Step-1 deals with selecting appropriate input parameters and providing data files. Step-2 relates to selecting appropriate probability models, scan analysis and area. Step-3 addresses to prescribing the result formate while Step-4 provides the output of the analysis.

The SatScan output gives a summary of data (here, number of locations, total number of cases and population under risk, Analysis (here, purely discrete spatial Poisson, scaning for high rates), type of Spatial Window (here, circular with maximum cluster size of 50 per cent of population at risk), Data checking & Inference (here, to ensure that all cases and controls are within the specified temporal study period and ensure that all observations (cases, controls and populations) are within the specified geographical area) and Most Likely Cluster (here, districts, their coordinates/radius, their Relative risk, Log likelihood ratio, Monte Carlo rank, P-value, etc.). Districts Supaul, Madhepura, Saharsa, Darbhanga, Khagaria, Munger, Araria, Begusarai, Bhagalpur and Samastipur constitute the hot-spot for getting WPV cases in Bihar as per 2009 data as presented in Fig. 4. The detailed output is provided at Appendix-1.

Fig. 4. WPV Hot-Spot districts in Bihar, 2009

6. Findings and Analysis

The study evaluates the performance of AFP surveillance strategy for eradicating polio in state of Bihar. AFP Surveillance Indicators like total AFP cases, Total WPV, Adequate Stool Rate, Polio compatible cases and Non-Polio AFP Rate for year 2009 were analyzed using exploratory data analysis tools. The AFP data for these districts were categorized as per Table 4 below into four mutually exclusive and exhaustive classes based upon the severity of polio in the districts.

Table 4. Categorization of polio endemic districts.

Category of WPV	Description	Endemic status
Category-1	No WPV cases	No risk Area
Category-2	1-10 WPV cases	Low Risk Area
Category-3	11-20 WPV cases	Moderate Risk Area
Category-4	Above 20 WPV cases	High Risk Area

AFP Surveillance indicators for the year 2009, we observe in Table. 5 below that there are 22 districts without any WPV case and only 6 districts having more than 10 WPV cases during the year. We also observe that overall average of Adequate Stool Rate in all the four Categories of districts have been higher than the benchmark of 80 per cent.

Table 5. Analysis for AFP indicators, 2009.

Variable	Category	Districts	Per cent	Mean	Minimum	Median	Maximum
Total AFP	1	22	57.89	297.30	68	302.50	597
	2	10	26.32	329.50	143	337.50	486
	3	5	13.16	648.00	371	683.00	952
	4	1	2.63	414.00	414	414.00	414
Adequate Stool Rate	1	22	57.89	86.99	79.95	86.78	93.44
	2	10	26.32	88.97	76.98	90.21	95.80
	3	5	13.16	81.85	80.05	81.50	84.66
	4	1	2.63	81.40	81.40	81.40	81.40

The analysis of AFP cases in Table 6 below indicate that Patna has disproportionately high AFP case with respect to only 14 WPV cases, and closely followed by 806 AFP cases in Samastipur (WPV-16), 683 AFP cases in Darbhanga (WPV-14). We observe that AFP surveillance has been sub-optimal in high–risk districts of Khagaria (with 371 AFP cases for 16 WPV) and Saharsa (with 414 AFP Cases for 23 WPV cases). We also observe that the only Category-4 district, Saharsa has only 414 AFP cases for 23 WPV cases indicating a weak surveillance in Saharsa resulting into a relatively higher WPV cases in year 2009. We also note that Adequate Stool Rate in Category-3 and 4 districts, although above the benchmark figure of 80 per cent, is much lower than those in Category-1 and 2 districts. It indicates that although AFP surveillance has improved in all polio endemic districts, the priority area need to be further strengthened to stop WPV transmission.

Table 6. AFP surveillance indicators of districts, 2009.

Category	District	AFP	WPV	Adequate Stool Rate	Polio Compatible Cases	NPAFP Rate
1	Arwal	134	0	90.30	2	46.84
	Aurangaabad	277	0	81.95	3	27.23
	Banka	190	0	93.16	1	22.58
	Bhojpur	275	0	91.27	0	26.13
	Buxar	183	0	93.44	2	26.26
	champaran(e)	358	0	83.24	5	17.42
	champaran(w)	355	0	82.82	8	20.86
	Gaya	532	0	89.66	1	30.62
	Gopalganj	312	0	86.22	2	25.04
	Jehanabad	244	0	90.98	2	48.63
	Kaimur	166	0	93.37	0	23.73
	Kishanganj	141	0	83.69	4	22.40
	Madhubani	597	0	81.91	14	29.70
	Munger	305	0	84.26	3	57.97
	Nawada	300	0	87.33	6	33.16
	Purnia	237	0	86.08	2	16.85
	Rohtas	319	0	84.64	3	25.20
	Saran	432	0	81.71	1	25.76
	Sheohar	68	0	88.24	1	23.22
	Sitamarhi	429	0	79.95	11	32.57
	Siwan	343	0	90.67	4	24.13
	Vaishali	343	0	88.92	4	24.31
2	Araria	342	2	86.55	4	32.35
	Bhagalpur	374	2	85.29	3	31.99
	Jamui	278	1	95.32	1	42.96
	Katihar	278	2	76.98	6	24.53
	Lakhisarai	267	1	92.51	0	72.00
	Madhepura	416	7	91.35	8	50.78
	Muzaffarpur	378	1	85.45	5	20.89
	Nalanda	486	1	90.33	6	41.69
	Sheikhpura	143	1	95.80	1	52.63
	Supaul	333	3	90.09	5	33.02
3	Begusarai	427	13	81.50	11	38.71
	Darbhanga	683	14	81.41	15	42.51
	Khagaria	371	16	80.05	7	54.14
	Patna	952	14	84.66	3	47.66
	Samastipur	806	16	81.64	13	47.42
4	Saharsa	414	23	81.40	16	46.35
	Total	13488	117	86.69	183	34.48

A combined study of the logistics & management scenario of the AFP surveillance using scan statistics and exploratory data analysis tools indicates that high endemic districts are primarily located in Central Bihar that is affected by perennial flood. These are hard-to-reach areas where almost all socio-economic development indicators like income, nutrition, education, sanitation, etc are far below the national average. The difficult terrain and seasonally inaccessible areas pose serious logistical & management challenges to implement polio eradication programme in these districts. The emerging pattern of polio transmission in Bihar suggests that such hot-spot districts are essentially falling in *Kosi* river-bed area that has conducive socio-economic indicators for WPV transmission in the area.

We observe that Patna (WPV-14) has been excluded from the spatial hot spot whereas low endemic districts like Munger (WPV-0), Supaul (WPV-3), Araria (WPV-2) and Bhagalpur (WPV-2) have been included in the hot-spot districts. Missing Patna with high AFP cases as well as high WPV cases indicates high dependence of spatial scan statistical tools on cases in adjoining districts for detecting a hot-spot zone. It may be construed as a limitation of spatial scan statistics tools to detect hot spots with respect to poliovirus transmission. Further, classification of low incidence districts like Munger, Araria, Bhagalpur and Supaul are detected as hot-spots, also suggests that pure spatial distributional assumption in the SaTScan may be confirmed with temporal and space-time analysis.

References

1. Field Guide: Surveillance of Acute Flaccid Paralysis, Govt of India, Ministry of Health & Family Welfare, New Delhi (2005).
2. K. Park: Park's Textbook of Preventive and Social Medicine, 18th Ed. (2005).
3. M. Kulldorff and N. Nagarwal: Spatial disease clusters: Detection and inference, Statistics in Medicine, 14, 799-810(1995).
4. WHO: Field guide for supplementary activities aimed at achieving polio eradication. National Polio Surveillance Project - India, AFP Surveillance bulletin (various) (1997).
5. WHO: SEARO: Immunization and vaccine development, VPD surveillance bulletins (2009).

Appendix-1

1. SUMMARY OF DATA

Study period..................: 2009/1/1 - 2009/12/31
Number of locations...........: 38
Total population..............: 13488
Total number of cases.........: 117
Annual cases / 100000.........: 868.0

2. Analysis

Type of Analysis : Purely Spatial
Probability Model : Discrete Poisson
Scan for Areas with : High Rates

3. Input

Case File : E:\SatScan data\WPVcases.txt
Population File : E:\SatScan data\AFP Cases.txt
Coordinates File : E:\SatScan data\Lat long.txt
Time Precision : Year
Start Time : 2009/1/1
End Time : 2009/12/31
Coordinates : Latitude/Longitude

4. Data Checking

Temporal Data Check : Check to ensure that all cases and controls are within the specified temporal study period.

Geographical Data Check: Check to ensure that all observations (cases, controls and populations) are within the specified geographical area.

5. Spatial Window

Maximum Spatial Cluster Size : 50 per cent of population at risk
Window Shape : Circular
Isotonic Scan : No

6. Inference

P-Value Reporting : Default Combination
Adjusting for More Likely Clusters : No
Number of Replications : 999

7. OUTPUT

MOST LIKELY CLUSTER
1.Location IDs included.: SUPAUL, MADHEPURA, SAHARSA, DARBHANGA,
KHAGARIA, MUNGER, ARARIA, BEGUSARAI,
BHAGALPUR, SAMASTIPUR
Coordinates / radius..: (26.070000 N, 86.360000 E) / 108.00 km
Population............: 4471
Number of cases.......: 96
Expected cases........: 38.78
Annual cases / 100000.: 2148.6
Observed / expected...: 2.48
Relative risk.........: 9.22
Log likelihood ratio..: 59.396587
Monte Carlo rank......: 1/1000
P-value...............: 0.001

On the Stochastic Modeling and Analysis of Bloom Caster System of Continuous Casting Shop Area of an Integrated Steel Plant

S. K. Singh
S. O. S. in Statistics
Pt. Ravishankar Shukla University
Raipur, Chhattisgarh - 492010, India.
profsksingh@hotmail.com

This paper deals with the stochastic modeling and analysis of Bloom caster system of continuous casting shop area of Bhilai steel plant. System consists of two tundishes along with one mold with rolling belt and gas cutter each. Initially one tundish is in preparation while other is kept in the cold standby. Once the functioning of all the required units gets started it is continued till the completion of the cycle, after this whole system is sent to scheduled maintenance. Blooms obtained are cut into the required sizes with the help of gas cutters and are sent to Bloom storage yard. Failure time distributions of all units are taken to be negative exponential whereas repair and maintenance time distributions are taken to be arbitrary. Using regenerative point technique, several system characteristics such as mean time to system failure (MTSF), availability, busy period analysis of the repairman which are useful to the system managers and engineers, are evaluated. At last some graphs are plotted in order to highlight the important results.

Keywords: Regenerative point technique; Mean time to system failure; Availability; Reliability; Repair facility.

1. Introduction

In the past, although the term “reliability” has not been used specifically, yet the consequences of the failure of the system have been considered. Today, with the increasing complexity and automation of systems, reliability has become inextricably linked with all major phases of the genesis and use of commercial, military and space systems at all levels of design, development, procurement, production, operation and maintenance. Where emphasis has been placed on these concepts, dramatic improvements have often been obtained in the form of reduction in costs, superior workmanship, a higher probability of satisfactory operation and better customer-vendor relationship.

Much of the early reliability work was confined to hypothetical work without taking real data into consideration. Some of the papers are available concerning industry. Very few authors have studied the industrial systems with real existing situation. Kececioglu and Jiang [1] considered the system which has a sensor with a non-zero failure rate, and a switching unit which has both cyclic and continuous failure rates. The reliability analysis of k out of n systems has been carried out by Chung [2] and Dhillon [3]. Recently Singh *et al.* [4] have studied a stone crushing system having one apron feeder, one grizzly, one primary gyratory crusher. This group of equipment is used to get iron ores from stones in mining crushing plants. They have obtained various parameters of the systems which are useful to the system managers and engineers. For the purpose of analyzing industrial models, system of Bhilai Steel Plant, India is selected for study. Bhilai Steel Plant being one of the leading steel plants of India is having the following major units:

1. Coke-ovens
2. Sintering plant No. 1
3. Sintering plant No. 2
4. Blast furnaces
5. Steel melting shop
6. Oxygen Blow Converter
7. Continuos Casting Plant
8. Blooming Mill
9. Billet Mill
10. Rail and Structural Mill
11. Merchant Mill
12. Wire rod mill
13. Plate Mill

The purpose of this model is to study a **Bloom Caster system** of the continuous casting Shop area of the Bhilai Steel Plant. Bloom caster system consists of two tundishes, one mold with rolling belt and gas cutter each. Initially one tundish is in preparation while other is kept in the cold standby. With the completion of the preparation of the previous tundish latter is brought into preparation which remains in the warm standby and does not fail. If the tundish in operation fails it is replaced by the one kept in the warm standby. Once the functioning of all the required units get started it is continued till the completion of the cycle, after this whole system is sent to scheduled maintenance. Blooms obtained are cut into the required sizes with the help of gas cutters and are sent to blooms storage yard. Besides regular repair; scheduled maintenance is also performed. Besides this, shut down repair concept is also imposed on the system to make the system ready as early as possible in case of major breakdowns. Failure time distributions are taken to be negative exponential, whereas repair time distributions are taken to be arbitrary. At last some graphs are plotted in order to highlight the important results. Using regenerative point technique, following measures of system effectiveness are obtained to carry out the profit analysis:

(i) Steady state transition probabilities and Mean sojourn times in different states;
(ii) Mean time to system failure;
(iii) Mean up time during (0,t];
(iv) Expected busy period of the repairman in simple repair (electrical, mechanical, onine) in (0,t] and in steady state;
(v) Expected busy period of the repairman in scheduled maintenance in (0,t] and in steady state;
(vi) Expected busy period of the repairman in shut down in (0,t] and in steady state;
(vii) Expected profit earned by the system in (0,t] and in steady state.

At last some particular cases are also discussed and graphs are plotted to highlight the important results.

1.1. *Description of the system*

1. Bloom Caster system consists of two tundishes, one mold with rolling belt and gas cutter each.
2. Initially one tundish remains in preparation and other is kept as a cold standby. With the completion of preparation, the tundish in standby is kept in preparation and after its preparation, it remains as a warm standby but it does not fail in warm standby condition.
3. If working tundish fails then warm standby immediately comes into operation and failed tundish is replaced by the new one and is immediately kept for preparation.
4. System undergoes for shut down:
   (i) if mold/rolling belt fails.
   (ii) tundish fails provided one tundish is under preparation/Under replacement.
5. Gas cutter is taken as statistically perfect. Initially gas cutter remains in good and non-operative condition and after a random time, they start working.
6. Once the functioning of all the units has been started, it is continued till the required blooms are obtained and with the completion of the cycle, whole system undergoes schedule maintenance.

7. All the failure and repair time distributions are taken exponentially distributed except shut-down repair time and scheduled maintenance time which are taken as arbitrary distributed.
8. After repair unit works as good as new.
   State Transition Diagram is shown in Figure 1.

1.2. *Notations*

E_1	: Set of regenerative states $\{S_0, S_1, S_2, S_3,, S_5, S_6, S_8,\}$
E_2	: Set of non-regenerative states $\{S_4, S_7\}$
ψ	: Scheduled Maintenance holding rate.
b(t)	: Scheduled Maintenance completion rate.
$g_1(t), G_1(t)$	: pdf and cdf of Shut -Down repair rate.
λ, α, β	: Constant failure rate of Tundish/Mold/Rolling Belt.
ϕ	: Preparation completion rate of Tundish.
r_2	: Constant repair rate of Tundish.
η	: Job arrival rate of Gas cutter.
$p_{i,j}$	: Transition from state i to j.
$B_i^j(t)$ $j = 1, 2, 3, 4$	: Probability that repairman is busy in simple repair/ scheduled Maintenance/ shut down repair/ preparation at epoch t starting from state $S_i \in E$.
$W_i(t)$	: Probability that repairman is busy initially in simple repair for (i= 6,8) and remains busy at epoch t without transiting to any other regenerative state.
$\square$	Symbol for Stieltjes convolution
$\boxed{c}$	Symbol for ordinary convolution

1.3. *Symbols used for the states of the system*

$T_0 / T_p / T_s / T_r / T_{ws}$ Tundish in Operation/ under preparation/ under Standby/ under repair/ under warm standby.

$M_0 / M_g / R_0 / R_g$ Mold in operation/ good and non-operative/ Rolling belt in operation/ good and non-operative state.

G_0 / G_g Gas cutter under operation/ good and non-operative.

SM/SD System under scheduled maintenance/ System under shut down repair.

Possible transitions between states are shown in Fig. 1.

2. Transition Probabilities and Sojourn Times

Simple probabilistic considerations yields the following expressions for non-zero transition probabilities p_{ij}.

$$p_{01}=\int_0^\infty \phi e^{-\phi t}dt\ ;\ p_{12}=\int_0^\infty \phi e^{-X_1 t}dt;\ p_{13}=\int_0^\infty \eta e^{-X_1 t}dt$$

$$p_{14}=\int_0^\infty (\lambda+\alpha+\beta)e^{-X_1 t}dt; \qquad p_{25}=\int_0^\infty \eta e^{-X_2 t}dt; \qquad p_{26}=\int_0^\infty \lambda e^{-X_2 t}dt$$

$$p_{24}=\int_0^\infty (\alpha+\beta)e^{-X_2 t}dt\ ;\ p_{35}=\int_0^\infty \phi e^{-X_3 t}dt;\ p_{34}=\int_0^\infty (\lambda+\alpha+\beta)e^{-X_4 t}dt$$

$$p_{37}=\int_0^\infty \varphi e^{-X_3 t}dt\ ;\ p_{40}=\int_0^\infty g_1(t); \quad p_{54}=\int_0^\infty (\alpha+\beta)e^{-X_4 t}dt$$

$$p_{57}=\int_0^\infty \varphi e^{-X_4 t}dt\ ;\ p_{58}=\int_0^\infty \lambda e^{-X_4 t}dt; p_{64}=\int_0^\infty (\lambda+\alpha+\beta)e^{-X_5 t}dt$$

$$p_{61}=\int_0^\infty r_2 e^{-X_5 t}dt\ ;\ p_{68}=\int_0^\infty \eta(t)e^{-X_5 t}dt\ ; \qquad p_{70}=\int_0^\infty b(t)\,dt$$

$$p_{84}=\int_0^\infty (\lambda+\alpha+\beta)e^{-X_6 t}dt\ ;$$

$$p_{87}=\int_0^\infty \varphi e^{-X_6 t}dt;\ p_{83}=\int_0^\infty r_2 e^{-X_6 t}dt \tag{1-21}$$

where

$X_1=\phi+\lambda+\alpha+\beta+\eta$; $X_2=\lambda+\alpha+\beta+\eta$; $X_3=\psi+\lambda+\alpha+\beta+\varphi$

$X_4=\varphi+\lambda+\alpha+\beta$; $X_5=\eta+\lambda+\alpha+\beta+r_2$; $X_6=\varphi+\lambda+\alpha+\beta+r_2$

Mean sojourn time μ_i in S_i is defined as the time that the system continues in state S_i before transiting to any other state.

$$\mu_0 = \int_0^\infty e^{-\phi t}\,dt\,;\ \mu_1 = \int_0^\infty e^{-X_1 t}dt\,;\ \mu_2 = \int_0^\infty e^{-X_2 t}dt$$

$$\mu_3 = \int_0^\infty e^{-X_3 t}dt\,;\ \mu_4 = \int_0^\infty \overline{G}_1(t)dt\,;\ \mu_5 = \int_0^\infty e^{-X_4 t}dt$$

$$\mu_6 = \int_0^\infty e^{-X_5}dt\,;\ \mu_7 = \int_0^\infty \overline{B}(t)dt\,;\ \mu_8 = \int_0^\infty e^{-X_6 t}dt \qquad (22\text{-}30)$$

3. Mean Time to System Failure

Time to system failure can be regarded as first passage time to the failed state. To obtain it, we regard the down states as absorbing states. Using arguments as for the regenerative process we obtain the following recursive relations for $\pi_j(t)$.

$$\pi_j(t) = \sum_j Q_{ij}(t)\boxed{s}\,\pi_j(t) + \sum_k Qik(t) \qquad (31)$$

Where j is an un-failed regenerative state to which the given regenerative state I can transit k is failed state to which I can transit directly.Taking Laplace Steltjes transform of above relation (31) and solving for $\tilde{\pi}_0(s)$, we have

$$E(T) = -\frac{d}{ds}\tilde{\pi}_0(s)\Big|_{s=0} = \frac{D_1^{/}(0) - N_1^{/}(0)}{D_1(0)} \qquad (32)$$

$$\text{MTSF} = E(T) =$$

$$\frac{\mu_0 a_0 + \mu_1 a_1 + \mu_2 a_2 + \mu_3 a_3 + \mu_5 a_5 + \mu_6 a_6 + \mu_8 a_8}{1 - p_{12}p_{26}p_{61} - p_{35}p_{58}p_{83} + p_{12}p_{26}p_{61}p_{35}p_{58}p_{83}} \qquad (33)$$

Where:

$$(1 - p_{12}p_{26}p_{61})(1 - p_{35}p_{58}p_{83}) = a_0\,;$$

$$p_{01}(1 - p_{35}p_{58}p_{83}) = a_1$$

$$p_{01}p_{12}(1 - p_{35}p_{58}p_{83}) = a_2\,;$$

$$p_{01}p_{13} + p_{01}p_{12}p_{83}(p_{25}p_{58} + p_{26}p_{68}) = a_3$$

$$p_{01}p_{12}p_{25} + p_{01}p_{13}p_{35} + p_{01}p_{12}p_{26}p_{35}p_{68}p_{83} = a_5\,;$$

$$p_{01}p_{12}p_{26}(1-p_{35}p_{58}p_{83}) = a_6$$

$$p_{01}p_{12}(p_{25}p_{58}+p_{26}p_{68})+p_{01}p_{13}p_{35}p_{58} = a_8$$

4. Availability Analysis

Let $M_i(t)$ be the probability that the system, initially in regenerative state S_i, is up at time t without passing through any other regenerative state or returning to itself through one or more non-regenerative state, i.e. either it continues to remain in regenerative state S_i or a non-regenerative state including itself.

By probabilistic arguments we have:

$$M_0(t) = e^{-\phi t};\ M_1(t) = e^{-X_1 t};\quad M_2(t) = e^{-X_2 t}$$

$$M_3(t) = e^{-X_3 t};\ M_5(t) = e^{-X_4 t};\ M_6(t) = e^{-X_5 t}$$

$$M_8(t) = e^{-X_6 t};\ M_4(t) = \bar{G}_1(t);\qquad M_7(t) = \bar{B}(t) \qquad (34\text{-}42)$$

Let $A_i(t)$ be the probability that the system is in upstate at instant 't' given that the system entered regenerative state at t=0 . The recursive relations $A_i(t)$ are given as:

$$A_i(t) = M_i(t) + \sum_j q_{ij}(t)\ \boxed{c}\ A_j(t) \qquad (43)$$

Where j is any successive regenerative state to which the regenerative state i can transit through transitions. Taking laplace transforms of [43], and after solving for $A_0^*(s)$, using equations [1-21] of Section 3, the steady state availability of the system when the system starts from $S_i \in E$ is obtained as follows:

$$A_0(\infty) = \lim_{s\to 0} s\ A_0^*(s) = \frac{N_2(0)}{D_2'(0)} \qquad (44)$$

Where

$$N_2(0) = \mu_0 a_0 + \mu_1 a_1 + \mu_2 a_2 + \mu_3 a_3 + \mu_5 a_5 + \mu_6 a_6 + \mu_8 a_8$$

$$D_2'(0) = \mu_0 a_0 + \mu_1 a_1 + \mu_2 a_2 + \mu_3 a_3 + \mu_4 a_4 + \mu_5 a_5 + \mu_6 a_6 + \mu_7 a_7 + \mu_8 a_8 \qquad (45\text{-}46)$$

Where $a_0, a_1, a_2, a_3, a_5, a_6, a_8$ are same as given for MTSF and

$$a_4 = p_{01}p_{12}p_{24}(1-p_{35}p_{58}p_{83}) + p_{01}p_{13}(1+q_{35}q_{54}+q_{35}q_{58}q_{84}) + p_{01}p_{14}(1-p_{35}p_{58}p_{83}) + p_{01}p_{12}p_{25}(p_{54}+p_{58}p_{84}+p_{35}p_{58}p_{83})$$

$$a_7 = p_{01}p_{13}(p_{57}+p_{58}p_{87}) + p_{01}p_{13}p_{37} + p_{01}p_{12}p_{26}p_{68}p_{83}(p_{37}+p_{35}p_{57}) + p_{01}p_{12}p_{25}p_{37}p_{58}p_{83} + p_{01}p_{12}p_{25}p_{57} + p_{01}p_{12}p_{25}p_{58}p_{87} + p_{01}p_{12}p_{26}p_{68}p_{87}$$

Using [20-21] in [19], we get the expression for $A_0(\infty)$.

5. Busy Period Analysis

5.1. *Expected busy period analysis of the repairman in simple repair (electrical, mechanical, on-line) in (0,t]*

By probabilistic arguments we have

$$W_6(t) = e^{-X_5 t} \quad ; \quad W_8(t) = e^{-X_6 t} \tag{47-48}$$

We define $B_i^1(t)$, the probability that repairman is busy in simple repair at epoch t starting from state $S_i \in E$. By probabilistic arguments, we have

$$B_i^1(t) = W_i(t) + \sum_j q_{ij}(t) \boxed{c} B_j^1(t) \tag{49}$$

Where j is any successive regenerative state to which the regenerative state i can transit through transitions. Taking laplace transforms of (49), and using relations (1-21) of section 3 we get:

$$N_3(0) = \mu_6 p_{01}p_{12}p_{26}(1 - p_{35}p_{58}p_{83}) + \mu_8[p_{01}p_{12}(p_{25}p_{58} + p_{26}p_{68}) + p_{01}p_{13}p_{35}p_{58}] \tag{50}$$

Therefore, in the long run,the fraction of time for which the system is under repair is given by:

$$B_0^1(\infty) = \lim_{t\to\infty} B_0^1(t) = \lim_{s\to 0} sB_0^{*1}(s) = \frac{N_3(0)}{D_2'(0)} \tag{51}$$

5.2. *Expected busy period analysis of the repairman in scheduled maintenance in (0,t]* is given by

$$B_0^2(\infty) = \lim_{t\to\infty} B_0^2(t) = \lim_{s\to 0} sB_0^{21}(s) = \frac{N_4(0)}{D_2'(0)} \tag{52}$$

5.3. *Expected busy period analysis of the repairman in shut-down repair in (0,t]* is given by

$$B_0^3(\infty) = \lim_{t\to\infty} B_0^3(t) = \lim_{s\to 0} sB_0^3(s) = \frac{N_5(0)}{D_2'(0)} \tag{53}$$

5.4. ***Expected busy period analysis of the repairman in tundish preparation in (0,t]*** is given by

$$B_0^4(\infty) = \lim_{t\to\infty} B_0^4(t) = \lim_{s\to 0} sB_0^4(s) = \frac{N_6(0)}{D_2^{/}(0)} \quad (54)$$

Where

$$N_4(0) = \mu_7[p_{01}p_{13}(p_{57} + p_{58}p_{87}) + p_{01}p_{13}p_{37} + p_{01}p_{12}p_{68}p_{83}p_{25}p_{37}p_{58} + p_{01}p_{12}p_{25}p_{57} + p_{01}p_{12}p_{25}p_{87}p_{58} + p_{01}p_{12}p_{26}p_{87}p_{68}]$$

$$N_5(0) = \mu_4[p_{01}p_{12}p_{24}(1 - p_{35}p_{58}p_{83}) + p_{01}p_{13}(1 + p_{35}p_{54} + p_{35}p_{58}p_{84}) + p_{01}p_{14}(1 - p_{35}p_{58}p_{83}) + p_{01}p_{12}p_{25}(p_{54} + p_{58}p_{84} + p_{34}p_{58}p_{83})]$$

$$N_6(0) = \mu_0(1 - p_{12}p_{26}p_{61})(1 - p_{35}p_{58}p_{83}) + \mu_1 p_{01}(1 - p_{35}p_{58}p_{83}) + \mu_3[p_{01}p_{13} + p_{01}p_{12}p_{83}(p_{25}p_{58} + p_{25}p_{68})]$$

6. Particular Cases

When all time distributions arc taken to be negative exponential, then steady state transition probabilities are:

$$\begin{aligned}
&p_{01} = 1; \quad p_{12} = \phi / X_1; \quad p_{13} = \eta / X_1; \quad p_{14} = (\lambda + \alpha + \beta) / X_1 \\
&p_{25} = \eta / X_2; \quad p_{26} = \lambda / X_2; \quad p_{24} = (\alpha + \beta) / X_2; \quad p_{35} = \phi / X_3 \\
&p_{34} = (\lambda + \alpha + \beta) / X_3; \quad p_{37} = \varphi / X_3; \quad p_{40} = 1; \quad p_{54} = (\alpha + \beta) / X_4 \\
&p_{57} = \varphi / X_4; \quad p_{58} = \lambda / X_4; \quad p_{64} = (\lambda + \alpha + \beta) / X_5 \\
&p_{68} = \eta / X_5; \quad p_{61} = r_2 / X_5; \quad p_{70} = 1; \quad p_{84} = (\lambda + \alpha + \beta) / X_6 \\
&p_{83} = r_2 / X_6; \quad p_{87} = \varphi / X_6
\end{aligned} \quad (55\text{-}75)$$

7. Cost Analysis

The cost benefit analysis of the system can be carried out by considering the expected busy period of the repairman in repair in (0,t]. Therefore, G(t) = expected revenue earned by the system in (0,t] - expected repair cost of the repair facility in (0,t]

$$= C_1\mu_{up}(t) - C_2\mu_b^1(t) - C_3\mu_b^2(t) - C_4\mu_b^3(t) - C_5\mu_b^4(t) \quad (76)$$

$$\mu_{up}(t) = \int_0^t A_0(t)dt; \quad \mu_b^1(t) = \int_0^t B_0^1(t)dt; \quad \mu_b^2(t) = \int_0^t B_0^2(t)dt$$

$$\mu_b^3(t) = \int_0^t B_0^3(t)dt; \quad \mu_b^4(t) = \int_0^t B_0^4(t)dt$$

The expected profit per unit of time in steady state is

$$G = \lim_{t\to\infty} \frac{G(t)}{t} = \lim_{s\to 0} s^2 G^*(s) =$$

$$C_1\mu_{up}(t) - C_2\mu_b^1(t) - C_3\mu_b^2(t) - C_4\mu_b^3(t) - C_5\mu_b^4(t) \quad (77)$$

where C_1 is the revenue per unit up time and C_2, C_3, C_4, C_5 are the repair cost, scheduled maintenance cost, shut down repair cost and preparation cost respectively.

Acknowledgement

The authors are grateful to Dept. of Science & Technology, New Delhi, India, for the financial assistance to the research project under which the work is initiated and carried out.

References

1. D. Kececioglu and S. Jiang, Reliability of repairable standby system with imperfect sensing and switching, Proceedings of Annual Reliability and Maintainability Symposium, 261–267, 1990.
2. W. K. Chung, Reliability analysis of a k-out-of-n: G parallel redundant system with multiple critical errors, Microelectron. and reliability, vol. 30, 585–589, 1990.
3. B. S. Dhillon, Stochastic modelling of k-out-of-n units family of systems, Int. Journal of system science vol. 23, 1277–1287, 1992.
4. S. K. Singh and Sheeba G. Nair, Stochastic modelling and analysis of stone crushing system used in iron ore mines, Journal of Ravishankar University, vol. 8, 101–104, 1995.

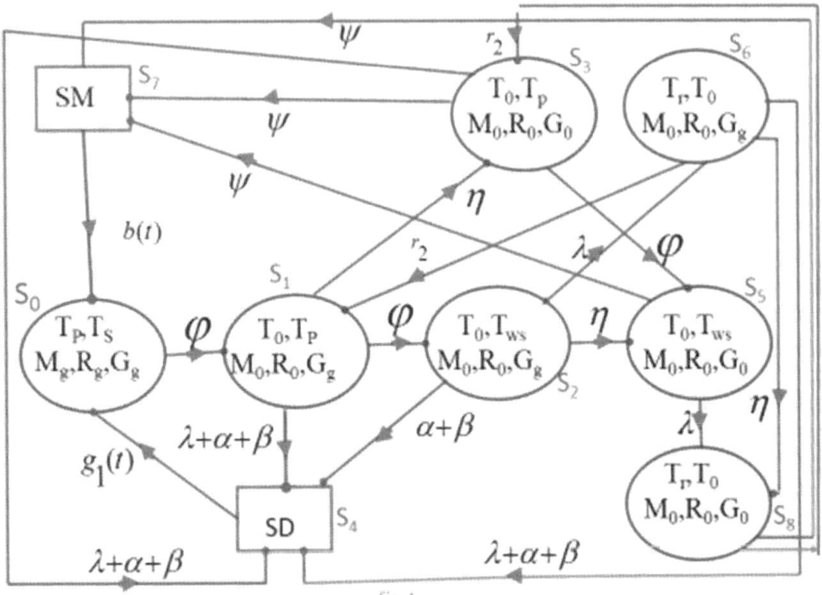

Fig. 1
State transition diagram of Bloom caster system

up state ◯ down state ▭ regenerative point ●

A Generalized Exponential-Lindley Distribution

A. Mishra

Department of Statistics, Patna University, Patna, India
E-mail: mishraamar@rediffmail.com

Binod Kumar Sah

R.R.M. Campus, Janakpur, Tribhuvan University
Kathmandu, Nepal

A two-parameter generalized exponential-Lindley distribution, of which the exponential distribution and the Lindley (1958) distribution are particular cases, has been introduced. Its moments, failure rate function, mean residual life function have been discussed. The maximum likelihood method and the method of moments have been discussed for estimating its parameters. The distribution has been fitted to some data-sets to test its goodness of fit.

Keywords: Exponential distribution, Lindley distribution, moments, failure rate function, meansresidual life function, stochastic ordering, estimation of parameters, goodness of fit.

1. Introduction

The exponential distribution occupies a central place among the continuous probability distributions and plays an important role in statistical theory with many interesting properties and having a wide range of applications in various fields. Its probability density function (pdf) is given by;

$$f(x;\theta) = \theta e^{-\theta x} \quad ; \; x > 0, \; \theta > 0 \tag{1}$$

One of the interesting properties of this distribution is its 'memory loss' which is expressed as;

$$P(X \geq s+t) = P(X \geq s).P(X \geq t) \tag{2}$$

Its rth moment about origin is given by

$$\mu_r' = \Gamma(r+1)/\theta^r \, , \; r = 1,2,3,\ldots. \tag{3}$$

Lindley (1958) introduced a one-parameter distribution, known as Lindley distribution, given by its probability density function

$$f(x;\theta) = \frac{\theta^2}{\theta+1}(1+x)e^{-\theta x} \quad ; \; x > 0, \; \theta > 0 \tag{4}$$

The first four moments about origin of the Lindley distribution have been obtained as

$$\mu_1' = \frac{\theta+2}{\theta(\theta+1)},\ \mu_2' = \frac{2(\theta+3)}{\theta^2(\theta+1)},\ \mu_3' = \frac{6(\theta+4)}{\theta^3(\theta+1)},\ \mu_4' = \frac{24(\theta+5)}{\theta^4(\theta+1)} \quad (5)$$

And its central moments have been obtained as

$$\mu_2 = \frac{\theta^2+4\theta+2}{\theta^2(\theta+1)^2},\quad \mu_3 = \frac{2(\theta^3+6\theta^2+6\theta+2)}{\theta^3(\theta+1)^3}, \quad (6)$$

$$\mu_4 = \frac{3(3\theta^4+24\theta^3+44\theta^2+32\theta+8)}{\theta^4(\theta+1)^4}$$

Ghitany *et al.* (2008) have discussed various properties of this distribution and showed that in many ways (1.4) provides a better model for some applications than the exponential distribution. Mazucheli and Achcar (2011), Ghitany *et al.* (2009, 2011) and Bakouchi *et al.* (2012) are some among others who discussed its various applications. Zakerzadah and Dolati (2009), Shanker and Mishra (2013a, 2013b) obtained generalized Lindley distributions and discussed their various properties and applications. Sankaran (1970) obtained a Lindley mixture of Poisson distribution.

In this paper, a generalized exponential-Lindley distribution of which, both the exponential distribution and the Lindley distribution are the particular cases, has been suggested. Its first four moments and some of the related measures have been obtained. Its failure rate, mean residual rate and stochastic ordering have also been studied. Estimation of its parameters has been discussed and for testing its goodness of fit the distribution has been fitted to some of those data sets where the Lindley distribution has earlier been fitted by others.

2. A Generalized Exponential-Lindley Distribution

A generalized exponential-Lindley distribution with two parameters α and θ is defined by its probability density function (pdf)

$$f(x;\alpha,\theta) = \frac{\theta^2(1+\alpha x)e^{-\theta x}}{(\theta+\alpha)};\ x>0,\theta>0,\alpha\geq 0 \quad (7)$$

It can easily be seen that at $\alpha=0$, the distribution (2) reduces to the exponential distribution (1) and at $\alpha=1$, it reduces to the Lindley distribution (4).

The first derivative of (7) is obtained as;

$$f'(x) = \frac{\theta^2}{\theta+\alpha}e^{-\theta x}(\alpha-\theta-\theta\alpha x) \quad (8)$$

And so $f'(x)=0$ gives $x=(\alpha-\theta)/\theta\alpha$. From this it follows that;

(i) For $\alpha > \theta$, $x_0 = (\alpha - \theta)/\theta\alpha$ is the unique critical point at which $f(x)$ is maximum.

(ii) $f'(x) \le 0$ i.e. $f(x)$ is decreasing in x.

Therefore, the mode of the distribution is given by

$$\text{Mode} = (\alpha - \theta)/\theta\alpha, \ \alpha > \theta \tag{9}$$

$$= 0, \text{ otherwise.}$$

The cumulative distribution function of the distribution is given by

$$F(x) = 1 - \frac{\theta + \alpha + \theta\alpha x}{\theta + \alpha}; x > 0, \ \alpha > \theta > 0. \tag{10}$$

Behavior of the distribution with varying values of the additional parameter α: A large number of distributions associated with various combinations of θ and α, by preparing several tables of ordinates of the generalized exponential-Lindley distribution have been studied. The additional parameter α introduced in the generalized exponential-Lindley distribution has the flexibility to encompass a wide variety of distributions within its range. A graph of the pdf of the distribution for θ= 0.5 and α = 0, 0.2, 0.5, 0.8, 1.0, 1.5, 2.5 and 3.5 is given below.

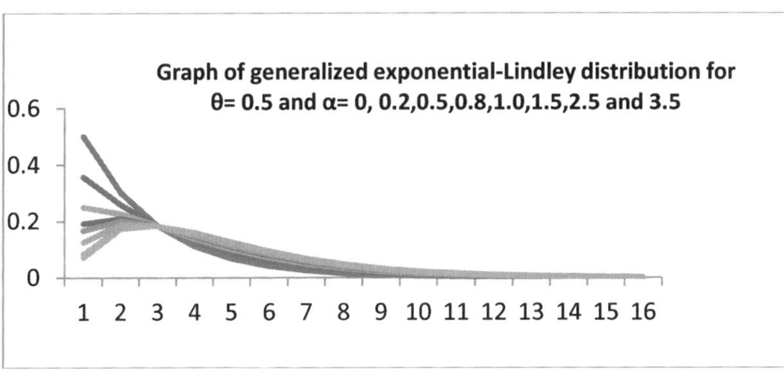

It can be seen that it is the value of α that makes changes in the originating points of the distribution. The higher values of α give lower ordinates for $x = 0$ that increase to certain points and then taper off. It is the parameter α that takes into account any departure of the observed distributions from the exponential and the Lindley distribution.

3. Moments and Related Measures

The rth moment about origin of the generalized exponential-Lindley distribution has been obtained as;

$$\mu_r^/ = \frac{\theta^2}{(\theta+\alpha)} \int_0^\infty x^r (1+\alpha x) e^{-\theta x} \, dx \tag{11}$$

Taking r = 1, 2, 3 and 4 in it the first four moments about origin are obtained as

$$\mu_1^/ = \frac{\theta+2\alpha}{\theta(\theta+\alpha)}, \ \mu_2^/ = \frac{2(\theta+3\alpha)}{\theta^2(\theta+\alpha)}, \ \mu_3^/ = \frac{6(\theta+4\alpha)}{\theta^3(\theta+\alpha)}, \ \mu_4^/ = \frac{24(\theta+5\alpha)}{\theta^4(\theta+\alpha)} \tag{12}$$

It can be easily verified that at $\alpha = 0$, the moments about origin of the distribution reduce to the respective moments of the exponential distribution and at α =1 these reduce to the respective moments of the Lindley distribution. Further, since the mean of the distribution is always greater than the mode, the distribution is positively skewed.

The central moments of this distribution have been obtained as

$$\mu_2 = \frac{\theta^2 + 4\theta\alpha + 2\alpha^2}{[\theta(\theta+\alpha)]^2} \tag{13}$$

$$\mu_3 = \frac{2\theta^3 + 12\theta^2\alpha + 12\theta\alpha^2 + 4\alpha^3}{[\theta(\theta+\alpha)]^3} \tag{14}$$

$$\mu_4 = \frac{3\theta^4 + 24\theta^3\alpha + 48\theta^2\alpha^2 + 48\theta\alpha^3 + 8\alpha^4}{[\theta(\theta+\alpha)]^4} \tag{15}$$

It can be easily verified that at $\alpha = 0, 1$, the central moments of the distribution reduce to the respective moments of the exponential and Lindley distributions respectively.

The moment generating function of the generalized exponential-Lindley distribution is obtained as

$$M_X(t) = \frac{\theta+\alpha-t}{(\theta-t)^2} = \left(1 - \frac{t}{\theta+\alpha}\right)\left(1-\frac{t}{\theta}\right)^{-2} \tag{16}$$

The coefficients of variation, (γ) skewness $\left(\sqrt{\beta_1}\right)$ and the kurtosis (β_2) of the generalized exponential-Lindley distribution are given by

$$\gamma = \frac{\sigma}{\mu_1^/} = \frac{\sqrt{(\theta^2 + 4\theta\alpha + 2\alpha^2)}}{(\theta+2\alpha)} \times 100 \tag{17}$$

$$\sqrt{\beta_1} = \frac{2\theta^3 + 12\theta^2\alpha + 12\theta\alpha^2 + 4\alpha}{(\theta^2 + 4\theta\alpha + 2\alpha^2)^{3/2}} \tag{18}$$

$$\beta_2 = \frac{3\theta^4 + 24\theta^3\alpha + 48\theta^2\alpha^2 + 48\theta\alpha^3 + 8\alpha^4}{(\theta^2 + 4\theta\alpha + 2\alpha^2)^2} \tag{19}$$

4. Failure Rate and Mean Residual Life

For a continuous distribution with pdf $f(x)$ and cdf $F(x)$, the failure rate function (also known as the hazard rate function) and the mean residual life function are respectively defined as;

$$h(x) = \lim_{\Delta x \to 0} \frac{P(X < x + \Delta x \mid X > x)}{\Delta x} = \frac{f(x)}{1 - F(x)} \tag{20}$$

and

$$m(x) = E[X - x \mid X > x] = \frac{1}{1 - F(x)} \int_x^\infty [1 - F(t)]\, dt$$

The corresponding failure rate function, $h(x)$ and the mean residual life function, $m(x)$ of the distribution are thus given by

$$h(x) = \frac{\theta^2 (1 + \alpha x)}{(\theta + \alpha + \theta\alpha x)} \tag{21}$$

and

$$m(x) = \frac{1}{(\theta + \alpha + \theta\alpha x)e^{-\theta x}} \int_x^\infty (\theta + \alpha + \theta\alpha t)e^{-\theta t}\, dt \tag{22}$$

$$= \frac{(\theta + 2\alpha + \theta\alpha x)}{(\theta + \alpha + \theta\alpha x)\theta}$$

It can be easily verified that $h(0) = \theta^2/(\theta + \alpha) = f(0)$ and $m(0) = (\theta + 2\alpha)/\theta(\theta + \alpha) = \mu_1'$. It is also obvious that $h(x)$ is an increasing function of x, α and θ, whereas $m(x)$ is a decreasing function of x, α and θ. For $\alpha = 0, 1$ (4.3) and (4.4) reduce to the corresponding measures of the exponential and the Lindley distributions respectively.

5. Stochastic Orderings

Stochastic ordering of positive continuous random variables is an important tool for judging the comparative behavior. A random variable X is said to be smaller than a random variable Y in the

(i) stochastic order $(X \leq_{st} Y)$ if $F_X(x) \geq F_Y(x)$ for all x

(ii) hazard rate order $(X \leq_{hr} Y)$ if $h_X(x) \geq h_Y(x)$ for all x

(iii) mean residual life order $(X \leq_{mrl} Y)$ if $m_X(x) \leq m_Y(x)$ for all x

(iv) likelihood ratio order $\left(X \leq_{lr} Y\right)$ if $\dfrac{f_X(x)}{f_Y(x)}$ decreases in x.

The following results due to Shaked and Shanthikumar (1994) are well known for establishing stochastic ordering of distributions

$$X \leq_{lr} Y \Rightarrow X \leq_{hr} Y \Rightarrow X \leq_{mrl} Y \tag{23}$$

$$\Downarrow$$

$$X \leq_{st} Y$$

The generalized exponential-Lindley distribution is ordered with respect to the strongest 'likelihood ratio' ordering as shown in the following theorem:

Theorem: *Let* $X \sim$ *gen.exp-Lindley* $\left(\alpha_1, \theta_1\right)$ *and* $Y \sim$ *gen.exp-Lindley* $\left(\alpha_2, \theta_2\right)$ *If* $\alpha_1 = \alpha_2$ *and* $\theta_1 \geq \theta_2$ *(or if* $\theta_1 = \theta_2$ *and* $\alpha_1 \geq \alpha_2$*), then* $X \leq_{lr} Y$ *and hence* $X \leq_{lr} Y$, $X \leq_{mrl} Y$ *and* $X \leq_{st} Y$.

Proof: We have

$$\frac{f_X(x)}{f_Y(x)} = \left(\frac{\theta_1}{\theta_2}\right)^2 \left(\frac{1+\alpha_1 x}{1+\alpha_2 x}\right)\left(\frac{\theta_2+\alpha_2}{\theta_1+\alpha_1}\right) e^{-(\theta_1-\theta_2)x} \quad , x>0$$

Now;

$$\log \frac{f_X(x)}{f_Y(x)} = 2\log\left(\frac{\theta_1}{\theta_2}\right) + \log\left(1+\alpha_1 x\right) - \log\left(1+\alpha_2 x\right) + \log\left(\frac{\theta_2+\alpha_2}{\theta_1+\alpha_1}\right) - \left(\theta_1-\theta_2\right)x \; .$$

Thus

$$\frac{d}{dx}\frac{f_X(x)}{f_Y(x)} = \frac{\alpha_1}{1+\alpha_1 x} - \frac{\alpha_2}{1+\alpha_2 x} + \left(\theta_2-\theta_1\right)$$

$$= \frac{\alpha_1-\alpha_2}{(1+\alpha_1 x)(1+\alpha_2 x} + \left(\theta_2-\theta_1\right) \tag{24}$$

(i) If $\alpha_1 = \alpha_2$ and $\theta_1 \geq \theta_2$, then $\dfrac{d}{dx}\dfrac{f_X(x)}{f_Y(x)} < 0$. This means that $X \leq_{lr} Y$ and hence $X \leq_{lr} Y$, $X \leq_{mrl} Y$ and $X \leq_{st} Y$.

(ii) If $\theta_1 = \theta_2$ and $\alpha_1 \geq \alpha_2$, then $\dfrac{d}{dx}\dfrac{f_X(x)}{f_Y(x)} < 0$. This means that $X \leq_{lr} Y$ and hence $X \leq_{lr} Y$, $X \leq_{mrl} Y$ and $X \leq_{st} Y$.

This theorem shows the flexibility of the generalized exponential-Lindley distribution over exponential and Lindley distributions.

6. Estimation of Parameters

6.1. *Maximum likelihood estimates*

Let $(x_1, x_2, \ldots, x_n)$ be a random sample of size n from a two-parameter generalized exponential-Lindley distribution and let f_x be the observed frequency in the sample corresponding to $X = x$; x = 1, 2, 3, …k such that $\sum_{x=1}^{k} f_x = n$, where k is the largest observed value having non-zero frequency.

The likelihood function, L of the generalized exponential-Lindley distribution (2.1) is given by

$$L = \left(\frac{\theta^2}{\theta+\alpha}\right)^n \prod_{x=1}^{n} \left(1+x\alpha\right)^{f_x} e^{-n\theta\bar{x}} \tag{25}$$

and so the log likelihood function is obtained as

$$\log L = n\log\theta^2 - n\log\left(\theta+\alpha\right) + \sum_{i=1}^{k} f_X \log\left(1+\alpha x\right) - n\theta\bar{x} \tag{26}$$

The two log likelihood equations are thus obtained as

$$\frac{\partial \log L}{\partial\theta} = \frac{2n}{\theta} - \frac{n}{\theta+\alpha} - n\bar{x} = 0 \tag{27}$$

$$\frac{\partial \log L}{\partial\alpha} = \frac{n}{\theta+\alpha} + \sum_{x=1}^{k} \frac{x f_x}{1+x\alpha} = 0 \tag{28}$$

Equation (27) gives $\vec{X} = \dfrac{\theta+2\alpha}{\theta\left(\theta+\alpha\right)}$, which is the mean of the generalized exponential-Lindley distribution. The two equations (27) and (28) do not seem to be solved directly. However, the Fisher's scoring method can be applied to solve these equations. For applying this method we have

The following equations for $\hat{\theta}$ and $\hat{\alpha}$ can be solved;

$$\begin{bmatrix} \dfrac{\partial^2 \log L}{\partial\theta^2} & \dfrac{\partial^2 \log L}{\partial\theta\,\partial\alpha} \\ \dfrac{\partial^2 \log L}{\partial\theta\,\partial\alpha} & \dfrac{\partial^2 \log L}{\partial\alpha^2} \end{bmatrix}_{\hat{\theta}=\theta_0,\,\hat{\alpha}=\alpha_0} \begin{bmatrix} \hat{\theta} - \theta_0 \\ \hat{\alpha} - \alpha_0 \end{bmatrix} = \begin{bmatrix} -\dfrac{\partial \log L}{\partial\theta} \\ -\dfrac{\partial \log L}{\partial\theta} \end{bmatrix}_{\hat{\theta}=\theta_0,\,\hat{\alpha}=\alpha_0} \tag{29}$$

where, θ_0 and α_0 are the initial values of θ and α respectively. These equations are solved iteratively till sufficiently close estimates of $\hat{\theta}$ and $\hat{\alpha}$ are obtained.

6.2. *Estimates by method of moments*

Using the first two moments about origin, we have

$$\mu_2'/\mu_1' = K\,(\text{say}) = 2(\theta+3\alpha)/\theta(\theta+2\alpha) \tag{30}$$

This gives

$$\alpha = \theta(K\theta-2)/(6-2K\theta) \tag{31}$$

Substituting this value of α in the expression for mean of the distribution, we get after some simplification

$$\mu_1' = 2/\left(4\theta - K\theta^2\right) \tag{32}$$

Replacing the first and the second moments μ_1' and μ_2' by the respective sample moments, $\overline{X}$ and m_2', an estimate k of K, can be obtained. Using this in the equation (32) an equation in θ can be obtained as

$$f(\theta) = k\theta^2 - 4\theta + \left(2/\overline{X}\right) = 0 \tag{33}$$

which gives an estimate of θ as

$$\theta = \left[1 \pm \left(\frac{2\overline{X}-k}{2\overline{X}}\right)^{1/2}\right] \tag{34}$$

Substituting this estimate of θ in (31), an estimate of α can be obtained.

7. Goodness of Fit

The generalized exponential-Lindley distribution has been fitted to a number of data-sets to which earlier the exponential distribution and the Lindley distribution have been fitted by others and to almost all these data-sets the generalized exponential-Lindley distribution provides closer fits than these two distributions.

The fitting of the generalized exponential-Lindley distribution to onesuch data-set has been presented in the following table. The data set given in Table 1 is the data set reported by Bzerkedal (1960) respectively. The expected frequencies according to the exponential distribution and the Lindley distribution have also been given. The estimates of the parameters have been obtained by the method of maximum likelihood.

Table 1. Survival times (in days) of 72 guinea pigs infected with virulent tuberclebacilli.

Survival Time (In days)	Observed frequency	Expected frequency Exponential	Lindley	Gen. Exp-Lindley
0 – 80	8	14.3	16.1	12.3
80 – 160	30	20.6	21.9	24.0
160 – 240	18	13.3	15.4	16.9
240 – 320	8	8.5	9.0	8.8
320 – 400	4	5.5	5.5	5.1
400 – 480	3	3.5	1.8	3.2
480 – 560	1	6.3	2.3	1.7
Total	72	72.0	72.0	72.0

Estimates of parameters $\hat{\theta}$=0.00552, $\hat{\theta}$= 0.011 $\begin{matrix}\hat{\theta} = 0.010968 \\ \hat{\alpha} = 0.800146\end{matrix}$.

The value of χ^2 for exponential, Lindley and generalize exponential-Lindley distributions have been obtained as 12.60, 7.77 and 3.55 respectively. The respective probabilities of thevalues of χ^2 to be exceeded by these values at 4, 3 and 2 degrees of freedom respectively are 0.021, 0.051 and 0.176. This indicates that the generalized exponential- Lindley distribution gives much closer fits than the exponential distribution and the Lindley distribution and thus provides a better alternative to these distributions.

References

1. Bakouch, H. S., Al-Zahrani, B. M., Al-Shomrani, A. A., Marchi, V. A. A., Louzada, F., An extended Lindley distribution, *Journal of the Korean Statistical Society,* 41 (1), 75–85 (2012).
2. Bzerkedal, T., Acquisition of resistance in guinea pigs infected with different doses of virulent tubercle bacilli, *American Journal of Epidemiology,* 72 (1), 130–148 (1960).
3. Ghitany, M. E., Atieh, B., Nadarajah, S., Lindley distribution and its Applications, *Mathematics and Computers in Simulation,* 78 (4), 493–506 (2009).
4. Ghitany, M. E. Al-qallaf, F., Al-Mutairi, D. K., Hussain, H. A., A two parameter weighted Lindley distribution and its applications to survival data, *Mathematics and Computers in Simulation,* 81 (6), 1190–1201 (2011).
5. Lindley, D. V., Fiducial distributions and Bayes' theorem, Journal of the *Royal Statistical Society*, Series B, 20, 102–107 (1958).

6. Mazucheli, J. and Achcar, J. A., The Lindley distribution applied to competing risks lifetime data, *Computer Methods and Programs in Biomedicine*, 104 (2), 188–192 (2011).
7. Sankaran, M., The discrete Poisson-Lindley distribution, *Biometrics*, 26, 145–149 (1970).
8. Shanker, R and Mishra, A., (A quasi Lindley distribution, *African Journal of Mathematics and Computer Science Research*, Vol. 6 (4), 64–71 (2013a).
9. Shanker, R and Mishra, A., A two-parameter Lindley distribution, *Statistics in Transition*, (New Series), Vol. 14 (1), 45–56 (2013b).
10. Shaked, M. and Shanthikumar, J. G., *Stochastic Orders and Their Applications*, Academic Press, New York (1994).
11. Zakerzadah, H. and Dolati, A., Generalized Lindley distribution, *Journal of Mathematical Extension*, 3(2), 13–25 (2009).

On Estimating the Urban Populations Using Minimum Information

Arun Kumar Sinha

Dept. of Statistics, Central University of South Bihar,
B I T Patna Campus, Patna 800 014, Bihar, India
E-mail: arunkrsinha@yahoo.com

Vijay Kumar

Rajendra Memorial Research Institute of Medical Sciences (ICMR),
Agamkuan, Patna 800 007, Bihar, India
E-mail: vijay_stat2007@rediffmail.com

Ravi B. P. Verma

Statistics Canada
Ottawa, Ontario KIG 5X3, Canada
E-mail: ravi2verma@yahoo.com

For implementing the programs and policies of the urban planning department we need to know the current populations of the urban areas of interest. Many times the required information is not easily available. In view of these scenarios the paper deals with the estimation of population totals based on ranked set sampling (RSS), which is relatively a new sampling method that utilizes only ranking information about the randomly selected sampling units with respect to the characteristic of interest without using their exact measurements. The estimator is more efficient than that of a simple random sample with the same size. Two illustrations based on reported data sets are given. The findings are expected to be of much help to the policy and decision makers, and also to those who look for a cost-effective sampling method to estimate the urban populations.

Keywords: Simple Random Sample; Relative Precision; Stratified Sampling; Unequal Allocation.

1. Introduction

For estimating the population total of some specific areas on the basis of sampling Rao (1997) proposed an estimator for the population total of urban areas of West Bengal, a state of India. The areas could be urban, rural, industrial, non-industrial, or of particular professionals, etc. This is a modified form of its simple random sample (SRS) estimator. But it is based on the assumption that a selected sample unit may be unusually very large. For example, while estimating the total population of West Bengal the selected sampling units may include an urban area like Kolkata (Calcutta) and this could jeopardize the usual estimate of the population total. As the sample mean possesses many other statistical properties, the proposed estimator is justified. Besides, the estimator is based on the principle of post-stratification. It is the motivation behind the proposed ranked set sampling (RSS) estimator for estimating the population total of some specific type of areas more efficiently and conveniently as compared with the usual SRS or Rao's estimator. RSS is a cost-effective sampling method proposed by McIntyre (1952). For obtaining a ranked set sample, a fairly large number of randomly identified sampling units are partitioned into small subsets of the same size. Next, the units of each subset are ranked separately with respect to the characteristic of interest without using their actual measurements. The ranking information is supposed to be easily available and inexpensive as compared with the quantification of the units. Exactly one unit of each subset with a specified rank is quantified after ranking is accomplished. As the ranking induces stratification on the population it provides a more structured sample than a simple random sample does with the same size. This sample, in turn, yields more efficient estimators of many parameters of interest than a simple random sample of the same size does. Moreover, its implementation needs only the ranking of the randomly selected units, which does not depend upon the method employed for determining the ranking. Thus, one can use any or all the available information (in absence of actual quantification) including subjective judgment for the purpose. Interestingly, by taking advantage of the experience and expertise of the field personnel it exploits the auxiliary information that is not effectively utilized by standard probability survey sampling designs. Even in the presence of ranking error it provides unbiased and more efficient estimators of various population parameters. It has been successfully used in the several areas on interest. Recently, it has been used under the parametric setting for estimating the parameters of several known distributions. See Patil, Sinha and Taillie (1994) and Sinha (2005) for a more detailed coverage of RSS. As illustrations of the purposed method we have considered the estimation of the total population of urban areas using published data sets, See Sinha and Sinha (2014), and Sinha (2016). This is required for implementing various government and non-government plans of immediate interest.

2. Ranked Set Sampling

In order to illustrate the technique, suppose that we wish to select a ranked set sample of size 12. We first of all select a simple random sample following simple random sampling with replacement (SRSWR) method of size 16. These 16 units are randomly put in four sets, each consisting of four units. Now, on the basis of visual perception or any other crude and inexpensive method, the four units of each set are ranked with respect to the main characteristic of interest without quantifying them. This ranking of units is performed in each set separately. For convenience, suppose the rank one is accorded to the unit having the smallest value of the variable of interest, the rank two to the unit with the next higher value, the rank three to the unit possessing the next higher value and the rank four is given to the unit with the highest value. The unit with the rank one is selected for quantification from the first set. The unit possessing the second rank is selected from the second set for measurement. This process is continued until the units having ranks three and four are chosen for the third and the fourth set respectively for quantification. The whole process yields four measurements representing rank one to rank four. Thus, out of 16 units only four units are measured. As we need to have an RSS of size 12, we draw two more sets of 16 units following SRSWR from the population. From each set of 16 units, we repeat the process of measurement of four units adopting the earlier method. Thus, we obtain an RSS of 12 measurements and this sample consists of three measurement of each rank order. In general, m denotes the set size and r represents the number of cycles (replications). In this illustration, we have $m = 4$ and $r = 3$. For obtaining an RSS of size mr, we need to draw a random sample of size m^2r in the first phase of sampling. Next, we consider RSS with equal and unequal allocation. The former uses the same number of quantification from each rank order while the latter takes care of the variation within the measurements of each rank order.

2.1. *RSS with equal allocation*

Let $X_{(i:m)j}$ denote the unit possessing the ith rank in the jth cycle. Here, $i = 1,\ldots, m$ and $j = 1,\ldots, r$. The estimator of the population mean μ is defined as

$$\hat{\mu}_{RSS} = \frac{\sum_{i=1}^{m}\sum_{j=1}^{r} X_{(i:m)j}}{mr} \quad \text{or,} \quad \hat{\mu}_{RSS} = \frac{1}{m}\sum_{i=1}^{m}\hat{\mu}_{(i:m)}$$

where

$$\hat{\mu}_{(i:m)} = \frac{\sum_{j=1}^{r} X_{(i:m)j}}{r} \tag{1}$$

Here, $E(\hat{\mu}_{RSS}) = \mu$ since $E(\hat{\mu}_{(i:m)}) = \mu_{(i:m)}$ and $\mu = \frac{1}{m}\sum_{i=1}^{m}\mu_{(i:m)}$

Now, $\text{var}(\hat{\mu}_{RSS}) = \frac{1}{m^2}\sum_{i=1}^{m}\frac{\sigma^2_{(i:m)}}{r}$ or, $\text{var}(\hat{\mu}_{RSS}) = \frac{1}{m^2 r}\sum_{i=1}^{m}\sigma^2_{(i:m)}$ (2)

Here, $\mu_{(i:m)}$ and $\sigma^2_{(i:m)}$ denote the population mean and variance respectively of the ith order statistic. The estimator of $\text{var}(\hat{\mu}_{RSS})$ is defined as follows:

$$\hat{\text{var}}(\hat{\mu}_{RSS}) = \frac{1}{m^2 r}\sum_{i=1}^{m} s^2_{(i:m)}$$

where

$$s^2_{(i:m)} = \frac{1}{r-1}\sum_{j=1}^{r}(X_{(i:m)j} - \hat{\mu}_{(i:m)})^2 \text{ and } E(s^2_{(i:m)}) = \sigma^2_{(i:m)}.$$

The estimate of σ^2 for simple random sampling (SRS) is given by,

$$\hat{\sigma}^2_{RSS} = \left[\frac{mr - m + 1}{m^2 r(r-1)}\right]\sum_{i=1}^{m}\sum_{j=1}^{r}(X_{(i:m)j} - \hat{\mu}_{(i:m)})^2 + \left(\frac{1}{m}\right)\sum_{i=1}^{m}(\hat{\mu}_{(i:m)} - \mu_{RSS})^2 \quad (3)$$

See Norris, Patil and Sinha (1995).

For comparing the performance of $\hat{\mu}_{RSS}$ as compared with SRS estimator, $\hat{\mu}_{SRS}$ the relative precision (*RP*) is defined as below:

$$RP = \frac{\text{var}(\hat{\mu}_{SRS})}{\text{var}(\hat{\mu}_{RSS})} = \frac{\sigma^2 / mr}{\sum_{i=1}^{m}\sigma^2_{(i:m)} \Big/ m^2 r} \text{ or, } RP = \frac{\sigma^2}{\frac{1}{m}\sum_{i=1}^{m}\sigma^2_{(i:m)}} \quad (4)$$

Here, σ^2 denotes the population variance.

Obviously, $RP \geq 1$ as long as $\sigma^2 \geq \frac{1}{m}\sum_{i=1}^{m}\sigma^2_{(i:m)}$

Relative cost (RC) = $\frac{1}{RP}$ and relative savings (RS) = $\left(1 - \frac{1}{RP}\right)$X100 (5)

In general, $1 \leq RP \leq \frac{m+1}{2}$ and $0 \leq RS \leq \frac{m-1}{m+1}$

Note that the performance of RSS is never worse than that of SRS.

2.2. *RSS with unequal allocation*

For increasing the performance of RSS, we use the Neyman's criterion as used in stratified sampling. According to this principle,

$$\mathrm{r_i} = \frac{\mathrm{n}\sigma_{(\mathrm{i:m})}}{\sum_{i=1}^{m} \sigma_{(i:m)}}, \; i = 1, 2, 3, \ldots, m. \tag{6}$$

Here, $n = \sum_{i=1}^{m} r_i$ and $\mathrm{r_i}$ denotes the number of quantifications of the units with rank i.

The estimator of the population mean is defined as

$$\hat{\mu}_{\mathrm{RSSUA}} = \frac{1}{m}\sum_{i=1}^{m}\frac{\mathrm{T_i}}{\mathrm{r_i}} \quad \text{and} \tag{7}$$

$$\mathrm{var}(\hat{\mu}_{RSSUA}) = \frac{1}{m^2}\sum_{i=1}^{m}\frac{\sigma^2_{(i:m)}}{r_i} \tag{8}$$

where $\mathrm{T_i}$ shows the sum of the quantifications of the units having the ith rank. On putting the value of r_i into the expression for $\mathrm{var}(\hat{\mu}_{RSSUA})$, we have

$$\mathrm{var}(\hat{\mu}_{RSSUA}) = \frac{(\bar{\sigma})^2}{n}, \text{ where } \bar{\sigma} = \frac{\sum_{i=1}^{m}\sigma_{(i:m)}}{m}.$$

The estimate of σ^2 for simple random sampling in case of unequal allocation is given by

$$\sigma^2_{\mathrm{RSSUA}} = \sum_{\mathrm{i=1}}^{\mathrm{m}} \frac{\mathrm{m}\,(\mathrm{r_i}-1)+1}{\mathrm{m}^2\,\mathrm{r_i}\,(\mathrm{r_i}-1)} \sum_{\mathrm{j=1}}^{\mathrm{r_i}} \left(\mathrm{X}_{(\mathrm{i:m})\mathrm{j}} - \hat{\mu}_{(\mathrm{i:m})}\right)^2 + \frac{1}{\mathrm{m}}\sum_{\mathrm{i=1}}^{\mathrm{m}}\left(\hat{\mu}_{(\mathrm{i:m})} - \hat{\mu}_{\mathrm{RSSUA}}\right)^2 \tag{9}$$

The relative precision (RP_{ua}) of $\hat{\mu}_{\mathrm{RSSUA}}$ relative to $\hat{\mu}_{\mathrm{SRS}}$ with the same number of quantification is given by

$$RP_{ua} = \frac{\mathrm{var}(\hat{\mu}_{SRS})}{\mathrm{var}(\hat{\mu}_{RSSUA})} = \frac{\sigma^2/n}{\sum_{i=1}^{m}\frac{\sigma^2_{(i:m)}}{r_i}\Big/ m} \tag{10}$$

Relative cost (RC) = $\dfrac{1}{RP_{ua}}$ and

relative savings (RS) = $\left(1 - \dfrac{1}{RP_{ua}}\right) X100$ (11)

In general, $0 \le RP_{ua} \le m$ and $RP_{ua} \ge RP$.

See Patil, Sinha and Taillie (1994) and Sinha (2005).

3. Illustration

The males, females and the total populations of Class I, II and III of 130 urban places of Bihar, India in 2011 are 5867548, 5229560 and 11097108 respectively according to the data sets of Census of Bihar 2011 Reports **(Source: Primary Census Abstract: Directorate of Census Operations, Bihar)**. The numbers of urban places are 26, 27 and 77 respectively in Class I, Class II and Class III. Class I has 100,000 and above populations, Class II has populations between 50,000 and 99,999 and Class III has populations between 20,000 and 49,999. For estimating the total population of these places one could either estimate the total population directly or estimate the males and females population separately and then by adding these two one could find of the total population. As ranked set sampling uses ranking with respect to the variable of interest, the ranking of the urban places based on the male population may not be the same as that based on the female population if the sex ratio fluctuates from place to place. Considering this fact, we have shown both the estimation methods of the total population. Obviously, if the sex ratio does not differ much from place to place the two methods would provide almost the same value. In this section, we illustrate both the methods.

3.1. *Estimation of total population*

We estimate their total population with the help of a sample of 27 urban places. In sampling terminology, here N = 130 and n = 27. For obtaining an RSS, we need to specify the set size and the number of replications in the beginning. As the ranking of three places with respect to their population is supposed to be convenient for the field personnel, we take the set size as 3 in this illustration. Theoretically, one could consider any number for the set size. Note that, the efficiency of an RSS estimator increases with the increase in the set size provided the ranking is perfect. In this illustration for getting a sample of size 27, the number of replications is determined as 9 because the set size is taken as

3. For this, sample 81 out of 130 places are located first following SRSWR. Thus, we have m = 3 and r = 9. The population of these 81 places for total, males and females are presented in 9 sets, i.e. 27 row each having three places in Table 1. Here, the total population is not based on the sum of male and female populations, as male and female populations are drown from a separate population of male and female.

Table 1. Randomly selected 81 urban places separately for total population, male population and female population.

	Total, Males, Females	Total, Males, Females	Total, Males, Females
	67885, 13855, 9831	51469, 139319, 48213	24286, 18876, 9980
Set 1	43113, 20011, 17318	29888, 13460, 16749	49320, 82589, 220252
	27286, 15845, 10086	36949, 12343, 220252	43113, 18389, 31890
	37736, 12466, 16749	398138, 37132, 18935	468940, 16065, 12751
Set 2	37756, 26897, 10086	51469, 14895, 15294	32409, 26985, 789071
	31904, 70381, 12309	27286, 23507, 18423	44298, 18389, 46628
	32739, 26897, 12751	280547, 13468, 35616	31961, 154915, 12116
Set 3	23447, 97129, 139146	23567, 28163, 28400	36949, 97129, 165693
	78888, 20132, 18423	240565, 17110, 12632	48298, 15161, 14572
	29888, 186145, 141974	23649, 21947, 165693	21266, 77687, 21051
Set 4	48401, 11009, 63788	21266, 25492, 10201	31904, 17110, 9980
	33732, 14037, 22919	71531, 16783, 32023	35257, 25611, 20355
	147396, 32041, 38265	46083, 26622, 48990	86720, 20803, 85112
Set 5	294116, 20566, 9509	25576, 42089, 24572	296889, 248688, 24629
	48406, 11105, 11388	49320, 11065, 99928	30270, 25492, 11365
	26528, 35910, 14399	251136, 21983, 14641	35707, 12515, 69709
Set 6	28477, 77687, 85112	52340, 212005, 789071	61107, 20056, 72586
	37756, 15939, 12993	40659, 14105, 11104	25678, 24302, 28158
	50928, 139319, 12116	40659, 16783, 24047	296889, 26401, 14727
Set 7	43387, 11009, 9697	351838, 14881, 28158	67678, 20555, 72586
	30578, 186145, 12265	132896, 41671, 16749	38727, 24603, 22568
	26219, 25030, 23288	26528, 23507, 12478	31961, 27487, 16749
Set 8	62712, 126834, 9831	1683275, 10644, 9640	39018, 21983, 9697
	40659, 41671, 15945	280547, 24603, 23235	78887, 19792, 69709
	66624, 82859, 18890	39240, 11009, 14727	147126, 78561, 13219
Set 9	25678, 113173, 9980	26515, 212005, 220252	24286, 13735, 11365
	51002, 22993, 47588	52792, 23032, 17451	32409, 31268, 23943

Now, we rank the three places of each row separately on the basis of their populations. The rank one is given to the place having the lowest population, the rank two to the place possessing the next higher population and the rank three is accorded to the place with the highest population. After ranking, we select the place with the minimum population from the first set, the place with the second rank from the second set, and finally the place with the highest population from the third set. This process is continued for each row of 9 sets. The selected places with their total populations, male populations and female populations and their ranks are mentioned in Table 2.

Table 2. Ranked set sample of 27 urban places for total populations, male populations and female populations.

Set	Rank 1	Rank 2	Rank 3
Set 1	24286, 13855, 9831	43113, 20011, 17318	43113, 18389, 220252
Set 2	37736, 12466, 12751	37756, 26897, 15294	44298, 70381, 46628
Set 3	31961, 13468, 12116	23567, 97129, 139146	240565, 20132, 18423
Set 4	21266, 21947, 21051	31904, 17110, 10201	71531, 25611, 32023
Set 5	46083, 20803, 38265	294116, 42089, 24572	49320, 25492, 99928
Set 6	26528, 12515, 14399	52340, 77687, 85112	40659, 24302, 28158
Set 7	40659, 16783, 12116	67678, 14881, 28158	132896, 186145, 22658
Set 8	26219, 23507, 12478	62712, 21983, 9697	280547, 41671, 69709
Set 9	39240, 11009, 13219	26515, 113173, 11365	52792, 31263, 47588

The arithmetic means of the urban places having ranks one, two and three are obtained as 32664, 71078 and 106191 respectively, their standard deviations are obtained as 8608, 85036 and 92575 respectively while their variances are computed as 74093320, 7231095915 and 8570080454 respectively.

Thus, using equation (1) the mean population of the urban places for Class I, II and III of Bihar in 2011 is estimated as $\hat{\mu}_{\text{RSS}}$ = 69977.7 and the estimate of the total population is obtained as $N\hat{\mu}_{\text{RSS}}$ = 9097097. The difference between the actual and the RSS based estimated population is 2000011, which is 18.02% as compared with the actual population. On using equation (2), the value of the $\text{var}(\hat{\mu}_{\text{RSS}})$ = 195990984.

On applying equation (3), the estimate of σ^2 for simple random sampling (SRS) is given by,

$$\hat{\sigma}^2_{RSS} = 5801416584 \text{ and } \text{var}(\hat{\mu}_{SRS}) = 214867281.$$

Thus, on using equation (4) and (5), we obtain RP = 1.10 and RS = 9.09%.

For implementing RSS with unequal allocation we need to determine first the number the sampling units to be quantified for each rank order separately unlike equal allocation where the same number of units is quantified for each rank order. This is carried out following the Neyman's criterion, which is also used in the case of stratified sampling with unequal allocation.

Here, on using equation (6), we obtain r_1 = 2, r_2 = 12 and r_3 = 13. We, then, note down the populations of two urban places with rank one, twelve urban places with rank two and thirteen urban places with rank three. The selected places with their populations are given in Table 3.

Table 3. Ranked set sample (using unequal allocation) of 27 urban places for total populations, male populations and female populations.

Set	Rank 1	Rank 2	Rank 3
Set 1	24286, 13855, 9831	43113, 20011, 17318	43113, 18389, 220252
Set 2	37736, 12466, 12751	37756, 26897, 15294	44298, 70381, 46628
Set 3		32739, 26897, 12751	36949, 97129, 165693
Set 4		78888, 17110, 14572	29888, 186145, 165693
Set 5		31904, 17110, 10201	71531, 25611, 32023
Set 6		86720, 26622, 48990	296889, 248688, 24629
Set 7		48406, 11105, 11388	251136, 35910, 69709
Set 8		52340, 77687, 85112	40659, 24302, 28158
Set 9		50928, 26401, 14727	351838, 20555, 72586
Set 10		38727, 41671, 16749	31961, 27487, 23288
Set 11		62712	280547, 126834, 9831
Set 12		66624	26515, 41671, 69709
Set 13			52792, 82589, 18890
Set 14			212005, 220252
Set 15			31268, 47588

The arithmetic means of urban places with ranks one, two and three are computed as 31011, 52571 and 119855 respectively while their variances are obtained as 90451250, 318199350 and 15360150184 respectively. Finally, we compute an estimate of the population mean of cities/towns classified according to three ranks as $\hat{\mu}_{\text{RSSUA}} = 67812.3$ (using equation (7)) and an estimate of the population total as N $\hat{\mu}_{\text{RSSUA}} = 8815603$. The difference between the actual and the estimated population is 2281505, which comes to be 20.55% with respect to the actual population. Further, after applying equation (8), we obtain $\text{var}(\hat{\mu}_{\text{RSSUA}}) = 139254695$ and on using equation (9), the estimate of σ^2 for simple random sampling in case of unequal allocation is given by

$$\sigma^2_{\text{RSSUA}} = 6409925667 \text{ and } \text{var}(\hat{\mu}_{SRS}) = 237404654.$$

Thus, on using equation (10) and (11), we obtain $\hat{RP}_{ua} = 1.70$ and $RS =$ 41%.

Next, we estimate the male and the female populations separately to obtain the estimate of the total population. This investigation takes into account the varying sex ratios in different urban places, which could influence the ranking of a place in a set. Also, this, in turn, could help obtain a better estimate of the total population.

3.2. *Estimation of male population*

Similarly the male populations of 81 urban places are randomly selected from 130 urban places and placed in Table 1 and proceed as Section 3.1, we quantified 27 urban places for male population with rank one, two and three which is placed in Table 2. Thus, we have the arithmetic means for the first, second and the third ranks are computed as 16261, 47884 and 49265 respectively, their standard deviations are obtained as 4682, 37970 and 53751 respectively and their variances are obtained as 21918136, 1441747695 and 2889204797 respectively. Thus, on using equation (1), RSS estimate of the population mean of the male population is computed as 37803.3 and that of the estimated total male population is obtained as 4914433, which differs from the estimated total by 953115 (16%). After applying equation (2), we obtain the value of the $\text{var}(\hat{\mu}_{\text{RSS}}) = 53739144$.

On applying (3), the estimate of σ^2 for simple random sampling (SRS) is given by,

$$\hat{\sigma}^2_{RSS} = 1575832512 \text{ and, } \text{var}(\hat{\mu}_{SRS}) = 58364167.$$

Thus, on using equation (4) and (5), $RP = 1.09$ and $RS = 8.26\%$.

Using the Neyman criterion for unequal allocation and using equation (6), we obtain the value of $r_1 = 2$, $r_2 = 10$ and $r_3 = 15$ and proceed similar as Section 3.1, we quantified two urban places for rank one, ten urban places for rank two and fifteen urban places for rank three. The selected places with their populations are given in Table 3. Thus, we have the arithmetic mean of male population for the first, second and the third ranked urban places as 13161, 29151 and 83264 respectively while their variances are obtained as 964661, 359531350 and 5822104452 respectively. On applying equation (7), the RSS estimator of the population mean under unequal allocation, $\hat{\mu}_{\text{RSSUA}} = 41858.7$, which helps obtain the estimate of the male population as 5441631. The difference between the actual and the estimated male population is 425917, which comes to be 7.26% with respect to the actual population of 130 places. Further, using equation (8), we obtain $\text{var}(\hat{\mu}_{\text{RSSUA}}) = 47175085$ and on applying equation (9), the estimate of σ^2 for simple random sampling in case of unequal allocation is given by

$$\sigma^2_{\text{RSSUA}} = 2872686997 \text{ and } \operatorname{var}(\hat{\mu}_{SRS}) = 106395815.$$

Thus, on using equation (10) and (11), we obtain $\hat{\text{RP}}_{\text{ua}} = 2.26$ and RS = 56%.

Next, we consider the estimation of total female population.

3.3. *Estimation of female population*

Similarly, the female populations of 81 urban places are randomly selected from 130 urban places placed in Table 1 and proceed as Section 3.1, we quantified 27 urban places for female population with rank one, two and three which is placed in Table 2. Thus, we have the RSS means of the female population of the places possessing the first, second and the third rank are computed as 16247, 37874 and 65041 respectively, their standard deviations are obtained as 8819, 44657 and 63633 respectively and their variances are obtained as 77772173, 1994207670 and 4049136740 respectively. Thus, on using equation (1), we get $\hat{\mu}_{RSS}$ = 39730.7 and the estimate of the total female population based on the RSS mean as 5164991, which differs from the actual female population by 64569 which is 1.23% with respect to the actual female population of 130 places. After applying equation (2), the value of the $\operatorname{var}(\hat{\mu}_{\text{RSS}}) = 75569341$.

On using equation (3), the estimate of σ^2 for simple random sampling (SRS) is given by,

$$\hat{\sigma}^2_{MRSS} = 2287747775 \text{ and } \operatorname{var}(\hat{\mu}_{SRS}) = 84731399.$$

Thus, on using, equation (4) and (5), RP = 1.12 and RS = 11%.

Further, RSS with unequal allocation is considered with the replications of the first, second and the third rank as $r_1 = 2$, $r_2 = 10$ and $r_3 = 15$ as the earlier situations on using equation (6). In fact, these numbers could also be computed following the Neyman's criterion. Now, proceeding similar as Section 3.1, we quantified two urban places for rank one, ten urban places for rank two and fifteen urban places for rank three. The selected places with their population are given in Table 3. Thus, the RSS means of the first, second and third ranks under unequal allocation are computed as 11291, 24710 and 80995 respectively. On applying equation (7), the RSS estimator under unequal allocation of the population mean is obtained as $\hat{\mu}_{RSSUA}$ = 38998.7, The estimate of the population total on this basis is obtained by 5069831, which differs from the actual population by 159729 (3.05%). Further, using equation (8), we obtain the estimate of $\text{var}(\hat{\mu}_{RSSUA}) = 47038845$ and on applying equation (9), the estimate of σ^2 for simple random sampling in case of unequal allocation is given by

$$\sigma^2_{\text{RSSUA}} = 3039974356 \text{ and, } \text{var}(\hat{\mu}_{SRS}) = 112591643.$$

Thus, on using equation (10) and (11), we obtain $\hat{\text{RP}}_{\text{ua}} = 2.39$ and RS =58%.

Table 4. RP, RS and estimated and actual population with % difference under equal and unequal allocation for male, female and total populations.

Allocation	Relative Precision (RP)	Relative Savings (RS)	Estimated Population and Actual Population with % Difference
Equal Allocation (Direct Method)			
($m = 3, r = 9, n = 27$)			
Total Population	1.10	9.09%	9097097, 11097108 (18.02%)
Male Population	1.09	8.26%	4914433, 5867548 (16%)
Female Population	1.12	11%	5164991, 5229560 (1.23%)
Total Population (Indirect Method)	---	---	10079424, 11097108 (9.17%)
Unequal allocation (Direct Method)			
Total Population			
($m = 3, r_1 = 2, r_2 = 12, r_3 = 13, n = 27$)	1.70	41%	8815603, 11097108 (20.55%)
Male Population			
($m = 3, r_1 = 2, r_2 = 10, r_3 = 15, n = 27$)	2.26	56%	5441631, 5867548 (7.26%)
Female Population			
($m = 3, r_1 = 2, r_2 = 10, r_3 = 15, n = 27$)	2.39	58%	5069831, 5229560 (3.05%)
Total Population (Indirect Method)	---	---	10511462, 11097108 (5.28%)

The approach of indirect method appears better because of the variation in sex ratios, and the smaller difference between the actual total population and new estimated total population. For the discussion on the population estimation by RSS see Swanson and Tayman (2012) and Verma (2015). Though all the estimates are based on specific samples, the performance of RSS with unequal allocation is found better for estimating the total population, see Table 4. This is true when the total population is estimated directly on the basis of the total populations of the sampled places as well as when the estimated total population is obtained as the sum of the separately estimated male and female populations.

From Table 4, it is obvious that in all the categories of populations, the relative saving is about 10% in equal allocation but on applying unequal allocation we obtain the relative saving more than 50%.

References

1. C. Radhakrishna Rao, Statistics and Truth: Putting Chance to Work, Second Edition. World Scientific Publishing Co. Pte. Ltd. Singapore (1997).
2. G. A. McIntyre, A method of unbiased selective sampling using ranked sets. *Australian Journal of Agricultural Research,* 3, 385–390 (1952).
3. G. P. Patil, A. K. Sinha and C. Taillie, Ranked set sampling. In: *Handbook of Statistics,* 12 (G. P. Patil and C. R. Rao eds.), 167–200. North-Holland, Elsevier Science B. V., Amsterdam (1994).
4. Arun K. Sinha, On some recent developments in ranked set sampling. *Bulletin of Informatics and Cybernatics,* 37, 137–160 (2005).
5. Ravi R. Sinha and Arun K. Sinha, Sampling of urban places, *Journal of Patna Science* College (JPSC), 2, 165–176 (2014).
6. Arun K Sinha, Ranked set sampling: As a cost effective and more efficient data collection method. To appear in *Statistical Journal of the International Association for Official Statistics* (2016).
7. R. C. Norris, G. P. Patil and A. K. Sinha, Estimation of multiple characteristics by ranked set sampling methods. *Coenoses,* 10, 95–111 (1995).
8. Census of India, Provisional Population Totals – Bihar. Published by the Registrar General of India, Govt. of India, New Delhi "Appendix – 1 to Table – 6", pp. 330–334 (2011).
9. D. A. Swanson and J. Tayman, Subnational Population Estimates, *Springer* (2012).
10. Ravi B. P. Verma, Review of book, Subnational population estimates by David Swanson and Jeff Tayman, Springer, 2012, published in *Canadian Studies in Population*, Vol. 42, No. 3–4, 86–90, (2015).

Fitting of Some Statistical Distributions of Daily Precipitation Data on North West India (NWI) Regions

Dr. Ranjan Kumar Sahoo
P. G. Dept. of Statistics, Utkal University
India-751004
E-mail: ranjansahoostatistics@ gmail.com

Analysis of rainfall data strongly depends on its distribution pattern. It has long been a topic of interest in the fields of meteorology in establishing a probability distribution that provides a good fit to rainfall. In this study three major statistical distributions are fitted to the daily precipitation data for nine meteorological subdivisions of NWI regions such as, Punjab, Haryana, Chandigarh &Delhi, Himachal Pradesh, Uttarakhand, East and West Uttar Pradesh, Gujarat, East and West Rajasthan, Jammu and Kashmir for the period 1982 to 2013. These distributions are Log-normal, Gamma and Weibull distributions. To study the goodness of fit, Chi-square test is taken over the DJF data among with individual December, January and February data for 91 days. From the Chi-square test it shows that lognormal distribution is more suitable for Daily Precipitation on NWI region among these three distributions.

KeyWords: Daily Precipitation on **NWI** (North West India); Statistical distributions; Weibull distribution; Gamma distribution; Log-normal distributions; DJF: December, January and February.

1. Introduction

The amount of rainfall received over an area is an important factor in assessing the amount of water available to meet the various demands of agriculture, industry, and other human activities. Therefore, the study of the distribution of rainfall in time and space is very important for the welfare of the national economy. Many applications of rainfall data are enhanced by knowledge of the actual distribution of rainfall rather than relying on simple summary statistics.

Rainfall data from arid and semi-arid regions are best fit to one of several probability distribution functions such as Normal (Gaussian), Log-Normal, Gamma, Weibull, and Gumbel distributions (Maliva and Missimer, 2012).

Ozturk (1981) reviewed some of probability distribution models for precipitation totals and their applications. Thegeneral properties of a probability distribution model which was a mixture of Gamma and Poisson distribution were discussed. A new approximation was given for the solution of likelihood equations and the efficiency of the estimators was obtained. An application was also made using monthly precipitation totals.

Sen and Eljadid (1999) studied the rainfall distribution function for Libya and rainfall prediction. Libyan monthly rainfall distributions were found to abide by Gamma probability distribution function which was confirmed on the basis of chi-square tests. Almost all the rainfall sequences recorded for at least the last 20 years in Libya were investigated statistically and Gamma distribution parameters were calculated at existing stations. The shape and scale parameters were then regionalized. Predictions of 10, 25, 50 and 100 mm rainfall amounts were achieved by this probability function.

Al-Mansory (2005) identified the proper theoretical statistical distributions for the data set of highest monthly rainfall for Basrah station, south of Iraq, for 75-given years with missing data. Six distributions were selected: Normal, Log-Normal, Log-Normal type III, Pearson type III, Log-Person type III, and Gumbel. Maximum likelihood method was selected to estimate theoretical distribution parameters. Adequacy test was conducted using Chi-square test. Results indicated that Person type III and Gumbel distributions were the proper for describing maximum monthly rainfall in the area being examined.

Husak *et al.* (2006) studied the use of Gamma distribution to represent monthly rainfall in Africa for drought monitoring applications. They demonstrated the feasibility of fitting cell-by-cell probability distributions to grids of monthly interpolated, continent-wide data Gamma distribution was well suited to these applications because it was fairly familiar to African scientists, and capable of representing a variety of distribution shapes. This study tested the goodness-of-fit using the Kolmogorov–Smirnov test, and compared these results against Weibull distribution which was commonly used in rainfall events. They found that Gamma distribution was suitable for roughly 98% of the locations over all months. The techniques and results presented in this study provided a foundation for use of the

Gamma distribution to generate drivers for various rain related models.

Olumide *et al.* (2013) fitted various probability distribution models to various rainfall and runoff for the Tagwai dam site in Minna, Niger State, Nigeria to evaluate the model that was best suitable for the prediction of their values and subsequently using the best model to predict for both the expected yearly maximum daily-rainfall and yearly maximum daily-runoff at some specific return periods. The distributions Gumbel, log Gumbel, Normal and Log-Normal distribution were used in the study. The Normal distribution model was found most appropriate for the prediction of yearly maximum daily-rainfall and the Log-Gumbel distribution model was the most appropriate for the prediction of yearly maximum daily-runoff.

2. Probability Distributions of Rainfall Data

Analysis of rainfall data strongly depends on its distribution pattern. It has long been a topic of interest in the fields of meteorology in establishing a probability distribution that provides a good fit to rainfall. Papalexiou *et al.* (2012) proposed a typical procedure for selecting a distribution law of rainfall with three steps:

1. Select a priori some of the many parametric families of distributions.
2. Estimate the parameters according to one of the many fitting methods.
3. Choose the one best fitted according to some metric or fitting test.

In this study three probability distributions are selected, *i.e.,* Normal, Gamma (2 Parameters) and Weibull (2 Parameters) distributions.

2.1. *Gamma distribution*

Gamma distribution is widely used in hydrologic analysis. The probability distribution function of a random variable *x* having Gamma distribution is:

$$f(x) = \mathrm{P}(\mathrm{X} \leq \mathrm{x}) == \int_0^x f(x)\, dx = \frac{1}{\beta^\alpha \gamma \alpha} \int_0^{\mathrm{X}} \mathrm{x}^{\alpha-1} \mathrm{e}^{-\frac{\mathrm{x}}{\beta}}\; x \geq 0,\ \alpha \geq 0,\ \beta \geq 0; \quad (1)$$

Where: α: shape parameter β: scale parameter $\gamma(\alpha)$: Gamma function.

The two parameter Gamma distribution has a lower bound at zero, which is disadvantage for application to hydrologic variables that have a lower bound larger than zero (Chow, 1988). Two parameters of Gamma distribution (α and β) are estimated by method of moments as following (Forbes *et al.* 2011): $\alpha = \frac{\bar{x}}{s^2}$, $\beta = \frac{s^2}{\bar{x}}$.

2.2. *Chi-square goodness of fit test (Q)*

The chi-square (Q) test is a technique that check if a specific distribution of a certain observed event's frequency in a sample is suitable for that sample or not. The first step in the chi-square test is to arrange the number of observations *N* into a set of *k* cells (class intervals), and then calculate the chi-square statistic from the statistical formula (Guizani, 2010):

$$\chi^2 = \sum_{i=1}^{k} \frac{(O_i - E_i)^2}{E_i} \quad (2)$$

O_i: observed frequency in the *ith* cell, E_i: expected frequency in the same cell, ***k***: number of intervals. The expected frequency can be computed by: $Ei = Np_i$, where N: total number of observations, p_i: probability of the *ith* cell, the degrees of freedom can be computed using $V = K - S - 1$, denotes the number of parameters used in fitting distribution. The null hypotheses of the chi-square test is that there is no difference between observed and estimated value and a tested distribution is describing the observed data, while the alternative hypothesis is that the tested distribution doesn't describe the observed data. The null hypothesis will be rejected if Q is greater than the critical value.

Goodness of Fit Tests

The goodness of fit tests provides objective procedures to determine whether an assumed theoretical distribution provides an adequate description of the observed data, these tests are valid only for rejecting an inadequate model; they cannot prove that the model is correct (Almog, 1979). Two types are used in this study, these are: Chi-square goodness of fit test, Kolmogorov-Smirnov goodness of fit test.

2.3. *Weibull distribution*

The two parameter Weibull distribution is commonly applied in hydrology since many hydrological variables are bounded by zero (Tallaksen and Lanen, 2004). The cumulative distribution function is given by (Reddy, 1997):

$$f(X) = 1 - e^{\left(-\frac{X}{\beta}\right)^{\alpha}} \tag{3}$$

where α: shape parameter, β: scale parameter. The mean and variance for Weibull distribution are:

$$\mu = \beta\Gamma\left(1 + \frac{1}{\alpha}\right) \sigma^2 = \beta^2\left[\Gamma\left(1 + \frac{2}{\alpha}\right) - \left[\Gamma\left(1 + \frac{1}{\alpha}\right)\right]^2\right]$$

Since the basic Weibull model has two parameters, estimates of the model parameters can be obtained using the sample mean $\overline{x}$ and sample variance s^2. By equating μ with $\overline{x}$ and σ^2 with β is obtained as follows (Murthy *et al.* 2004):

$$\frac{\sigma^2}{\mu^2} = \frac{S^2}{\overline{X}^2} = \frac{\Gamma\left(1+\frac{2}{\alpha}\right)}{\Gamma^2\left(1+\frac{1}{\alpha}\right)} - 1$$

by comparing Equation (11) with a coefficient of variation equation which is: $C_V = \frac{S}{\overline{X}}$ finding that (Al-Fawzan, 2000):

$$C_V = \frac{\sqrt{\Gamma\left(1+\frac{2}{\alpha}\right) - \Gamma^2\left(1+\frac{1}{\alpha}\right)}}{\Gamma\left(1+\frac{1}{\alpha}\right)},$$

α is obtained by trial and error from the above equation and β is given by the following equation

$$\beta = \frac{\overline{x}}{\Gamma\left(1+\frac{1}{\alpha}\right)}$$

2.4. *Lognormal distribution*

Random variable X has the *lognormal distribution* with parameters $\mu \in R$ and $\sigma \in (0,\infty)$ if ln(X) has the normal distribution with mean μ and standard deviation σ. The parameter σ is the *shape parameter* of X, while e^{μ} is the scale parameter of X.

Equivalently, $X = e^{Y}$ where Y is normally distributed with mean μ and standard deviation σ. We can write Y = μ+σZ where Z has the standard normal distribution. Hence we can write $X=e^{\mu+\sigma z}=e^{\mu}(e^{z})^{\sigma}$.

Random variable e^{z} has the lognormal distribution with parameters 0 and 1, and naturally enough, this is the *standard lognormal distribution*. The lognormal distribution is used to model continuous random quantities when the distribution is believed to be skewed, such as certain income and lifetime variables.

The probability density function of the lognormal distribution with parameters μ and σ is given by

$$f(x) = \frac{1}{\sqrt{2\pi\sigma x}} exp\left(-\frac{[\ln(x)-\mu]^2}{2\sigma^2}\right), x \in (0,\infty) \tag{4}$$

a. *f* increases and then decreases with mode at $x=\exp(\mu-\sigma^2)$.
b. f is concave upward then downward then upward again, with inflection points at $x=\exp(\mu-3/2\sigma^2 \pm 1/2\sigma^2+\sqrt{\sigma^2+4})$.
c. *f*(x)→0 as x↓0 and as x→∞.

PROBABILITY DISTRIBUTION TABLE: RAINFALL FOR DJF									
	Weibull Distribution			Gamma Distribution			Log-normal Distribution		
Rainfall (in mm)	Days	Obs. Prob.	Exp. Prob.	Days	Obs. Prob.	Exp. Prob.	Days	Obs. Prob.	Exp. Prob.
0.090	380	0.3133	0.0429	380	0.3133	0.0412	380	0.3133	0.0544
0.544	191	0.1575	0.0305	191	0.1575	0.0272	191	0.1575	0.0486
0.998	106	0.0874	0.0338	106	0.0874	0.0223	106	0.0874	0.0354
1.452	86	0.0709	0.0256	86	0.0709	0.0223	86	0.0709	0.0322
1.906	56	0.0462	0.0231	56	0.0462	0.0223	56	0.0462	0.0322
2.360	49	0.0404	0.0272	49	0.0404	0.0198	49	0.0404	0.0190
2.814	31	0.0256	0.0256	31	0.0256	0.0223	31	0.0256	0.0124
3.268	33	0.0272	0.0132	33	0.0272	0.0173	33	0.0272	0.0206
3.722	18	0.0148	0.0099	18	0.0148	0.0214	18	0.0148	0.0107
4.176	29	0.0239	0.0198	29	0.0239	0.0091	29	0.0239	0.0132
4.630	28	0.0231	0.0107	28	0.0231	0.0107	28	0.0231	0.0132
5.084	18	0.0148	0.0157	18	0.0148	0.0190	18	0.0148	0.0165
5.538	22	0.0181	0.0132	22	0.0181	0.0091	22	0.0181	0.0082
5.992	20	0.0165	0.0181	20	0.0165	0.0140	20	0.0165	0.0107

PROBABILITY DISTRIBUTION TABLE: RAINFALL FOR DJF									
	Weibull Distribution			Gamma Distribution			Log-normal Distribution		
Rainfall (in mm)	Days	Obs. Prob.	Exp. Prob.	Days	Obs. Prob.	Exp. Prob.	Days	Obs. Prob.	Exp. Prob.
6.446	11	0.0091	0.0099	11	0.0091	0.0132	11	0.0091	0.0124
6.900	15	0.0124	0.0124	15	0.0124	0.0165	15	0.0124	0.0082
7.354	14	0.0115	0.0157	14	0.0115	0.0099	14	0.0115	0.0148
7.808	8	0.0066	0.0157	8	0.0066	0.0124	8	0.0066	0.0082
8.262	11	0.0091	0.0115	11	0.0091	0.0148	11	0.0091	0.0082
8.716	7	0.0058	0.0165	7	0.0058	0.0165	7	0.0058	0.0115
9.170	6	0.0049	0.0157	6	0.0049	0.0115	6	0.0049	0.0148
9.624	8	0.0066	0.0124	8	0.0066	0.0173	8	0.0066	0.0091
10.078	6	0.0049	0.0157	6	0.0049	0.0165	6	0.0049	0.0082
10.532	6	0.0049	0.0148	6	0.0049	0.0132	6	0.0049	0.0107
10.986	4	0.0033	0.0107	4	0.0033	0.0157	4	0.0033	0.0082
11.440	6	0.0049	0.0107	6	0.0049	0.0165	6	0.0049	0.0107
11.894	5	0.0041	0.0132	5	0.0041	0.0091	5	0.0041	0.0041
12.348	5	0.0041	0.0148	5	0.0041	0.0165	5	0.0041	0.0066
12.802	1	0.0008	0.0107	1	0.0008	0.0148	1	0.0008	0.0082
13.256	5	0.0041	0.0124	5	0.0041	0.0115	5	0.0041	0.0082
13.710	2	0.0016	0.0165	2	0.0016	0.0124	2	0.0016	0.0082
14.164	1	0.0008	0.0157	1	0.0008	0.0140	1	0.0008	0.0099
14.618	4	0.0033	0.0099	4	0.0033	0.0181	4	0.0033	0.0049
15.072	2	0.0016	0.0115	2	0.0016	0.0157	2	0.0016	0.0066
15.526	1	0.0008	0.0124	1	0.0008	0.0115	1	0.0008	0.0066
15.980	2	0.0016	0.0091	2	0.0016	0.0115	2	0.0016	0.0066
16.434	1	0.0008	0.0107	1	0.0008	0.0107	1	0.0008	0.0099
16.888	3	0.0025	0.0280	3	0.0025	0.0148	3	0.0025	0.0099
17.342	0	0.0000	0.0140	0	0.0000	0.0280	0	0.0000	0.0091
17.796	0	0.0000	0.0132	0	0.0000	0.0140	0	0.0000	0.0074
18.250	1	0.0008	0.0099	1	0.0008	0.0132	1	0.0008	0.0049
18.704	3	0.0025	0.0099	3	0.0025	0.0107	3	0.0025	0.0082
19.158	0	0.0000	0.0132	0	0.0000	0.0107	0	0.0000	0.0033
19.612	0	0.0000	0.0058	0	0.0000	0.0107	0	0.0000	0.0066
20.066	3	0.0025	0.0140	3	0.0025	0.0140	3	0.0025	0.0066
20.520	0	0.0000	0.0074	0	0.0000	0.0082	0	0.0000	0.0058
20.974	0	0.0000	0.0132	0	0.0000	0.0140	0	0.0000	0.0049
21.428	0	0.0000	0.0115	0	0.0000	0.0124	0	0.0000	0.0058
21.882	1	0.0008	0.0099	1	0.0008	0.0107	1	0.0008	0.0099
22.336	0	0.0000	0.0082	0	0.0000	0.0082	0	0.0000	0.0190
22.790	0	0.0000	0.0082	0	0.0000	0.0074	0	0.0000	0.0082
23.244	0	0.0000	0.0074	0	0.0000	0.0099	0	0.0000	0.0074
23.698	1	0.0008	0.0107	1	0.0008	0.0115	1	0.0008	0.0082
24.152	0	0.0000	0.0058	0	0.0000	0.0082	0	0.0000	0.0058
24.606	1	0.0008	0.0091	1	0.0008	0.0107	1	0.0008	0.0074
25.060	1	0.0008	0.0124	1	0.0008	0.0099	1	0.0008	0.0041
25.514	0	0.0000	0.0099	0	0.0000	0.0124	0	0.0000	0.0058
25.968	0	0.0000	0.0115	0	0.0000	0.0099	0	0.0000	0.0049

PROBABILITY DISTRIBUTION TABLE: RAINFALL FOR DJF									
	Weibull Distribution			Gamma Distribution			Log-normal Distribution		
Rainfall (in mm)	Days	Obs. Prob.	Exp. Prob.	Days	Obs. Prob.	Exp. Prob.	Days	Obs. Prob.	Exp. Prob.
26.422	0	0.0000	0.0074	0	0.0000	0.0041	0	0.0000	0.0099
26.876	0	0.0000	0.0058	0	0.0000	0.0049	0	0.0000	0.0033
27.330	0	0.0000	0.0033	0	0.0000	0.0058	0	0.0000	0.0066
27.784	0	0.0000	0.0074	0	0.0000	0.0082	0	0.0000	0.0082
28.238	0	0.0000	0.0091	0	0.0000	0.0115	0	0.0000	0.0041
28.692	0	0.0000	0.0132	0	0.0000	0.0115	0	0.0000	0.0082
29.146	0	0.0000	0.0091	0	0.0000	0.0082	0	0.0000	0.0066
29.600	0	0.0000	0.0099	0	0.0000	0.0091	0	0.0000	0.0082
30.054	0	0.0000	0.0066	0	0.0000	0.0033	0	0.0000	0.0066
30.508	0	0.0000	0.0041	0	0.0000	0.0066	0	0.0000	0.0082
30.962	0	0.0000	0.0099	0	0.0000	0.0091	0	0.0000	0.0066
31.416	0	0.0000	0.0074	0	0.0000	0.0082	0	0.0000	0.0058
31.870	0	0.0000	0.0058	0	0.0000	0.0058	0	0.0000	0.0132
32.324	0	0.0000	0.0082	0	0.0000	0.0074	0	0.0000	0.0132
32.778	0	0.0000	0.0049	0	0.0000	0.0058	0	0.0000	0.0140
33.232	0	0.0000	0.0124	0	0.0000	0.0091	0	0.0000	0.0091
33.686	0	0.0000	0.0058	0	0.0000	0.0049	0	0.0000	0.0165
34.140	0	0.0000	0.0049	0	0.0000	0.0074	0	0.0000	0.0173
34.594	0	0.0000	0.0033	0	0.0000	0.0016	0	0.0000	0.0247
35.048	0	0.0000	0.0049	0	0.0000	0.0049	0	0.0000	0.0247
35.502	0	0.0000	0.0074	0	0.0000	0.0066	0	0.0000	0.0247
35.956	1	0.0008	0.0082	1	0.0008	0.0082	1	0.0008	0.0742

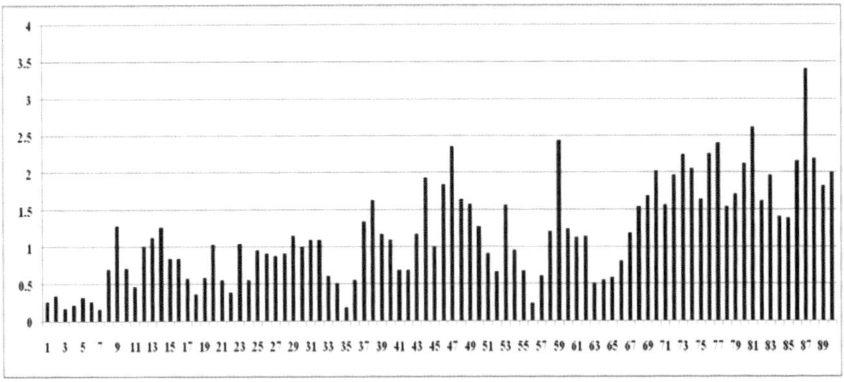

Fig 1. Graphic representation of daily precipitation of NWI.

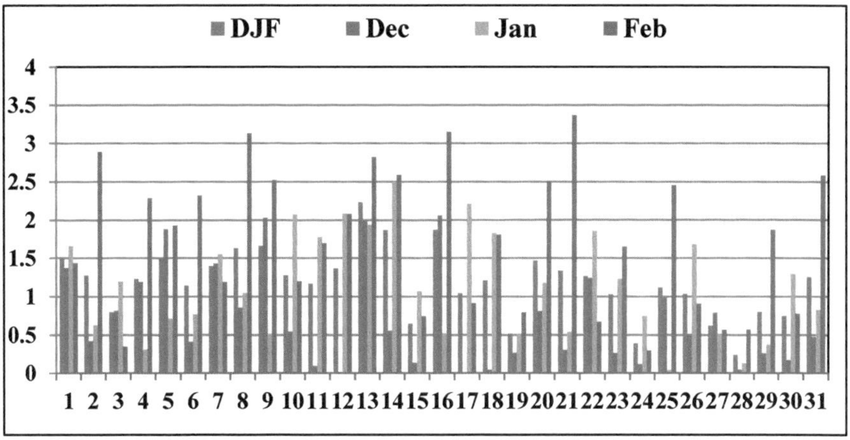

Fig 2. Year to year inter annual variation (1982–2013).

Fitting of statistical distributions on DJF pPrecipitation data over NWI (North West India) regions: For DJF

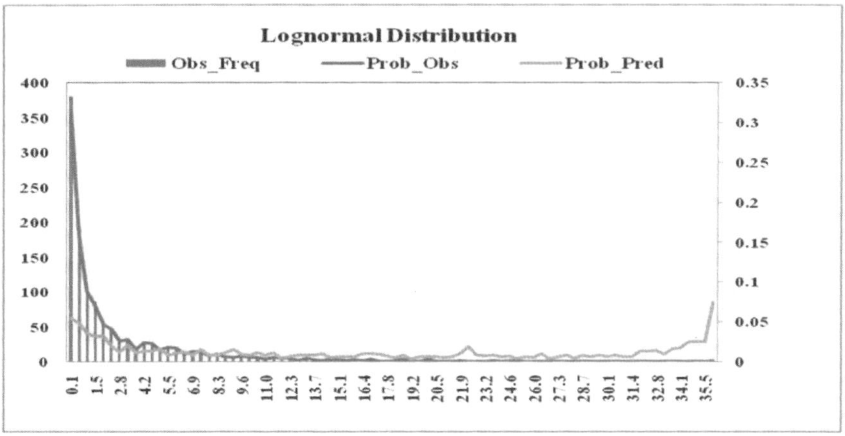

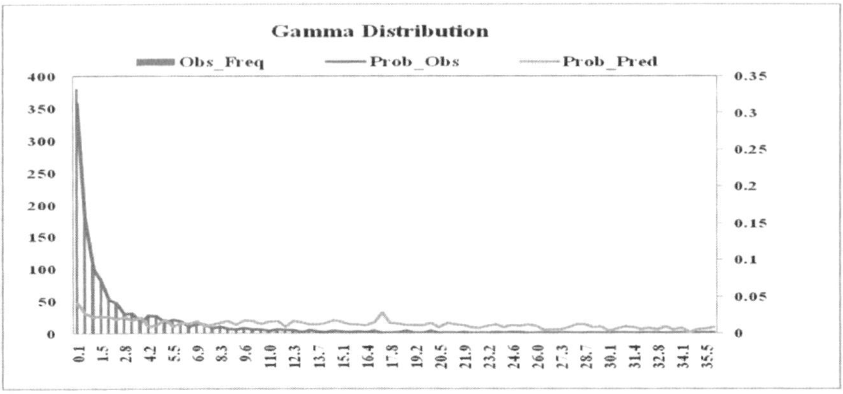

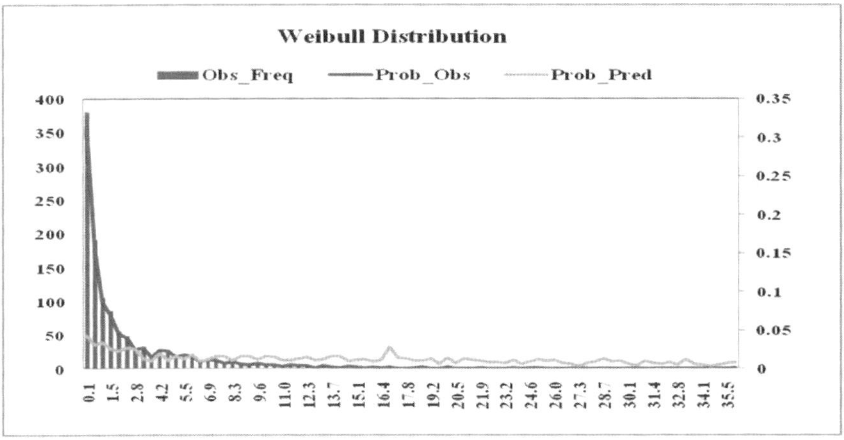

The chi square value for Weibull, Gamma, and Log-normal distribution are 2.9737, 3.3595 and 2.1920 respectively.

Fitting of statistical distributions on December precipitation data over NWI (North West India) regions:

For Dec

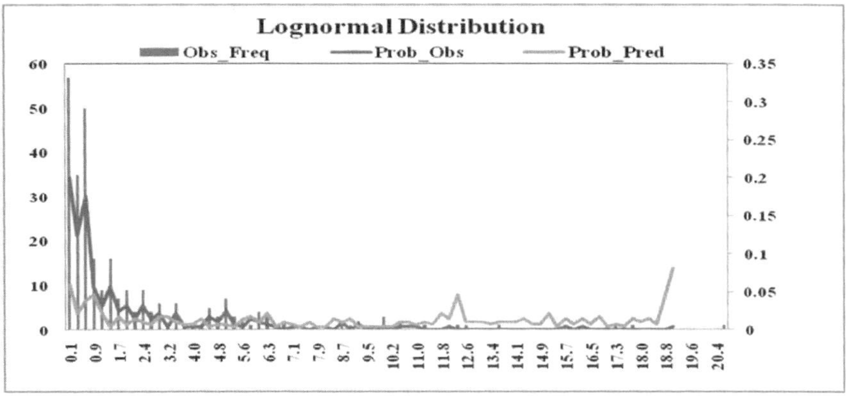
Lognormal Distribution
Obs_Freq
Prob_Obs
Prob_Pred

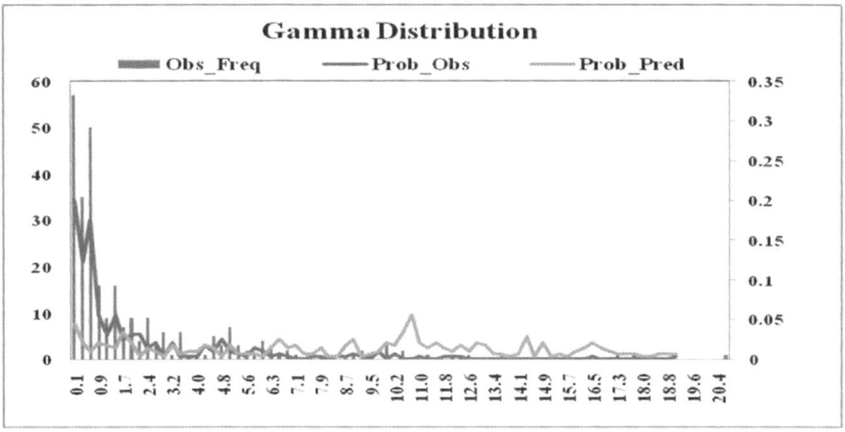
Gamma Distribution
Obs_Freq
Prob_Obs
Prob_Pred

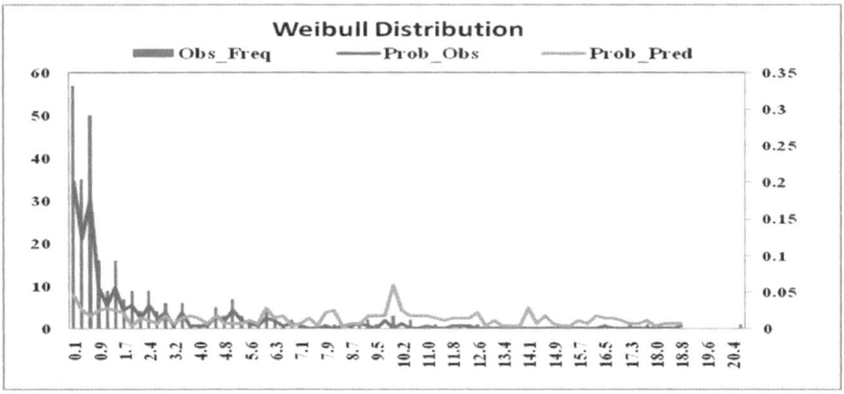
Weibull Distribution
Obs_Freq
Prob_Obs
Prob_Pred

The chi square value for Weibull, gamma, and Log-normal distributions are 3.2860, 4.7557 and 2.9270 respectively.

Fitting of statistical distributions on January precipitation data over NWI (North West India) regions:

For Jan

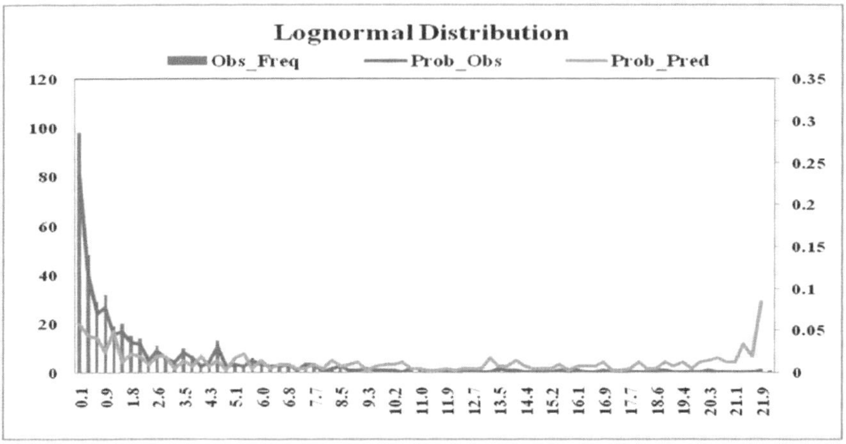

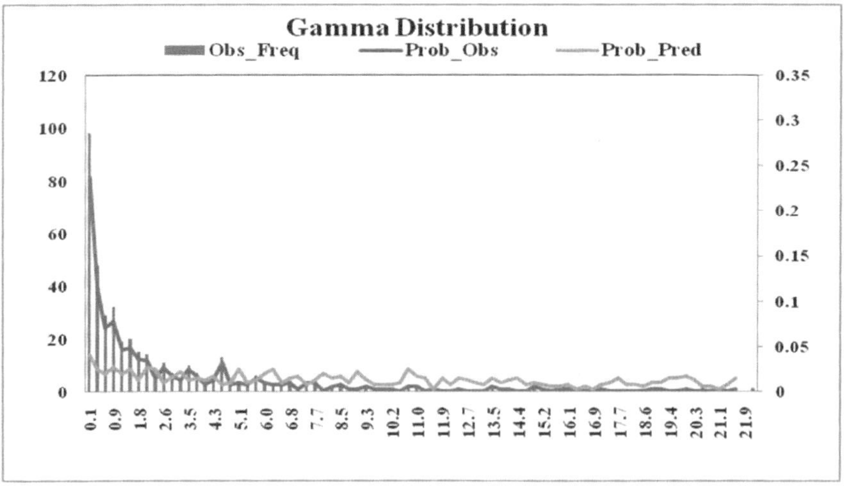

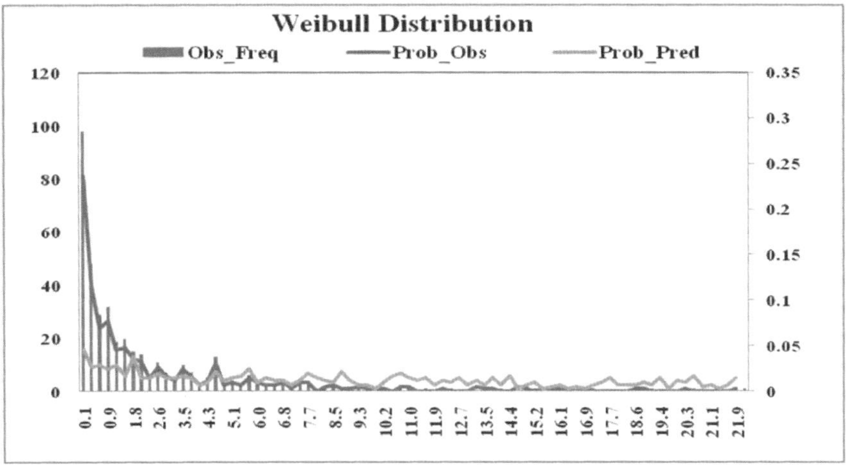

The chi square values for Weibull, Gamma, and Log-normal distributions are 1.7773, 2.1799 and 1.4653 respectively.

Fitting of statistical distributions on February precipitation data over NWI (North West India) regions:

For Feb

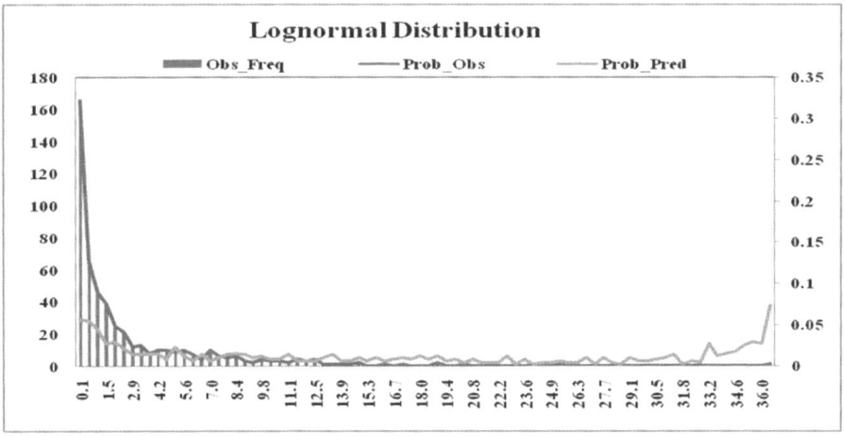

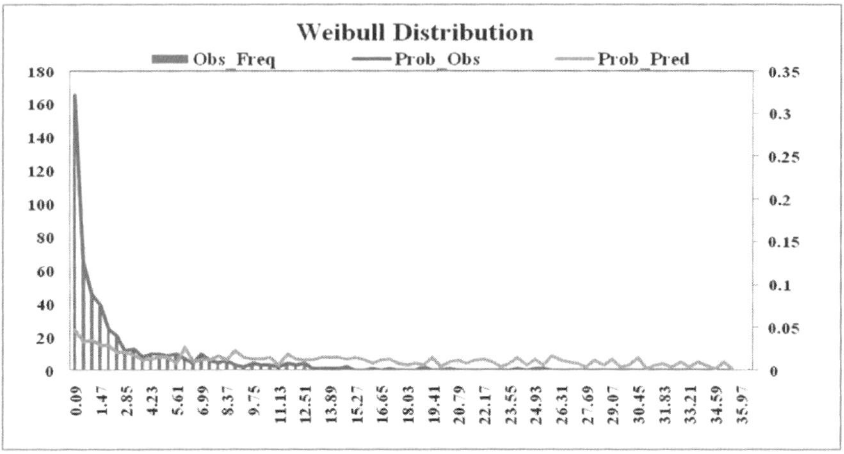

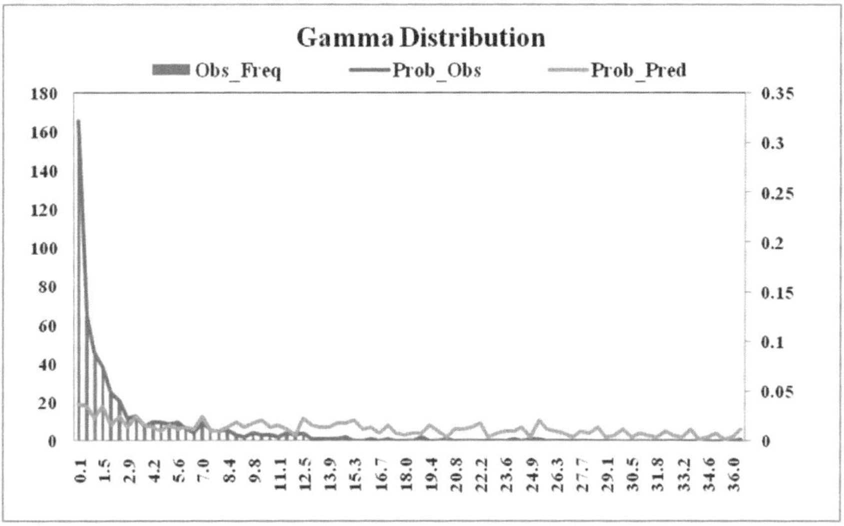

The chi square value for Weibull, Gamma, and Log-normal distribution are 2.6051, 3.2948 and 2.0958 respectively.

Chi-Square Table:

	Weibull	Gamma	Lognormal
DJF	2.9737	3.3595	2.1920
DEC	3.2860	4.7557	2.9270
JAN	1.7773	2.1799	1.4653
FEB	2.6051	3.2948	2.0958

Conclusion

From the above three distributions we concluded that from the Chi-Square table Lognormal distribution is suitable for nine regions such as: East Uttar Pradesh, West Uttar Pradesh, Uttarakhand, Haryana Chandigarh and Delhi, Punjab, Himachal Pradesh, Jammu and Kashmir, West Rajasthan, East Rajasthan regions for the month of Dec, Jan and Feb, also it shows the significantly different from the other distribution like Gamma and Weibull for any regions. So in this way we neglecting two distributions such as Weibull and gamma and considering the only one distribution *i.e.* lognormal distribution because it shows highly significant result other than two distributions.

References

1. Al-Fawzan M. A. (2000), "Methods for Estimating the Parameters of the Weibull Distribution", King Abdulaziz City for Science and Technology.
2. Al-Mansory H. B. (2005), "Statistical Analysis of Extreme Monthly Rainfall in Basrah City, South of Iraq", Marina Mesopotamica, Basrah, Iraq, Vol. 20, 2, PP. (283–296).
3. Almong, R. (1979), "A Study of the Application of the Lognormal Distribution to Corrective Maintenance Repair Time", M. Sc. Thesis, Naval postgraduate school.
4. Chow V. T., Maidment D. R. and Mays L. W. (1988), "Handbook of applied hydrology", McGraw-Hill series inwater Resources and Environmental Engineering, New York, ISBN 0-07-010810-2.
5. Forbes C., Evans M., Hastings N. and Peacock B. (2011), "Statistical Distributions", John Wiley & Sons, Inc., 4th Edition, New Jersey.
6. Guizani M., Rayes A., Khan B. and Al-Fuqaha A. (2010), "Network Modeling and Simulation: a Practical Perspective", John Wiley & Sons Ltd., U. S. A.
7. Maliva R. and Missimer T. (2012), "Arid Lands and Water Evaluation and Management", Springer-Verlag Berlin Heidelberg, ISBN 978-3-642-29104-3.
8. Murthy D. P., Xie M. and Jiang R. (2004), "Weibull Models", John Wiley & Sons, Inc., New Jersey.
9. Olumide B. A., Saidu M., Oluwasesan A. (2013), "Evaluation of Best Fit Probability Distribution Models for the Prediction of Rainfall and Runoff Volume (Case Study Tagwai Dam, Minna-Nigeria)", International Journal of Engineering and Technology, U. K., Vol. 3, No. 2, PP. (94–98), ISSN: 2049-3444.

10. Ozturk A. (1981), "On the Study of a Probability Distribution for Precipitation Totals", Department of Statistics, Division of Mathematical Science, Mimeograph Series (81-3).

On Systematic Sampling Strategies for a Varying Sample Size

K. B. Panda

PG Department of Statistics, Utkal University,
Bhubaneswar, Odisha-751004, India
kunja.st@utkaluniversity.ac.in
www.utkaluniversity.ac.in

In this paper, strategies based on linear systematic sampling (lss) and circular systematic sampling (css) procedures, when the population size is not a multiple of sample size, have been compared under certain models. It is found that the usual sample mean in lss can be employed in preference to the other competing known estimators in lss and css. A method of lss which can be looked upon as an extension of centered systematic sampling has been suggested for the purpose of eliminating the effect of linear trend for the case when population size is not a multiple of sample size and it has been shown that, under a linear trend model, this method gives better results than the one due to Bell house and Rao [3] who have extended Yates' end corrections to css. An empirical investigation based on two real populations confirms that a conventional strategy involving lss and the usual estimator based thereon is superior to two other well-known strategies when the population units are arranged in ascending order of a related variable.

Keywords: Systematic sampling; linear trend; superpopulation model; bilinear systematic sampling.

1. Introduction

A sampling procedure in which out of a population of N units we select a unit from the first k units followed by selection of every kth unit is called linear systematic sampling (lss). The number$k$ is known as sampling interval, while the unit selected initially from the first k units is called random start. Thus, having decided about k, we can express

$$N = nk + r, \ (0 \leq r < k) \tag{1.1}$$

where n is the largest integer not exceeding N/k, i.e., $n = [N/k]$. Such a procedure of systematic sampling will render samples of size n or $n+1$. However, if we are aiming at a fixed sample size, a way-out is provided by circular systematic sampling (css) which consists in selecting a random start from 1 to N and thereafter selecting every kth unit cyclically until a sample of n units is obtained. Cyclic selection means assigning the number $(n+1)$ to the first unit, $(n + 2)$ to the second unit and so on, in order to continue the selection procedure when the Nth unit has been reached. For $r = 0$, css reduces to lss.

In this paper, we consider the general case of lss, i.e., $N = nk + r$, $(0 \leq r < k)$ and compare, uninhibited by the fact that the sample size is n or $n + 1$, certain estimators based on lss with the usual estimator in css and, to rationalize this comparison, we have introduced the notion of randomized sample size. Thus, we would examine whether, for special populations characterized by certain known models, lss strategies are preferable to css strategy. Besides, we would be interested in identifying as to which of the two customary estimators under lss would fare better for the special populations considered by us. A detailed account is available in Panda [7].

Under lss, a conceivable estimator of the population mean $\bar{Y}$ is the simple sample mean defined as

$$\bar{y}_{sy} = \frac{1}{n+1}\sum_{j=1}^{n+1} y_{ij} \; (i \leq r)$$

$$= \frac{1}{n}\sum_{j=1}^{n} y_{ij}, \; (i > r)$$

where i is the random start and y_{ij} is the jth unit in the ith systematic sample. It can be noted that the simple sample mean $\bar{y}_{sy}$ is not an unbiasedestimator of $\bar{Y}$ in the general case of lss under (1.1). However, the bias associated with $\bar{y}_{sy}$ is quite small and can be neglected if sample size is large. However, for lss, an unbiased estimator which is usually mentioned in the literature, when (1.1) holds, is given by

$$\bar{y}'_{sy} = \frac{k}{N} y_{i.}, \quad (i \leq k)$$

where $y_{i.}$ is the total of all y-measurements in the ith systematic sample. Under css, the usual estimator which is mean of units selected is unbiased for $\bar{Y}$ and we will denote it by $\bar{y}_{css}$. All the three estimators are identical when $r = 0$.

2. Comparison of Systematic Sampling Strategies

To compare the three estimators $\bar{y}_{sy}$, $\bar{y}'_{sy}$ and $\bar{y}_{css}$ under some special populations, we have their mean square errors given by

$$MSE(\bar{y}_{sy}) = \frac{1}{k}\sum_{i=1}^{k}(\bar{y}_{i.} - \bar{Y})^2 = V_1, \text{ say},$$

$$V(\bar{y}'_{sy}) = \frac{1}{k}\sum_{i=1}^{k}(\frac{k}{N} y_{i.} - \bar{Y})^2 = V_2, \text{ say},$$

$$\text{and } V(\bar{y}_{css}) = \frac{1}{N}\sum_{i=1}^{k}(\bar{y}_{i.} - \bar{Y})^2 = V_{3,n}, \text{ say},$$

where $\bar{y}_{i.}$ is the mean of the ith systematic sample. To be precise V_1 represents mean square error because $\bar{y}_{sy}$ is biased, while V_2 and $V_{3,n}$ signify the variances as both $\bar{y}'_{sy}$ and $\bar{y}_{css}$ are unbiased. It would be apt to clarify that $V_{3,n}$ is the variance of $\bar{y}_{css}$ based on a fixed sample size n. It may be noted that, under (1.1), the probability that both $\bar{y}_{sy}$ and $\bar{y}'_{sy}$ are based on $(n+1)$ unit is r/k and that they are based on n units is $(k-r)/k$. Hence, in order to rationalize the comparison of $\bar{y}_{css}$ with $\bar{y}_{sy}$ and $\bar{y}'_{sy}$ in terms of the same expected sample size, say, n^*. Let

$$n^* = \begin{cases} n+1 \; with\; probability\; r/k \\ n \; with\; probability \dfrac{k-r}{k}. \end{cases}$$

So that the expected value of n^* is N/k. Then the variance of $\bar{y}_{css}$ based on randomized sample size will become

$$V(\bar{y}^*_{css}) = \tfrac{r}{k} V_{3,n+1} + \tfrac{k-r}{k} V_{3,n} = V_3, \text{ say,}$$

where $\bar{y}^*_{css}$ is the mean based on n^* under css and $V_{3,n+1}$ is the variance under css for sample size $(n+1)$.

2.1. *Populations with random ordering*

When the finite population of N units is in random order, Madow and Madow [5] showed that, for $r = 0$ in (1.1), the mean square error of $\bar{y}_{sy}$ (or $\bar{y}'_{sy}$) averaged over the $N!$ permutations of the N population values $y_1, y_2, \dots\dots\dots, y_N$ yielded $\frac{N-n}{Nn} S^2$ which tantamounts to the variance of the customary estimator of $\bar{Y}$ under simple random sampling without replacement where

$$S^2 = \tfrac{1}{N-1} \textstyle\sum_{i=1}^{N} (y_i - \bar{Y})^2.$$

The linear model ξ that embodies the random permutation assumption is given by

$$y_i = \mu + e_i, \; (i = 1, 2, \dots., N) \tag{2.1}$$

where $E_\xi = 0$, $E_\xi(e_i^2) = \sigma^2$ and $E_\xi(e_i e_j) = -\frac{\sigma^2}{N-1} (i \neq j)$. Such a random permutation model has been considered by Rao and Bellhouse [7]. In the context of comparison of the three estimators under consideration, it would be apt to mention that the estimators $\bar{y}_{sy}$, $\bar{y}'_{sy}$ and $\bar{y}_{css}$ are $p\xi$ (design-cum-model) unbiased. Hence, with a view to comparing them, we have the average (model expected or anticipated) mean square errors given by

$$E_\xi(V_1) = \frac{nk(n+1)(k-1) + r(k-r)}{nk(n+1)(N-1)} \sigma^2$$

$$E_\xi(V_2) = \frac{(N^2 - n^2 k - 2nr - r)k}{N^2(N-1)} \sigma^2 + \frac{r(k-r)}{N^2} \mu^2$$

$$\text{and } E_\xi(V_3) = \frac{nk(n+1)(k-1)+r(k-r)}{nk(n+1)(N-1)}\sigma^2$$

Thus, under the model ξ, $\bar{y}_{sy}$ and $\bar{y}^*_{css}$ perform equally well. Besides, if we consider the usual sample mean based on the randomized sample size n^* in simple random sampling without replacement, we find that its average variance is equal to $E_\xi(V_1)$ and $E_\xi(V_3)$. However, it can be seen from the relevant average mean square errors that $\bar{y}_{sy}$ performs better than, $\bar{y}'_{sy}$ if

$$\frac{\sigma^2}{\mu^2} \leq \frac{nk(n+1)(N-1)}{N^2 + nkr(n+1)},$$

and such a condition will always hold if $\sigma^2/\mu^2 \leq 1$. This, in fact, would imply that, in a large variety of practical situations, $\bar{y}_{sy}$ will fare better than $\bar{y}'_{sy}$ provided the model given by (2.1) prevails, because the quantity σ/μ, representing the coefficient of variation in the superpopulation, often does not exceed unity.

2.2. *Populations generated under a classical model*

Let's consider that the population at hand has been drawn at random from a superpopulation ξ_0 in which

$$E_{\xi_0}(y_i) = \mu, E_{\xi_0}(y_i - \mu)^2 = \sigma^2 \quad (i = 1,2, \ldots\ldots..N)$$

$$E_{\xi_0}(y_i - \mu)(y_j - \mu) = 0 \quad (i \neq j) \tag{2.2}$$

Here, the crucial condition is that all y_i's have the same mean μ and there is no linear trend and linear correlation existing between y_i and $y_j (i \neq j)$ at any time point. Averaging over E_{ξ_0}, we have

$$E_{\xi_0}(V_1) = \frac{nk(n+1)(k-1) + r(k-r)}{n(n+1)kN}\sigma^2$$

$$E_{\xi_0}(V_2) = \frac{(k-1)}{N}\sigma^2 + \frac{r(k-r)}{N^2}\mu^2$$

$$E_{\xi_0}(V_3) = \frac{nk(n+1)(k-1) + r(k-r)}{n(n+1)kN}\sigma^2.$$

As in the case of random permutation model considered in subsection 2.1, we find that $\bar{y}_{sy}$ and $\bar{y}^*_{css}$ have the same average mean square error, while $\bar{y}_{sy}$ has a lower average mean square error than that of $\bar{y}'_{sy}$ if

$$\frac{\sigma^2}{\mu^2} \leq \frac{nk(n+1)}{N},$$

which is tantamount to a sufficient condition that the square of the coefficient of variation in the superpopulation is smaller than the sample size n, i.e.,

$$\frac{\sigma^2}{\mu^2} \leq n.$$

It may be noted here that, in practice, one seldom encounters the coefficient of variation so large as to exceed $n^{1/2}$, meaning thereby that $\bar{y}_{sy}$, under the model (2.2), would always be better than $\bar{y}'_{sy}$.

2.3. *Populations with a linear trend*

Let's consider a fixed population linear trend model, viz.,

$$y_i = i \qquad (i = 1,2, \dots\dots\dots, N). \tag{2.3}$$

It may be noted that all the three estimators, viz., $\bar{y}_{sy}$, $\bar{y}'_{sy}$ and $\bar{y}^*_{css}$ are design-unbiased under (2.3). The variances of these three estimators under (2.3) are obtainable as

$$V_1 = \frac{k^2-1}{12} - \frac{r(k-r)}{4} = A_1, \text{ say}$$

$$V_2 = \frac{k}{N^2}\left[\frac{n^2k(k^2-1)}{12} + \frac{r(r+1)}{2}\left\{nk - \frac{1}{3}(n-1)(2r+1) - \frac{r(r+1)}{2k}\right\}\right] = A_2, say,$$

$$V_3 = \frac{k^2-1}{12} + \frac{r^2}{12} + \frac{r(nk+k-r)}{6n(n+1)} = A_3, \text{ say}.$$

A comparison of these variances yields

$$V_1 \leq V_2 \leq V_3,$$

thus signalling the use of $\bar{y}_{sy}$ in preference to the other estimators. Now, let the linear trend be modelled as

$$y_i = \alpha + \beta i \quad (i = 1,2, \dots\dots\dots, N) \tag{2.4}$$

wherein we assume that both the constants α and β are either positive or negative. We have deliberately chosen to allude to the model (2.4) because, unlike the other two estimators, $\bar{y}'_{sy}$ has a variance which is a function of α. The variances of the three estimators under (2.4) become

$$V_1 = A_1\beta^2$$

$$V_2 = A_2\beta^2 + \frac{r(k-r)}{N^2}\alpha[\alpha + (N+1)\beta]$$

$$V_3 = A_3\beta^2,$$

which, in turn, enable us to conclude once again that $\bar{y}_{sy}$ is superior to both $\bar{y}'_{sy}$ and $\bar{y}^*_{css}$, while, unless $\alpha = 0$, nothing can be said unconditionally regarding the performance of $\bar{y}'_{sy}$ vis-à-vis $\bar{y}^*_{css}$. Thus, $\bar{y}_{sy}$ is invariably the best in the presence of linear trend as characterized by (2.3) and (2.4).

Now, if we turn to superpopulation linear trend model (say, ξ_1), i.e.,

$$y_i = i + e_i \quad (i = 1, 2, \dots \dots \dots, N) \tag{2.5}$$

Where $E_{\xi_1}(e_i) = 0, E_{\xi_1}(e_i^2) = \sigma^2$ and $E_{\xi_1}(e_i e_j) = 0 \ (i \neq j)$, a comparison of the three estimators, viz., $\bar{y}_{sy}$, $\bar{y}'_{sy}$ and $\bar{y}^*_{css}$ that are design-cum-model unbiased will involve computation of their average variances given by

$$E_{\xi_1}(V_1) = A_1 + \frac{1}{k}[\frac{r}{n+1} + \frac{k-r}{n} - \frac{k}{N}]\sigma^2$$

$$E_{\xi_1}(V_2) = A_2 + \frac{k-1}{N}\sigma^2$$

and $E_{\xi_1}(V_3) = A_3 + \left[\frac{r}{k}\left(\frac{1}{n+1} - \frac{1}{n}\right) + \left(\frac{1}{n} - \frac{1}{N}\right)\right]\sigma^2,$

which can then be used to verify that, in the process of switch over from (2.3) to (2.5), the superiority of $\bar{y}_{sy}$ vis-à-vis $\bar{y}'_{sy}$ is not significantly affected due to superimposition of an error component provided the error variance is small and the sample size is large. Besides, both $\bar{y}_{sy}$ and $\bar{y}'_{sy}$ continue to perform better than $\bar{y}^*_{css}$ under (2.5). However, if the linear trend is modeled by ξ_2 as specified by

$$y_i = \alpha + \beta i + e_i \quad (i = 1, 2, \dots \dots \dots, N) \tag{2.6}$$

where $E_{\xi_2}(e_i) = 0, E_{\xi_2}(e_i^2) = \sigma^2$ and $E_{\xi_2}(e_i e_j) = 0 \ (i \neq j)$, the conclusions relating to the performance of $\bar{y}_{sy}$ relative to $\bar{y}'_{sy}$ and $\bar{y}^*_{css}$ are the same as stated under (2.5) while nothing can be said unconditionally (due to the presence of α in the average variance of $\bar{y}'_{sy}$) regarding the performance of $\bar{y}'_{sy}$ vis-à-vis $\bar{y}^*_{css}$.

2.4. *Elimination of the effect of linear trend*

In this subsection, the estimator $\bar{y}'_{sy}$ has been omitted as it is not easily amenable to manipulations for complete elimination of the effect of linear trend.

Bellhouse and Rao [3] have examined and established the efficacy of centred systematic sampling in the presence of a trend, while Agrawal and Jain [1] have concluded that the centred systematic sampling provides an efficiency robust estimator in the sense that it maintains its superiority vis-à-vis the other competing estimators in the face of a variety of conditions that are assumed to characterize the underlying trend in the population. The work of these authors centering round the case when $N = nk$ motivates us to extend it to the case when $N = nk + r \ (0 \leq r < k)$, i.e., we have to devise a systematic sampling method which is a generalization of centred systematic sampling method and aims at eliminating the effect of a linear trend. The extended method to be called modified centred systematic sampling consists in selecting

(a) the column $(r + k + 1)/2$ (comprising n units) if either r or k is odd, or

(b) the column $(r+1)/2$ (comprising $(n+1)$ units) if both r and k are odd, or

(c) the column $(r+k)/2$ or $(r+k+2)/2$ (comprising n units) if both r and k are even.

It may be noted here that, when $r=0$, it is the sampling scheme given in (a), which reduces to centred systematic sampling, that will be applicable. The estimation with recourse to (a) and (b) results in complete elimination of the effect of linear trend characterized by (2.3) and (2.4). However, the use of (c) entails invocation of the end corrections with a view to bringing about the complete elimination of the effect of linear trend, i.e., the first and the last observations in the column $(r+k)/2$ or $(r+k+2)/2$ are weighted by $1+\theta$ and $1-\theta$, respectively, while the remaining observations in the column are weighted by 1 before taking the average, the value of θ being

$$\theta = \begin{cases} -\dfrac{n}{2(n-1)k} \; if\ we\ choose\ the\ column\ \dfrac{r+k}{2} \\ \dfrac{n}{2(n-1)k} \; if\ we\ choose\ the\ column\ \dfrac{r+k+2}{2}. \end{cases}$$

Under the model (2.6), the average variance of $\bar{y}_{sy}$ is given by

$$E_{\xi_2}V(\bar{y}_{sy})_a = \left(\frac{1}{n}-\frac{1}{N}\right)\sigma^2 \tag{2.7}$$

for the scheme given in (a),

$$E_{\xi_2}V(\bar{y}_{sy})_b = \left(\frac{1}{n+1}-\frac{1}{N}\right)\sigma^2 \tag{2.8}$$

for the scheme given in (b) and

$$E_{\xi_2}V(\bar{y}_{sy})_c = \left(\frac{1}{n}-\frac{1}{N}\right)\sigma^2 + \frac{1}{2(n-1)^2k^2}\sigma^2 \tag{2.9}$$

for the scheme given in (c) after applying end corrections as indicated above.

Bellhouse and Rao [3] have extended Yates' end corrections to circular systematic sampling. The average variance of $\bar{y}_{css}$ based on a sample of n units under (2.6) after incorporating the Bellhouse-Rao end corrections is obtainable as

$$E_{\xi_2}V(\bar{y}_{css}) =$$
$$\frac{N-n}{Nn}\sigma^2 + \frac{\sigma^2}{6(N-k)^2}\left\{k^2+r^2+\frac{2rk}{n}-1+\frac{(N-2k-r)r(r+k)(r+k+1)(r+k-1)}{N(n-1)^2k^2}\right\} \tag{2.10}$$

which could be suitably modified for a sample of size $(n+1)$ if (2.10) is to be compared with (2.8).

Now, to examine the performance of $\bar{y}_{sy}$ relative to $\bar{y}_{css}$ under the model ξ_2, a comparison of (2.7), (2.8) and (2.9) with (2.10) is carried out after taking the involved sample size into consideration, and it is found that $\bar{y}_{sy}$ is invariably more efficient than $\bar{y}_{css}$.

It will be apt to point out that the average variance of $\bar{y}_{sy}$ for the sampling schemes in (a) and (b) under the model (2.6) assumes the minimum value of the Godambe-Joshi [4] lower bound given by $\left(\frac{1}{n}-\frac{1}{N}\right)\sigma^2$.

3. Strategies for Predetermined Sample sizes

In the preceding subsections, we compared different estimators, after starting with specified k, that were based on a sample size n or $n+1$. Here, we consider selection of samples of predetermined size n from a population of size $N = nk + r^*$ $(0 \le r^* < n)$. A known method of ensuring a sample of size n is circular systematic sampling. Agrawal *et al.* [2], in the context of linear trend, have proposed bilinear systematic sampling to obtain a predetermined sample size. For this purpose, they have mooted the following approach by splitting N as

$$\begin{aligned} N &= nk + r^* \\ &= (n-r^*)k + r^*(k+1) \\ &= n_1k_1 + n_2k_2 \\ &= N_1 + N_2, \end{aligned}$$

where $N_1 = n_1k_1, N_2 = n_2k_2, n_1 = (n-r^*), n_2 = r^*, k_1 = k$ and $k_2 = k+1$, the population of size N has now been split into two subpopulations, say, P_1 and P_2 of sizes N_1 and N_2, respectively, wherefrom samples of sizes n_1 and n_2 are drawn, thus making up a total of sample size n. It may be noted that, for each of the subpopulations P_1 and P_2, the usual systematic sampling methods, viz., linear systematic sampling, centred systematic sampling, modified systematic sampling, balanced systematic sampling etc. can be employed for the purpose of drawing samples with a view to eliminating the effect of linear trend. A sampling situation where one of the sampling methods indicated in the preceding line is applied to P_1 and /or P_2 will be termed as adjusted bilinear systematic sampling. Let $\bar{y}_b$ denote the mean based on bilinear systematic sampling. It would now be of interest to examine the performance of $\bar{y}_b$ vis-à-vis $\bar{y}_{css}$ in the presence of a linear trend as characterized by the model (2.6). The average variances of $\bar{y}_b$ and $\bar{y}_{css}$ under (2.6) are found to be

$$E_{\xi_2}V(\bar{y}_b) = \left(\frac{1}{n}-\frac{1}{N}\right)\sigma^2 + \frac{r^*(n-r^*)}{nN^2}\sigma^2 + \beta^2\frac{k(k+1)}{12N^2}\left[(n-r^*)^2k(k-1) + {r^*}^2(k+1)(k+2)\right]$$

and $E_{\xi_2}V(\bar{y}_{css}) = \left(\frac{1}{n}-\frac{1}{N}\right)\sigma^2 + \beta^2\left\{\frac{k^2-1}{12} + \frac{{r^*}^2}{12} + \frac{r^*k}{12}\right\}$, implying thereby that, for large populations, $\bar{y}_b$ will be better than $\bar{y}_{css}$.

If adjusted bilinear systematic sampling (described above) results in complete elimination of linear trend, then the average variance of $\bar{y}_b$ would be given by

$$E_{\xi_2}V(\bar{y}_b) = \left(\frac{1}{n} - \frac{1}{N}\right)\sigma^2 + \frac{r^*(n-r^*)}{nN^2}\sigma^2,$$

which, on comparison with (2.10), shows that $\bar{y}_b$ is more efficient than $\bar{y}_{css}$. Thus, in the case of a predetermined sample size, the above findings point to the fact that it would be worth being well-disposed to $\bar{y}_b$ (superseding $\bar{y}_{css}$) under models given in (2.5) and (2.6) that depict the presence of a linear trend. These conclusions will remain valid even under fixed population linear trend models considered in (2.3) and (2.4).

4. Empirical Study

With a view to comparing the three strategies, viz.,

(i) $S_1 = (\bar{y}_{sy}, P)$

(ii) $S_2 = (\bar{y}'_{sy}, P)$ and

(iii) $S_3 = (\bar{y}^*_{sy}, P^*)$,

where P and P^* are linear and circular systematic sampling designs, respectively, we consider the following two real populations taken from Murthy ([6], pp. 127–130 and p. 228):

(1) Population 1: 1961 census population of 128 villages arranged in increasing order of 1951 census population.

(2) Population 2: Output for 80 factories in a region arranged in increasing order of number of workers.

Table 4.1 (for population 1) and Table 4.2 (for Population 2) present the mean square errors $V_1, V_2 and\ V_3$ (as spelt out in section 2 in respect of strategies $S_1, S_2\ and\ S_3$) for various combinations of $n, k\ and\ r$.

Table 4.1. Population 1 (N=128), mean square errors of the competing strategies S_1, S_2 and S_3.

n	k	r	V_1	V_2	V_3
4	32	0	487733.00	487733.00	487733.00
5	25	3	264330.10	396524.90	400095.10
6	21	2	219934.10	291774.80	316839.10
7	18	2	126896.30	186482.80	190364.40
8	16	0	133174.30	133174.30	133174.30
9	14	2	86448.04	136935.20	148904.60
10	12	8	45783.33	80460.45	103755.40
11	11	7	31191.53	63669.90	79865.01
12	10	8	48383.24	81176.74	55861.88

Table 4.1. (*Continued*)

n	k	r	V_1	V_2	V_3
14	9	2	25559.73	46485.01	46837.50
16	8	0	39643.52	39643.52	39643.52
18	7	2	34034.84	58720.10	65442.19
21	6	2	24764.43	37084.22	38308.85
25	5	3	16746.92	36600.92	16186.82
32	4	0	5887.06	5887.06	5887.06

Table 4.2. Population 2 (N=80), mean square errors of the competing strategies S_1, S_2 *and* S_3.

n	k	r	V_1	V_2	V_{32}
2	40	0	970080.90	970080.90	970080.90
3	26	2	224256.10	365110.80	411250.30
6	13	2	71226.49	137495.70	155978.10
7	11	3	35906.24	107086.30	126035.60
10	8	0	54267.18	54267.18	54267.18
11	7	3	13458.56	77272.81	52137.16
13	6	2	7830.80	45878.28	37544.24
26	3	2	1080.02	12299.66	7426.23
40	2	0	630.64	630.64	630.64

Table 4.1 and Table 4.2 bear it out that, when the arrangement of units is in ascending order of a related variable, the strategy S_1 is the most efficient followed by S_2 and S_3 in order of efficiency. These findings are in keeping with the results obtained in subsection 2.3 relating to "population with a linear trend".

Acknowledgement

I thank Dr. M.C.Agrawal, former Professor of Statistics, University of Delhi for his help and support in the preparation of this paper.

References

1. Agrawal, M. C. and Jain, N.(1988). Comparison of some sampling strategies in the presence of a trend. Jour. Ind. Soc. Ag. Statistics, Vol. XL, 3, 191–201.
2. Agrawal, M. C., Bersales, Lisa Grace S. and Acusta, Andre P. (1991). Bilinear systematic sampling. Papers and Proceedings of joint conference on Statistics, National Statistical Coordination Board, Manila.
3. Bellhouse, D. R. and Rao, J. N. K. (1975). Systematic sampling in the presence of a trend. Biometrika, 62, 97.

4. Godambe, V. P. and Joshi, V. M. (1965). Admissibility and Bayes estimation in sampling finite populations I, Ann. Math. Stat., 36, 1707–1722.
5. Madow, W. G. and Madow, L. H. (1944). On the theory of systematic sampling, Ann. Math. Stat., 15, 1–24.
6. Murthy (1967). Sampling Theory and Methods, Statistical Publishing Society, Calcutta, India.
7. Panda, K. B. (1994), Some Contributions to the Theory of Survey Sampling, Ph.D. thesis submitted to the University of Delhi, Delhi, India.
8. Rao, J. N. K. and Bellhouse, D. R. (1978). Estimation of finite population mean under generalized random permutation model. J. Statist. Plann. Infer., 2, 125–141.

Estimation of Measurement Variance Under Two-Stage Sampling: Estimation of Population Mean

Professor Pulakesh Maiti
Indian Statistical Institute
Kolkata, West Bengal, India
E-mail: pulakesh@isical.ac.in, maiti.pulakesh@gmail.com

Considered is the mean square error decomposition model for estimation of population mean and the problem of estimation of measurement variance under two-stage sampling. The measurement technique considered here is that of employing personal interviewers for data collection. The sample mean has been defined taking care of interviewer's effects and the measurement variance associated with the estimator has been derived with an effort to estimating the measurement variance.

Keywords: Decomposition model; Two-stage sampling; Measurement variance.

1. Introduction to General Model for Measurement Variability

Considered is the set up of simple i.e., direct response on a quantity response variable in the context of a finite labeled population. It so happens that we need investigators which will have intervention on the response profile finally received by data collecting agency. The problem considered is to unbiasedly estimate the finite population mean and also to estimate measurement variance by incorporating a two stage sampling design and by administering the sampling design in a situation, where the above interventions are likely to be present.

The importance of non-sample or measurement errors arising due to interventions of survey management personnel, namely investigators, supervisors, coders etc., has long been recognized [for numerous references see e.g., the comprehensive papers by Mahalanobis (1946), Hansen *et al.* (1961), Bailar and Dalenius (1969)].

Non-sampling bias is a measure of difference between the expected value of repeated observations and the corresponding true value, whereas non-sampling variance/measurement variance measures variations of observed values for fixed samples in hypothetical repetitions, if it is agreed upon that the survey is conceptually repeatable under identical conditions. More precisely, it is assumed that a measurement observed has a well defined, though quite likely unknown probability distribution. Non-sampling variance has such components as (a) simple or uncorrelated response variance and (b) correlated response variance.

1.1. *Model description*

The models in the literature for the effect of measurement errors are expressed either as mean square error decomposition model or as mixed linear models. Sometimes, the net bias is assumed to be zero so, that models deal with variability only. The mean square error formulation and the linear model formulation tend to merge, as various models focus on a particular source of error– most usually of the interviewer. In such a case, both approaches can be expressed in terms of a linear model, although the assumptions are different under different approaches. The major difference between the two approaches is that decomposition approach often has a component attributable to the interaction between sampling and measurement errors and the linear model approach generally quits this component. In addition, the linear model approach usually defines response variability in terms of variation around the true value.

1.2. *Hansen-Hurwitz-Bershad model (1961)*

Let X_i be the true value of the variable X for the ith element, and Y_{it} be the measurement of ith element at t^{th} instant with

$$E_{s>i}\,E_{t|s>i}\,(Y_{it}) = Y_i\,,$$

where t stands for the random behavior of the ith response.

Another formulation of this model is by Bailar and Dalenius (1969) which states that "the survey procedure is such that the process of recording response for any unit is repeatable; it gives rise to a random variable whose value at trial t is not correlated with its value at any other trial".

2. Formulation of Mean Square Error of the Sample Mean Under the Model

2.1. *Without specifying measurement technique*

Assuming a simple random sampling of size n, and without specifying the measurement technique, Hansen-Hurwitz-Bershad (1961) derives the mean square error, of $\bar{y}_t$, the sample mean,

$$\bar{y}_t = \sum_{i=1}^{n} y_{it} \Big/ n \qquad (2.1)$$

With $V(\bar{y}_t)$ = Sample variance (SV) + Measurement variance (MV) + covariance of response and sampling distribution (CRS)

$$= SV + \frac{1}{n}\left[\sigma_d^2 + (n-1)\rho\sigma_d^2\right] + CRS \tag{2.2}$$

Where,

$$\sigma_d^2 = \sum_{i=1}^{N} Var(y_{it}) \Big/ N \,,\ \rho\sigma_d^2 = \sum_{i\#i'}\sum Cov(y_{it}, y_{i't}) \Big/ N(N-1).$$

Koch (1993) represents a response error model for the multivariate unequal probability case and he arrives at a general model in which the response or measurement variance is decomposed into three components with

$$V(\bar{y}_t) = SV + (SRV + CRV + IRV) + IV\,, \tag{2.3}$$

Where, the response variance now consists of three components namely simple response variance (SRV), the correlated response variance (CRV) and the interaction response variance.

2.2. *Specifying measurement technique*

A number of different models in the literature treat measurement errors for surveys that employ personal interviewers. In the most general finite population case, we have a finite population of elements $i = 1,2,...,N$. Also let us consider a finite population of interviewers indexed by $j = 1,2,...,K$. Now with the sampling designs of selecting a sample of elements, and a sample of interviewers, and with a survey design for assigning elements to the selected interviewers, an estimator of population mean is defined as

$$\bar{y}_t = \frac{1}{n}\sum_{j=1}^{K}\sum_{i=1}^{N} U_i V_j C_{ij}\, y_{ijt}\,, \tag{2.4}$$

where, U_i, V_j are indicator variables to denote presence/absence of the ith element, j^{th} investigator into the samples of elements and investigators and C_{ij} stands for assignment of ith element to the jth investigator.

Under the assumption of no interaction between the sampling and measurement errors, we have

$$V(\bar{y}_t) = \text{Measurement Variance} + \text{Sample Variance.} \tag{2.5}$$

The measurement variance has such components as,

a) Variance of the measurement of the ith element by the jth investigator given that both element i and interviewer j have been selected for the survey and element i has been assigned to interviewer j (2.6)

b) Measurement covariance between the elements measured by the same investigator.
c) Measurement covariance between interviewers within elements, and
(d) Between element between interviewer covariance. (Lesseler *et al.*, 1992)

The above results in (2.5), (2.6) were further extended by Maiti (2009) taking care of both investigator as well as supervisor effects. For discussion of other similar models we refer to Census Bureau Model/Cochran Model, Raj Model, Murthy Model (Lessler *et al.* 1992).

2.3. *Estimation of measurement variance*

Estimation of the effect of measurement variability on survey statistics requires that some types of experimental procedure (survey design) be introduced into the survey, so that there is replication. Three types of such approaches as (a) repetition on same population elements, (b) use of interpenetrated or replicated samples or (c) use of methods with a combination of the replicated samples and repeat methods.

2.3.1. *Based on the approach of repeat measurements*

The statistics defined by

$$BMWE = \frac{2}{n}\sum_{j=1}^{K}\sum_{i=1}^{N} C_{ij}\left(y_{ij1t} - y_{ij2t}\right)^2 \tag{2.7}$$

estimates simple measurement variance (SMV), with the assumption that repeat measurements are independent of the original once.

2.3.2. *Use of interpenetrating or replicated samples*

Let us consider a fixed population of K-interviewers and an independent sample of m elements for each interviewer.

Let

$$\bar{y}_{jt} = \sum_{i=1}^{N} C_{ij}\, y_{ijt} \Big/ m, \tag{2.8}$$

$$\bar{y}_t = \sum_{j=1}^{K} \bar{y}_{jt} \Big/ k$$

and $BIMS = \sum_{j} \left(\bar{y}_{jt} - \bar{y}_t\right)^2 \Big/ K(K-1);$

It may be shown that

$$E(BIMS) = V(\bar{y}_t) \tag{2.9}$$

Thus, BIMS estimates total variance, without estimating measurement variance separately. Another sample statistic for estimation of Measurement Variance (MV) is Within-Interviewer Mean Square:

$$WIMS = \frac{1}{k(m-1)} \sum_{j=1}^{K} V_j \sum_{i=1}^{N} C_{ij} \left(y_{ijt} - \bar{y}_{jt}\right)^2 \tag{2.10}$$

with,

$$E(WIMS) = MV + S_Y^2 = \sigma_d^2 (1-\rho) + S_Y^2 \tag{2.11}$$

Where a sample of *SRS (K,k)* and k-interpenetrating samples of size m for each investigator have been drawn, (Lesseler, 1992).

Remark

It may be noted that the previous results relate to estimation of measurement variance based on a single-stage sampling design. In the next section, what follows is the study of estimation of measurement variance under a two-stage sampling design.

3. Estimation of Measurement Variance under Two-Stage Sampling

Let MN number of population elements be grouped into N-number of first stage units of equal size of M-number of second stage units. An *SRS (N, n)* is selected followed by simultaneous application of *SRS (M, m)* to each of the selected n-number of first stage units. Let an *SRS (L, l)* sample of investigators be drawn.

Let,

$$U_i = \begin{cases} 1, if\ i^{th} \text{ first stage unit is selected;} \\ 0, \text{otherwise.} \end{cases}$$

$$V_i = \begin{cases} 1, if\ j^{th} \text{ investigator is selected;} \\ 0, \text{otherwise.} \end{cases}$$

$$\delta_{k(i)} = \begin{cases} 1, if\ k(i)^{th} \text{ second stage unit from } i^{th} \text{first stage unit is selected;} \\ 0, \text{otherwise.} \end{cases}$$

$$C_{jk(i)} = \begin{cases} 1, if\ k(i)^{th} \text{ second stage unit from } i^{th} \text{first stage} \\ \text{unit is assigned to } j^{th} \text{ selected investigator;} \\ 0, \text{otherwise.} \end{cases}$$

Let

$$\bar{y}_{jt} = \sum_{i=1}^{N} U_i \sum_{k(i)} \delta_{k(i)} C_{jk(i)} y_{jk(i)t} \Big/ m^*$$

Where, $\sum_{k(i)} \delta_{k(i)} C_{jk(i)} = m^*$, the equal number of elements allotted to the jth investigator ($j = 1,2,\ldots,l$);

$$s_{jt}^2 = \sum_{i=1}^{N} U_i \sum_{k(i)=1}^{M} \delta_{k(i)} C_{jk(i)} (y_{jk(i)t} - \bar{y}_{jt})^2 \Big/ (m^* - 1)$$

$$\bar{y}_t = \sum_{j=1}^{L} V_j \bar{y}_{jt} \Big/ l$$

Let the statistic WIMS be defined as

$$WIMS = \sum_{j=1}^{L} V_j s_{jt}^2 \Big/ l$$

$$= \sum_{j=1}^{L} V_j \sum_{i=1}^{N} U_i \sum_{k(i)=1}^{M} \delta_{k(i)} C_{jk(i)} (y_{jk(i)t} - \bar{y}_{jt})^2 \Big/ l(m^* - 1)$$

$$= \Bigg[\frac{(m^*-1)}{m^*} \sum_i \sum_j \sum_{k(i)} U_i V_j \delta_{k(i)} C_{jk(i)} y_{jk(i)t}^2$$

$$- \frac{1}{m^*} \Bigg\{ \sum_j \sum_i \sum_{k\#k'} U_i V_j \delta_{k(i)} \delta_{k'(i)} C_{jk(i)} C_{jk'(i)} y_{jk(i)t} y_{jk'(i)}$$

$$+ \sum_j \sum_{i\#i'} \sum_k U_i U_{i'} V_j \delta_{k(i)} \delta_{k(i')} C_{jk(i)} C_{jk(i')} y_{jk(i')t} y_{jk(i)t}$$

$$+ \sum_j \sum_{i\#i'} \sum_{k\#k'} U_i U_{i'} V_j \delta_{k(i)} \delta_{k'(i')} C_{jk(i)} C_{jk(i')} y_{jk(i)t} y_{jk'(i)t} \Bigg\} \Bigg] \frac{1}{l(m^*-1)}$$

The above general approach can be used to examine several different selection and interviewer assignment schemes.

4. Specific Situation

Firm simple random sampling without replacement at both the stages of two stage sampling and simple random sampling of interviewers, we have,

$$P_r(U_i = 1) = \frac{n}{N}; \quad P_r(U_i = 1, U_{i'} = 1) = \frac{n(n-1)}{N(N-1)};$$

$$P_r\left(\delta_{k(i)}=1\middle|\,U_i=1\right)=\frac{m}{M};\;P_r\left(\delta_{k(i)}=1,\delta_{k'(i)}=1\middle|\,U_i=1\right)=\frac{m(m-1)}{M(M-1)};$$

$P_r\left(\delta_{k(i)}=1,\delta_{k'(i')}=1\middle|\,U_i=1,U_{i'}=1\right)=\frac{m}{M}\frac{m}{M}$ (Since sampling of second stage units is done independently);

$P_r\left(\delta_{k(i)}=1,\delta_{k'(i)}=1\middle|\,U_i=1,U_{i'}=1\right)=0$ (Since the first stage units are non-over lapping);

$$P\left(V_j=1\right)=\frac{l}{L};\;P_r\left(V_j=1,V_{j'}=1\right)=\frac{l\,(l-1)}{L\,(L-1)}.$$

The assignment scheme is such that the elements of the sample are assigned to the interviewers so that each element is interviewed by a single interviewer. For this, given mn number of elements selected, l number of random groups of each size m^* is formed such that $lm^*=nm$. Then, we have,

$$P_r\left(C_{jk(i)}=1\middle|U_i=1\right)=\frac{m^*}{mn};$$

$$P_r\left(C_{jk(i)}=1,C_{jk'(i)}=1\;\middle|U_i=1\right)=\frac{m^*(m^*-1)}{mn(mn-1)};$$

$$P\left(C_{jk(i)}=1,C_{j'k(i)}=1\;\middle|U_i=1\right)=0$$

$$P_r\left(C_{jk(i)}=1,C_{j'k'(i')}=1\;\middle|U_{i'}=1,U_{i'}=1\right)=\frac{m^*}{mn}\,\frac{m^*}{mn-1}$$

$$P_r\left(C_{jk(i)}=1,C_{j'k'(i')}=1\;\middle|U_i=1\;\;U_{i'}=1\right)=0$$

$$P_r\left(C_{jk(i)}=1,C_{j'k'(i')}=1\;\middle|U_i=1,\,U_{i'}=0\right)=\frac{m^*(m^*-1)}{mn(mn-1)}$$

Theorem 4. 1. Under the situations as mentioned above and for large mn, we have,

$$E\,(WIMS)=E_1\,E_{2|1}\;E_t\,(WIMS)$$
$$=\text{Measurement Variance}+\left(S_w^2+S_b^2\right)$$

where,

$E_{2|1}(\cdot)$ = Expectation with respect to second stage sampling;

$E_1(\cdot)$ = Expectation with respect to first stage sampling, and;

$E_t(\cdot)$ = Expectation with respect to randomness of the response;

S_w^2 and S_b^2 are within and between sum of squares.

Proof:

We have,

$$E_t\left(y^2_{jk(i)t}\right)=\left[V_t\left(y_{jk(i)t}\right)+\left(y^2_{jk(i)}\right)\right];$$
$$E_t\left(y_{jk(i)t}\ y_{jk'(i)t}\right)=\left[Cov_t\left(y_{jk(i)t}, y_{jk'(i)t}\right)+y_{jk(i)}\ y_{jk'(i)}\right];$$
$$E_t\left(y_{jk(i)t}\ y_{j'k'(i)t}\right)=\left[Cov_t\left(y_{jk(i)t}, y_{jk'(i')t}\right)+y_{jk(i)}\ y_{jk'(i')}\right].$$

Thus, we have

$$E(WIMS)=\Bigg[\Bigg\{\sum_i\sum_j\sum_k \frac{V_t\left(y_{jk(i)t}\right)}{NLM}-\frac{(m-1)}{NLM(M-1)(mn-1)}\times$$
$$\sum_i\sum_j\sum_{k\#k'\in i} Cov_t\left(y_{jk(i)t}, y_{jk'(i)t}\right)$$
$$-\frac{(n-1)m}{(mn-1)M^2N(N-1)L}\sum_j\sum_{i\#i'}\sum_{k\#k'k\in i,k'\in i'} Cov_t\left(y_{jk(i)t}, y_{jk'(i')t}\right)\Bigg\}, \tag{4.1}$$
$$+\Bigg\{\frac{1}{NLM}\sum_i\sum_j\sum_k y^2_{jk(i)}-\frac{(m-1)}{NLM(M-1)(mn-1)}\sum_i\sum_j\sum_{k\#k'\in i} y_{jk(i)}\ y_{jk'(i)}$$
$$-\frac{(n-1)m}{(mn-1)M^2N(N-1)L}\sum_{i\#i'}\sum_j\sum_{k\#k'k\in i,k'\in i'} y_{jk(i)}\ y_{jk'(i')}\Bigg\}\Bigg]$$

= Measurement variance assuming no between interviewer covariance + $(S_w^2+S_b^2)$ (4.2)

This is very much similar to the expression in (2.11) under one stage sampling. For routine calculation see Appendix (A1).

Appendix A1

Assuming, for all j,

$$y_{jk(i)}=y_{k(i)},\ y_{jk'(i)}=y_{k'(i)} \text{ and } y_{jk'(i')}=y_{k'(i')}$$

We have,

$$\frac{1}{NLM}\sum_i\sum_j\sum_k y^2_{jk(i)}-\frac{(m-1)}{NLM(M-1)(mn-1)}\sum_i\sum_j\sum_{k\#k'} y_{jk(i)}\ y_{jk'(i)}$$

$$-\frac{(n-1)m}{(mn-1)M^2N(N-1)L}\sum_{i\#i'}\sum_j\sum_{k\#k'} y_{jk(i)}\,y_{jk'(i')}$$

$$=\frac{1}{MN}\sum_{i=1}^{N}\sum_{k=1}^{M} y_{k(i)}^2-\frac{(m-1)}{MN(M-1)(mn-1)}\times\sum_i\sum_{k\neq k'} y_{k(i)}\,y_{k'(i')}$$

$$-\frac{(n-1)m}{(mn-1)M^2N(N-1)}\sum_{i\#i'}\sum_j\sum_{k\#k'} y_{k(i)}\,y_{k'(i')} \qquad \text{(A1.1)}$$

Now,

(i) $$\frac{1}{MN}\sum_{i=1}^{N}\sum_{k=1}^{M} y_{k(i)}^2=\frac{1}{N}\sum_{i=1}^{N}\left\{\frac{1}{M}\sum y_{k(i)}^2-\overline{Y}_i^2+\overline{Y}_i^2\right\}$$

$$=\frac{1}{N}\sum_{i=1}\left(S_i^2+\overline{Y}_i^2\right);$$

(ii) $$\frac{(m-1)}{NM(M-1)(mn-1)}\sum_i\sum_{k\#k'} y_{k(i)}\;y_{k'(i)}$$

$$=\frac{(m-1)}{NM(M-1)(mn-1)}\sum_i\left\{\left(\sum_k y_{k(i)}\right)^2-\sum_k y_{k(i)}^2\right\}$$

$$=\frac{(m-1)}{NM(M-1)(mn-1)}\sum\left\{M^2\,\overline{y}_i^2-\left(\sum y_{ki}^2-M\,\overline{y}_i^2+M\,\overline{y}_i^2\right)\right\}$$

$$=\frac{(m-1)}{NM(M-1)(mn-1)}\sum\left(M^2\,\overline{y}_i^2-M\,S_i^2-M\,\overline{y}_i^2\right)$$

$$=\frac{(m-1)}{(mn-1)N}\sum\overline{Y}_i^2-\frac{(m-1)\sum S_i^2}{(mn-1)N(M-1)}$$

Observing that, for (i,i'),

$$\left(\sum_{k=1}^{M} y_{k(i)}\right)\left(\sum_{k'=1}^{M} y_{k'(i')}\right)=\sum y_{k(i)}\,y_{k(i')}+\sum_{k\neq k'}\sum y_{k(i)}\,y_{k'(i')}$$

Since the first stage units are non overlapping,

We have $\{K : k \in i \text{ and } k(i')\} = \Phi$,

$$\sum_{k\neq k'}\sum y_{k(i)}\, y_{k'(i')} = \left(\sum_{k=1}^{M} y_{k(i)}\right)\left(\sum_{k'}^{M} y_{k'(i')}\right)$$

$$= Y_i \; Y_{i'} .$$

Hence,

$$\frac{(n-1)m}{(mn-1)M^2N(N-1)}\sum_{i\neq i'}\sum_{k\neq k'} y_{k(i)}\, y_{k'(i')} = \frac{(n-1)m}{(mn-1)M^2N(N-1)}\sum_{i\neq i'} Y_i\, Y_{i'}$$

$$= \frac{(n-1)m}{(mn-1)M^2N(N-1)} M^2 \sum_{i\neq i'} \overline{Y}_i\, \overline{Y}_{i'}$$

$$= \frac{(n-1)m}{(mn-1)N(N-1)}\left\{\left(\sum \overline{Y}_i\right)^2 - \sum \overline{Y}_i^{\,2}\right\}$$

$$= -\frac{(n-1)m}{(mn-1)}\left\{\frac{\sum \overline{Y}_i^{\,2}}{N(N-1)} - \frac{N^2\,\overline{Y}^2}{N(N-1)}\right\},$$

$$= -\frac{(n-1)m}{(mn-1)}\frac{1}{N-1}\left\{\frac{1}{N}\sum \overline{Y}_i^{\,2} - N\,\overline{Y}^2\right\}, \text{ where } \overline{Y} = {}^{\Sigma Y_i}\!/_{N} .$$

Therefore, the expression in (A1.1) becomes,

$$\frac{1}{N}\sum\left(S_i^2 + \overline{Y}_i^{\,2}\right) - \frac{(m-1)}{(mn-1)N}\sum \overline{Y}_i^{\,2} + \frac{(m-1)\sum S_i^2}{(M-1)N(mn-1)}$$

$$+\frac{(n-1)m}{(mn-1)(N-1)}\left(\frac{\Sigma\overline{Y}_i^{\,2}}{N} - N\,\overline{y}^2\right)$$

$$= \left(1 + \frac{(m-1)}{(mn-1)(M-1)}\right)\frac{\Sigma S_i^2}{N} + \frac{1}{N}\left(1 - \frac{(m-1)}{(mn-1)}\right)\Sigma\overline{Y}_i^{\,2}$$

$$+\frac{(n-1)m}{(mn-1)(N-1)}\left(\frac{\Sigma\overline{Y}_i^{\,2}}{N} - \overline{Y}^2 + \overline{Y}^2 - N\overline{Y}^2\right)$$

$$= \left(1+\frac{(m-1)}{(mn-1)(M-1)}\right)\frac{\Sigma S_i^2}{N}+\frac{1}{N}\left(1-\frac{(m-1)}{(mn-1)}\right)\Sigma \overline{Y}_i^2$$
$$+\frac{(n-1)m}{(mn-1)(N-1)}\left(S_b^2-(N-1)\overline{Y}^2\right)$$
$$=\left(1+\frac{(m-1)}{(mn-1)(M-1)}\right)\frac{\Sigma S_i^2}{N}+\left(1-\frac{(m-1)}{(mn-1)}\right)\frac{\Sigma \overline{Y}_i^2}{N}$$
$$+\frac{(n-1)m}{(mn-1)}\frac{S_b^2}{(N-1)}-\frac{(n-1)m}{(mn-1)}\overline{Y}^2 \qquad (3)$$

Under the assumption $mn \to \infty$ i.e., for large sample size, the previous expression reduces to,

$$\frac{1}{N}\sum S_i^2+\frac{1}{N}\sum \overline{Y}_i^2+\frac{S_b^2}{N-1}1-\overline{Y}^2$$
$$=\frac{1}{N}\sum S_i^2+\frac{S_b^2}{N-1}\left(\frac{1}{N}\Sigma\overline{Y}_i^2-\overline{Y}^2\right)$$
$$=\frac{1}{N}\sum S_i^2+S_b^2\left(\frac{1}{N-1}+1\right)$$

$$=\frac{\sum S_i^2}{N}+S_b^2\left(\frac{N}{N-1}\right) \simeq \frac{\sum S_i^2}{N}+S_b^2, \text{(Assuming} \frac{N}{N-1}\simeq 1)$$
$$=S_w^2+S_b^2, \text{ where } S_w^2=\frac{1}{N}\sum_{i=1}^{N} S_i^2 \text{ and} S_b^2=\frac{1}{N}\sum \left(\overline{Y}_i-\overline{Y}\right)^2.$$

Now it follows that,

$$E(WIMS)=\left[\sum_i\sum_j\sum_k \frac{V_t(y_{jk(i)t})}{NLM}-\frac{(m-1)}{NLM(M-1)(mn-1)}\times\right.$$
$$\sum_i\sum_j\sum_{k\neq k'} Cov_t\,(y_{jk(i)t},\, y_{jk'(i)t})$$

$$-\frac{(n-1)m}{(mn-1)M^2N(N-1)}\sum_j\sum_{i\#i',}\sum_{k\#k'k\in i,k'\in i'}Cov_t(y_{jk(i)t},y_{jk'(i')t})\Bigg]+$$

$$\left(S_w^2+S_b^2\right)$$

$$= \text{Measurement variance} + \left(S_w^2+S_b^2\right)$$

$$= MV+\left(S_b^2+S_w^2\right).$$

Therefore, $M\hat{V} = WIMS-\left(\hat{S}_b^2+\hat{S}_w^2\right).$

Remark

Thus without repeat measurement, measurement variance can estimated if, $\hat{S}_b^2$ and $\hat{S}_w^2$ are available.

References

1. Bailar, Barbara A. and Tore Dalenius, *Estimating the Response Variance Components of the U.S. Bureau of the Census Survey Model*, Sankhya, 31B, pp. 341–360 (1969).
2. Hansen, Morris, H. and Max A. *Bershad Measurement Errors in Census and Surveys*, Buletin of International Statistical Institute, *38*, pp. 359–374 (1961).
3. Koch, Gory G, An Alternative *Approach to Multivariate Response Error Models for Sample Survey data with applications to estimators involving sub class mean*, JASA, 68, pp. 906–913 (1973).
4. Lesseler, Judith T. and Kalsbeek, William D, *Non-sampling Errors in Surveys,* John Wiley and Sons Inc. (1992).
5. Mahalanobis P. C., Recent Experiments in Statistical Sampling in the Indian Statistical Institute , JRSS 109, pp. 327–328 (1946).
6. Maiti, P, Estimation of Non-Sampling Variance Components Under the linear model approach Statistics in Transition, New Series, Volume 10, No. 2, pp. 193–222 (2009).

The Interior-Point Revolution in Mathematical Programming and its Place in Applied Mathematics

J. N. Singh
Prof. of Mathematics, Dept. of Mathematics & Computer Science,
Barry University 11300 NE, Second Avenue, Miami Shores
Florida 33161, USA
E-mail: jsingh@barry.edu

In this paper we present various developments that took place after seminal work of Dr. N. Karmarkar of Bell Laboratory and indicate several directions in which future progress can be achieved. Finally we present various consequences of the Interior-Point Revolution in different disciplines of applied mathematics.

Keywords: Linear Programming; Interior-Point Methods; Curvature of Central Path; Semi definite Programming; Computational Complexity 2010 Mathematics Subject Classifications: **49M37**, **65K05**, **90C05**, **90C22**, **90C25**, **90C30**, **90C46**.

1. Introduction

Linear Programming Problem and Notations

A primal-dual pair of linear programming problem in standard form can be stated as

$(P)\min imize\ \{c^T x : Ax = b, x \geq 0\}$ and

$(D)\max imize\ \{b^T y : A^T y + s = c\}$, where $A \in R^{m\times n}, b \in R^m\ and\ c, s \in R^n$.

Professor G. B. Dantzig of Stanford University formulated this problem in 1947 and proposed a method called the simplex algorithm to solve it. In 1972 Klee and Minty [12] constructed an exponential example of linear programming problem which forced the simplex method to take exponential number of iterations to find a solution. Since then 'to prove the existence of a polynomial algorithm for linear programming problem' remained as an open problem before mathematicians and computer scientists. In 1979 a Russian Mathematician L. G. Khachiyan [11] proposed a method called ellipsoid algorithm and showed how a linear programming problem with integer (or rational) data can be solved in polynomial time. Although the work of Khachiyan solved an open problem in theoretical computer science, but it could not prove an efficient algorithm to solve real life linear programming problems. In 1984 N. Karmarkar [10] of Bell Lab proposed another polynomial-time algorithm for linear programming which is very efficient in solving large linear programming problems. In fact, the US air force tested the Bell Lab Production code that implements Karmarkar's algorithm independently and they found that this algorithm was able to solve some extremely large linear programming problems, which were previously unsolved due to the limitations of the earlier methods.

Notations:

$A \in R^{m \times n}$ is an $m \times n$ matrix, $b \in R^m$ is called requirement vector and $c \in R^n$ is called price vector.

If $x \in R^n$ and (y, s)$\in R^{m+n}$ are feasible solutions of primal problem (P) and the dual problem (D) respectively then the duality gap $c^T x - b^T y = x^T s$ > 0.

If feasible solutions x of (P) and *(y, s)* of (D) satisfy $x > 0$ and $s > 0$ then they are called strictly feasible solutions (or interior feasible solutions) to the respective problems and lie in the interior of problems' feasible region.

Let X = diag $\{x_1, x_2, ..., x_n\}$ be an $n \times n$ diagonal matrix whose diagonal entries are precisely the components $x_1, x_2, ..., x_n$ of the vector $x \in R^n$ *and* $e = (1, 1, \cdots, 1)^T$ be a vector of ones, then

$$X^{-1} = \text{diag } \{1/x_1, 1/x_2, ..., 1/x_n\},$$

$$Xe = x \text{ and } X^{-1}e = x^{-1} = (1/x_1, 1/x_2, ..., 1/x_n).$$

2. Karmarkar's Algorithm for Linear Programming

Karmarkar's algorithm considers a linear programming problem in canonical form as:

$$\min imize \left\{ c^T x : Ax = 0, e^T x = 1, x \geq 0 \right\},$$

where $A \in Z^{m \times n}, c \in Z^n$, which can be obtained from a linear programming problem in standard form and Z is the set of integers.

2.1. *The algorithm*

The algorithm generates a sequence of points $x^{(0)}, x^{(1)}, ..., x^{(k)}, ...$ starting with $x^{(0)} = a_0 = (1/n)e$, which is the center of the simplex

$S = \{x \geq 0 : e^T x = 1\}$ in the following steps:

Step 0. Initialize $x^{(0)} = (1/n)e = a_0$

Step 1. At any $k\,th$ iteration, the current point $x^{(k)}$ is mapped to the center $a_0 = (1/n)e$ of the simplex *S* and a point *b* is obtained in the following substeps whose inverse image gives the next point $x^{(k+1)}$ of the sequence.

Projective Transformation T_k :

Let $D = diag\{x_1^{(k)}, x_2^{(k)}, ..., x_n^{(k)}\}$, where $x_1^{(k)}, x_2^{(k)}, ..., x_n^{(k)}$ are the components of the kth iterate $x^{(k)}$, then $x' = T_k(x) = \dfrac{D^{-1}x}{e^T D^{-1}x}$. This is an one-to one map and its inverse projective transformation is given as $x = T_k^{-1}(x') = \dfrac{Dx'}{e^T Dx'}$.

Substep 1.1. Let $B = \begin{bmatrix} AD \\ e^T \end{bmatrix}$.

Substep 1.2. Project Dc orthogonally on the null space of *B* i.e.,

$$c_p = [I - B^T(BB^T)B]Dc.$$

Substep 1.3. Normalize c_p

$$\hat{c} = \frac{c_p}{\|c_p\|}.$$

Substep 1.4. Take $b = a_0 - \alpha r\hat{c}$, where $r = (n(n-1))^{-1/2}$ is the radius of the largest sphere inscribed in the simples *S* and $\alpha \in (0,1)$ is a parameter which Karmarkar [10] takes as α =1/4, Singh and Singh [16] takes α =2/3 and Franklin [7] takes α =1/2.

Substep 1.5. Apply inverse transformation T_k^{-1} to b so that

$$x^{(k+1)} = \frac{Db}{e^T Db}.$$

There is one more technicality to which we are supposed to deal with. In fact linear functions are not invariant under projective transformation, but the ratio of two linear functions can be transformed to the functions of the same type. Keeping this fact in mind Karmarkar has introduced a potential function as the ratio of two linear functions as follows:

$$f(x) = \sum_j \ln \frac{c^T x}{x_j}, \tag{1}$$

This potential function helps to measure the progress of the algorithm. We expect certain decrease δ in potential function $f(x)$ at each iteration of the algorithm. Karmarkar [10] has achieved $\delta \approx 0.125$, Singh and Singh [15] has achieved $\delta \approx$ 0.23 and Frankline has achieved $\delta \approx 0.31$. The iteration stops when the current iterate $x = x^{(k)}$ meets the required convergence check $c^T x^{(k)} \leq 2^{-q}(c^T x^{(0)})$.

Further details of this algorithm can be found in Karmarkar [10] and Singh *et al.* [16, 17].

3. Interior Point Revolution

In view of fundamental theorem on linear programming there is one to one correspondence between the set of basic feasible solutions to a linear programming problem and the set of all extreme points of the polytope formed by the linear constraints defining the linear programming problem. The earlier methods (simplex and its variants) travel from one extreme point (vertex) to another extreme point through the boundary of the polytope. Karmarkar's algorithm and other interior-point methods starts algorithm at an interior point of the polytope makes several corrections of the path to reach to the optimal extreme point (vertex).

After seminal work of N. Karmarkar thousands of paper appeared on interior-point methods. Many researchers called it Interior-Point Revolution. Roughly, these methods can be classified into three categories: (1) Affine Scaling Methods; (2) Potential Reduction Methods; and (3) Central Path Methods.

3.1. *Affine scaling methods*

The affine scaling algorithm, first introduced by Dikin [4] in 1967 and latter reintroduced by Barnes [3] and Vanderbei *et al.* [22] as a variant of projective interior point algorithm, is one of the simplest and efficient interior point methods to solve linear programming problems. Let

$$F = \{x \in R^n : Ax = b, x \geq 0\} \tag{2}$$

be the feasible region for linear programming problem (P). Given a strictly interior point x_0 in F, affine scaling algorithm creates an ellipsoid with the center at x_0 which approximates the feasible region F and optimizes the objective function $c^T x$ over it. If $x_0 \in F$ is a strictly interior point, then Dikin ellipsoid centered at x_0 is denoted by Ex_0 and is defined as

$$Ex_0 = \{x \in R^n : Ax = b, (x - x_0)^T X_0^{-2}(x - x_0) \leq 1\},$$

where X_o^{-2} is the diagonal matrix whose diagonal entries are the reciprocals of the squares the components of x_0. It is easy to see that this ellipsoid Ex_0 is contained in F. In affine scaling algorithm we obtain affine scaling direction at x_0 as the solution of the following direction finding problem:

$$minimize\left\{c^T d : Ad = 0, d^T X_0^2 d \leq 1\right\} \tag{3}$$

It is important to note that (3) is a convex program with one convex quadratic constraint. Direction finding problem (3) is solved by obtaining Karush-Kuhn-Tucker conditions and then solving associated linear system of equations. Let d* be the solution to the direction finding problem (3), then the next point x^* in affine scaling algorithm is obtained by the formula

$$x^* = x_0 + \alpha\, d^*, \tag{4}$$

where α is chosen so that the new iterate x^* remains inside the feasible region F and ensures some improvement in the objective function value $c^T x$, i.e., $c^T x_0 > c^T x^*$.

The affine scaling methods are not only attractive because of their simple description but also because of their efficient performance in practice. Because of its theoretical and practical significance there are number of papers which studied its global and local convergence. It is general consensus that the proof of convergence of the algorithm in the absence of degeneracy is easy but under degeneracy, the proof of convergence is long and cumbersome. Recently, Dikin [4] and others proved global convergence of the long-step version of the affine scaling algorithm, in which the next iterate is obtained by taking a fixed fraction $\alpha \in (0, 1)$ of the whole step to the boundary of the feasible region F. Assuming α=0.5, Dikin [6] showed that the sequence of primal iterates converges to a point lying in the relative interior of the primal face and that the sequence of the dual iterates converges to the analytic center of the dual optimal face. To the best of our knowledge no general results on the bounds on the efficiency of the algorithm is available. But because of the analysis carried out in Megiddo and

Shub [15] about associated continuous trajectory, it is suspected that primal-dual affine scaling algorithms have exponential time complexity in the worst case. To prove or disprove this conjecture is one of the research problems in affine scaling methods which can be tried with some hope of success.

3.2. *Potential reduction methods*

In potential reduction methods consider the following optimization problem:

$$\textit{minimize}\, \phi(x,y,s) = q\ln(c^T x - b^T y) - \sum_{j=1}^{n} \ln(x_j)$$

such that $Ax = b,\, x \geq 0;\;\; A^T y + s = c,\, s > 0$

where q is a parameter (called search direction parameter). The potential or merit function ϕ (x, y, s) consists of two terms: the first term of ϕ (x, y, s) is the product of search direction parameter q and the logarithm of the duality gap, and we intend to drive this term to - ∞. The second term of the potential function is the logarithmic barrier function defined on the interior of the nonnegative orthant, which compels the feasible solutions to remain away from the boundary of the feasible region. Thus the potential function, in fact, is a tool to reduce the duality gap to zero, without allowing the feasible solution moving towards the boundary of the feasible region. With some small assumption on linear programming problem (P) and simple manipulation of potential function $\phi(x,y,s)$ gives the inequality

$$c^T x - b^T y \leq k_1 e^{\phi(x,y,s)/q}, \tag{5}$$

where k_1 is a constant, which depends on the problem (P). i.e. the duality gap is bounded above by $k_1 e^{\phi(x,y,s)/q}$ which is a function of potential function value $\phi(x,y,s)$. In inequality (5), as $\phi(x,y,s) \to -\infty$, the duality gap $(c^T x - b^T y) \to 0$.

For the execution of the potential reduction algorithm, we first select parameter q, and obtain new iterate $(x_1, y_1, s_1) \in R^{n+m+n}$, from the current iterate (x_0, y_0, s_0), which ensures some decrease in the potential function. This can be achieved by using one of the various existing methods like: (1) Newton's method, (2) Newton's method that only accounts for the Hessian of barrier function, $\sum \ln(x_j)$ etc.

In potential reduction method each iteration ensures the reduction of potential function by at least an amount δ. Corresponding to stepsize parameter

$\alpha = \dfrac{1}{4}$, Karmarkar's potential function decreases at each iteration of the algorithm by at least a number $\delta > \dfrac{1}{8}$. Various other guaranteed reductions in Karmarkar's potential function corresponding to different choices of stepsize parameter α can be found in Singh and Singh [16]. Starting with (x^0, y^0, s^0) as an initial solution, potential reduction algorithm obtains a near optimal solution within some optimality tolerance ε in at most

$$\left\lceil (q/\delta)\ln(c^T x^0 - b^T y^0)/\varepsilon) + k_2 \right\rceil \tag{6}$$

iterations, where k_2 is a constant, whose value depends on the problem (P) and $\lceil . \rceil$ is a ceiling function. During the execution of the algorithm it also provides dual information which is quite useful for the user for specifying an optimality tolerance to the computer.

In general all potential reduction methods attempt to drive potential function ϕ to $-\infty$. In such a circumstance the duality gap will tend to zero. i.e., $c^T x = b^T y$ and in view of fundamental duality theorem this current iterate will be the optimal solution of the primal-dual pair (P) and (D). Some potential function methods possess good and some excellent complexity bounds. Theoretical results in worst case, indicate that the convergence of the algorithm in $O(n^{0.5}L)$ iterations can be guaranteed if parameter q is chosen around $n + O(n^{0.5})$. However, its performance in practice indicates that much faster convergence is achieved when q is set around $n + 0(n^2)$.

In spite of good to excellent performance guarantee associated with almost all potential reduction methods, they have not received much attention in terms of computational testing. One can try some of the research problems related to the computational aspects of various potential function methods which have not been fully explored, so far. Further details of these methods can be found in Anstreicher [2] and Freund [8] and Singh *et al.* [20, 21].

3.3. *Central path methods*

The central path methods obtain central trajectory for a given linear programming problem and then solve it approximately along the central path for the diminishing value of the barrier parameter. In order to obtain central trajectory for the linear programming problem (P), we consider logarithmic barrier problem (P_μ) as follows:

$$(P_\mu)\ \min\left\{c^T x - \mu\sum_{j=1}^{n} x_j : Ax = b, x > 0\right\} \tag{7}$$

where μ>0 is called barrier parameter. This logarithmic barrier problem (P_μ) is obtained from (P) by adding barrier term $-\mu\sum_{j=1}^{n} x_j$ to the objective function $c^T x$ of (P) and restricting x to be strictly positive. It is important to note that when barrier parameter as $\mu \to 0, (P_\mu)$ reduces to (P).

Karush-Kuhn-Tucker condition for (P_μ) can be written as:

$$\begin{gathered} Ax = b, x > 0 \\ c - \mu X^{-1}e = A^T y \end{gathered} \tag{8}$$

If we introduce a slack variable vector $s = \mu X^{-1}e$, then these optimality conditions can be restated as follows:

$$Ax = b,\ x > 0;\ A^T y + s = c,\ s > 0;\ Sx - \mu e = 0, \tag{9}$$

where S = diag $\{s_1, s_2, ..., s_n\}$ is a diagonal matrix whose diagonal entries are slack variable $s_1, s_2, ..., s_n$ respectively. Last equation in (9) can also be written as:

$$SXe - \mu e = 0 \tag{10}$$

Let for $\mu > 0, (x(\mu), y(\mu), s(\mu)) \in R^{n+m+n}$ denote the solution of the system (8), then the set

$$\Gamma = \{(x(\mu), y(\mu), s(\mu) \in R^{n+m+n} : \mu > 0\} \tag{11}$$

is called central path or central trajectory for the problem (P). Now if $(x(\mu), y(\mu), s(\mu)) \in \Gamma$ then it satisfies first two equations of (9), which are primal-dual constraints; so $x(\mu)$ and $(y(\mu), s(\mu))$ are strictly feasible for the primal problem (P) and the dual problem (D) and the duality gap on the central trajectory is

$$x^T(\mu)s(\mu) = n\mu, \tag{12}$$

which follows from the last nonlinear equation in (9). Since the duality gap along the central trajectory is $x^T s = n\mu$ so the barrier parameter μ can be expressed as

$$\mu = (x^T s) / n . \tag{13}$$

The majority of central path methods adopt the strategy to solve for approximate solution along the central path for diminishing value of the barrier parameter μ that tends to zero in the limit. There are a number of ways in which this strategy can be adopted. For example, for given $\mu > 0$ one can choose to approximately optimize (p_μ) or to approximately solve (8) or some other equivalent characterization of the central path. Also there may be a number of ways to approximately solve the system of nonlinear equations (9). For example Newton's method or predictor-corrector methods or pre-conditioned conjugate gradient may be used for this purpose. As to the complexity of the central path methods are concerned, the best central path methods reduce the duality gap by a fixed proportion in $O(n^{0.5})$ iterations. Further details of central path methods can be found in Kojima, Mizuno and Yoshise [13], Jansen, Roos and Terlacky [9].

4. Applications of Linear Programming

Linear Programming models are used in: 1) Agriculture, 2) Contract Awards, 3) Diet and Menu planning, 4) Economic Analysis, Banking and Finance, 5) Energy and Environment, 6) Facility and Site Location, 7) Health Care and Medical, 8) Land Use Planning, 9) Marketing and Advertising, 10) Military, 11) Personnel Assignment and Workforce Planning, 12) Production Scheduling, Inventory Control and Planning, 13) Public Policy and Urban Affairs, 14) Structural Design, 15) Traffic Analysis, 16) Transportation and Network, 17) Travelling-Salesperson Problem, 18) University Management and Planning, 19) Regression Analysis, 20) Industry: a) Airlines Industries, b) Chemical Industry, c) Coal Industry, d) Communications and Computer Industries, e) Iron and Steel Industry, f) Paper Industry, g) Petroleum Industry, h) Railroad Industry etc. It has been reported in literature that extremely large linear programming problems with several millions to one billion variable can be solved using interior point methods. Some Interior Point Methods are extremely fast on large linear programming problems. For example, ***fit2p*** is one of the largest problems from NETLIB suit (a standard collection of test problems) has 3,000 constraints and 14,000 variables. By today's standard, this problem is considered smaller than medium size problems. An implementation of the simplex and a variant of interior point methods on ***fit2p*** has indicated that the simplex algorithm consumed more than 36.5 hours, whereas interior point method took less than 3 minutes to find the solution of this problem. Thus, interior point method is about 900 times faster than the simplex method on ***fit2p***. Such a big difference in solution time cannot be overlooked. It is expected that this Interior-Point Revolution in Mathematical Programming will play a very significant role in wide applications of linear programming models and can result in huge savings for various sectors of economy and business.

5. Semidefinite Programming (SDP) and Second Order Cone Programming (SOCP)

5.1. *Semidefinite programming (SDP)*

A semidefinite Programming Problem can be stated as

$$\text{(SDP)}\ Minimize\ \{C \bullet X : A_i \bullet X = b_i,\ X \succeq 0, i = 1, \cdots, m\},$$

where $C, A_1, A_2, \cdots, A_m \in SR^{n\times n}$ are $n \times n$ symmetric matrices, X is a PSD matrix, $b_1, b_2, \cdots, b_m$ are given scalars and $A \bullet X = trace\, A^T X$. The semidefinite programming is the most exciting development of the mid-nineties. All Interior pints methods and their variants can be used to solve a semidefinite programming problems. Recently SDP has got applications in structural optimization, signal processing, circuit design, algebraic geometry, communication and information theory, finance and quantum computing.

5.2. *Second order cone programming (SOCP)*

A second order cone programming problem can be stated as

(SOCP): Given problem data: $A_j \in R^{m\times(1+n_j)}$, $c_j \in R^{(1+n_j)}$ and $b \in R^m$

Minimize $c_1^T x_1 + c_2^T x_2 + \cdots + c_k^T x_k$

Such that $A_1 x_1 + A_2 x_2 + \cdots + A_k x_k = b$ and

$x_j \in S_2^{1+q} = \left\{ \begin{bmatrix} \xi \\ - \\ x \end{bmatrix} \in R^{1+q} : \|\overline{x}\| \le \xi \right\}$ is a second order cone. Like linear programming (LP) and semidefinite programming (SDP) the second order cone programming (SOCP) has got lots of applications in science and engineering. All the interior point methods developed for linear programming problems can be used to solve second order cone programming (SOCP). In fact $LP \subset SOCP \subset SDP$.

6. Concluding Remarks

There may be a large number of research problems associated with various aspects of different variants of interior-point methods such as convergence, termination, step-length, strong-polynomiality and computational complexity etc., for LP, SDP and SOCP. Dikin, Tsuchiya *et al.* proved global convergence of the long step version of affine scaling algorithm in which the next iterate is obtained by taking a fixed step-length $\alpha \in (0,1)$, one can examine the possibility to improve the efficiency of the algorithm by taking an optimal step-length at each iteration of the algorithm like Anstreicher [1] in the case of projective algorithm. In view of analysis carried out in Megiddo and Shub [15] about associated continuous trajectory, it is suspected that in the worst primal-dual affine scaling algorithms have exponential time complexity. To prove or disprove this conjecture might be an important research direction in affine scaling methods in linear programming. The worst case analysis of the potential reduction methods indicates that the convergence of the algorithm can be achieved in $o(\sqrt{n}L)$ iterations, if the parameter q is chosen around $n+o(\sqrt{n})$. One can examine the possibility of relatively better convergence results in less restrictive conditions. Recently some researchers have used the disjoint crossover events to study computational complexity of linear programming algorithms. This area is quite new and seems to be promising but it has received very little attention from the researchers so far. Recently several authors have studied the curvature of the central path to find upper and the lower bounds on the numbers of iterations to solve a linear programming problem. Some authors have tried to extend these results for SDP and SOCP under some restrictive conditions.It may be interesting to extend the results on curvature of the central path and disjoint crossover events to SDP and SOCP. Also "to prove the existence of strongly polynomial algorithm for linear programming problem with integer data" is still an open problem. Recently Vavasis and Ye [22] have obtained a complexity bound for a variant of interior point methods which depends on dimension n and some condition measure χ_A of the constraint matrix A. This result is significant because problem data is not restricted to integers and the price vector c and the requirement vector b do not play any role in the complexity bounds. Also the existence of polynomial (or strongly polynomial) algorithm for linear programming problem with real data is another open problem in this area. Ye [24] has obtained some improved complexity results for linear feasibility problems with real data.

References

1. K. M. Anstreicher, "The worst case step in Karmarkar'salgorithm", Mathematics of Operations Research, vol. 14 (2), pp. 294–302, (1989).
2. K. M. Anstreicher, "Potential Reduction Methods", in Interior Point Methods in Mathematical Programming,Terlaky, T. ed., Kluwer Academic Publisher (1996).
3. E. R. Barners, "A variation on Karmarkar's algorithm for solving linear programming problems," Mathematical Programming, 36 pp. 174–182, (1986).
4. I. I. Dikin, "Iterative solution of problems of linear and quadratic programming", Soviet Mathematics, Doklady, 8, pp. 675–679, (1967).
5. I. I. Dikin, "On the convergence of iterative process," Upravlyaemye Systemi, 12, pp. 59–60, (1974).
6. I. I. Dikin, "The Convergence of dual variables", Technical Report Siberian Energy Institute, Irkutsk, Russia (Dec. 1991).
7. Joel Franklin, "Convergence in Karmarkar's Algorithm", SIAM Journal of Numerical Analysis, volume 24(4), pp. 928–945, (1987).
8. R. M. Freund and S. Mizuno, "Interior Point Methods: Current Status and Future Directions", Optima No 51, pp. 1–9, (October 1996).
9. B. Jansen, C. Roos and T. Terlaky, "A short survey on ten years of interior point methods", Report 95-45, Delft University of Technology, Delft, The Netherlands (1995).
10. N. Karmarkar, "A new polynomial-time algorithm for linear programming, "Combinatorica, volume 4(4), pp. 373–395, (1984).
11. L. G. Khachiyan, "A polynomial algorithm in linear programming," USSR Computational Mathematics and Mathematical Physics, 20, pp. 53–72, (1980).
12. V. Klee and G. L. Minty, "How good is the simplex algorithm?", in O shisha ed., Inequality III, Academic Press, New York, pp. 159–175, (1972).
13. M. Kojima, S. Mizuno and A. Yoshise, "A primal-dual interior point algorithm for linear programming", in Progress in Mathematical Programming, ed. N. Megiddo, pp. 29–47, (1989).
14. C. J. H. McDiarmid, "On improvement per iteration in Karmarkar's method for linear programming", Math. Programming. volume 40, pp. 290–320, (1990).

15. N. Megiddo and M. Shub "Boundary behavior of interior point algorithms in linear programming", Mathematics of Operations Research, volume 14 pp. 97–114, (1989).
16. J. N. Singh and D. Singh, "Convergence in Karmarkar's Algorithm: A Review", International Journal of Mathematical Education in Science and Technology, Vol. 25(3) (1994) pp. 333–341, (1994).
17. J. N. Singh, D. Singh and R. Shashi, "A modified termination rule for Karmarkar's algorithm", Journal of Information and Optimization Sciences, volume 16(3) pp. 585–591 (1995).
18. J. N. Singh, "Some remarks on the construction of central path for linear programming" Paper presented at the Sixth SIAM Conf. on optimization at Atlanta USA, May 10–12, (1999).
19. J. N. Singh, M. Shakil and D. Singh, "Disjoint Crossover Events and Computational Complexity of Linear Programming", PNG journal of Mathematics, Computing and Education, volume 6 (2001).
20. J. N. Singh, "Some Recent Advances and new Directions in Interior Point Methods in Linear Programming" Nonlinear Forum, volume 6 pp. 233–248, (2001).
21. J. N. Singh, D. Singh, " Interior point Methods in Linear Programming: a review", International Journal of Mathematical Education in Science and Technology, volume 33(3) pp. 405–423, (1994).
22. R. J. Vanderbei, J. C. Lagarias, M. S. Meketon and B. A. Freedman, "A modification of Karmarkar's linear programming algorithm", Algorithmica, volume 1(4) pp. 395–907, (1986).
23. S. Vavasis and Yinue Ye, "A primal-dual interior point method whose running time depends only on the constraint matrix", Mathematical programming, volume74 pp. 79–120, (1996).
24. Yinue Ye, "Improved complexity results on solving real number linear feasibility problems" Working Paper of the Department of Management Science University of Iowa, (March 2002).

Combined Exponential Type Estimators of Population Mean in Stratified Random Sampling

R. Pandey and K. Yadav

Department of Statistics, University of Delhi,
New Delhi-110007, India
E-mail: ranjitapandey111@gmail.com, kalpana22yadav@gmail.com
www.du.ac.in

N. S. Thakur

Department of Mathematics and Statistics, Banasthali University,
Banasthali-304022, Rajasthan, India
E-mail: nst_stats@yahoo.co.in
www.banasthali.org

The present work proposes combined exponential ratio-product type estimators for the study variable to estimate the population mean under non-response, by using auxiliary information in a stratified random sampling setup. The mean squared error (M.S.E.) and minimum M.S.E of suggested estimators are derived under Proportional and Neyman allocation. An empirical study illustrates relative efficiency performance of the constructed estimators.

Keywords: Auxiliary variable, Combined exponential ratio-product type estimator, Mean Squared Error, Neyman allocation, Proportional allocation.

1. Introduction

Stratified sampling often improves the precision over un-stratified sampling strategies for finite population estimates. Practitioners use auxiliary information to improve efficiency of the estimators at the estimation stage. Kadilar and Cingi [5] have extended estimators of Upadhyaya and Singh [16] for stratified random sampling. To improve the estimators, Kadilar and Cingi [6], Shabbir and Gupta [11, 12] have suggested new estimators in stratified sampling. Koyuncu and Kadilar [8] have proposed families of estimators for estimating population mean in stratified sampling by considering the estimators proposed in Khoshnevisan *et al.* [7]. Koyuncu and Kadilar [9] have suggested a family of estimators in stratified sampling following Diana [3] and Kadilar and Cingi [5]. Motivated by the exponential ratio and product type estimators for population mean proposed by Bahl and Tuteja [1], exponential ratio type estimator of Srivastava [14] and combined exponential ratio and product type estimators of Singh *et al.* [13], we develop combined exponential type estimators of population mean in stratified random sampling using Hansen and Hurwitz [4] technique under non-response. Our purpose in developing new exponential type estimators is to propose some imputation strategies corresponding to stratified adaptations of some well-known sample mean estimators. Efficiency of the proposed strategies are compared with stratified adaptation of Hansen and Hurwitz estimator under non response.

2. Sampling Strategy and Estimation Procedure

A finite population $U(U_1, U_2, ..., U_k)$ of size N divided into k non-overlapping homogeneous layers (strata) is considered. The size of the ith stratum is assumed to be N_i, (i=1,2,...,k) and a sample of size n_i is drawn from the ith stratum using simple random sampling without replacement such that $\sum_{i=1}^{k} n_i = n$. The non-response is observed on the study variable and the auxiliary variable is free from non-response. Let n_{i1} and n_{i2} denote respondent and non-respondent units in the sample of n_i units for the ith stratum regarding study variable, respectively. Using Hansen and Hurwitz [4] procedure, a sub-sample of size m_i units is selected out of n_{i2} non-respondent units with the help of simple random sampling without replacement scheme such that $n_{i2}=L_i m_i$, $L_i \geq 1$ and the information is recorded on all the m_i units. Let X_0 and X_1 be the study and auxiliary variables with respective population means $\overline{X}_0$ and $\overline{X}_1$. The Hansen and Hurwitz estimator of population mean $\overline{X}_{0i}$ of study variable X_0 for the ith stratum is $T_{0i}^{*} = \frac{n_{i1}\overline{x}_{0i1} + n_{i2}\overline{x}_{0mi}}{n_i}$, i= 1...,k, where $\overline{x}_{0i1}$ and $\overline{x}_{0mi}$ are the sample means based on n_{i1} respondent units and m_i respondent units, respectively, in the ith stratum for the study variable. Obviously, T_{0i}^{*} is an unbiased estimator of $\overline{X}_{0i}$. Therefore, an unbiased estimator of population mean $\overline{X}_0$ of the study variable is given by $T_{0st}^{*} = \sum_{i=1}^{k} p_i T_{0i}^{*}$ where $p_i = \frac{N_i}{N}$ and the variance of the estimator is given by

$$V\left[T_{0st}^{*}\right] = \sum_{i=1}^{k}\left(\frac{1}{n} - \frac{1}{N}\right) p_i^2 S_{0i}^2 + \sum_{i=1}^{k} \frac{L_i - 1}{n_i} W_{i2} p_i^2 S_{0i2}^2 \tag{1}$$

The estimator of population mean $\overline{X}_1$ of auxiliary variable is $T_{1st} = \sum_{i=1}^{k} p_i \overline{x}_{1i}$ where $\overline{x}_{1i}$ is the sample mean based on n_i units of ith stratum for the auxiliary variable. It can easily be seen that T_{1st} is an unbiased estimator of $\overline{X}_1$.

3. Proposed Strategies

We suggest the following stratified adaptations of some existing estimators for estimating the population mean $\overline{X}_0$ under non-response. Motivated by Bahl and Tuteja [1], we postulate

$$T_{\exp 1} = T_{0st}^{*} \exp\left(\frac{\overline{X}_1 - T_{1st}}{\overline{X}_1 + T_{1st}}\right) \tag{2}$$

and

$$T_{\exp 2} = T_{0st}^{*} \exp\left(\frac{T_{1st} - \overline{X}_1}{T_{1st} + \overline{X}_1}\right) \tag{3}$$

Following Srivastava [14], an estimator $T_{\exp 3}$ in stratified random sampling

$$T_{\exp 3} = T_{0st}^{*} \exp\left(\frac{\overline{X}_1 - T_{1st}}{\overline{X}_1 + T_{1st}}\right)^{l} \tag{4}$$

is proposed where l is a suitably chosen constant. For l=1, $T_{\exp 3}$ is similar to the conventional exponential ratio type estimator whereas for l=−1, it is reduced to the conventional exponential product type estimator.

Singh *et al.* [13] proposed an estimator which is linear combination of exponential ratio and product type estimator for estimating the population mean of study variable in simple random sampling. Adapting Singh *et al.* [13] estimator, we propose an estimator $T_{\exp 4}$ in stratified random sampling as follows:

$$T_{\exp 4} = T_{0st}^{*}\left[\theta \exp\left(\frac{\overline{X}_1 - T_{1st}}{\overline{X}_1 + T_{1st}}\right) + (1-\theta)\exp\left(\frac{T_{1st} - \overline{X}_1}{T_{1st} + \overline{X}_1}\right)\right] \tag{5}$$

where θ is the constant and suitably chosen by minimizing mean squared error of the estimator $T_{\exp 4}$.

3.1. *Properties of the suggested estimators*

Let $T_{0st}^{*} = \overline{X}_0(1+e_0)$, $T_{1st} = \overline{X}_1(1+e_1)$ such that $E[T_{0i}^{*}] = \overline{X}_{0i}$; $E[T_{0st}^{*}] = \overline{X}_0$; $E[T_{1st}] = \overline{X}_1$.

Using the mechanism of missing completely at random (MCAR) where the probability of a response being missing is unrelated to the observed and unobserved data values on a specific unit. Hence, the observed data is deemed as true representative of the population and for given values of r and n, under stratified random sampling design, we have

$E[e_0]=E[e_1]=0;$

$E\left[e_0^2\right]=\frac{1}{\overline{X}_0^2}Var\left(T_{0st}^*\right);Var\left(T_{0st}^*\right)=\sum_{i=1}^{k}\left(\frac{1}{n_i}-\frac{1}{N_i}\right)p_i^2S_{0i}^2+\sum_{i=1}^{k}\left(\frac{L_i-1}{n_i}\right)W_{i2}p_i^2S_{0i2}^2$

$E\left[e_1^2\right]=\frac{1}{\overline{X}_1^2}\sum_{i=1}^{k}\left(\frac{1}{n_i}-\frac{1}{N_i}\right)p_i^2S_{1i}^2$; $Var\left(T_{1st}\right)=\sum_{i=1}^{k}\left(\frac{1}{n_i}-\frac{1}{N_i}\right)p_i^2S_{1i}^2$; $E\left[e_0e_1\right]=\frac{1}{\overline{X}_0\overline{X}_1}Cov\left(T_{0st}^*,T_{1st}\right)$;

$Cov\left(T_{0st}^*,T_{1st}\right)=\sum_{i=1}^{k}\left(\frac{1}{n_i}-\frac{1}{N_i}\right)p_i^2\rho_{01i}S_{0i}S_{1i}$; $R=\frac{\overline{X}_0}{\overline{X}_1}$; where S_{0i}^2 and S_{1i}^2 are the mean squares of study and auxiliary variables respectively in the ith stratum; S_{0i2}^2 is the population mean square of the non-response group in the ith stratum; ρ_{01i} is the population correlation coefficient between $\overline{X}_0$ and $\overline{X}_1$ in the ith stratum; $W_{i2}=\frac{N_{i2}}{N_i}$ is the non-response rate of the ith stratum in the population.

Theorem 3.1:[a] The estimator $T_{\exp 1}$ in terms of e_0 and e_1 is

$$T_{\exp 1}=\overline{X}_0\left\{1+e_0-\frac{1}{2}e_1+\frac{3}{8}e_1^2-\frac{1}{2}e_0e_1\right\}$$

[b] The M.S.E. of $T_{\exp 1}$ is given by

$$M\left(T_{\exp 1}\right)=\sum_{i=1}^{k}\left(\frac{1}{n_i}-\frac{1}{N_i}\right)p_i^2S_{0i}^2+\sum_{i=1}^{k}\left(\frac{L_i-1}{n_i}\right)W_{i2}p_i^2S_{0i2}^2$$
$$+\frac{1}{4}R^2\sum_{i=1}^{k}\left(\frac{1}{n_i}-\frac{1}{N_i}\right)p_i^2S_{1i}^2-R\sum_{i=1}^{k}\left(\frac{1}{n_i}-\frac{1}{N_i}\right)p_i^2\rho_{01i}S_{0i}S_{1i} \quad (6)$$

Theorem 3.2:[a] The estimator $T_{\exp 2}$ in terms of e_0 and e_1 is expressible as

$$T_{\exp 2}=\overline{X}_0\left\{1+e_0+\frac{1}{2}e_1-\frac{3}{8}e_1^2+\frac{1}{2}e_0e_1\right\}$$

[b] The M.S.E. of $T_{\exp 2}$ is given by

$$M\left(T_{\exp 2}\right)=\sum_{i=1}^{k}\left(\frac{1}{n_i}-\frac{1}{N_i}\right)p_i^2S_{0i}^2+\sum_{i=1}^{k}\left(\frac{L_i-1}{n_i}\right)W_{i2}p_i^2S_{0i2}^2$$
$$+\frac{1}{4}R^2\sum_{i=1}^{k}\left(\frac{1}{n_i}-\frac{1}{N_i}\right)p_i^2S_{1i}^2+R\sum_{i=1}^{k}\left(\frac{1}{n_i}-\frac{1}{N_i}\right)p_i^2\rho_{01i}S_{0i}S_{1i} \quad (7)$$

Theorem 3.3:[a] The estimator $T_{\exp 3}$ in terms of e_0 and e_1 expressible as

$$T_{\exp 3}=\overline{X}_0\left\{1+e_0-\frac{1}{2}le_1+\frac{1}{8}l(2+l)e_1^2-\frac{l}{2}e_0e_1\right\}$$

[b] The minimum M.S.E. of $T_{\exp 3}$ is obtainable when $l=\frac{2}{R}\frac{\sum_{i=1}^{k}\left(\frac{1}{n_i}-\frac{1}{N_i}\right)p_i^2\rho_{01i}S_{0i}S_{1i}}{\sum_{i=1}^{k}\left(\frac{1}{n_i}-\frac{1}{N_i}\right)p_i^2S_{1i}^2}$ is

given by

$$\left[M\left(T_{\exp 3}\right)\right]_{\min}=\sum_{i=1}^{k}\left(\frac{1}{n_i}-\frac{1}{N_i}\right)p_i^2S_{0i}^2+\sum_{i=1}^{k}\left(\frac{L_i-1}{n_i}\right)W_{i2}p_i^2S_{0i2}^2-\frac{\left[\sum_{i=1}^{k}\left(\frac{1}{n_i}-\frac{1}{N_i}\right)p_i^2\rho_{01i}S_{0i}S_{1i}\right]^2}{\sum_{i=1}^{k}\left(\frac{1}{n_i}-\frac{1}{N_i}\right)p_i^2S_{1i}^2} \tag{8}$$

Theorem 3.4:[a] The estimator $T_{\exp 4}$ in terms of e_0 and e_1 is expressible as

$$T_{\exp 4}=\frac{1}{\bar{X}_0}\left\{\theta\left(1+e_0-\frac{e_1}{2}+\frac{3}{8}e_1^2-\frac{1}{2}e_0e_1\right)+(1+\theta)\left(1+e_0+\frac{e_1}{2}-\frac{e_1^2}{8}+\frac{1}{2}e_0e_1\right)\right\}$$

[b] The minimum M.S.E. of $T_{\exp 4}$ is obtained when $l=\frac{1}{2}+\frac{\sum_{i=1}^{k}\left(\frac{1}{n_i}-\frac{1}{N_i}\right)p_i^2\rho_{01i}S_{0i}S_{1i}}{R\sum_{i=1}^{k}\left(\frac{1}{n_i}-\frac{1}{N_i}\right)p_i^2S_{1i}^2}$

and is given by

$$\left[M\left(T_{\exp 4}\right)\right]_{\min}=\sum_{i=1}^{k}\left(\frac{1}{n_i}-\frac{1}{N_i}\right)p_i^2S_{0i}^2+\sum_{i=1}^{k}\left(\frac{L_i-1}{n_i}\right)W_{i2}p_i^2S_{0i2}^2-\frac{\left[\sum_{i=1}^{k}\left(\frac{1}{n_i}-\frac{1}{N_i}\right)p_i^2\rho_{01i}S_{0i}S_{1i}\right]^2}{\sum_{i=1}^{k}\left(\frac{1}{n_i}-\frac{1}{N_i}\right)p_i^2S_{1i}^2} \tag{9}$$

The proofs of the above theorems are based on the standard techniques from sampling theory and hence are omitted. The $T_{\exp q}$'s , *q=1,...,4* are expressed in terms of the e_j's, *j=0,1.* Using the Taylor series approximation, the mean squared errors are obtained by using the properties of the e_j's.

4. M.S.E. of Suggested Estimators under Proportional and Neyman Allocation

[a] **Under Proportional Allocation:** The allocation of n_i's to various stratais called proportional if the sample fraction is constant for each stratum such that each stratum is represented according to its size,

i.e. $\frac{n_1}{N_1}=\frac{n_2}{N_2}=\ldots\frac{n_k}{N_k}=\frac{\sum n_i}{\sum N_i}=\frac{n}{N}=C;\ \frac{n_i}{N_i}=C=\frac{n}{N}\Rightarrow n_i\propto N_i\cdot$

Hence, $\left[Var\left(T_{0st}^{*}\right)\right]_{pro}=\left(\frac{1}{n}-\frac{1}{N}\right)\sum_{i=1}^{k}p_i S_{0i}^2+\frac{1}{nN}\sum_{i=1}^{k}\left(\frac{n_{i2}-m_i}{m_i}\right)N_{i2}S_{0i2}^2$;

$\left[Var\left(T_{1st}\right)\right]_{pro}=\sum_{i=1}^{k}\left(\frac{1}{n}-\frac{1}{N}\right)p_i S_{1i}^2$; $\left[Cov\left(T_{0st}^{*},T_{1st}\right)\right]_{pro}=\left(\frac{1}{n}-\frac{1}{N}\right)\sum_{i=1}^{k}p_i\rho_{01i}S_{0i}S_{1i}$

Theorem 4.1: The M.S.E. of estimator $T_{\exp1}$ under proportional allocation is given by

$$M\left[\left(T_{\exp1}\right)\right]_{pro}=\left(\frac{1}{n}-\frac{1}{N}\right)\sum_{i=1}^{k}p_i\left\{S_{0i}^2+\frac{1}{4}R^2S_{1i}^2-R\rho_{01i}S_{0i}S_{1i}\right\}+\frac{1}{nN}\sum_{i=1}^{k}\left(\frac{n_{i2}-m_i}{m_i}\right)N_{i2}S_{0i2}^2 \qquad (10)$$

Theorem 4.2: The M.S.E. of estimator $T_{\exp2}$ under proportional allocation is given by

$$M\left[\left(T_{\exp2}\right)\right]_{pro}=\left(\frac{1}{n}-\frac{1}{N}\right)\sum_{i=1}^{k}p_i\left\{S_{0i}^2+\frac{1}{4}R^2S_{1i}^2+R\rho_{01i}S_{0i}S_{1i}\right\}+\frac{1}{nN}\sum_{i=1}^{k}\left(\frac{n_{i2}-m_i}{m_i}\right)N_{i2}S_{0i2}^2 \qquad (11)$$

Theorem 4.3: The minimum M.S.E. obtainable at $l=\frac{2}{R}\frac{\sum_{i=1}^{k}\left(\frac{1}{n_i}-\frac{1}{N_i}\right)p_i^2\rho_{01i}S_{0i}S_{1i}}{\sum_{i=1}^{k}\left(\frac{1}{n_i}-\frac{1}{N_i}\right)p_i^2S_{1i}^2}$ of estimator $T_{\exp3}$ under proportional allocation are given by

$$\left[M\left[\left(T_{\exp3}\right)\right]_{pro}\right]_{\min}=\left(\frac{1}{n}-\frac{1}{N}\right)\left[\sum_{i=1}^{k}p_i S_{0i}^2-\frac{\left\{\sum_{i=1}^{k}p_i\rho_{01i}S_{0i}S_{1i}\right\}^2}{\sum_{i=1}^{k}p_i S_{1i}^2}\right]+\frac{1}{nN}\sum_{i=1}^{k}\left(L_i-1\right)N_{i2}S_{0i2}^2 \qquad (12)$$

Theorem 4.4: The minimum M.S.E. at $l=\frac{1}{2}+\frac{\sum_{i=1}^{k}\left(\frac{1}{n_i}-\frac{1}{N_i}\right)p_i^2\rho_{01i}S_{0i}S_{1i}}{R\sum_{i=1}^{k}\left(\frac{1}{n_i}-\frac{1}{N_i}\right)p_i^2S_{1i}^2}$ of estimator $T_{\exp4}$ under proportional allocation are given by

$$\left[M\left[\left(T_{\exp4}\right)\right]_{pro}\right]_{\min}=\left(\frac{1}{n}-\frac{1}{N}\right)\left[\sum_{i=1}^{k}p_i S_{0i}^2-\frac{\left\{\sum_{i=1}^{k}p_i\rho_{01i}S_{0i}S_{1i}\right\}^2}{\sum_{i=1}^{k}p_i S_{1i}^2}\right]+\frac{1}{nN}\sum_{i=1}^{k}\left(L_i-1\right)N_{i2}S_{0i2}^2 \qquad (13)$$

[b] **Under Neyman Allocation:** Neyman allocation is a special case of optimal allocation strategy which is used in stratified samples to obtain maximum survey precision, given a fixed total sample size. The best sample size for stratum i would be $n_i\propto N_iS_i\Rightarrow n_i=\frac{nN_iS_i}{\sum_{i=1}^{k}N_iS_i}$. $\left[Var\left(T_{1st}\right)\right]_{ney}==\frac{1}{n}\left(\sum_{i=1}^{k}p_iS_{1i}\right)^2-\frac{1}{N}\sum_{i=1}^{k}p_iS_{1i}^2$

$$\left[Var\left(T_{0st}^{*}\right)\right]_{ney}=\frac{1}{n}\left(\sum_{i=1}^{k}p_iS_{0i}\right)^2-\frac{1}{N}\sum_{i=1}^{k}p_iS_{0i}^2+\frac{1}{nN}\sum_{i=1}^{k}(L_i-1)\sum_{i=1}^{k}p_iN_{i2}S_{0i2}^2$$

$$\left[Cov\left(T_{0st}^{*},T_{1st}\right)\right]_{ney}=\frac{1}{n}\sum_{i=1}^{k}p_i\rho_{01i}\sum_{i=1}^{k}p_iS_{0i}S_{1i}-\frac{1}{N}\sum_{i=1}^{k}p_i\rho_{01i}S_{0i}S_{1i}$$

Theorem 4.5: The M.S.E. of estimator $T_{\exp 1}$ under Neyman allocation is given by

$$M\left[\left(T_{\exp 1}\right)\right]_{ney}=\frac{1}{n}\left\{\left(\sum_{i=1}^{k}p_iS_{0i}\right)^2+\frac{1}{4}R^2\left(\sum_{i=1}^{k}p_iS_{1i}\right)^2-R\sum_{i=1}^{k}p_i\rho_{01i}\sum_{i=1}^{k}p_iS_{0i}S_{1i}\right\}$$

$$-\frac{1}{N}\left\{\sum_{i=1}^{k}p_iS_{0i}^2+\frac{R^2}{4}\sum_{i=1}^{k}p_iS_{1i}^2-R\sum_{i=1}^{k}p_i\rho_{01i}S_{0i}S_{1i}\right\}+\frac{1}{nN}\sum_{i=1}^{k}(L_i-1)\sum_{i=1}^{k}p_iN_{i2}S_{0i2}^2 \quad (14)$$

Theorem 4.6: The M.S.E. of estimator $T_{\exp 2}$ under Neyman allocation is given by

$$M\left[\left(T_{\exp 2}\right)\right]_{ney}=\frac{1}{n}\left\{\left(\sum_{i=1}^{k}p_iS_{0i}\right)^2+\frac{1}{4}R^2\left(\sum_{i=1}^{k}p_iS_{1i}\right)^2+R\sum_{i=1}^{k}p_i\rho_{01i}\sum_{i=1}^{k}p_iS_{0i}S_{1i}\right\}$$

$$-\frac{1}{N}\left\{\sum_{i=1}^{k}p_iS_{0i}^2+\frac{R^2}{4}\sum_{i=1}^{k}p_iS_{1i}^2+R\sum_{i=1}^{k}p_i\rho_{01i}S_{0i}S_{1i}\right\}+\frac{1}{nN}\sum_{i=1}^{k}(L_i-1)\sum_{i=1}^{k}p_iN_{i2}S_{0i2}^2 \quad (15)$$

Theorem 4.7: The minimum M.S.E. is obtained when $l=\frac{2}{R}\frac{\sum_{i=1}^{k}\left(\frac{1}{n_i}-\frac{1}{N_i}\right)p_i^2\rho_{01i}S_{0i}S_{1i}}{\sum_{i=1}^{k}\left(\frac{1}{n_i}-\frac{1}{N_i}\right)p_i^2S_{1i}^2}$ of estimator $T_{\exp 3}$ under Neyman allocation are given by

$$\left[M\left[\left(T_{\exp 3}\right)\right]_{ney}\right]_{\min}=\left(\frac{1}{n}\left(\sum_{i=1}^{k}p_iS_{0i}\right)^2-\frac{1}{N}\sum_{i=1}^{k}p_iS_{0i}^2\right)$$

$$-\frac{\left\{\frac{1}{n}\sum_{i=1}^{k}p_i\rho_{01i}\sum_{i=1}^{k}p_iS_{0i}S_{1i}-\frac{1}{N}\sum_{i=1}^{k}p_i\rho_{01i}S_{0i}S_{1i}\right\}^2}{\frac{1}{n}\left(\sum_{i=1}^{k}p_iS_{1i}\right)^2-\frac{1}{N}\sum_{i=1}^{k}p_iS_{1i}^2}+\frac{1}{nN}\sum_{i=1}^{k}(L_i-1)\sum_{i=1}^{k}p_iN_{i2}S_{0i2}^2 \quad (16)$$

Theorem 4.8: The minimum M.S.E. at $l=\frac{1}{2}+\frac{\sum_{i=1}^{k}\left(\frac{1}{n_i}-\frac{1}{N_i}\right)p_i^2\rho_{01i}S_{0i}S_{1i}}{R\sum_{i=1}^{k}\left(\frac{1}{n_i}-\frac{1}{N_i}\right)p_i^2S_{1i}^2}$ of estimator $T_{\exp 4}$ under Neyman allocation are given by

$$\left[M\left[\left(T_{\exp 4}\right)\right]_{ney}\right]_{\min}=\left(\frac{1}{n}\left(\sum_{i=1}^{k}p_iS_{0i}\right)^2-\frac{1}{N}\sum_{i=1}^{k}p_iS_{0i}^2\right)$$

$$-\frac{\left\{\frac{1}{n}\sum_{i=1}^{k} p_i\rho_{01i}\sum_{i=1}^{k} p_i S_{0i} S_{1i} - \frac{1}{N}\sum_{i=1}^{k} p_i \rho_{01i} S_{0i} S_{1i}\right\}^2}{\frac{1}{n}\left(\sum_{i=1}^{k} p_i S_{1i}\right)^2 - \frac{1}{N}\sum_{i=1}^{k} p_i S_{1i}^2} + \frac{1}{nN}\sum_{i=1}^{k}(L_i - 1)\sum_{i=1}^{k} p_i N_{i2} S_{0i2}^2 \quad (17)$$

5. Empirical Study

In this section, we have considered the data from Sarndal *et al.* [10]. Population in the year 1985 is considered as the study variable and that in the year 1975 as auxiliary variable. There are 284 municipalities which have been divided randomly into four strata of sizes i.e. 73, 70, 97 and 44. Table 1 shows the values of the parameters of the population under consideration.

Table 1. Parameters of the population.

i	N_i	$\overline{X}_{0i}$	$\overline{X}_{1i}$	S_{0i}^2	S_{1i}^2	ρ_i	S_{0i2}^2
1	73	40.85	39.56	6369.10	6624.44	0.999	618.89
2	70	27.83	27.57	1051.07	1147.01	0.998	240.91
3	97	25.79	25.44	2014.97	2205.40	0.999	265.52
4	44	20.64	20.36	538.47	485.26	0.997	83.69

Table 2. Allocation of sample.

i	Size of Samples									
	Proportional Allocation					Neyman Allocation				
	n_i	n_{i1}	n_{i2}	m	L_i	n_i	n_{i1}	n_{i2}	m	L_i
1	15	10	5	3	1.67	26	16	10	5	2.00
2	15	9	6	3	2.00	10	7	3	2	1.50
3	21	15	6	4	1.50	19	13	6	3	2.00
4	9	5	4	2	2.00	5	3	2	2	1.00
Total	60	39	21	12	1.75	60	39	21	12	1.75

We calculate the Percent Relative Efficiencies (P.R.E.'s) of different suggested estimators with respect to the Hansen and Hurwitz unbiased estimator T_{0st}^* by using the formula: $P(*, T_{0st}^*) = \frac{V(T_{0st}^*)}{[MSE(*)]} \times 100$; (18)

* defines different M.S.E.'s of suggested estimators.

Table 3. Percentage relative efficiency (P.R.E.) of different suggested estimators w.r. to T_{0st}^*.

Proportional allocation				
L_i	$P(T_{\exp 1}, T_{0st}^*)$	$P(T_{\exp 2}, T_{0st}^*)$	$P(T_{\exp 3}, T_{0st}^*)$	$P(T_{\exp 4}, T_{0st}^*)$
1.50	119.56	12.13	2750.00	2750.00
1.67	119.35	12.23	2220.00	2220.00
2.00	118.95	12.42	1614.29	1614.29

Table 3. (*Continued*)

Neyman allocation				
1.00	1528.57	146.35	192600.0	192600.00
1.50	1122.73	144.66	3800.00	3800.00
2.00	896.46	143.08	1986.28	1986.28

It is observed that the relative efficiency of product type estimators under proportional allocation increases as L_i is increased. For the remaining proposed estimators under both types of allocations relative efficiency decreases with increase in L_i.

6. Conclusion

The present paper considers the problem for estimating the population mean of the study variable in presence of non-response which uses the auxiliary information under stratified random sampling. The performance of the proposed estimators are compared with Hansen and Hurwitz [4] procedure of sub-sampling, and their properties are studied. The proposed estimators $T_{\exp 1}, T_{\exp 3}$ and $T_{\exp 4}$ under both Proportional and Neyman allocations are found to be more efficient than the usual Hansen-Hurwitz estimator T^*_{ost}. The proposed estimator $T_{\exp 2}$ maximizes survey precision relative to T^*_{ost} under Neyman allocation only. The suggested estimators $T_{\exp 3}$ and $T_{\exp 4}$ are best among all the proposed estimators in terms of precision and hence are recommended as a suitable alternative under stratified sampling setup with missing data (Table 3).

Acknowledgments

Authors are grateful to the editor and a referee for careful reading of the manuscript and for their constructive suggestions which led to an improved version of the article. The first author acknowledges with thanks R&D Grant from University of Delhi and DU-DST PURSE Grant.

References

1. S. Bahl, and R.K. Tuteja, Ratio and Product type Exponential estimators, *Journal of Information and Optimization Sciences*, **159–164**, 12(1), (1991).
2. M. K. Chaudhary, V. K. Singh and R. K. Shukla, Combined-type family of estimators of population mean in stratified random sampling under non-

response, *Journal of Reliability and statistical studies*, **133–142**, 5(2), (2012).

3. G. Diana, A class of estimators of the population mean in stratified sampling, *Statistica*, **59–66**, 53(1), (1993).
4. M. H. Hansen and W. N. Hurwitz, The problem of non-response in sample surveys, *Journal of American Statistical Association*, **517–529**, 41, (1946).
5. C. Kadilar and H. Cingi, Ratio estimator in stratified sampling, *Biometrical Journal*, **218–225**, 45(2), (2003).
6. C. Kadilar and H. Cingi, A new estimator in stratified random sampling, *Communication in Statistics Theory and Methods*, **597–602**, 34, (2005).
7. M. Khoshnevisan, R. Singh, P. Chauhan, N. Sawan and F. Smarandache, A general family of estimators for estimating population mean using known value of some population parameter(s), *Far East Journal of Theoretical Statistics*, **181–191**, 22, (2007).
8. N. Koyuncu and C. Kadilar, Ratio and product estimators in stratified random sampling, *Journal of Statistical Planning and Inference*, **2552–2558**, 139(8), (2008).
9. N. Koyuncu and C. Kadilar, On the family of estimators of population mean in stratified random sampling, *Pakistan Journal of Statistics*, **427–443**, 26(12), (2010).
10. C. E. Sarndal, B. Swensson and J. Wretman, Model Assisted Survey Sampling, *Springer-Verlag*, New York, Inc., (1992).
11. J. Shabbir and S. Gupta, Improved ratio estimators in stratified sampling, *American Journal of Mathematical and Management Sciences*, **293–311**, 25, (2005).
12. J. Shabbir and S. Gupta, A new estimator of population mean in stratified sampling, Communication in Statistics Theory and Methods, **1201–1209**, 35, (2006).
13. R. Singh, P. Chauhan and N. Sawan, On linear combination of ratio-product type exponential estimator for estimating finite population mean. *Statistics in Transition*, **105–115**, 9(1), (2008).
14. S. K. Srivastava, An estimator using auxiliary information in sample surveys. Cal. Statist. Ass. Bull., **121–132**, 6, (1967).
15. P. V. Sukhatme and B. V. Sukhatme, *Sampling theory of surveys with applications* (Iowa State University Press, Ames, U.S.A, 1970).
16. L. N. Upadhyaya and H. P. Singh, Use of transformed auxiliary variable in estimating the finite populations mean, Biometrical Journal, **627–636**, 41, (1999).

An Analytical Study on Fractional Fokker-Planck Equation by Homotopy Analysis Transform Method

Jitendra Singh†

Department of Mathematics, Central University of South Bihar,
Patna, Bihar - 800014, India
E-mail: jitendraprof@gmail.com
www.cub.ac.in

Rajeev Kumar

Department of Mathematics, Central University of South Bihar
Patna, Bihar-800014, India
E-mail: rajeevkumar@cub.ac.in

In this article, we illustrate an analytical method, namely Homotopy Analysis Transform Method (HATM) which is a combination of Homotopy Analysis Method (HAM) and Laplace Decomposition Method (LDM). This scheme is simple to apply linear and nonlinear fractional differential equations and having less computational work in comparison of other exiting methods. The most useful advantage of this method is to solve the fractional nonlinear differential equation without using Adomain polynomials and He's polynomials for the computation of nonlinear terms. The proposed method has no linearization and restrictive assumptions throughout the process. Presently, homotopy analysis transform method is used to solve time fractional Fokker-Planck equation and similar equations. The series solution obtained by HATM converges very fast. A good agreement between the obtained solution and some well-known results has been demonstrated.

Keywords: Homotopy analysis transform method, Fractional Fokker-Planck equation, Closed form solution.

1. Introduction

Fractional differential equations have long history. These equations have demonstrated a considerable interest both in mathematics and in application in recent years. They have been used in modeling of many physical and chemical processes in engineering[1-2]. The aim of using fractional calculus facilities the description of some problems which are not easily described by ordinary calculus due to modeling error. The advantage of the fractional order system is

---

† This work is partially supported by UGC-BSR Research Start-Up Grant project No. F.20-6(20)/2012(BSR).

that they allow greater degree of freedom in the model. To find the solution of fractional differential equation is not easy due to many fundamental law of calculus having integer order derivative is not hold good in case of fractional derivatives such as product rule etc.

Hesameddini *et al.*[3] solved nonlinear fractional differential equation with a coupling method of Homotopy technique and Laplace transform. Partial differential equations of fractional order are often very complicated to be exactly solved and even if an exact solution is obtainable, the required calculations may be too complicated to be practical. Finding accurate and efficient method for solving fractional differential equation becomes an important task.

The Fokker-Planck equation manly generates in natural science like quantum optics, chemical physics, fluid dynamic, theoretical biology, solid-state physics and circuit theory. Fokker and Planck[4] used this equation to describe the Brownian motion of particles. The fractional order system of the Fokker-Planck equation are obtained by replacing the first order time derivative term by fractional order α (0 < α ≤ 1). A generalized time fractional Fokker-Planck equation in general form to n variables $x_1, x_2, \dots, x_n$ may be stated as,

$$\frac{\partial^\alpha u}{\partial t^\alpha} = \left[- \sum_{i=1}^{n} \frac{\partial}{\partial x_i} A_i(x,t) + \sum_{i=1}^{n} \sum_{j=1}^{n} \frac{\partial}{\partial x_i \partial x_j} B_{i,j}(x,t) \right] u(x,t) (0 < \alpha \leq 1), \quad (1)$$

with the initial condition:

$$u(x,0) = f(x) \qquad x = (x_1, x_2, \dots, x_n) \in \mathbb{R}^n. \quad (2)$$

Equations (1−2) is an equation of motion for the distribution function $u(x,t)$, which is called the time fractional forward Kolmogorov equation, where $B_{i,j}(x,t) > 0$ is called the diffusion coefficient and $A_i(x,t)$ is the drift coefficient. The similar partial differential equation,which might know as the time fractional backward Kolmogorov equation, can be written as[4],

$$\frac{\partial^\alpha u}{\partial t^\alpha} = \left[-A_i(x,t) \sum_{i=1}^{n} \frac{\partial}{\partial x_i} + B_{i.j}(x,t) \sum_{i=1}^{n} \sum_{j=1}^{n} \frac{\partial^2}{\partial x_i \partial x_j} \right] u(x,t) (0 < \alpha \leq 1). \quad (3)$$

The nonlinear Fokker-Planck equation provides useful mathematical models in different areas such as plasma physics, surface physics, population dynamics, biophysics, engineering, nonlinear hydrodynamic, polymer physics, psychology and marketing[5]. A generalized nonlinear time fractional Fokker-Planck equation in n-variables, $x_1, x_2, \dots, x_n$ may be written as

$$\frac{\partial^\alpha u}{\partial t^\alpha} = \left[- \sum_{i=1}^{n} \frac{\partial}{\partial x_i} A_i(x,t,u) + \sum_{i=1}^{n} \sum_{j=1}^{n} \frac{\partial^2}{\partial x_i \partial x_j} B_{i,j}(x,t,u) \right] u(x,t)$$

$$(0 < \alpha \leq 1), \quad (4)$$

The numerical and approximate analytical solution of the Fokker-Planck equation have been studied since the work of Risken[4]. There are several methods in literature which are used to solve this equation[6-9]. Some methods are based on transformation technique and some other give the solution in series form which converges to the exact solution. Recently S. Hesam *et al.*[10] successfully solved the Fokker-Planck equation by differential transform method.

The advantage of the proposed method HATM is its capability of combining two powerful methods for obtaining rapid convergent series for the solution of fractional nonlinear partial differential equation. Khan *et al.*[11] used this method to solve the famous Blasius flow equation on a semi-infinite domain. Motivated by the above work, in this article, authors are concerned to solve time fractional Fokker-Planck equations and time fractional Kolmogorov equation.

This paper is organized as follows: In Section 2, the basic idea of the HATM, Section 3, contains basic definition of Caputo order fractional derivative, Section 4, application of HATM on some examples with analytical solutions in one and two dimensional cases, in last, Section 5 have conclusion.

2. Homotopy Analysis Transform Method (HATM)

Consider equation $N\big(u(x,t)\big) = g(x,t)$, where N represents a general nonlinear ordinary or partial differential operator including both linear and nonlinearterms. The linear term is decomposed into $L + R$, where L is the highest order linear operator and R is the remaining of the linear operator. Thus, the equation can be written as

$$Lu + Ru + Nu = g(x,t), \tag{5}$$

where Nu indicates the nonlinear terms. By applying Laplace transform on both sides of Eq. (5), we get

$$£[Lu + Ru + Nu = g(x,t)], \tag{6}$$

Using the differential property of Laplace transform, we have

$$S^n £[u] - \sum_{k=1}^{n} S^{k-1}\, u^{(n-k)}(0) + £[Ru] + £[Nu] = £[g(x,t)], \tag{7}$$

On simplifying

$$£[u] - \frac{1}{S^n}\sum_{k=1}^{n} S^{k-1} u^{(n-k)}(0) + \frac{1}{S^n}(£[Ru] + £[Nu]) - \frac{1}{S^n}(£[g(x,t)]) = 0, \tag{8}$$

We define the nonlinear operator

$$N[\phi(x,t;q)] = £[\phi(x,t;q)] - \frac{1}{S^n}\sum_{k=1}^{n} S^{k-1}\, \phi^{(n-k)}(x,0;q)$$

$$+\frac{1}{s^n}\left(£\left[R\left(\phi(x,t;q)\right)\right]+£\left[N\left(\phi\,(x,t;q)\right)\right]\right)$$

$$-\frac{1}{s^n}\left(£[g(x,t)]\right), \tag{9}$$

where $q \in [0,1]$ and $\phi(x,t;q)$ is real function of x,t and q. We construct a homotopy as follows,

$$(1-q)£[\phi(x,t;q)-u_0(x,t)] = \hbar H(x,t)qN[\phi(x,t;q)], \tag{10}$$

where £ denotes the Laplace transform, $q \in [0,1]$ is the embedding parameter, $\hbar$ is a non-zero auxiliary parameter, $H(x,t) \neq 0$ is an auxiliary function, $u_0(x,t)$ is an initial guess of $u(x,t)$ and $\phi(x,t;q)$ is an unknown function. Obviously, when $q = 0$ and $q = 1$, it holds,

$$\phi(x,t;0) = u_0(x,t), \qquad \phi(x,t;1) = u(x,t). \tag{11}$$

respectively. Thus, as q increases from 0 to 1, the solution $\phi(x,t;q)$ varies from the initial guess $u_0(x,t)$ to the solution $u(x,t)$. Expanding $\phi(x,t;q)$ in Taylor series with respect to q, we have

$$\phi(x,t;q) = u_0(x,t) + \sum_{m=1}^{\infty} u_m\,(x,t)q^m, \tag{12}$$

where

$$u_m\,(x,t) = \frac{1}{m!}\frac{\partial^m \phi(x,t;q)}{\partial q^m} \quad at \quad q = 0. \tag{13}$$

If the auxiliary linear operator (L), the initial guess $u_0(x,t)$, the auxiliary parameter $\hbar$ and the auxiliary function $H(x,t)$ are properly chosen, series Eq. (12) converges at q=1, then we have

$$u(x,t) = u_0(x,t) + \sum_{m=1}^{\infty} u_m, \tag{14}$$

which must be one of solution of original nonlinear equation. According to definition (14), the governing equation can be deduced from the zero-order deformation Eq. (10). Define the vector

$$\vec{u}(x,t) = \{u_0(x,t), u_1(x,t), u_2(x,t), \ldots, u_m(x,t)\}. \tag{15}$$

Differentiating Eq. (10) $m-times$ with respect to the embedding parameter q and then setting $q = 0$ and finally dividing them by $m!$, we have the so called $m^{th}-order$ deformation equation

$$£[u_m(x,t) - \chi_m u_{m-1}(x,t)] = hH(x,t)R_m(\vec{u}_{m-1}). \tag{16}$$

Applying inverse Laplace transform,we have

$$u_m(x,t) = \chi_m u_{m-1}(x,t) + \hbar £^{-1}[H(x,t)R_m(\vec{u}_{m-1})], \tag{17}$$

where

$$R_m(\vec{u}_{m-1}) = \frac{1}{(m-1)!}\frac{\partial^{m-1} N(\emptyset(x,t;q))}{\partial q^{m-1}}, \quad at\, q = 0, \tag{18}$$

and

$$\chi_m = \begin{cases} 0, & m \le 1 \\ 1, & m > 1 \end{cases}. \tag{19}$$

3. Basic Definition of Fractional Derivative in Caputo Sense

There are various definitions of fractional integration and differentiation, such as Grunwald-Letnikov's definition and Riemann-Liouville's definition. The Riemann-Liouville derivative has certain disadvantages when trying to model real-world phenomena with fractional differential equations. Therefore, we shall introduce a modified fractional differential operator D^α proposed by Caputo in his work on the theory of viscoelasticity[12].

Definition 3.1 The Caputo definition of the fractional-order derivative is defined as

$$D^\alpha f(x) = \frac{1}{\Gamma(n-\alpha)} \int_0^x \frac{f^{(n)}(t)}{(x-t)^{\alpha+1-n}} dt, \qquad n-1 < \alpha \le n, \ n \in \mathbb{N}, \tag{20}$$

where α>0 is the order of the derivative and *n* is the smallest integer greater than α. For the Caputo derivative, we have

$$D^\alpha C = 0 \qquad (C \text{ is a constant}) \tag{21}$$

$$D^\alpha x^\beta = \begin{cases} 0 & , \ for\ \beta \in \mathbb{N}_0\ and\ \beta < \lceil \alpha \rceil \\ \frac{\Gamma(1+\beta)}{\Gamma(1+\beta-\alpha)} x^{\beta-\alpha} & , \ for\ \beta \in \mathbb{N}_0\ and\ \beta \ge \lceil \alpha \rceil\ or\ \beta \notin \mathbb{N} \\ & \quad and\ \beta > \lfloor \alpha \rfloor \end{cases} \tag{22}$$

Where $\lceil \alpha \rceil$ is ceiling function and $\lfloor \alpha \rfloor$ is the floor function. $\mathbb{N} = \{1,2,3,\dots\}$ and $\mathbb{N}_0 = \{0,1,2,\dots\}$. Recall that for $\alpha \in \mathbb{N}$, the Caputo differential operator coincides with the usual differential operator of an integer order. Similar to the integer-order differentiation, the Caputo fractional differentiation is a linear operation

$$D^\alpha(\lambda f(x) + \mu g(x)) = \lambda D^\alpha f(x) + \mu D^\alpha g(x) \tag{23}$$

Where λ and μ are constants. In this paper, the fractional derivatives are considered in the Caputo sense.

Definition 3.2 The Laplace transform $\pounds[f(t); s]$ of the Caputo fractional derivative is given by

$$\pounds[D_t^\alpha f(t); s] = s^\alpha \pounds[f(t)] - \sum_{k=0}^{m-1} s^{\alpha-k-1} f^k(0), \ m-1 < \alpha \le m \tag{24}$$

4. Numerical Experiments

In this section, we shall illustrate the Homotopy analysis transform technique by some examples.

Example 4.1 Consider the backward Kolmogorov Eq. (3) with $n = 1, x_1 = x$,

$$A_1(x,t) = -(x+1),$$
$$B_{1,1}(x,t) = e^t x,^2$$
$$f(x) = x+1, \qquad x \in \mathbb{R}.$$

Applying the HATM, we have

$$R_m(\overrightarrow{u_{m-1}}) = \text{£}[u_{m-1}] - \frac{(1-\chi_m)}{s}u(x,0) - \frac{(1+x)}{s^\alpha}\text{£}[(u_{m-1})_x]$$
$$-\frac{x^2}{s^\alpha}\text{£}[e^t(u_{m-1})_{xx}], \tag{25}$$

The $m^{th} - order$ deformation equation, using auxiliary function $H(x,t) = 1$, we have

$$u_m(x.t) = \chi_m u_{m-1}(x,t) + \hbar \text{£}^{-1}[R_m(\vec{u}_{m-1})]. \qquad (m \geq 1). \tag{26}$$

Using the Eqs. (25–26), we have the following approximations

$$u_0(x,t) = u(x,0) = (1+x), \tag{27}$$
$$u_1(x,t) = -\frac{\hbar(x+1)t^\alpha}{\Gamma(1+\alpha)}, \tag{28}$$
$$u_2(x,t) = \frac{\hbar(x+1)(\hbar t^\alpha\Gamma(\alpha+1) - \hbar\Gamma(2\alpha+1) - \Gamma(2\alpha+1))t^\alpha}{\Gamma(\alpha+1)\Gamma(2\alpha+1)}, \tag{29}$$
$$u_3(x,t) = -\frac{\hbar(1+x)t^\alpha}{\Gamma(1+\alpha)\Gamma(2\alpha+1)\Gamma(3\alpha+1)}(\hbar^2 t^{2\alpha}\Gamma(\alpha+1)\Gamma(2\alpha+1)$$
$$-2\hbar^2 t^\alpha\Gamma(\alpha+1)\Gamma(3\alpha+1) + \hbar^2\Gamma(2\alpha+1)\Gamma(3\alpha+1)$$
$$-2\hbar t^\alpha\Gamma(\alpha+1)\Gamma(3\alpha+1) + 2\hbar\Gamma(2\alpha+1)\Gamma(3\alpha+1)$$
$$+\Gamma(2\alpha+1)\Gamma(3\alpha+1), \tag{30}$$

Now, we obtain the series solution from Eq. (14), using the Eqs. (27–30). For getting the solution in closed form taking $\hbar = -1$ and $\alpha = 1$, we have

$$u(x,t) = (1+x)e^t. \tag{31}$$

Which is exact solution of the given problem, when $\alpha = 1$.

Example 4.2 In this example, we consider Eqs. (1–2) with $n = 2,\ x_1 = x, x_2 = y$ (Two dimensional case)

$$A_1(x,y,t) = x,$$
$$A_2(x,y,t) = 5y,$$
$$B_{1,1}(x,y,t) = x^2,$$
$$B_{1,2}(x,y,t) = 1,$$
$$B_{2,1}(x,y,t) = 1,$$
$$B_{2,2}(x,y,t) = y^2,$$

$$f(x) = x, \qquad x, y \in \mathbb{R}.$$

Applying the HATM, we have

$$R_m(\vec{u}_{m-1}) = \text{£}[u_{m-1}] - \frac{(1-\chi_m)}{s} u(x,y,0) + \frac{2}{s^\alpha}\text{£}[u_{m-1}] + \frac{y}{s^\alpha}\text{£}\left[(u_{m-1})_y\right]$$
$$-\frac{y^2}{s^\alpha}\text{£}\left[(u_{m-1})_{yy}\right] - \frac{3x}{s^\alpha}\text{£}[(u_{m-1})_x] - \frac{2}{s^\alpha}\text{£}\left[(u_{m-1})_{xy}\right]$$
$$-\frac{x^2}{s^\alpha}\text{£}[(u_{m-1})_{xx}], \tag{32}$$

The $m^{th} - order$ deformation equation, using auxiliary function $H(x,y,t) = 1$, we have

$$u_m(x,y,t) = \chi_m u_{m-1}(x,y,t) + \hbar \text{£}^{-1}[R_m(\vec{u}_{m-1})]. \qquad (m \geq 1). \tag{33}$$

Using the Eqs. (32–33), we have the following approximations

$$u_0(x,y,t) = u(x,y,0) = x, \tag{34}$$

$$u_1(x,y,t) = -\frac{\hbar x t^\alpha}{\Gamma(1+\alpha)}, \tag{35}$$

$$u_2(x,y,t) = \frac{\hbar x(\hbar t^\alpha \Gamma(\alpha+1) - \hbar\Gamma(2\alpha+1) - \Gamma(2\alpha+1))t^\alpha}{\Gamma(\alpha+1)\Gamma(2\alpha+1)}, \tag{36}$$

$$u_3(x,y,t) = -\frac{\hbar x t^\alpha}{\Gamma(1+\alpha)\Gamma(2\alpha+1)\Gamma(3\alpha+1)}(\hbar t^{2\alpha}\Gamma(\alpha+1)\Gamma(2\alpha+1)$$
$$-2\hbar^2 t^\alpha \Gamma(\alpha+1)\Gamma(3\alpha+1) + \hbar^2\Gamma(2\alpha+1)\Gamma(3\alpha+1)$$
$$-2\hbar t^\alpha \Gamma(\alpha+1)\Gamma(3\alpha+1) + 2\hbar\Gamma(2\alpha+1)\Gamma(3\alpha+1)$$
$$+\Gamma(2\alpha+1)\Gamma(3\alpha+1) \tag{37}$$

Now, we obtain the series solution from Eq. (14), using the Eqs. (34–37). The solution in closed form taking $\hbar = -1$ and $\alpha = 1$, we have

$$u(x,y,t) = xe^t \tag{38}$$

Which is exact solution of the given problem, when $\alpha = 1$.

Example 4.3 Finally, we apply the HATM on the nonlinear Fokker-Planck equation, consider Eq. (4) $n = 1, x_1 = x$,

$$A_1(x,t,u) = 4\frac{u}{x} - \frac{x}{3},$$

$$B_{1,1}(x,t,u) = u,$$

$$f(x) = x^2, \qquad x \in \mathbb{R}.$$

Using the HATM, we have

$$R_m(\vec{u}_{m-1}) = \text{£}[u_{m-1}] - \frac{(1-\chi_m)}{s} u(x,0) + \frac{8}{s^\alpha x}\text{£}\left[\sum_{k=0}^{m-1} u_{m-1-k}(u_k)_x\right]$$

$$-\frac{2}{s^{\alpha}}£\left[\sum_{k=0}^{m-1} u_{m-1-k}(u_k)_{xx}\right]-\frac{2}{s^{\alpha}}£\left[\sum_{k=0}^{m-1}(u_{m-1-k})_x(u_k)_x\right]$$

$$-\frac{x}{3s^{\alpha}}£[(u_{m-1})_x]-\frac{4}{x^2s^{\alpha}}£\left[\sum_{k=0}^{m-1} u_{m-1-k}u_k\right]$$

$$-\frac{1}{3s^{\alpha}}£[u_{m-1}] \tag{39}$$

The $m^{th}-order$ deformation equation, using auxiliary function $H(x,t)=1$, we have

$$u_m(x,t)=\chi_m u_{m-1}(x,t)+\hbar £^{-1}[R_m(\vec{u}_{m-1})]. \qquad (m\geq 1) \tag{40}$$

Using the Eqs. (39–40), we have the following approximations

$$u_0(x,t)=u(x,0)=x^2, \tag{41}$$

$$u_1(x,t)=-\frac{\hbar x^2t^{\alpha}}{\Gamma(1+\alpha)}, \tag{42}$$

$$u_2(x,t)=\frac{\hbar x^2(\hbar t^{\alpha}\Gamma(\alpha+1)-\hbar\Gamma(2\alpha+1)-\Gamma(2\alpha+1))t^{\alpha}}{\Gamma(\alpha+1)\Gamma(2\alpha+1)}, \tag{43}$$

$$\begin{aligned}u_3(x,t)=&-\frac{\hbar x^2t^{\alpha}}{\Gamma(1+\alpha)\Gamma(2\alpha+1)\Gamma(3\alpha+1)}(\hbar^2t^{2\alpha}\Gamma(\alpha+1)\Gamma(2\alpha+1)\\&-2\hbar^2t^{\alpha}\Gamma(\alpha+1)\Gamma(3\alpha+1)+\hbar^2\Gamma(2\alpha+1)\Gamma(3\alpha+1)\\&-2\hbar t^{\alpha}\Gamma(\alpha+1)\Gamma(3\alpha+1)+2\hbar\Gamma(2\alpha+1)\Gamma(3\alpha+1)\\&+\Gamma(2\alpha+1)\Gamma(3\alpha+1)\end{aligned} \tag{44}$$

Now, we obtain the series solution from Eq. (14), using the Eqs. (41–44). Now the solution in closed form taking $\hbar=-1$ and $\alpha=1$, we have

$$u(x,t)=x^2e^t \tag{45}$$

5. Conclusion

The present work is easily shown that the series solution obtained by HATM converges rapidly. The solution obtained by HATM is in good agreement with $\alpha=1$, and had a better explanation in the fractional cases. The present study on fractional Fokker-Plank equation affirmed that HATM is easy to apply and have less computational work with high accuracy. Therefore, this method may be applied to many complicated fractional nonlinear differential equations without any assumptions, transformations discretization and perturbation.

Acknowledgments

Authors are very thankful to U.G.C. for the scheme "Start-Up Grant for Newly Recruited Faculty" and to grant the fund UGC-BSR Research Start-Up Grant project No. F.20-6(20)/2012(BSR).

References

1. I. Podlubny, *Fractional Differential Equation, in: Mathematics in Science and Engineering*, Academic Press Inc., San Diego, CA, **vol. 198**, 1999.
2. D. Kusnezov, A. Bulgac and G. D. Dang, Quantum Levy processes and fractional kinetics, *Phys. Rev. Lett.* **82**, 1136–1139, (1999).
3. E. Hesameddini, M. Riahi and H. Latifizadesh, A coupling method of homotopy technique and Laplace transform for nonlinear fractional differential equations, *Int. J. Adv. Appl. Sci.* **1**, 159–170, (2012).
4. H. Risken, *The Fokker-Planck Equation Method of Solution and Applications,* Springer Verlag, Berlin, Heidelberg (1989).
5. T. D. Frank, Stochastic feedback, nonlinear families of Markov processes and nonlinear Fokker-Planck equations, *phy. Lett. A***331**, 391–408, (2004).
6. G. Harrison, Numerical solution of the Fokker-Planck equation using moving finite elements, *Numer. Methods Partial Differential Equations* **4**, 219–232, (1988).
7. M. Lakestani and M. Dehghan, Numerical solution of Fokker-Planck equation using the cubic B-spline scaling function, *Numer. Methods Partial Differential Equations* **25**, 418–429, (2009).
8. V. Palleschi and M. de Rosa, Numerical solution of the Fokker-Planck equation II. Multidimensional case, *Phys. Lett. A.***163**, 381–391, (1992).
9. M. P. Zorzano, H. Mais and L. Vazquez, Numerical solution of two dimensional Fokker-Planck equations, *Appl. Math. Comput.* **98**, 109–117, (1999).
10. S. Hesam, A. R. Nazemi and A. Haghbin, Analytical solution for the Fokker-Planck equation by differential transform method, *Sci. Iranica* **19**, 1140–1145, (2012).
11. M. Khan, M. A. Gondal, I. Hussain and S. KarimiVanani, A new comparative study between homotopy analysis transform method and homotopy perturbation transform method on a semi-infinite domain, *Math. Compu. Mod.* **55**, 1143–1150, (2012).
12. M. Caputo, Linear models of dissipation whose Q is almost frequency independent. Part II, *J. Roy Austral. Soc.* **13**, 529–539, (1967).

L-Primitive Words in Submonoids of a Free Monoid

Shubh Narayan Singh
Department of Mathematics, Central University of South Bihar,
Patna, Bihar-800014, India
E-mail: shubh@cub.ac.in
www.cub.ac.in

K. V. Krishna
Department of Mathematics, IIT Guwahati,
Guwahati, Assam-781039, India
E-mail: kvk@iitg.ac.in
www.iitg.ac.in

This work considers a natural generalization of primitive words with respect to a language. Given a language L, a nonempty word $w$is said to be L-primitive if w is not a proper power of any word in L. After ascertaining the number of primitive words in submonoids of a free monoid, the work proceeds to count L-primitive words in submonoids of a free monoid. The work also studies the distribution of L-primitive words in certain subsets of a free monoid.

Keywords: Free monoids, Primitive words, Numerical monoids.

1. Introduction

A nonempty word which is not a power of any other word is called a primitive word. It is well known that every nonempty word can be uniquely expressed as a power of a primitive word [9]. The study of primitive words is often the first step towards the understanding of words and plays an important role in the theory of languages. Ito *et al.* have investigated the number of primitive words inthe languages accepted by automata [6]. Shyr and Tseng have proved that any noncommutative submonoid of a free monoid contains infinitely many primitive words [11]. In the literature, there are various types of generalizations/ extensions of classic definition of primitive words [3, 4, 5, 7]. We propose yet another generalization of primitive word, viz. L-primitive word — a nonempty word that is not a proper power of any word in given L.

In this paper, we first investigate the primitive words in submonoids of free monoid. We could ascertain that the number of primitive words in a submonoid of free monoid is either at most one or infinity. Further, we study the distribution of L-primitive words in the certain subsets of free monoid. In particular, wecount L-primitive words in submonoids of free monoid.

The paper is organized as follows. In Section 2, we present some necessary preliminaries of the paper. We introduce the concept of L-primitive words in Section 3 and study some basic properties. In Section 4, we count the number of L-primitive words in a given language L as well as in submonoids of free monoid. Finally, Section 5 concludes the paper.

2. Preliminaries

In this section, we present some basic definitions and fix notations which are useful in the work. For more details one may refer [1, 8, 12].

Let A be a nonempty finite set called an *alphabet* with its elements as *letters*. The free monoid over A is denoted by A^* whose elements are called words, and ε denotes the identity element of A^* — the empty word. The set of all nonempty words over A is denoted by A^+. A *language* is a subset of A^*.

A nonempty word u is said to be a *prefix* of a nonempty word v if there exists a nonempty word t such that $ut = v$. A set X of nonempty words is called a *prefix set* if no element of X is a prefix of another word of X. A *power* of a nonempty word u is a word of the form u^k for some number $k \in \mathbb{N} = \{0, 1, 2, \dots\}$. It is convenient to set $u^0 = \varepsilon$ for each word$u$. If $k \in \mathbb{N} - \{0, 1\}$, we say that u^k is a *proper power* ofnonempty word$u$. A nonempty word x is said to be a *primitive word* if it is not a proper power of another nonempty word in A^*, i.e., for $u \in A^*$, if $x = u^k$ implies $k = 1$. For an alphabet A, the set of all primitive words $x \in A^*$ is denoted by$Q(A^*)$. For a subset X of A^*, we denote the set of all primitive words in X by $Q(X)$. We recall the following well known property of primitive words from [9].

Proposition 2.1. For every nonempty word w, there exists a unique primitive word u and a unique integer $k \geq 1$ such that $w = u^k$.

The unique primitive word u obtained in Proposition 2.1 is called the *primitive root* of w, denoted by $\sqrt{w}$. The *root* of a language L, denoted by $\sqrt{L}$, is the set of all primitive root of all nonempty words in L, i.e.,

$$\sqrt{L} = \left\{\sqrt{w} \in Q(A^*) \mid w \in L \setminus \{\epsilon\}\right\}.$$

A language L is said to be *commutative* if $uv = vu$ for all $u, v \in L$. Note that a language L is commutative if and only if there exists a word w such that $L \subseteq \{w\}^*$.

We recall some properties of numerical monoids from [10]. A *numerical monoid* is a submonoid of the additive monoid $(\mathbb{N}, +)$ whose complement in $\mathbb{N}$

is finite. For a nonempty subset X of $\mathbb{N}$, the submonoid of the additive monoid $(\mathbb{N},+)$ generated by X is denoted by $\langle X\rangle$, i.e.,

$$\langle X\rangle = \{\lambda_1 x_1 + \lambda_2 x_2 + \cdots + \lambda_n x_n \mid n,\ \lambda_i \in \mathbb{N}, x_i \in X,\ \forall i\ (1 \leq i \leq n)\}.$$

Theorem 2.2. Let X be a nonempty subset of $\mathbb{N}$. The submonoid $\langle X\rangle$ is a numerical monoid if and only if gcd $(X) = 1$.

Theorem 2.3. Every numerical monoid admits a unique finite minimal set of generators.

Theorem 2.4. Any nontrivial submonoid of the additive monoid $(\mathbb{N},+)$ is isomorphic to a numerical monoid.

3. *L*-Primitive Words

In this section, we introduce the notion of primitive words relative to a language L, called L-primitive words, and obtain some properties related to L-primitive words. We prove that every primitive word is an L-primitive word so that the latter notion is a generalization of the former one. Unless it is specified otherwise, in what follows, L is an arbitrary language.

Definition 3. 1. A nonempty word x is said to be an L-*primitive word* if x is not a proper power of any word in L, i.e., for $u \in L$,

$$x = u^k \Rightarrow k = 1.$$

For a subset X of A^*, we denote the set of all L-primitive words in X by $LQ(X)$.

We begin with some basic properties of L-primitive words in a subset of A^*.

Remark 3.2.

(i) If $L = \emptyset$, then $LQ(A^*) = A^+$.

(ii) If $L = A^*$, then $LQ(A^*) = Q(A^*)$.

Proposition 3.3. If L and L' are two languages such that $L \subseteq L'$, then $L'Q(A^*) \subseteq LQ(A^*)$.

Proof. On the contrary, let us assume that $L'Q(A^*) \nsubseteq LQ(A^*)$. Then there exists $w \in L'Q(A^*)$, but $w \notin LQ(A^*)$. Since $w \notin LQ(A^*)$, there exists $u \in L$ such that $w = u^k$ for some $k > 1$. In view of hypothesis, we have $u \in L'$, and so that $w \notin L'Q(A^*)$; a contradiction.

In view of Remark 3.2(ii), we have the following corollary of Proposition 3.3.

Corollary 3.4. Every primitive word is an L-primitive word. In particular, if $|A| \geq 2$, then $|LQ(A^*)| = \infty$.

Remark 3.5. An L-primitive word need not be a primitive word. For instance, let $L = \{abab\} \subseteq \{a,b\}^*$. Clearly, the word $abab$ is an L-primitive word, but not a primitive word.

Definition 3.6. For $w \in A^+$, we define the set of *L-primitive roots* of w, denoted $\sqrt[L]{w}$, is defined as

$$\sqrt[L]{w} = \{x \in LQ(A^*) \mid x^k = w \text{ for some } k \geq 1\}.$$

Further, for $X \subseteq A^*$, the *L-primitive root* of X, denoted $\sqrt[L]{X}$, is the set of L-primitive roots of all nonempty words in X.

Remark 3.7. The primitive root of a nonempty word is an L-primitive root of the word. Thus, if $w \neq \epsilon$, then $\sqrt[L]{w} \neq \emptyset$.

4. *L*-Primitive Words in Various Subsets

This section is divided into two subsections. We investigate L-primitive words in some sets related to L itself in Subsection 4.1. Then we carry on the investigations on submonoids of a free monoid in Subsection 4.2.

4.1. *L-primitive words in L*

In this subsection, we make an attempt to investigate L-primitive words in L and also in L^c -- complement of L in A^*. In this connection, we provide some sufficient conditions and characterizations. In fact we give a relation between L-primitive words and L-primitive roots in L.

Theorem 4.1. Let L be a subset of A^+. Then $L \neq \emptyset$ if and only if $LQ(L) \neq \emptyset$.

Proof. Let us assume that $L \neq \emptyset$ and choose $w \in L$. If $w \in LQ(L)$, then we are through. Otherwise there exists $u \in L$ such that $w = u^k$ for some $k > 1$. Clearly $|u| < |w|$. If $u \in LQ(L)$, then we are through. Otherwise we continue to choose shorter words in L whose power is w. This process of choosing a shorter word in L terminates at a finite stage and eventually we get a word $x \in LQ(L)$

such that $w = x^m$ for some $m > 1$. This shows that $LQ(L) \neq \emptyset$. The converse is straightforward.

Theorem 4.2. If L is a prefix subset of A^+, then $L = LQ(L)$.

Proof. Clearly, $LQ(L) \subseteq L$. Let $x \in L$, but $x \notin LQ(L)$. Then there exists a nonempty word $u \in L$ such that $x = u^k$ for some $k > 1$. This implies that the word $u \in L$ is a prefix of the word $x \in L$. This contradicts that L is a prefix set. Hence $L = LQ(L)$.

Remark 4.3. The converse of Theorem 4.2 is not necessarily true. For instance, let $L = \{a, ab\} \subseteq \{a, b\}^+$. Clearly $L = LQ(L)$ but L is not a prefix set.

It is clear that $LQ(L) \subseteq \sqrt[L]{L}$. Now we explore the possibilities so that $LQ(L) = \sqrt[L]{L}$. We need the notion of power of a subset of A^* introduced by Calbrix and Nivat [2]. The *power* of a subset $X$of A^*, denoted by pow(X), is defined as

$$\text{pow}(X) = \{x^k \mid x \in X \text{ and } k \geq 1\}.$$

Remark 4.4. $\text{pow}\big(LQ(A^*)\big) = A^+$.

Theorem 4.5.

(i) $\sqrt[L]{L} \subseteq L \Leftrightarrow LQ(L) = \sqrt[L]{L}$.

(ii) $L^c = \text{pow}(L^c) \Rightarrow LQ(L) = \sqrt[L]{L}$.

(iii) $L \subseteq Q(A^*) \Rightarrow LQ(L) = L = \sqrt[L]{L}$.

Proof. We first note that $LQ(L) \subseteq \sqrt[L]{L}$.

(i) $(:\Leftarrow)$ Since $LQ(L) \subseteq L$, from the hypothesis, we have $\sqrt[L]{L} \subseteq L$.
$(\Rightarrow:)$ Let $x \in \sqrt[L]{L}$; then x is L-primitive word. Also from the hypothesis, we have $x \in L$ and so $x \in LQ(L)$. Hence we have the part (i).

(ii) Let us assume that $x \in \sqrt[L]{L} \setminus LQ(L)$. Since x is an L-primitive word and $x \notin LQ(L)$, we have $x \notin L$. From the hypothesis, we have $x^k \in L^c$ for all $k \geq 1$. Since $x \in \sqrt[L]{L}$, we have $x \in \sqrt[L]{w}$ for some $w \in L$. That means there is a number $t \geq 1$ such that $x^t = w(\in L)$; a contradiction. Hence $LQ(L) = \sqrt[L]{L}$.

(iii) Clearly $LQ(L) \subseteq L$. Let $x \in L$; from the hypothesis, we have $x \in Q(A^*)$. By Corollary 3.5, since every primitive word is an L-primitive word, we have $x \in LQ(L)$. Thus $LQ(L) = L$.
It is clear that for $w \in Q(A^*)$ we have $\sqrt[L]{w} = \{w\}$. Since $L \subseteq Q(A^*)$, we have $\sqrt[L]{L} = \text{L}$. Hence if $L \subseteq Q(A^*) \Rightarrow LQ(L) = L = \sqrt[L]{L}$.

Remark 4.6. The converse of Theorem 4.5(ii) is not necessarily true. For instance, consider $L = \{a, b, a^6\} \subseteq \{a, b\}^+$. We observe that $LQ(L) = \sqrt[L]{L} = \{a, b\}$. Since $a^2 \in L^c$, we have $a^6 \in pow(L^c)$. But $a^6 \notin L^c$ and therefore $L^c \neq \text{pow}(L^c)$.

Theorem 4.7. $L = \text{pow}(L) \Leftrightarrow L^c = LQ(L^c)$.

Proof. ($\Rightarrow$:) Clearly $LQ(L^c) \subseteq L^c$. Let $x \in L^c$, but $x \notin LQ(L^c)$. Then there exists a word $w \in L$ such that $x = w^k$ for some $k > 1$. Since $w \in L$, we have $w^k \in \text{pow}(L)$. It follows that $x \in \text{pow}(L)$. But $L = \text{pow}(L)$, we have $x \in L$. This is a contradiction.
(:$\Leftarrow$) Clearly $L \subseteq \text{pow}(L)$. Let $x \in \text{pow}(L)$ but $x \notin L$. Then there exists a word $u \in L$ such that $x = u^k$ for some $k > 1$. Since $x \notin L$, we have $x \in L^c$. But $L^c = LQ(L^c)$, it follows that x is an L-primitive word; which is a contradiction.

4.2. *L-primitive words in submonoids*

In this subsection, for comparison, we first investigate the number of primitive words in submonoids of a free monoid. Further, we study the L-primitive words in submonoids of a free monoid. Also we count the L-primitive words in submonoids of the free monoid over a unary alphabet. In this case, when L is finite, we prove that a submonoid has either at most one or infinitely many L-primitive words. Finally, we leave certain remarks on estimating the number of L-primitive words in submonoids over an arbitrary alphabet. We require the following theorem from [11].

Theorem 4.9. A submonoid H of $A^*$is noncommutative if and only if H contains infinitely many primitive words.

Now we observe that a submonoid of A^* contains either at most one primitive word or infinitely many primitive words.

Theorem 4. 10. A submonoid H of A^* contains either at most one or infinitely many primitive words.

Proof. If $H = \{\varepsilon\}$, then clearly H has no primitive words. Let us assume that $H \neq \{\varepsilon\}$. If H is a noncommutative submonoid, then by Theorem 4.9, the submonoid H contains infinitely many primitive words. Otherwise, we have $H \subseteq \{w\}^*$ for some nonempty word w. Without loss of generality, assume that

w is a primitive word. Thus, according to $w \in H$or $w \notin H$, the primitive words in submonoid H is either 1 or 0.

Corollary 4.11. If H is nontrivial submonoid of A^*, then either $|\sqrt{H}| = 1$ or ∞.

Let A be a unary alphabet, say $A = \{a\}$. It is known that the free monoid A^* is isomorphic to the additive monoid $(\mathbb{N}, +)$ under the isomorphism given by $a^k \mapsto k$. Thus, each word a^k of A^* is characterized by its length $k \in \mathbb{N}$. Hence we count the L-primitive words in a submonoid of the additive monoid $\mathbb{N}$, instead of A^*. In what follows, H is a nontrivial submonoid of the additive monoid $\mathbb{N}$ and L is a nonempty subset of $\mathbb{N}$. Now, we count the number of L-primitive words in H. We begin with the following remark.

Remark 4.12. If $1 \in L$, then according to $1 \in H$ or $1 \notin H$, the L-primitve words in H is either 1 or 0, respectively.

Let us assume further that $1 \notin L$. In view of Theorem 2.3 and Theorem 2.4, let Y be the finite minimal generating set of H.

Theorem 4.13. If $\gcd(Y) = 1$, then the number of L-primitive words in H is infinite.

Proof. If $\gcd(Y) = 1$, by Theorem 2.2, the submonoid H is a numerical monoid so that $|\mathbb{N} - H| < \infty$. Therefore H contains infinitely many prime numbers. Since $1 \notin L$, every prime number is a L-primitive word. Hence the number of L-primitive words in H is infinite.

Theorem 4.14. If L is a finite set and $\gcd(Y) > 1$, then the number of L-primitive words in H is either at most one or infinite.

Proof. Let $\gcd(Y) = d(> 1)$. For all $l \in L$, we first assume that $l \nmid d$ and claim that the number of L-primitive words in H is infinite. Since $d \neq 1$, by Proposition 2.2, the submonoid H is not a numerical monoid. We define the function $f: H \longrightarrow \mathbb{N}$ by $f(h) = \frac{h}{d}$. The function f is a monomorphism and so the image of f, $\mathrm{Im}(f)$, is isomorphic to H. By Theorem 2.2, the submonoid $\mathrm{Im}(f)$ is a numerical monoid.

The numerical monoid $\mathrm{Im}(f)$ has infinitely many prime numbers. Let $p \in \mathrm{Im}(f)$ be a prime number such that $p > \max(L)$, then $pd \in H$. By

Euclid's lemma, $l \nmid pd$ for all $l \in L$. Since $\mathrm{Im}(f)$ has infinitely many such prime numbers, therefore the number of L-primitive words in H is infinite.

Now, we assume that $l \mid d$ for some $l \in L$. We determine the number of L-primitive words in H with respect to $d \in L$or $d \notin L$. If $d \notin L$, then clearly the submonoid H contains no L-primitive words. If $d \in L$, we consider both the cases $d \in H$ and $d \notin H$.

If $d \notin H$, then clearly the submonoid H contains no L-primitive words. In the case $d \in H$, if there is an $l'(\neq d)$ such that $l' \mid d$, then H contains no L-primitive words; otherwise H contains exactly one L-primitive word.

Remark 4.15. If L is an infinite subset of $\mathbb{N}$, then the number of L-primitive words in H need not satisfy the Theorem 4.14. For instance, let H be the submonoid of the additive monoid $\mathbb{N}$ generated by the set $\{4,6\}$. Consider the infinite subset $L = \{4\} \cup \{\wp - \{2,5\}\}$, where $\wp$ is the set of prime numbers. We observe that the numbers 4 and 10 are only L-primitive words in H. Thus the number of L-primtive words in H is exactly two.

In the following, we make remarks on the number of L-primitive words in submonoids of a free monoid over an alphabet of size at least two. We observe that if H is the trivial submonoid, then H contains no L-primitive words. If H is a nontrivial submonoid, then by Corollary 3.4, we have the following remark.

Remark 4.16. If H is a noncommutative submonoid of A^*, where $|A| \geq 2$, then the number of L-primitive words in H is infinite.

5. Conclusion

Motivated by the work of Ito *et al.* in [6], we have considered a study on the number of primitive words in the languages of semi-flower automata (SFA). An SFA precisely accepts a finitely generated submonoid of a free monoid [12]. We extended the study to submonoids of a free monoid and observed that the number of primitive words in a submonoid is either at most one or infinite. Further, we have initiated a study on the number of L-primitive words in submonoids of a free monoid. If L is finite, we have counted the number of L-primitive words in submonoids of free monoid over a unary alphabet. When L is infinite, the problem appears to be more complicated and a systematic study in this regard is necessary. In case the alphabet size is at least two, we could remark only on the number of L-primitive words in noncommutative submonoids. One can consider the problem in commutative submonoids.

References

1. J. Berstel and D. Perrin, Theory of Codes, volume 117 of Pure and Applied Mathematics (Academic Press Inc., 1985).
2. H. Calbrix and M. Nivat, Prefix and period languages of rational ω-languages, *In Developments in languages theory, II* (Magdeburg, 1995), pages 341–349 (World Sci. Publ., River Edge, NJ, 1996).
3. E. Czeizler, L. Kari and S. Seki, On a special class of primitive words, *Theoret. Comput. Sci.*, 411(3):617–630 (2010).
4. M. Domaratzki, Trajectory-based operations, PhD thesis (Queen's University, Canada, 2004).
5. H. K. Hsiao, C. C. Huang and S. S. Yu, Word operation closure and primitivity of languages, *J. UCS*, 8(2):243–256 (2002).
6. M. Ito, M. Katsura, H. J. Shyr and S. S. Yu, Automata accepting primitive words, *Semigroup Forum*, 37(1): 45–52 (1988).
7. L. Kari and G. Thierrin, Word insertions and primitivity, *Util. Math.*, 53:49–61 (1998).
8. M. Lothaire, Combinatorics on words, volume 17 of Encyclopedia of Mathematics and its Applications (Addison-Wesley Publishing Co., Reading, Mass., 1983).
9. R. C. Lyndon and M. P. Schützenberger, The equation $a^M = b^N c^P$ in a free group, *Michigan Math. J.*, 9: 289–298 (1962).
10. J. C. Rosales and P. A. García-Sánchez, Numerical semigroups, volume 20 of Developments in Mathematics (Springer, New York, 2009).
11. H. J. Shyr and D. C. Tseng, Some properties of dense languages, *Soochow J. Math.*, 10:127–131(1984).
12. S. N. Singh, Semi-Flower Automata, PhD thesis (IIT Guwahati, India, 2012), URL: http://www.iitg.ac.in/kvk/papers/sns-phd-thesis.pdf.

Comparison of the Performance of Ranked Set Sampling with the Linear Regression Estimation

Rahbar Ali
Department of Statistics, Patna University,
Patna - 800 005, India
E-mail: rahbar.007@gmail.com
www.patnauniversity.ac.in

Arun Kumar Sinha
Department of Statistics,
Central University of South Bihar,
Patna - 800 014, India
E-mail: arunkrsinha@yahoo.com

We intend to examine and explore the capability of ranked set sampling, a relatively new sampling technique for the data collection that provides more efficient estimators of some parameters of interest than the corresponding estimators based on simple random sampling. The main focus of the investigation is to develop a well-designed and cost-effective sampling technique. RSS is one of the sampling methods that can help accomplish such objectives. In this paper, the performance of RSS is compared with that of the linear regression method based on SRS. These procedures are illustrated using real data sets consisting of height and weight of sugarcane provided by the Government of Bihar, India and the data set on tuber referred to by Kumar (2013). The numerical investigation, in turn, suggests that the RSS method performs better than the linear regression method based on SRS in most of the real- life situations.

Keywords: Simple random sampling, Auxiliary variable, Relative precision.

1. Introduction

Ranked set sampling (RSS) was first suggested by McIntyre (1952) to increase the efficiency of the estimator of the population mean. It has been shown that RSS provides more efficient estimator than simple random sampling (SRS) does with the same sample size. There has been a considerable development and many modifications were suggested to this method. Comparing RSS estimator of the population mean with the SRS based linear regression estimator Patil, Sinha and Taillie (1993) showed that the RSS estimator was considerably more efficient than the regression estimator unless correlation is very high. They considered the regression estimator with and without double sampling while for

RSS, they assumed perfect and concomitant ranking scenarios. They considered that the main and auxiliary variables jointly followed a bivariate normal distribution. Note that, this model is favorable to the regression estimator since the latter is unbiased under the bivariate normality. Muttlak (1995) used the method to estimate the slope and the intercept of a simple linear regression equation, which are more efficient than the corresponding estimators based on SRS. Further, Muttlak (1996) obtained more efficient estimators of the parameters of the multiple regression model based on this sampling technique. Muttlak (1997) reported that the extreme RSS estimator of the population mean was more efficient than the regression estimator unless the correlation between the variable of interest and the auxiliary variable was more than 0.80. Yu and Lam (1997) proposed a more efficient RSS regression estimator than usual RSS and SRS estimators unless the correlation between the variable of interest and the concomitant variable is low ($|\rho| < 0.4$). Muttlak (1998) also showed that the median RSS estimator showed a better performance than the equal allocation based on RSS under perfect and concomitant ranking scenarios. Sinha (2005) mentions the recent developments in this method with same new results. Further, for the applications of RSS in vegetation research, see Sinha (2014).

2. RSS Selection

2.1. *RSS under perfect ranking scenario*

For studying a ranked set sample m^2 units are randomly selected from an infinite population and the selected units are arranged in m sets, each consisting of m units. After ranking the units of each set separately with respect to the variable of interest, the unit with the lowest rank, i.e., with rank one is quantified from the first set, the unit with the second lowest rank, i.e., with rank two is measured from the second set and the process is continued until the unit possessing rank m is quantified from the mth set. This process results in m quantifications out of the m^2 units selected. In order to get a larger sample size n the whole process is repeated r times. This provides a ranked set sample of size $n = mr$ units from m^2r units randomly selected. These mr units form the RSS. In this scheme, m denotes the set size while r refers to the number of cycles.

Suppose that $X_{(i:m)j}$ denotes the ith order statistic under perfect ranking in jth cycle, for $i = 1, 2, \ldots, m$ and $j = 1, 2, \ldots, r$. Also, $E(X_{(i:m)j}) = \mu_{(i:m)}$ and $var(X_{(i:m)j}) = \sigma^2_{(i:m)}$. The McIntyre's estimator, $\hat{\mu}_{(RSS)}$, of the population mean, μ, is defined as follows:

$$\hat{\mu}_{RSS} = \frac{1}{mr}\sum_{i=1}^{m}\sum_{j=1}^{r} X_{(i:m)j} \quad (1)$$

Also, as

$$\hat{\mu}_{(i:m)} = \frac{1}{r}\sum_{j=1}^{r} X_{(i:m)j},\ \hat{\mu}_{RSS} = \frac{1}{m}\sum_{i=1}^{m} \hat{\mu}_{(i:m)},\ E(\hat{\mu}_{(i:m)}) = \mu_{(i:m)}.$$

Here,

$$E(\hat{\mu}_{RSS}) = \frac{1}{m}\sum_{i=1}^{m} \mu_{(i:m)} = \mu, \text{ and, } var(\hat{\mu}_{RSS}) = \frac{1}{m^2 r}\sum_{i=1}^{m} \sigma_{(i:m)}^2 \quad (2)$$

This expression is also expressed as

$$var(\hat{\mu}_{RSS}) = \frac{1}{mr}\left\{\sigma^2 - \frac{1}{m}\sum_{i=1}^{m}\left(\mu_{(i:m)} - \mu\right)^2\right\} = \frac{\sigma^2}{mr}\left\{1 - \frac{1}{m}\sum_{i=1}^{m}\left(\frac{\mu_{(i:m)}-\mu}{\sigma}\right)^2\right\} \quad (3)$$

2.1.1. *RSS under concomitant ranking scenario*

While dealing with real life situations many times it may not be possible to rank the randomly selected units with respect to the main variable of interest. In this situation the ranking is carried out with respect to an auxiliary variable usually referred to as the concomitant variable, which is highly correlated with the main variable of interest. In order to investigate the effect of the concomitant variable Y on the performance of RSS, we presume that (a) the regression of X on Y is linear, and (b) the standardized variables $(X - \mu_X)/\sigma_X$ and $(Y - \mu_Y)/\sigma_Y$ follow the same distribution.

The linear regression relationship is maintained when we rank according to the concomitant variable. The assumption (a) implies that

$$X_{[i:m]j} = \mu_X + \frac{\rho\sigma_X}{\sigma_Y}(Y_{(i:m)j} - \mu_Y) + \epsilon \text{ or, } E\left[\frac{X_{[i:m]j}-\mu_x}{\sigma_x}\right] = \rho E\left[\frac{Y_{(i:m)j}-\mu_y}{\sigma_y}\right] \quad (4)$$

where the error term ϵhas expectation zero and $Y_{(i:m)}$ denotes the correctly ranked ordered statistics on Y.

3. Linear Regression Estimator

In this section, we show how to use the linear regression estimator on the ranked set sampling to estimate the mean and variance. The linear regression estimator $\hat{\mu}_l$ is defined,

$$\hat{\mu}_l \equiv \hat{\bar{X}}_l = \bar{x} + \hat{\beta}(\bar{Y} - \bar{y}) \quad (5)$$

where $\bar{Y}$, $\bar{y}$ and $\hat{\beta}$ indicate the population mean of the auxiliary variable Y, the sample mean of Y based on *n* observations and the least square estimator of the population regression coefficient β respectively.

As we know that the regression estimator is biased in general and its precision would be assessed by MSE $(\hat{\mu}_l)$. Further, when the sampling is from a bivariate normal population, the regression estimator is known to be unbiased.

Sukhatme and Sukhatme (1970, p.196) show that the variance of the linear regression estimator with $n = mr$ quantification is given by

$$var(\hat{\mu}_l) = \frac{\sigma_X^2}{mr}(1-\rho^2)\left(1+\frac{1}{mr-3}\right) \quad (6)$$

3.1. *Comparison of the linear regression estimator with the RSS estimator under perfect ranking*

The relative precision of RSS estimator as compared with the regression estimator is given by

$$RP = \frac{var(\hat{\mu}_l)}{var(\hat{\mu}_{RSS})} = \frac{\frac{\sigma_X^2}{mr}(1-\rho^2)\left(1+\frac{1}{mr-3}\right)}{\frac{\sigma_X^2}{mr}\left\{1-\frac{1}{m}\sum_{i=1}^{m}\left(\frac{EX_{(i:m)}-\mu_X}{\sigma_X}\right)^2\right\}} = \frac{(1-\rho^2)\left(1+\frac{1}{mr-3}\right)}{\left\{1-\frac{1}{m}\sum_{i=1}^{m}\left(\frac{EX_{(i:m)}-\mu_X}{\sigma_X}\right)^2\right\}} \quad (7)$$

3.2. *RSS estimator of the population variance*

Norris, Patil and Sinha (1995) mentioned an unbiased estimator of the population variance σ^2 based on a ranked set sample when the number of cycle is more than one which is given below:

$$\hat{\sigma}_{RSS}^2 = \left[\frac{mr-m+1}{m^2r(r-1)}\right]\sum_{i=1}^{m}\sum_{j=1}^{r}\left(X_{(i:m)j}-\hat{\mu}_{(i:m)}\right)^2 + \frac{1}{m}\sum_{i=1}^{m}\left(\hat{\mu}_{(i:m)}-\mu_{RSS}\right)^2$$

A balanced ranked set sampling uses equal number of allocation because each rank order consists of the same number of measurements, i.e., r observations. This is also termed as RSS with equal allocation.

3.3. *RP under perfect ranking scenario*

On replacing σ_x^2 by $\hat{\sigma}_{RSS}^2$ we obtain an estimator of relative precision ($\widehat{RP}$) and it is mentioned below:

$$\widehat{RP} = \frac{\frac{\hat{\sigma}_{RSS}^2}{mr}(1-\rho^2)\left(1+\frac{1}{mr-3}\right)}{\frac{\hat{\sigma}_{RSS}^2}{mr}\left\{1-\frac{1}{m}\sum_{i=1}^{m}\left(\frac{EX_{(i:m)}-\mu_X}{\hat{\sigma}_{RSS}}\right)^2\right\}} = \frac{(1-\rho^2)\left(1+\frac{1}{mr-3}\right)}{1-\frac{1}{m}\sum_{i=1}^{m}\left(\frac{EX_{(i:m)}-\mu_X}{\hat{\sigma}_{RSS}}\right)^2} \quad (8)$$

4. RP under Concomitant Ranking Scenario

We also compare the linear regression estimator with the RSS estimator when the ranking of the randomly drawn units is accomplished by the concomitant variable Y. In this case, equations (3) and (4) are applied and the effect is to insert a factor of ρ^2 into the denominators of equation (8).

$$RP = \frac{(1-\rho^2)\left(1+\frac{1}{mr-3}\right)}{1-\frac{\rho^2}{m}\sum_{i=1}^{m}\left(\frac{EY_{(i:m)}-\mu_X}{\sigma_X}\right)^2} \tag{9}$$

On replacing σ_x^2 by $\hat{\sigma}_{\text{RSS}}^2$ in equation (9), we obtain an estimator of relative precision ($\widehat{RP}$), which is given below:

$$\widehat{RP} = \frac{\frac{\hat{\sigma}_{\text{RSS}}^2}{mr}(1-\rho^2)\left(1+\frac{1}{mr-3}\right)}{\frac{\hat{\sigma}_{\text{RSS}}^2}{mr}\left\{1-\frac{\rho^2}{m}\sum_{i=1}^{m}\left(\frac{EX_{(i:m)}-\mu_X}{\hat{\sigma}_{\text{RSS}}}\right)^2\right\}} = \frac{(1-\rho^2)\left(1+\frac{1}{mr-3}\right)}{1-\frac{\rho^2}{m}\sum_{i=1}^{m}\left(\frac{EX_{(i:m)}-\mu_X}{\hat{\sigma}_{\text{RSS}}}\right)^2} \tag{10}$$

5. Applications under Perfect Ranking and Concomitant Ranking

(a) Sugarcane Data

The procedures are illustrated using a real data set consisting of 150 observations of height and weight of sugarcane provided by the Government of Bihar, India. We examine the performance of RSS as compared with the linear regression method based on SRS for set size m = 3, 4, 5 and 6. In general, the regression estimator is biased and its precision would be assessed by MSE ($\hat{\mu}_l$).

Table 1. Sugarcane weight on their height presuming their perfect ranking.

When $m = 3$, $r = 20$, $n = mr = 3*20 = 60$					
Lowest		Middle		Highest	
Height	Weight	Height	Weight	Height	Weight
1.86	0.75	1.84	0.70	2.12	0.98
1.97	0.80	2.16	1.00	1.86	0.76
1.84	0.75	1.96	0.76	1.99	0.67
1.81	0.73	1.94	0.74	2.18	0.96
1.95	0.74	2.16	1.01	2.20	0.97
1.75	0.74	2.02	0.98	2.13	0.97
1.74	0.74	2.00	0.83	2.07	0.96
1.95	0.77	2.07	0.95	2.19	1.01
1.89	0.77	2.13	0.98	2.02	0.98
1.88	0.72	2.10	1.01	2.18	1.01
1.88	0.69	2.07	0.97	1.86	0.72
1.94	0.76	2.03	0.95	2.16	1.00
1.74	0.75	1.90	0.69	2.16	1.02
1.92	0.75	2.01	0.99	2.06	0.95
1.75	0.71	1.92	0.71	2.11	1.01
1.85	0.74	2.04	0.95	2.18	0.99
1.90	0.72	1.92	0.84	2.00	0.82
1.80	0.76	1.90	0.95	2.18	0.98
1.91	0.79	1.86	0.83	2.06	1.01
1.88	0.70	1.98	0.84	1.88	0.98
Mean	0.744	-	0.884	-	0.938

$\sigma_X^2 = 0.0148$, where σ_X^2 is calculated through 60 observations of sugarcane weight given in above Table 1. See Ali and Sinha (2015).

(i) When $m = 3$, $r = 20$, $n = mr = 3*20 = 60$
From equation (1), $\mu_{(1:3)} = 0.744$, $\mu_{(2:3)} = 0.884$ and $\mu_{(3:3)} = 0.938$.
We get $\hat{\mu} = 0.8552$, $\hat{\rho} = 0.8250$, where $\hat{\rho}$ denotes the correlation co-efficient between the height and the weight of sugarcane with the same sample size.
From equation (6),$\text{var}(\hat{\mu}_l) = 0.00008016126$, $var(\hat{\mu}_{RSS}) = 0.000135$.

From equation (3), we get $RP = \dfrac{\frac{\sigma_X^2}{mr}(1-\rho^2)\left(1+\frac{1}{mr-3}\right)}{\frac{\sigma_X^2}{mr}\left\{1-\frac{1}{m}\sum_{i=1}^{m}\left(\frac{EX_{(i:m)}-\mu_x}{\sigma_x}\right)^2\right\}} = 0.59145.$

The value of RP suggests that the regression estimator is better than RSS estimator, because the correlation coefficient is very high.

The expression for an unbiased estimator of the population variance σ^2 based on a ranked set sample when the number of cycle is more than one according to Norris, Patil and Sinha (1995) is given below:

$$\hat{\sigma}_{RSS}^2 = \left[\frac{mr-m+1}{m^2r(r-1)}\right]\sum_{i=1}^{m}\sum_{j=1}^{r}\left(X_{(i:m)j}-\hat{\mu}_{(i:m)}\right)^2 + \frac{1}{m}\sum_{i=1}^{m}\left(\hat{\mu}_{(i:m)}-\mu_{RSS}\right)^2 = 0.01465.$$

Putting the value of $\hat{\sigma}_{RSS}^2 = 0.0147$ in equation (3) we get the value of $\widehat{var}(\hat{\mu}_{RSS})$ as given below:

$$\widehat{var}(\hat{\mu}_{RSS}) = 0.00013456,\ \widehat{RP} = \frac{\frac{\hat{\sigma}_{RSS}^2}{mr}(1-\rho^2)\left(1+\frac{1}{mr-3}\right)}{\frac{\hat{\sigma}_{RSS}^2}{mr}\left\{1-\frac{1}{m}\sum_{i=1}^{m}\left(\frac{EX_{(i:m)}-\mu_x}{\hat{\sigma}_{RSS}}\right)^2\right\}} = 0.595746$$

The value of RP indicates that the regression estimator is better than RSS estimator, because the correlation coefficient is very high.

Table 2. Mean, variance and relative precision (RP and $\widehat{RP}$) for set size m.

m	μ_x	σ_x^2	$\hat{\rho}$	$var(\hat{\mu}_l)$	$var(\hat{\mu}_{RSS})$	RP	$\hat{\sigma}_{RSS}^2$	$\widehat{var}(\hat{\mu}_{RSS})$	$\widehat{RP}$
	Under Perfect Ranking								
3	0.8552	0.0148	0.8250	0.000080	0.000136	0.591	0.0147	0.000134	0.596
4	0.8655	0.0135	0.7850	0.000088	0.000096	0.915	0.0134	0.000094	0.928
5	0.8085	0.0112	0.7777	0.000075	0.000077	0.976	0.0111	0.000074	0.989
6	0.8288	0.0099	0.6730	0.000092	0.000066	1.406	0.0098	0.000064	1.427
	Concomitant Ranking								
3	0.8552	0.0148	0.8250	0.000080	0.000171	0.468	0.0147	0.000169	0.471
4	0.8655	0.0135	0.7850	0.000088	0.000146	0.604	0.0134	0.000144	0.607
5	0.8085	0.0112	0.7777	0.000075	0.000121	0.624	0.0111	0.000119	0.628
6	0.8288	0.0099	0.6730	0.000092	0.000120	0.766	0.0098	0.000118	0.777

Next, we explore the impact of the value of ρ presuming its different values on RP and $\widehat{RP}$. For this investigation, we consider $\rho = 0.100\ (0.05)\ 0.950$.

Table 3. Relative precision corresponding to different assumed values of ρ .

$\hat{\rho}$	RP				$\widehat{RP}$			
	Set size (m)				Set size (m)			
	3	4	5	6	3	4	5	6
0.100	1.831	2.360	2.440	2.544	1.847	2.390	2.474	2.582
0.200	1.775	2.289	2.366	2.467	1.791	2.318	2.399	2.504
0.300	1.683	2.169	2.243	2.338	1.697	2.197	2.274	2.373
0.400	1.553	2.003	2.070	2.158	1.567	2.028	2.099	2.191
0.500	1.387	1.788	1.848	1.927	1.399	1.811	1.874	1.956
0.600	1.183	1.526	1.577	1.644	1.194	1.545	1.599	1.669
0.700	0.943	1.216	1.257	1.310	0.951	1.231	1.274	1.330
0.800	0.666	0.858	0.887	0.925	0.671	0.869	0.899	0.939
0.900	0.351	0.453	0.468	0.488	0.354	0.459	0.475	0.496
0.950	0.180	0.232	0.240	0.251	0.182	0.235	0.244	0.254

(b) Case Study on Tuber Data Set

Further, we examine the performance of RSS on another 96 observations of tuber data set consisting of four characteristics haulm fresh weight (kg) (HFW), root fresh weight (gm) (RFW), tuber weight (kg) (TW) and chlorophyll measure (CM). We compare RSS estimator for these data set with the linear regression estimator based on SRS for the set size $m = 3$, 4, 5 and 6.

Table 4. Pearson correlation co-efficient (ρ) and p-value of tuber data set.

Haulm Fresh Weight and other Characteristics	Pearson Correlation co-efficient (ρ)	p-Value
Haulm Fresh Weight (kg) and Root Fresh Weight (gm)	0.085	0.410
Haulm Fresh Weight (kg) and Tubers Weight (kg)	0.219	0.032
Haulm Fresh Weight (kg) and Chlorophyll Measure	0.049	0.637

The plots are ranked on the basis of Haulm Fresh Weight without using their actual weights. This is carried out by the past experiences of field personnel.

Table 5. Ranked on the basis of Haulm fresh weight of root fresh weight.

When $m = 3$, $r = 20$, $n = mr = 3*20 = 60$					
Lowest		Middle		Highest	
HFC	RFW	HFC	RFW	HFC	RFW
0.064	8.2621	0.174	20.4146	0.207	15.2992
0.081	5.1844	0.115	8.1362	0.205	26.4916
0.121	8.6215	0.205	26.4916	0.174	20.4146
0.090	8.0957	0.205	26.4916	0.156	26.9758
0.084	6.3070	0.183	8.9363	0.174	20.4146

Table 5. (*Continued*)

When $m = 3, r = 20, n = mr = 3*20 = 60$					
Lowest		Middle		Highest	
HFC	RFW	HFC	RFW	HFC	RFW
0.082	5.2725	0.156	26.9758	0.205	26.4916
0.058	5.2842	0.136	10.4908	0.174	12.4350
0.081	5.1844	0.207	15.2992	0.302	18.2810
0.144	8.0898	0.150	6.9436	0.207	15.2992
0.136	10.4908	0.180	9.4231	0.205	26.4916
0.144	10.2321	0.184	11.4502	0.210	9.8271
0.082	5.2725	0.213	10.5131	0.302	18.2810
0.084	6.3070	0.169	9.1624	0.161	15.0083
0.093	11.6110	0.169	9.1624	0.302	18.2810
0.084	6.3070	0.137	10.6332	0.146	8.0406
0.182	8.0240	0.156	7.5437	0.122	11.4822
0.054	6.2142	0.151	12.6285	0.131	14.2409
0.134	15.2509	0.132	7.5556	0.302	18.2810
0.094	9.0728	0.160	8.9316	0.197	11.6631
0.145	11.1028	0.161	15.0083	0.174	20.4146
Mean	8.0090	-	13.1090	-	17.7057

$\sigma_x^2 = 43.192$, here σ_x^2 is calculated through 60 observations of Root Fresh Weight of Tuber given in above Table 5.

(i) When $m = 3, r = 20, n = mr = 3*20 = 60$

From equation (1), $\mu_{(1:3)} = 8.009, \mu_{(2:3)} = 13.109$ and $\mu_{(3:3)} = 17.7057$.

We get $\hat{\mu} = 12.94, \hat{\rho} = 0.5950$, where $\hat{\rho}$ denotes the correlation coefficient between Halum Fresh Weight and Root Fresh Weight with the same sample size. From the equation (6) and (3).$var(\hat{\mu}_l) = 0.4732$, $var(\hat{\mu}_{RSS}) = 0.4585$.

$$RP = \frac{\frac{\sigma_x^2}{mr}(1-\rho^2)\left(1+\frac{1}{mr-3}\right)}{\frac{\sigma_x^2}{mr}\left\{1-\frac{1}{m}\sum_{i=1}^{m}\left(\frac{EX_{(i:m)}-\mu_x}{\sigma_x}\right)^2\right\}} = 1.032$$

The value of RP suggests that the RSS estimator is better than the regression estimator with the same sample size. The expression for an unbiased estimator of the population variance σ^2 based on a ranked set sample when the number of cycle is more than one according to Norris, Patil and Sinha (1995) is given below:

$$\hat{\sigma}_{RSS}^2 = \left[\frac{mr-m+1}{m^2r(r-1)}\right]\sum_{i=1}^{m}\sum_{j=1}^{r}\left(X_{(i:m)j} - \hat{\mu}_{(i:m)}\right)^2 + \frac{1}{m}\sum_{i=1}^{m}\left(\hat{\mu}_{(i:m)} - \mu_{RSS}\right)^2 = 42.942$$

Putting the value of $\hat{\sigma}_{RSS}^2 = 42.942$ in equation (3), we get the value of $\widehat{var}(\hat{\mu}_{RSS}) = 0.4543$ and $\widehat{RP} = 1.036$.

The value of $\widehat{RP}$ indicates that the RSS estimator is better than the regression estimator with the same sample size.

Table 6a. Mean, variance and relative precision (RP and $\widehat{RP}$) of set size m.

m	Under Perfect Ranking								
	μ_x	σ_x^2	$\hat{\rho}$	$var(\hat{\mu}_l)$	$var(\hat{\mu}_{RSS})$	RP	$\hat{\sigma}_{RSS}^2$	$\widehat{var}(\hat{\mu}_{RSS})$	$\widehat{RP}$
3	12.9420	43.192	0.5950	0.4732	0.4585	1.032	42.942	0.4543	1.036
4	11.3020	28.853	0.5410	0.3461	0.2519	1.374	28.640	0.2483	1.383
5	11.2072	25.791	0.5401	0.3098	0.2242	1.382	25.598	0.2209	1.392
6	12.9650	21.673	0.5220	0.2674	0.1801	1.485	21.505	0.1773	1.501
Concomitant Ranking									
3	12.9420	43.192	0.5950	0.4732	0.6273	0.754	42.942	0.6232	0.755
4	11.3020	28.853	0.5410	0.3461	0.4139	0.836	28.640	0.4100	0.837
5	11.2072	25.791	0.5401	0.3098	0.3700	0.838	25.598	0.3670	0.839
6	12.9650	21.673	0.5220	0.2674	0.3120	0.857	21.505	0.3091	0.865

Now, we need to investigate the impact of different assumed value of $\rho = 0.100$ (0.05) 0.950. The results are shown in Table 6b.

Table 6b. RP corresponding to different values of ρ based on regression model.

ρ	RP				$\widehat{RP}$			
	Set size (m)				Set size (m)			
	3	4	5	6	3	4	5	6
0.100	1.582	1.923	1.932	2.021	1.587	1.936	1.945	2.037
0.200	1.534	1.864	1.873	1.959	1.539	1.878	1.886	1.975
0.300	1.454	1.767	1.776	1.857	1.459	1.780	1.788	1.872
0.400	1.342	1.631	1.639	1.714	1.347	1.643	1.651	1.728
0.500	1.198	1.457	1.463	1.531	1.202	1.467	1.474	1.543
0.600	1.023	1.243	1.249	1.306	1.026	1.252	1.258	1.317
0.700	0.815	0.990	0.995	1.041	0.818	0.997	1.002	1.049
0.800	0.575	0.699	0.702	0.735	0.577	0.704	0.707	0.741
0.900	0.304	0.369	0.371	0.388	0.305	0.372	0.373	0.391
0.950	0.156	0.189	0.190	0.199	0.156	0.191	0.192	0.201

Table 7. Ranked on the basis of Haulm fresh weight of tubers weight.

When m=3, r = 20, n = mr = 3*20 = 60					
Lowest		Middle		Highest	
HFW	TW	HFW	TW	HFW	TW
0.064	10.050	0.174	26.010	0.207	23.122
0.081	37.948	0.115	34.928	0.205	28.665
0.121	15.640	0.205	28.665	0.174	26.010
0.090	18.542	0.205	28.665	0.156	12.048
0.084	17.456	0.183	22.130	0.174	26.010
0.082	29.462	0.156	12.048	0.205	28.665
0.058	24.604	0.136	16.652	0.174	27.469

Table 7. (*Continued*)

When m=3, r = 20, n = mr = 3*20 = 60					
Lowest		Middle		Highest	
HFW	TW	HFW	TW	HFW	TW
0.081	37.948	0.207	23.122	0.302	31.861
0.144	10.322	0.150	16.782	0.207	23.122
0.136	16.652	0.180	17.868	0.205	28.665
0.144	25.912	0.184	34.970	0.210	27.908
0.082	29.462	0.213	25.258	0.302	31.861
0.084	17.456	0.169	25.862	0.161	11.840
0.093	21.642	0.169	25.862	0.302	31.861
0.084	17.456	0.137	8.960	0.146	25.760
0.182	28.566	0.156	15.302	0.122	21.918
0.054	12.950	0.151	35.628	0.131	24.814
0.134	26.262	0.132	26.472	0.302	31.861
0.094	19.440	0.160	41.226	0.197	20.714
0.145	16.882	0.161	11.840	0.174	26.010
Mean	21.733		23.913		25.509

Here $\sigma_x^2 = 58.888$ is calculated above mentioned Tuber Weight in Table 7.

(i) When $m = 3$, $r = 20$, $n = mr = 3*20 = 60$
From equation (1), $\mu_{(1:3)} = 21.733, \mu_{(2:3)} = 23.913$ and $\mu_{(3:3)} = 25.509$
We get $\hat{\mu} = 23.718$, $\hat{\rho} = 0.316$, where $\hat{\rho}$ denotes the correlation co-efficient between Halum Fresh and Tuber Weight with the same sample size.
From equation (6) and (3) we get, $var(\hat{\mu}_l) = 0.899$, $var(\hat{\mu}_{RSS}) = 0.941$

$$RP = \frac{\frac{\sigma_x^2}{mr}(1-\rho^2)\left(1+\frac{1}{mr-3}\right)}{\frac{\sigma_x^2}{mr}\left\{1-\frac{1}{m}\sum_{i=1}^{m}\left(\frac{EX_{(i:m)}-\mu_x}{\sigma_x}\right)^2\right\}} = 0.955$$

The expression for an unbiased estimator of the population variance σ^2 based on a ranked set sample when the number of cycle is more than one mention by Norris, Patil and Sinha (1995) is given below:

$$\hat{\sigma}_{RSS}^2 = \left[\frac{mr - m + 1}{m^2 r(r-1)}\right]\sum_{i=1}^{m}\sum_{j=1}^{r}\left(X_{(i:m)j} - \hat{\mu}_{(i:m)}\right)^2 + \frac{1}{m}\sum_{i=1}^{m}\left(\hat{\mu}_{(i:m)} - \mu_{RSS}\right)^2 = 58.881$$

Putting the value of $\hat{\sigma}_{RSS}^2 = 58.881$ in equation (3) we get the value of $\widehat{var}(\hat{\mu}_{RSS}) = 0.940$ and $\widehat{RP} = 0.956$

Table 8a. Mean, variance and relative precision (RP and $\hat{RP}$) of set size m.

m	Under Perfect Ranking								
	μ_x	σ_x^2	$\hat{\rho}$	$var(\hat{\mu}_l)$	$var(\hat{\mu}_{RSS})$	RP	$\hat{\sigma}^2_{RSS}$	$\widehat{var}(\hat{\mu}_{RSS})$	$\widehat{RP}$
3	23.718	58.89	0.316	0.899	0.941	0.955	58.88	0.940	0.956
4	23.368	42.10	0.324	0.639	0.661	0.966	42.08	0.660	0.968
5	23.112	41.06	0.315	0.627	0.584	1.073	41.01	0.583	1.075
6	23.016	36.89	0.250	0.587	0.473	1.240	36.79	0.471	1.245
Concomitant Ranking									
3	23.718	58.89	0.316	0.899	0.978	0.915	58.88	0.977	0.916
4	23.368	42.10	0.324	0.639	0.697	0.916	42.08	0.697	0.917
5	23.112	41.06	0.315	0.627	0.674	0.930	41.01	0.673	0.930
6	23.016	36.89	0.250	0.587	0.606	0.968	36.79	0.604	0.971

Next, we examine the impact of the value of ρ presuming its different values on RP and $\widehat{RP}$. For this investigation, we consider $\rho = 0.100\ (0.05)\ 0.950$. The results are shown in Table 8b.

Table 8b. Relative precision corresponding to different values of assume ρ .

ρ	RP				$\widehat{RP}$			
	Set size (m)				Set size (m)			
	3	4	5	6	3	4	5	6
0.100	1.050	1.069	1.180	1.310	1.050	1.069	1.180	1.311
0.200	1.018	1.037	1.144	1.270	1.018	1.037	1.144	1.271
0.300	0.965	0.983	1.084	1.204	0.965	0.983	1.085	1.205
0.400	0.891	0.907	1.001	1.111	0.891	0.907	1.001	1.112
0.500	0.796	0.810	0.894	0.992	0.796	0.810	0.894	0.993
0.600	0.679	0.691	0.763	0.847	0.679	0.691	0.763	0.847
0.700	0.541	0.551	0.608	0.675	0.541	0.551	0.608	0.675
0.800	0.382	0.389	0.429	0.476	0.382	0.389	0.429	0.477
0.850	0.294	0.300	0.331	0.367	0.294	0.300	0.331	0.367
0.900	0.202	0.205	0.226	0.251	0.202	0.205	0.226	0.252
0.950	0.103	0.105	0.116	0.129	0.103	0.105	0.116	0.129

References

1. Arun K. Sinha, On some recent developments in ranked set sampling. *The Bulletin of Informatics and Cybernatics*, **37**, 137−160(2005).
2. Arun K. Sinha, Ranked set sampling methods, for vegetation research. *Int. J. Mendel*, ***31(1-2)***, 13−22(2014).
3. H. A. Muttlak, Parameters Estimation in a Simple Linear Regression Using Rank Set Sampling. *Biometrical Journal*, **37**, 799−810(1995).

4. H. A. Muttlak, Estimation of parameters in a multiple regression model using ranked set sampling. *Journal of Information & Optimization Sciences*, **17**(3), 521–533(1996).
5. H. A. Muttlak, Median ranked set sampling with concomitant variables and a comparison with ranked set sampling and regression estimators. *Environmetrics*, **9**, 255–267(1998).
6. H. A. Muttlak, Median ranked set sampling. *Journal of Applied Statistical Science* 6: 245–255(1997).
7. G. A. McIntyre, A method of unbiased selective sampling using ranked sets. *Australian Journal of Agricultural Research*, **59**, 385–390(1952).
8. P. L. H. Yu and K. Lam, Regression estimator in ranked set sampling. *Biometrics*, **53**, 1070–1080(1997).
9. P. V. Sukhatme and B. V. Sukhatme, Sampling Theory of Surveys with Applications. *Iowa State University Press*, Ames, Iowa (1970).
10. Rahbar Ali and Arun K. Sinha, Estimating the Weight of Sugarcanes by Ranked Set Sampling. *International journal of Statistics and Systems* (IJSS) 0973-2675. Vol. 10 No. **2**(2015).
11. G. P. Patil, A. K. Sinha and C. Taillie, Relative precision of ranked set sampling: A comparision with the regression estimator. *Environmetrics*, 4 (4), 339–412(1993).
12. R. C. Norris, G. P. Patil and A. K. Sinha, Estimation of multiple characteristics by ranked set sampling methods. *Coenoses*, **10(2-3)**, 95–111(1995).
13. V. Kumar (2013), Some Contributions to Theory and Designs of Survey Sampling for Estimating Multiple Characteristics. *Ph.D. Thesis, Department of Statistics, Patna University, Patna, India.*

Optimal Selection of Logistics Operating Channels for a Sustainable Reverse Supply Chain

Vernika Agarwal[1], Jyoti Dhingra Darbari[2], P. C. Jha[3]

Department of Operational Research, University of Delhi, Delhi, India-110007

E-mail: [1]vernika.agarwal@gmail.com, [2]jydbr@hotmail.com, [3]jhapc@yahoo.com

Recent years have witnessed reverse logistics (RL) gaining immense importance as a profitable and sustainable business strategy. A systematic decision-making model is required to guide organization wanting to engage in the reverse logistics business for market competence and environmental obligation. The main reverse logistics functions can be categorized as: collection of the returns; dismantling of the unwanted products; repair and refurbishment of the recovered products and components. In this study, a mathematical model is developed which can assist the decision makers of a company to assess the possibility of choosing one of the following three logistics operating channels (LOC) for managing each of the above reverse processes: the company takes control of the process itself; it completely outsources the services related to the process to third party service provider (3PRL); it coordinates the process through collaborative partnership with 3PRL. To evaluate the above three operating channels, the Analytical Hierarchy Process-Quality Function Deployment (AHP-QFD) methodology is used to incorporate the voice of the stakeholders. The requirements of the stakeholders determine the evaluating criteria using a series of house of quality (HOQ) in which the weights of the evaluating criteria are achieved using AHP. The criteria for assessment are therefore based on the strategic goal of implementing a sustainable RL system. The result from the AHP-QFD approach along with other quantitative data is used to parameterize the mathematical model which determines the most appropriate alternative for each of the RL activity. A case study is included to validate the proposed method.

Keywords: AHP; QFD; Sustainability; Reverse Logistics; Logistics Operating channels.

1. Introduction

Reverse supply chain management (RSCM) is defined as the effective and efficient management of the series of activities required to retrieve a product from a customer and either dispose of it or recover value [1]. Over the last few years reverse supply chain management has become an important area for gaining sustainable image for electronics manufacturers. Due to shortening of product life cycles and increased government pressure for enhancing take back laws, there has been tremendous increase in product returns. It has been a significant management concern for manufacturing companies to handle the overwhelming flow of returns and consequently the design strategies for RSC requires focused attention. Strategic planning of a supply chain network is one of the most challenging elements of managing reverse logistics (RL) [2]. A RSC integrates many processes out of which the basic three considered in the study are: (1) Collection and Inspection Process, (2) the Dismantling Process and (3) Repair and Refurbishment. Each process of RL requires different strategic focus; however the main underlying purpose is to retrieve maximum value in a sustainable fashion. Due to resource, capital and labor limitation, the management may collaborate either partially or fully with logistics providers and accordingly choose an appropriate operating channel (OC) for implementing all or some of the value added RL activities. As the number of returns are uncertain and untimely, reverse supply chain management becomes more critical and complex. It is hence essential to create a network of strategic interdependence whereby the company and its partners (logistics providers) work towards their sustainable network strategies and gain mutual benefits. Significant benefits can be attained by constructing collaborative relationships but a careful analysis is required before entering such a strategic alliance [3]. Selection of Logistics Operating Channel (LOC) to determine which reverse process should be allocated to which OC, plays a very important role in an effective and efficient supply chain. Which reverse process is to be undertaken by which LOC is an important decision as each process infers different cost and different expertise [4]. A perfect mix of complete or partial outsourcing increases flexibility and improvement in achievement level.

To engage in decision making at the organizational level, the decision makers need to be equipped with necessary information about all the reverse processes to make effective and effective decisions. A knowledge based decision support system can hence be developed by involving of stakeholders with different roles, from manufacturers, distributors, retailers, dismantlers, customers. Incorporating the conflicting interests and priorities of the various stakeholders increases the complexity of the decision making process. Multi criteria decision making (MCDM) is a powerful tool to deal with conflicting factors and goals and its variants such as analytical hierarchy process (AHP) and analytical network process (ANP) have been extensively used to derive priorities for the alternatives [5]. However, quality function deployment (QFD) is efficient planning tool to consider the voice of the company stakeholders and translate their expectations into appropriate strategic priorities. Furthermore, AHP-QFD has advantage over QFD in preventing inconsistent decisions due to biases and subjectivity of the decision makers [6][7][8]. Goal programming and other multi objective programming techniques can be used as a solution approach to solve the multi objective model to select the final alternatives while considering resource limitations [9].

The main objective of the paper is to analyze the impact of implementation of each RL process through one of the following OCs: the company takes control of the process itself; it completely outsources the services related to the process to third party service provider (3PRL); partially outsourcing activities under a collaborative partnership. To the best of our knowledge, evaluation and selection of OC for reverse supply chain has not been addressed in the literature till now. This paper proposes a novel multi criteria optimization model which deploys AHP-QFD methodology for evaluation of LOCs for each reverse process and Goal Programming is used as a solution approach to select the most appropriate (maximum weight) OC which gives the lowest cost for the system. The rest of the paper is organized as follows: Section 2 elaborated the solution methodology used in the paper. Section 3 presents the mathematical formulation of the proposed model. The results are discussed in Section 4 and Section 5 concludes the paper.

2. Proposed Methodology

The proposed model uses combines AHP-QFD approach to evaluate the three LOCs for RL- the company takes control of the process itself; it completely outsources the services related to the process to third party service provider (3PRL); it coordinates the process through collaborative partnership with 3PRL for collection of the returns; dismantling of the unwanted products; repair and refurbishment of the recovered products and components. The stakeholder requirements and company's sustainable strategies are identified based on interactions with decision makers and literature survey. The model consists of three interrelated HOQ, including HOQ1- linking sustainable stakeholder requirements to company's sustainable strategies (Steps 1 to 3), HOQ2 relating company's sustainable strategies to sustainable assessment criteria's (Steps 4 to 7) and HOQ3 assessing sustainable assessment criteria to LOC for each of the three RL activities (Steps 8 to 10) [8][10].

Step 1: The importance of each sustainable stakeholder requirements are determined using AHP [11].

Step 2: The relationship weights of company's sustainable strategies (i) and sustainable stakeholder requirements (k) are determined using AHP, that is C_{ik}^2.

Step 3: The importance rating of company's sustainable strategies is computed as $w_i^1 = \sum_{k \epsilon S} c_k^1 c_{ik}^2$; where c_k^1 denotes the importance rating of stakeholder requirements; R denotes the set of sustainable strategies and S denotes the set of stakeholder requirements $S = \{1,2..m\}$.

Step 4: For HOQ2 the sustainable assessment criteria's are identified.

Step 5: Copy the Company's sustainable strategies and their importance rating (as computed in step 3) into HOQ2.

Step 6: The relationship weights of sustainable assessment criteria's (i) and company's sustainable strategies (k) are determined using AHP, that is C_{ik}^3.

Step 7: The importance rating of sustainable assessment criteria's is computed as $w_i^2 = \sum_{k\epsilon S} w_k^1 c_{ik}^3$; where w_k^1 denotes the importance rating of company's sustainable strategies; R denotes the set of sustainable assessment criteria and S denotes the set of company's sustainable strategies $S = \{1,2..m\}$.

Step 10: Copy the sustainable assessment criteria's and their importance rating (as computed in step 3) into HOQ3 along with the three alternatives of LOCs.

Step 11: The relationship weights of LOC alternatives (i) and sustainable assessment criteria's (k) are determined using AHP , that is C_{ik}^4.

Step 12: The importance rating of LOCs is computed as $w_i^3 = \sum_{k\epsilon S} w_k^2 c_{ik}^4$; where w_k^2 denotes the importance rating of sustainable assessment criteria's; R denotes the set of alternate LOCs and S denotes the set of sustainable criteria's $S = \{1,2 \ldots m\}$.

3. Mathematical Modeling

Sets:

i: RL activity (collection, dismantling and repair and refurbishment), i=1,2,3.
j: LOC, j=1,2,3
k: sub cost of ith RL activity

Parameters

W_j^i	Weight of jth LOC option for i^{th} RL activity obtained from QFD
C_{jk}^i	k^{th} sub-cost of jth LOC option for i^{th} RL activity
P_j^i	Benefit/Profit obtained from j^{th} LOC
B^i	Total budget for i^{th} RL activity
W_j^{min}	Minimum threshold of weight for i^{th} RL activity

Decision Variable $X_j^i = \begin{cases} 1 \; when \; i^{th} RL \; activity \; is \; performed \; by \; j^{th} \; LOC \\ 0 \; otherwise \end{cases}$

Mathematical Model

$$Maximize \sum_i \sum_j W_j^i X_j^i \qquad (F1)$$

$$Minimize \sum_j \sum_k X_j^i \left(C_{jk}^i - P_j^i\right) \qquad (F2)$$

Subject to constraints

$$\sum_j X_j^i = 1 \qquad \forall i \qquad (1)$$

$$\sum_j \sum_k X_j^i (C_{jk}^i - P_j^i) \leq B^i \qquad \forall i \tag{2}$$

$$W_j^i X_j^i \geq W_{min}^i \qquad \forall i,j \tag{3}$$

$$X_j^i \in \{0,1\} \qquad \forall i,j \tag{4}$$

The first objective maximizes the performance of LOC of reverse logistics and the second objective minimizes the total cost of the reverse logistics network, while selecting LOC for each of the three RL processes considered in the study. The first constraint ensures that only one OC is selected for each RL operation. The second constraint the budget constraint which ensures that the total cost of each RL activity is less than the budget allocated for that activity. The third constraint is the threshold constraint which states that minimum threshold should be achieved by for each RL activity. Finally the fourth constraint is the binary restriction.

3.1. *Goal programming*

Goal programming is an effective mathematical technique for optimising multi objective models. An optimal solution is defined as one that minimizes both positive and negative deviations from set goals simultaneously or minimizes the amount by which each goal can be violated. The multi objective problem formulated above can be solved using GP following the steps:

Step 1: Solve problems (P1) and (P2) considering one objective at a time to obtain the target (optimal values) for each objective, say f_1^*andf_2^*.

Problem P1:

$$Maximize \sum_i \sum_j W_j^i X_j^i \tag{F1}$$

Subject to constraints (1)–(4)

Problem P2:

$$Minimize \sum_j \sum_k X_j^i (C_{jk}^i - P_j^i) \tag{F2}$$

Subject to constraints (1)–(4).

Step 2: Solve problem (P3) with the objective to minimize the non-achievement of the rigid constraints by minimizing specific deviational variables under the constraints $f(X) + \eta - \rho = b$ as follows:

if $f(X) <= b$, ρ is minimized

if $f(X) >= b$, η is minimized

if $f(X) = b$, $\eta + \rho$ is minimized

Problem P3:

$$Minimize \quad \sum_i \sum_j \sum_k \left(\rho_{jk_i} + \eta_{ij}\right) \tag{C1}$$

Subject to

$$\sum_j X_j^i = 1 \; ; \; \forall i \tag{C2}$$

$$\sum_j \sum_k X_j^i (C_{jk}^i - P_j^i + \eta_{jk_i} - \rho_{jk_i}) = B^i \; ; \; \forall i \tag{C3}$$

$$W_j^i X_j^i + \eta_{ij} - \rho_{ij} \geq W_{min}^i \ ; \ \forall i,j \qquad \text{(C4)}$$

$$X_j^i \in \{0,1\} \ ; \ \forall i,j \qquad \text{(C5)}$$

Problem P4:

Problem (P4) can be formulated by incorporating the optimal solution of the problem (P3) and assigning the weights λ_i (i=1,2) to the ith goal as follows:

Subject to

$$Minimize\ \lambda_i \eta 1 + \lambda_2 \rho 2 \qquad \text{(P1)}$$

$$\left(\frac{f1}{f1^*}\right) + \eta 1 - \rho 1 = 1 \qquad \text{(P2)}$$

$$\left(\frac{f2}{f2^*}\right) + \eta 2 - \rho 2 = 1 \qquad \text{(P3)}$$

$$\textstyle\sum_j X_j^i = 1 \qquad \forall i \qquad \text{(P4)}$$

$$\textstyle\sum_j \sum_k X_j^i \left(C_{jk}^i - P_j^i + \eta_{jk_i}\right) = B^i \ \forall i \qquad \text{(P5)}$$

$$W_j^i X_j^i - \rho_{ij} \geq W_{min}^i \qquad \forall i,j \qquad \text{(P6)}$$

$$X_j^i \in \{0,1\} \qquad \forall i,j \qquad \text{(P7)}$$

$$\lambda_i + \lambda_2 = 1 \qquad \text{(P8)}$$

$\left(\frac{f1}{f1^*}\right)$ and $\left(\frac{f2}{f2^*}\right)$ are used as normalized objectives functions to make these scale free.

4. Result Discussion

The company considered is an electronics manufacturing company which has to deal with the a huge number of returns of varying quality. The value retrieving processes therefore may vary significantly. To ensure minimum wastage and maximum recovery, a well planned and collaborative perspective can create an effective and sustainable reverse network. The options (LOCs) for the manufacturer is to either take full control of the reverse process itself or completely or partially outsources the services related to the process to 3PRL. Each LOC has its own level of complexity when implemented for each reverse process resulting in different costs, environmental impact and social effect. All these parameters influence the judgment for each LOC in RL processes as per the criteria considered. AHP-QFD is deployed to incorporate the voice of stakeholders into decision making process of evaluating the LOCs in reverse process.Following criteria are considered for HOQs for deployment of QFD for ranking of LOCs in reverse processes [12][13][14]:

Sustainable Stakeholder Requirements

Minimum logistics and operational cost (SR1); Affordable and quality processes (SR2); Effective collection and distribution network (SR3); Low environmental impact (SR4); Waste reducing reverse processes (SR5); Socially responsible supply chain (SR6); Legal compliance (SR7).

Company Sustainability Strategies

Safe and healthy practices (CS1); Use of energy efficient and value added processes (CS2); Optimize routing and work flow (CS3); Provide customer incentives for increase in returns (CS4); Priority to quality retrieval (CS5); Labor education in subjects like technology use and processes (CS6); Decrease fuel consumption and waste (CS7).

Sustainable Assessment Criteria

Low price, low cost (SA1); Socially responsible services (SA2); Retrieving products and components with high standards of safety and quality (SA3); Use of innovative and green practices (SA4); Effective logistics planning (SA5).

Alternatives (LOCs)

Company (A1); 3PRL (A2); Collaborative partnership (A3).

In the first phase company's sustainable strategies can be determined based on sustainable stakeholder requirements as given by Table 2 (HOQ1). The second phase is about translating company's sustainable strategies into sustainable stakeholder requirements Table 3 (HOQ 2). In the third phase based on each of the three RL process the LOCs are evaluated Table 4 (HOQ 3) (Following steps 1–12).

Table 1. HOQ1 for company sustainable strategies.

	Importance Rating	CS1	CS2	CS3	CS4	CS5	CS6	CS7
SR1	0.20		0.15	0.47	0.23	0.10	0.05	
SR2	0.10		0.43			0.43		0.14
SR3	0.15			0.43	0.14			0.43
SR4	0.15	0.07	0.40	0.13				0.40
SR5	0.13		0.33					0.67
SR6	0.10	0.56			0.12		0.32	
SR7	0.18		0.44		0.11			0.44
	Importance Rating	0.07	0.25	0.18	0.10	0.06	0.04	0.30

Table 2. HOQ2 for sustainable assessment criteria.

	Importance Rating	SA1	SA2	SA3	SA4	SA5
CS1	0.066		0.124	0.359	0.517	
CS2	0.253	0.163		0.297	0.540	
CS3	0.178	0.558	0.122			0.320
CS4	0.099	0.667	0.333			
CS5	0.062	0.126		0.416	0.458	

Table 2. (*Continued*)

	Importance Rating	SA1	SA2	SA3	SA4	SA5
CS6	0.042	0.122	0.320		0.558	
CS7	0.300	0.108	0.125		0.613	0.153
	Importance Rating	0.252	0.114	0.124	0.406	0.103

Table 3. HOQ3(i) for ranking of LOCs with respect to the three operations.

		Collection			Dismantling			Repair and Refurbish		
	Importance Rating	A1	A2	A3	A1	A2	A3	A1	A2	A3
SA1	0.252	0.163	0.540	0.297	0.630	0.218	0.151	0.260	0.413	0.327
SA2	0.114	0.630	0.218	0.151	0.413	0.260	0.327	0.674	0.226	0.101
SA4	0.406	0.400	0.200	0.400	0.687	0.186	0.127	0.630	0.218	0.151
SA5	0.103	0.149	0.474	0.376	0.550	0.210	0.240	0.163	0.297	0.540
	Importance Rating	0.291	0.291	0.293	0.515	0.193	0.189	0.287	0.278	0.332
	Normalized Values	0.332	0.332	0.335	0.574	0.215	0.211	0.320	0.310	0.370

Weight obtained from QFD for collection of returns (i1) by company, 3PRL and collaborative partnerships are 0.332, 0.332 and 0.335 respectively. While in case of dismantling of the unwanted products (i2) are 0.574, 0.215 and 0.211 and Repair and refurbishment of the recovered products and components (i3) 0.320, 0.310 and 0.370 when performed by company, 3PRL and collaborative partnerships. There are no benefit/profit obtained from collection of the returns; in case of dismantling of the unwanted products profit of the company is Rs.60 and in collaborative partnership is Rs.40 per product; while in case of repair and refurbishment of the recovered products and components benefit the company is Rs.400 and in collaborative partnership is Rs.200 per product. Total budget and minimum threshold weight for collection of the returns; dismantling of the unwanted products; repair and refurbishment of the recovered products and components are Rs.300, Rs.500 and Rs.250; and 0.3, 0.2 and 0.3 respectively. Collection of returns includes four sub costs viz. hiring/transportation cost (C1); handling, inspection and sorting cost (C2); man-power cost (C3) and penalty cost for uncollected returns (C4). When the company performs operations on its own the sub costs are Rs.0 as the hiring/transportation is done by the trucks/vehicles that deliver the products in the forward supply chain, Rs.150, Rs.180 and no penalty cost for uncollected returns. In case the company decides to take up 3PRL, hiring cost of Rs.300 is accrued and a penalty of Rs.50 is imposed for uncollected returns. Further in case the company decides to be in a collaborative partnership the costs will be Rs.100, Rs.90, Rs.70 and no penalty will be incurred. Dismantling of the unwanted products includes three sub costs viz. disassembly cost (D1); packaging cost (D2) and man-power cost (D3). When the dismantling is performed by the company the sub costs are Rs.200, Rs.30 and Rs.250. In case of outsourcing it to 3PRL the hiring cost will be Rs.500 and all the other costs are bearded by 3PRL and in case the company undergoes collaborative partnership the sub costs include Rs.100, Rs.10 and Rs.125. Repair and refurbishing operation includes repair and refurbishing cost (R1) and man-power cost (R2). When the company will perform the operations itself the costs will be Rs.300 and Rs.150. The hiring cost of 3PRL is Rs.400 which includes the manpower cost. In case of collaborative partnership the costs will be Rs.150 and Rs.150 respectively. The total budget per unit for collection of returns, dismantling and repair and refurbishment of products and components by the company are Rs.300, RS.500 and Rs.250.

The mathematical model is solved using LINGO 11.0. In order to validate the mathematical model equal weights are given to both the objectives, these weights can be changed based on the interactions with the decision makers. The model selects 3PRL to perform collection of returns; collaborative partnership should be carried out for dismantling of products while repair and refurbishment of recovered products and components should be performed by company.

5. Conclusion

The study provides an analytical framework to decision makers of an electronic manufacturing company for implementation of RL through various operating channels. The paper adopts methodology based on AHP-QFD to evaluate the options of implementing the various RL activities either in-house or by outside logistics providers. The criteria for evaluation takes into account the voice of the stakeholders of the company and economic, environmental and social aspects are also incorporated. A multi-optimization model is developed to select the OCs for carrying out the RL activities, under company's budgetary constraints and expectations. Future research can include Fuzzy AHP and Fuzzy QFD to handle the imprecise judgments of the decision makers and the model can be extended by considering additional tiers of the reverse supply chain.

References

1. S. Levi, P. Kaminsky and E.S. Levi, Designing and managing the supply chain: Concepts, strategies, and case studies. McGraw-Hill (2003).
2. K. K. Pochampally and S. M. Gupta, Strategic planning of a reverse supply chain network, Intern. J. of Integr. Supply Manag., 1, 4, 421, (2005).

3. V. Agarwal, J. D. Darbari and P. C. Jha, A sustainable framework for selection and evaluation of third party reverse logistics providers using Fuzzy AHP-TOPSIS, Analytics in Operations/Supply Chain Management, I. K. International Publishing House Pvt Ltd, 181 (2015).
4. Senthil, B. Srirangacharyulu and A. Ramesh, A decision making methodology for the selection of reverse logistics operating channels, Procedia Eng., 38, 418, (2012).
5. Y. Dou and J. Sarkis, A joint location and outsourcing sustainability analysis for a strategic offshoring decision, Intern. J. of Prod. Res., 48, 2, 567, (2010).
6. G. Büyüközkan and C. Berkol, Designing a sustainable supply chain using an integrated analytic network process and goal programming approach in quality function deployment, Expert Syst. with Appl., 38, 11, 13731, (2011).
7. R. Chadawada, A. Sarfaraz, K. Jenab and H. Pourmohammadi, Integration of AHP-QFD for selecting facility location, Benchmarking: An Intern. J., 22, 3, 411, (2015).
8. W. Ho, D. J. Bennett, K. L. Mak, K. B. Chuah, C. K. M. Lee and M. J. Hall, Strategic logistics outsourcing: An integrated QFD and AHP approach, in Ind. Eng. and Eng. Manag., IEEE, December, 1434, (2009).
9. W. Ho, X. Xu and P. K. Dey, Multi-criteria decision making approaches for supplier evaluation and selection: A literature review. Eur. J. of Oper. Res., 202, 1, 16, (2010).
10. W. Ho, T. He, C. K. M. Lee and A. Emrouznejad, Strategic logistics outsourcing: An integrated QFD and fuzzy AHP approach, Expert Syst. with Appl., 39, 12, 10841, (2012).
11. T. L. Saaty, The Analytic Hierarchy Process, 3rd ed., McGraw-Hill, New York, NY (1980).
12. S. Perçin and H. Min, A hybrid quality function deployment and fuzzy decision-making methodology for the optimal selection of third-party logistics service providers, Intern. J. of Logist. Res. and Applications, 16, 5, 380, (2013).
13. C. Y. Chiou, H. C. Chen, C. T. Yu and C. Y. Yeh, Consideration Factors of Reverse Logistics Implementation-A Case Study of Taiwan's Electronics Industry, Procedia-Soc. and Behav. Sci., 40, 375, (2012).
14. G. Büyüközkan and C. Berkol, Designing a sustainable supply chain using an integrated analytic network process and goal programming approach in quality function deploymen, Expert Syst. with Appl., 38, 11, 13731, (2011).

Reliability Measures of a Parallel-Unit System with Arbitrary Distributions of Random Variables

Jitender Kumar* and M. S. Kadyan#

Department of Statistics & O. R.,
Kurukshetra University, Kurukshetra (India)
*E-mail: *khatkarjitu@gmail.co, #mskadian@kuk.ac.in*

S. C. Malik

Department of Statistics,
M.D. University, Rohtak-124001 (Haryana)
E-mail: sc_malik@rediffmail.com

A parallel system of two identical units has been studied by considering arbitrary distributions for the random variables associated with failure time, repair time and replacement time. There is a single server who visits the system immediately to carry out repair activities. The unit does not work as new after repair and so called degraded unit. The degraded unit at its failure is replaced by new one in order to avoid unnecessary expenses on its repair. The system is considered in up-state if any one or two of new and/or degraded unit(s) are operative. Some measures of system effectiveness such as mean sojourn times, mean time to system failure (MTSF), availability, busy period of the server, expected number of visits by the server and profit incurred to the system model are obtained using semi-Markov process and regenerative point technique. Giving particular values to various parameters and costs, the numerical results for MTSF and profit function are obtained by considering exponential distribution for all random variables.

Keywords: Parallel-Unit System; Degradation; Replacement; Arbitrary Distributions.

1. Introduction

Redundancy and repair maintenance are some of the well known methods by which the availability and profit of the systems can be improved. Two-units parallel systems have been studied by various researchers such as Subramanian and Sarma [1981], Gupta and Goel [1990], Barros *et al.* [2006], Singh and Singh [2007], Kishan and Kumar [2009], Malik and Gitanjali [2012], Bashir and Joorel [2014] and Rathee and Malik [2014] due to their applicability in their respective fields.

But, no attention was paid to performance evaluation of parallel system due to degradation after its failure with Arbitrary Distributions of Random

Variables. Mokaddis *et al.* [1997] have proposed reliability model for two-unit warm standby systems subject to degradation.

Also, sometimes repair of the degraded unit is not feasible due to its excessive use and increased cost of maintenance. In such cases, the failed degraded unit is replaced by new one in order to avoid the unnecessary expenses of repair. Further, hazard rates of many systems are of linearly increased nature due to wear out under mechanical stress and so their failure times and repair times follow arbitrary distributions.

Keeping above facts in view, this paper is devoted to the study of a parallel system of two-identical units. A single server is provided immediately to carry out repair activities. The unit works with reduce capacity after its repair and so called degraded unit. The degraded unit at its failure is replaced by new one so that unnecessary expenses on its repair may be avoided. The system is considered in up-state if any one or two of new and/or degraded unit(s) are operative. The distribution of failure, replacement and repair times are arbitrary. Some measures of system effectiveness such as mean sojourn times, mean time to system failure (MTSF), availability, busy period of the server, expected number of visits by the server and profit incurred to the system model are obtained using semi-Markov process and regenerative point technique. The results for a particular case are also obtained.

2. Notations

E	:	Set of regenerative states.
No	:	The unit is new and operative.
Do	:	The unit is degraded and operative.
a(t)/ A(t)	:	p.d.f./c.d.f of failure rate of new unit.
b(t)/ B(t)	:	p.d.f./c.d.f of failure rate of degraded unit.
f(t)/ F(t)	:	p.d.f./c.d.f of failure rate of original unit when both are available to use.
$f_1(t)/ F_1(t)$	:	p.d.f./c.d.f of failure rate of degraded unit when both are available to use.
g(t)/G(t)	:	pdf/cdf of repair time for new unit.
$g_1(t)/G_1(t)$	:	pdf/cdf of replacement time of the degraded unit.
$NF_{ur}/NF_{UR}/NF_{wr}$	:	New unit is failed and under repair/under continuous repair from previous state/waiting for repair.
$DF_{ure}/DF_{URe}/DF_{wre}$	:	Degraded unit is failed and under replacement/under continuous replacement from previous state/waiting for replacement.

$q_{ij}(t), Q_{ij}(t)$	: pdf and cdf of first passage time from regenerative state i to a regenerative state j or to a failed state j without visiting any other regenerative state in (0,t].
$q_{ij.kr}(t), Q_{ij.kr}(t)$	: pdf and cdf of first passage time from regenerative state i to a regenerative state j or to a failed state j visiting state k, r once in (0,t].
$M_i(t)$	: Probability that system up initially in state $S_i \in E$ is up at time t without visiting any other regenerative sate.
$W_i(t)$	: Probability that server is busy in the state S_i up to time t without making any transition to any other regenerative state or returning to the same via one or more non-regenerative states.
m_{ij}	: Contribution to mean sojourn time in state $S_i \in E$ and non regenerative state if occurs before transition to $S_j \in E$.
®/©	: Symbols for Stieltjes convolution/Laplace convolution.
~|*	: Symbols for Laplace Stieltjes Transform (LST)/Laplace Transform (LT).

3. System Description

In this section, a reliability model for the system is developed as per the assumptions. The following are the possible transition states of the system:

State S_0: Both new units are working in parallel.

State S_1: One new unit is under repair after its failure and other new unit is operative.

State S_2: New unit and degraded unit are operative.

State S_3: New unit is operative and degraded unit is under replacement after its failure.

State S_4: Degraded unit is operative and new unit is under repair after its failure.

State S_5: New unit is waiting for repair after its failure and degraded unit is continuously under replacement after its failure.

State S_6: Both degraded units are working in parallel.

State S_7: New unit is continuously under repair after its failure and degraded unit is waiting for replacement after its failure.

State S_8: One degraded unit is under replacement after its failure and other degraded unit is operative.

State S_9: One degraded unit is continuously under replacement after its failure and other degraded unit is waiting for replacement after its failure.

State S_{10}: New unit is continuously under repair after its failure and other new unit is waiting for repair after its failure.

The states S_0, S_1, S_2, S_3, S_4, S_6 and S_8 are regenerative while the states S_5, S_7, S_9 and S_{10} are non-regenerative.

4. Transition Probabilities and Mean Sojourn Times

Simple probabilistic considerations yield the following expressions for the non-zero elements $p_{ij} = Q_{ij}(\infty) = \int q_{ij}(t)\,dt$ as

$$p_{01} = \int_0^\infty f(t)dt,\ p_{12} = \int_0^\infty g(t)\bar{A}(t)dt,\ p_{13} = \int_0^\infty a(t)\bar{G}(t)dt,\ p_{23} = \int_0^\infty b(t)\bar{A}(t)dt,$$

$$p_{24} = \int_0^\infty a(t)\bar{B}(t)dt,\ p_{30} = \int_0^\infty g_1(t)\bar{A}(t)dt,\ p_{35} = \int_0^\infty a(t)\overline{G_1}(t)dt,$$

$$p_{46} = \int_0^\infty g(t)\overline{B(t)}dt,\ p_{47} = \int_0^\infty b(t)\overline{G(t)}dt,\ p_{68} = \int_0^\infty f_1(t)dt,$$

$$p_{89} = \int_0^\infty b(t)\,\overline{G_1(t)}dt,\ p_{82} = \int_0^\infty g_1(t)\,\overline{B(t)}dt,$$

$$p_{14.3} = p_{13} \circledR p_{34},\ p_{31.5} = p_{35} \circledR p_{51},\ p_{48.7} = p_{47} \circledR p_{78},\ p_{83.9} = p_{89} \circledR p_{93} \qquad (1)$$

The probability law holds for these transition probabilities.

The mean sojourn times μ_i in state S_i are given by

$$\mu_0 = \int_0^\infty \overline{F(t)}\,dt,\ \mu_1 = \int_0^\infty \overline{A(t)G(t)}\,dt,\ \mu_2 = \int_0^\infty \overline{A(t)B(t)}\,dt,\ \mu_3 = \int_0^\infty \overline{A(t)G_1(t)}\,dt,$$

$$\mu_4 = \int_0^\infty \overline{b(t)G(t)}\,dt,\ \mu_6 = \int_0^\infty \overline{F_1(t)}\,dt,\ \mu_8 = \int_0^\infty \overline{G_1(t)B(t)}\,dt \qquad (2)$$

The unconditional mean time taken by the system to transit from any state S_i when time is counted from epoch at entrance into state S_j is stated as:

$$m_{ij} = \int t\,dQ_{ij}(t) = -q_{ij}^{*\prime}(0) \text{ and } \mu_i = E(T) = \int_0^\infty P(T > t)dt = \sum_j m_{ij} \qquad (3)$$

5. Mean Time to System Failure

Let $\phi_i(t)$ be the cdf of the first passage time from regenerative state S_i to a failed state. Regarding the failed state as absorbing state, we have the following recursive relations for $\phi_i(t)$:

$$\phi_i(t) = \sum_j Q_{i,j}(t) ® \phi_j(t) + \sum_k Q_{i,k}(t) \tag{4}$$

where S_j is an operative regenerative state to which the given regenerative state S_i can transit and S_k is a failed state to which the state S_i can transit directly. Taking LST of relations (6) and solving for $\tilde{\phi}_0(s)$. Using this, we have

$$R^*(s) = (1 - \tilde{\phi}_0(s))/s \tag{5}$$

The reliability R(t) can be obtained by taking Laplace inverse transform of (5). The mean time to system failure can be given by

$$\text{MTSF} = \lim_{s \to 0} R^*(s) = \frac{N_{11}}{D_{11}} \tag{6}$$

Where, $N_{11}=(1-p_{24}p_{82}p_{46})\ (\mu_0+\mu_1)+\ p_{12}\ (\mu_2\ +\mu_3p_{23}+p_{24}\mu_4)+\ p_{12}\ p_{24}p_{46}(\mu_6+\mu_8)$ and $D_{11}=1-p_{46}p_{24}p_{82}-p_{12}p_{30}\ p_{23}$.

6. Availability Analysis

Let $A_i(t)$ be the probability that the system is in up-state at instant t given that the system entered regenerative state S_i at $t=0$. The recursive relations for $A_i(t)$ are given by:

$$A_i(t) = M_i(t) + \sum_j q_{i,j}^{(n)}(t) © A_j(t) \tag{7}$$

Where S_j is any successive regenerative state to which the regenerative state S_i can transit through n transitions. We have,

$$M_0 = \int_0^\infty \overline{F(t)}\, dt,\ M_1 = \int_0^\infty \overline{A(t)G(t)}\, dt,\ M_2 = \int_0^\infty \overline{A(t)B(t)}\, dt,$$

$$M_3 = \int_0^\infty \overline{A(t)G_1(t)}\, dt,\ M_4 = \int_0^\infty \overline{b(t)G(t)}\, dt,$$

$$M_6 = \int_0^\infty \overline{F_1(t)}\, dt,\ M_8 = \int_0^\infty \overline{G_1(t)B(t)}\, dt \tag{8}$$

Taking LT of relations (7) and solving for $A_0^*(s)$. The steady-state availability of the system can be given by

$$A_0(\infty) = \lim_{s \to 0} sA_0^*(s) = \frac{N_{12}}{D_{12}} \tag{9}$$

where

$N_{12} = (\mu_0 + \mu_1)(1-p_{82}\, p_{24}) - (p_{12}p_{23} + p_{12}\, p_{24}p_{839} + p_{1410}\, p_{839} + p_{82}\, p_{24}\, p_{1410})$ $(p_{315}\mu_0-\mu_3) + \mu_2(p_{12}+p_{82}\, p_{1410}) + (p_{1410} + p_{12}\, p_{24})\,(\mu_4 + \mu_8 + \mu_6 p_{46})$.

$D_{12} = (\mu_0\, p_{30} + \mu^1{}_3)(p_{12}- p_{12}\, p_{24}p_{82} + p_{1410}p_{839} + p_{82}\, p_{24}p_{1410}) + (p_{23} + p_{839}p_{24})$ $\mu^1{}_1 + \mu_2\,(p_{12} + p_{82}p_{1410}) + (\mu_6\, p_{46} + \mu^1{}_4)(p_{82}\, p_{24} + p_{12}\, p_{24}p_{839} + p_{1410}p_{839} +$ $p_{82}\, p_{24}\, p_{1410}) + \mu^1{}_8\,(p_{24} + p_{23}p_{1410})$.

7. Busy Period Analysis for Server Due to Repair

Let $BR_i(t)$ be the probability that the server is busy in repair at an instant t given that the system entered regenerative state S_i at $t = 0$. The following are the recursive relations for $BR_i(t)$

$$BR_i(t) = W_i(t) + \sum_j q_{i,j}^{(n)}(t) © BR_j(t) \tag{10}$$

where S_j is a subsequent regenerative state to which state S_i transits through n transitions. Taking LT of relations (10) and solving for $BR_0^*(s)$ and using this, we can obtain the fraction of time for which the repairman is busy in steady state

$$BR_0 = \operatorname*{Lim}_{s \to 0} sBR_0^*(s) = \frac{N_{13}}{D_{12}} \tag{11}$$

$N_{13} = W_1(1-p_{82}\, p_{24}) + (p_{1410} + p_{12}\, p_{24})\, W_4$ and D_{12} is already mentioned.

8. Busy Period Analysis for Server Due to Replacement

Let $BP_i(t)$ be the probability that the server is busy in replacement at an instant t given that the system entered regenerative state S_i at $t = 0$. The following are the recursive relations for $BP_i(t)$

$$BP_i(t) = W_i(t) + \sum_j q_{i,j}^{(n)}(t) © BP_j(t) \tag{12}$$

where S_j is a subsequent regenerative state to which state S_i transits through n transitions. Taking LT of relations (12) and solving for $BP_0^*(s)$ and using this, we can obtain the fraction of time for which the repairman is busy in steady state

$$BR_0 = \mathit{Lim}_{s \to 0}\, sBP_0^*(s) = \frac{N_{14}}{D_{12}} \tag{13}$$

$N_{14} = (p_{12}p_{23} + p_{12}\, p_{24}p_{839} + p_{1410}p_{839} + p_{82}\, p_{24}\, p_{1410})W_3 + (p_{1410} + p_{12}\, p_{24})\, W_8$ and D_{12} is already mentioned.

9. Expected Number of Visits Due to Repair

Let $NR_i(t)$ be the expected number of visits by the server due to repair in (0,t] given that the system entered the regenerative state S_i at $t=0$. We have the following recursive relations for $NR_i(t)$:

$$NR_i(t) = \sum_j Q_{i,j}(t) \circledR [\delta_j + NR_j(t)] \tag{14}$$

where S_j is any regenerative state to which the given regenerative state S_i transits and $\delta_i=1$, if j is the regenerative state where the server does job afresh otherwise $\delta_i = 0$. Taking LST of relations (14) and solving for $\widetilde{NR}_0(s)$. The expected number of visits per unit time are given by

$$NR_0 = \lim_{s \to 0} s\,\widetilde{NR}_0(s) = \frac{N_{15}}{D_{12}}, \tag{15}$$

where $N_{15} = (1\text{-}p_{82}\ p_{24})\text{-}(p_{12}p_{23} + p_{12}\ p_{24}p_{839} + p_{1410}\ p_{839} + p_{82}\ p_{24}\ p_{1410})\ p_{315}$ and D_{12} is already specified.

10. Expected Number of Visits Due to Replacement

Let $NP_i(t)$ be the expected number of visits by the server due to replacement in (0,t] given that the system entered the regenerative state S_i at $t=0$. We have the following recursive relations for $NP_i(t)$:

$$NP_i(t) = \sum_j Q_{i,j}(t) \circledR [\delta_j + NP_j(t)] \tag{16}$$

where S_j is any regenerative state to which the given regenerative state S_i transits and $\delta_i=1$, if j is the regenerative state where the server does job afresh otherwise $\delta_i=0$. Taking LST of relations (16) and solving for $\widetilde{NP}_0(s)$. The expected number of visits per unit time are given by

$$NP_0 = \lim_{s \to 0} s\,\widetilde{NP}_0(s) = \frac{N_{16}}{D_{12}} \tag{17}$$

Where, $N_{16}= p_{23}(p_{12} + p_{82}\ p_{1410}) + (p_{1410} + p_{12}\ p_{24})p_{46}$ and D_{12} is already specified.

11. Profit Analysis

Profit incurred to the system model in steady state is given by

$$P = K_1A_0 - K_2BR_0 - K_3BP_0 - K_4NR_0 - K_5NP_0$$

where

K_1: Revenue per unit up-time of the system

K_2: Cost per unit time for which server is busy in repair

K_3: Cost per unit time for which server is busy in replacement

K_4: Cost per visit by the server for repair
K_5: Cost per visit by the server for replacement.

12. Interpretation and Conclusion

By taking exponential distribution for all random variables, Table 1 shows that MTSF decreases with the increase of failure rates (λ and λ_1) for fixed values of other parameters. It means that the system will become more reliable as industry/users pay more attention towards decreasing the failure rates of the units. However, MTSF increases as repair rate (θ) and replacement rate (θ_1) increase. From Table 2, it is observed that profit goes on decreasing as failure rates (λ and λ_1) increase. However, its value increases if repair rate (θ) and replacement rate (θ_1) increase for fixed values of other parameters. It means that replacement of the degraded unit at its failure will be economically beneficial under stated conditions.

Table 1. MTSF vs. failure rate of new unit.

λ ↓	Mean Time to System Failure (MTSF)			
	λ_1=0.13	λ_1=0.16	θ=2.7	θ_1=3.6
0.01	3841.918	3776.34	4393.618	4057.398
0.02	1111.822	1071.353	1269.026	1172.361
0.03	562.8267	533.6799	641.5636	592.5364
0.04	356.0203	333.089	405.4667	374.219
0.05	253.5953	234.5716	288.6495	266.1369
0.06	194.3351	177.9913	221.1204	203.6261
0.07	156.4247	142.0321	177.9532	163.649
0.08	130.4072	117.4987	148.3484	136.2212
0.09	111.6022	99.8602	126.9637	116.4021

Table 2. Profit of the system vs. failure rate of new unit.

λ ↓	Profit of the System			
	λ_1=0.13	λ_1=0.16	θ=2.7	θ_1=3.6
0.01	4659.178	4601.178	4914.073	4763.808
0.02	4100.486	4050.486	4644.756	4311.616
0.03	3738.172	3688.172	4413.242	4007.849
0.04	3468.871	3418.871	4197.777	3773.142
0.05	3264.845	3214.845	4005.953	3590.413
0.06	3106.297	3054.297	3837.979	3445.546
0.07	2980.118	2930.118	3691.561	3328.484
0.08	2877.585	2827.585	3563.807	3232.214
0.09	2792.767	2742.767	3451.94	3151.805

Hence, on the basis of the results obtained for a particular case it is concluded that a parallel unit system can be made more reliable and profitable to use by decreasing failure rates of the units and immediate replacement of the degraded unit at its failure.

Acknowledgement

The authors are very grateful to anonymous referees for their valuable suggestions and this work is a part of UGC-BSR Research Start-Up-Grant [No. F.20-9(5)/2012 (BSR)], supported by University Grant Commission (UGC), Government of India.

References

1. R. Subramanian and Y. V. S Sarma, A complex two-unit parallel system, *Microelectronics Reliability* **21**, 2 (1981).
2. R. Gupta and R. Goel, Cost-benefit analysis of two-unit parallel system with administrative delay in repair, *International Journal of Systems Science* **21**, 7 (1990).
3. G. S. Mokaddis, S. W. Labib and A. M. Ahmed, Analysis of a two-unit warm standby system subject to degradation, *Microelectronics Reliability* **37**, 4 (1997).
4. A. Barros, C. Berenguer and A. Grall, A maintenance policy for two-unit parallel systems based on imperfect monitoring information, *Reliability Engineering & System Safety* **91**, 2 (2006).
5. S. N. Singh, S. K. Singh, A study on a two unit parallel system with Erlangen repair time, *STATISTICA* **LXVII**, 1 (2007).
6. R. Kishan and M. Kumar, Stochastic analysis of a two-unit parallel system with preventive maintenance, *Journal of Reliability and Statistical Studies* **22**, (2009).
7. S. C. Malik and Gitanjali, Analysis of a Parallel System with Maximum Repair Time and Single Server who Appears and Disappears Randomly, *International Journal of Mathematical Archive* **3**, 11 (2012).
8. R. Bashir and J. P. S. Joorel, Reliability Measures of a Three Non-Identical Repairable Redundant Complex System with Load Sharing Units Working in Parallel, International Journal of Statistics and Reliability Engineering, **1**, 2 (2014).
9. Rathee, S. C. Malik, A Parallel System with Priority to Preventive Maintenance over Replacement Subject To Maximum Operation and Repair Times, *International Journal of Engineering Development and Research (IJEDR)*, **2**, 3 (2014).

Adoption and Evolution of FOSS: Key Factors in the Development of the Apache Web Server

Ranjan Kumar

Dept. of Computer Science, Aryabhatta College, (University of Delhi), New Delhi, 110021, India
E-mail: ranjan301@gmail.com

Subhash Kumar[†] and Sukanta Deb

Dept. of Physics, Acharya Narendra Dev College, (University of Delhi) New Delhi, 110019, India
[†]E-mail: subhashkumar@andc.du.ac.in
www.andcollege.du.ac.in

The paper reviews the practice and conceptual framework of Free and Open Software (FOSS) along with its evolution and global status. It is argued that Open Source software has come as a relief for programmers especially when technologies have evolved and are rapidly changing in the current environment of business, enterprise and academia. In this respect the concept of "FOSS Adoption Index (FAI)" is examined vis-a-vis the developing world with special reference to India. It is clear from the specifics of the FAI model which includes its parameters and the computational procedures involved that this index may serve us well in the understanding of the evolution of FOSS. The FAI model shows a medium to low levels of FOSS adoption amongst Indian organizations including the academia, which is particularly revealing because of the fact that sharing of information has been a mainstay in the academic environment and in the development of science and technology. The paper further investigates the above issues with respect to Apache web server's development as a robust and popular web server which is currently being used by 60% of all web servers. This shows that there may exist some key factors in the development of Apache web server which contribute to the popularity of open source software among various software developers on a FOSS platform.

Keywords: History of Computing; FOSS; web server.

1. Introduction

The commercial propriety products have dominated the software market in the world during the past many decades. Although these products have offered a myriad of functionalities, they by their intrinsic nature do suffer from many limitations. These limitations include lack of access to the program codes, unknown security threats, bugs, vendor lock-in, high cost of upgrades, etc. These limitations have led to the growth of an alternative economy in software, which consists of software that are free as well as open - the FOSS (Free Open Source Software). The FOSS or sometimes just called the Open Source Software (OSS) software provides the source code – the open code. Thus more programmers are able to have access to the code and thereby facilitate creation of new functionalities, fixing of bugs and development of robust security features. This is analogous to the same organic mode that science has grown over a period of time. Most of the FOSS product carries licenses which is associated with a legal liability, which requires any modifications such as enhancements and fixes to be made available under the same license. This invokes trust and confidence in developers as well as users which in turns create large, sustainable communities that drives the growth of the software over a period of time. In the last quarter century the FOSS is steadily getting better and robust. It not only provides more functionality, security and capability but also sustainability at much lesser cost. Given the democratic framework and the host of benefit it offers, FOSS should be increasingly the first option for both individuals and organizations alike. Despite the above mentioned positives about FOSS, the rate of adoption and evolution of FOSS products in the developed nations as well as in the developing countries is well below its potential. This clearly indicates to a need to discern the evolution and adoption model of the OSS and devise strategies for accelerated adoption of FOSS in the world in general and India in particular. This paper tries to address this issue by examining the chief attributes of a successful OSS product – the Apache web server, which may be replicated in the development matrix of other OSS products for their rapid evolution and adoption. The paper aims to study the macro-level evolution patterns of FOSS, with the assumption that the Apache web server is a representative sample of relevant FOSS projects. The paper is organized as follows. In Section II the history and salient features of a FOSS product is discussed, while the next section (Sec. III) gives an account of the adoption of FOSS at various levels. Section IV outlines the salient features in the development, adoption and growth strategies of Apache web server as a successful OSS product. The last section (Sec. V) provides the discussion and concluding remarks.

2. The Free Open Software

The origin of the present FOSS movement, in the most accepted narrative is associated to Richard M. Stallman of Massachusetts Institute of Technology (MIT, USA), who started the GNU project in 1984; mainly to launch a UNIX like powerful operating system. This was followed by the setting up of the Free

Software Foundation [1]. The other milestone that heralded the FOSS movement is the writing and its wide distribution of a Unix-like kernel by Linus Trovalds (a second year graduate student at the University of Helsinki) in 1991. By 1993 both 386BSD and GNU/Linux had become very reasonable stable platforms. Therefore it can be said that in spite of being in existence since the1960's it is only in late 1990's that FOSS had begun to change the face of the Software and IT industry. The phenomenal growth and expansion of FOSS has been mainly aided by the massive growth and expansion across the worlds of the Internet and technologies related to World Wide Web. The advent of FOSS has provided a new praxis to the way software is construed which includes its philosophy, usage, development and deployment in various areas [2]. The Free Software Foundation (FSF) embodies Richard Stallman's vision of free software by entitling its end users the following four freedoms –

- Freedom to use the code/program for all times at any geographical location and for any objective.
- Freedom to adapt/modify/ or customize the program to suit one's need. The availability of open code is critical for this.
- Freedom to make multiple copies and redistribute them.
- Freedom to enhance the code and to release the same in the public domain.

Thus "Free" in FOSS is not construed in the economic sense but, in the words of Stallman it incorporates the spirit of free essentially same as is understood for "freedom" or "free speech" and not as in "free beer" [3]. From the legal point of view, software distribution may follow the FOSS model. A different FOSS license entitles the user with a myriad of rights and permissions [4]. Stallman scripted the GNU General Public License (GPL) which guarantees the end-user availability of the source code and all the four freedoms mentioned above incorporating the FSF philosophy. Netscape followed Trovalds in allowing access to its browser code to all - the public and developers from around the world. The email client *Thunderbird* and the famous browser *Mozilla* are essentially products of this endeavor.

Besides the license or the philosophical aspect of FOSS another aspect of FOSS focuses on the process employed to create the software. This was clearly brought out in the book *"The Cathedral and the Bazaar"* by Raymond wherein the modus operandi for the development of proprietary software is likened to that of construction of a space – say a cathedral by a group of trained people in a confined environment. This is compared to the development process of FOSS which mimics the bazaar involving large group of people working in an environment which is open, porous and sans boundaries [5]. In fact this paradigm of collaborative development can be traced to Linus Trovalds when he released the Linux kernel written by him to the community which were essentially hackers and exhorted them to contribute to its further enhancement and development. Fundamentally FOSS is seen as a collaborative venture wherein members transcend the barriers of geography and politics from across the globe to participate and contribute in the development process of the software. The role of Internet is essentially of an enabler since most members are unpaid volunteers working together over the internet. It is no surprise that a significant amount of the technical infrastructure of the Internet itself is sourced from the FOSS [6].

3. Adoption and Evolution of FOSS

FOSS has attracted attention and gained popularity among various government, organizations and academic institutions primarily because of its unique features of affordability and the invaluable freedom it guarantees to its end-users. FOSS is not only about development of computer scripts/codes and a matter left alone to the computer scientists. In fact FOSS can be considered as an ecosystem with the academia, Government, Industry and Community as its essential components with bi-directional relationship among each of them [7]. The culture of collaboration it breeds makes it particularly important for the academia for which sharing of information has been a mainstay in the academic environment and in the development of science and technology. The academia while evangelizing the cause of FOSS, also produce FOSS products/technologies and generates FOSS trained human resource (HR). However it remains dependent on the Government for funding and creates enabling policies. It also needs patronage from the Industry to provide appropriate placements for the FOSS human resource. The role of the community is paramount to extend support and evangelize.

In specific spheres like academia and research, e-governance, Small and Medium Enterprises (SMEs) etc FOSS has witnessed growth. A survey done a decade ago has shown adoption of FOSS to the extent of 78% by the public administrations of 13 European countries [8], while at the same time 87% of the institutions surveyed in the US at the same time have reported the use of FOSS [9]. However, these surveys which show encouraging trends for FOSS adoption are mainly concerning the public institutions. The usage of FOSS in private enterprise, industry and personal space is still limited.

Despite there being awareness of the benefits of FOSS, India is not a significant player in the FOSS world till today, with few sporadic, unorganized, undirected and individual dependent developments [7]. There is a paucity of reliable and authorized quantitative primary data which may depict the extent of FOSS adoption and its usage. In a recent study [7], a FOSS Adoption Index (FAI) model is envisaged which attempts to quantify an organization's FOSS friendliness and FOSS maturity by its level of usage. Such models may serve as a tool in devising effective policy to enhance FOSS adoption and its usage.

Another way to devise means and ways for enhanced adoption of FOSS is to identify salient attributes of some successful FOSS product and attempt to replicate it in the development, usage and adoption of other FOSS products. This attempt has been made in the next section where the developmental matrix of a very successful FOSS – the Apache web server is analyzed.

4. Apache Web Server: A Successful Foss

In 1995 a motley group of people got together mainly with purpose of exchanging software patches for the NCSA web server. Later that year this group who called themselves the Apache Group, accomplished the task of developing their own web server - the Apache HTTP server; which was released in the public domain. In less than 6 months the Apache server surpassed the NCSA web server's adoption volume. Its unique modular design facilitated other sister projects for web server modules to gravitate and cluster around the Apache group. Once the Apache group acquired the position of the market leader, the fear of abuse and lack of legal framework crept in. This necessitated the creation of the Apache Software Foundation (ASF) being institutionalized in 1999. Its market shares in February 2007 i. e, in a period of 10 years since its inception reached a whooping 61-73%. Till recently, the Apache web server continues its lead in the market share of all web servers. The success story of the Apache web server is essentially inter-woven around the following features of the ASF:

4.1. *Philosophy*

i. Open Source i.e. making the software available to general public and usable for any purpose they deem fit.
ii. High quality and collaborative software development.
iii. Interaction being technical-based, respectful and honest.
iv. Faithful implementation of standards, wherein security being a major part of software development.

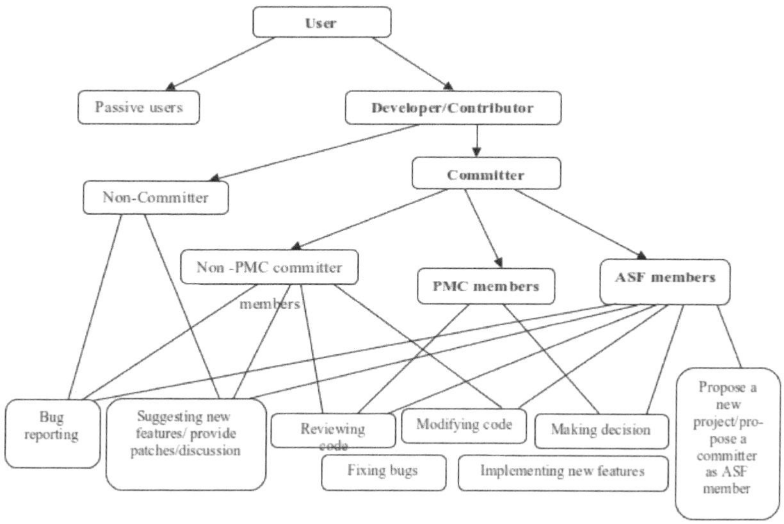

Fig. 1. Schematic representation: Role of Apache users and developers.

4.2. *Structure of ASF organization*

Unlike most open source project ASF has not been organized around a single person or primary contributors. Rather the unsaid underpinning of the organization is *meritocracy* - the governance of merit [10]. It means the more merit one has the more power she gets. The openness does not allow it to reduce into an old-men club. The chain of merit consists of the user, the committers, the project members (PMC) and the ASF member as shown schematically in Fig. 1 above. In the ASF project a total of around 835 committers out of which 111 are involved in Apache HTTP server project and out of 111, 63 ASF members (committers) are also there. PMC of Apache HTTP server project have 44 members. Decisions are taken by the voting and for code change at least 3 positive votes and no negative votes is a mandatory requirement. PMC members are bound to voting. A veto must be accompanied by either a suggestion of an alternative or a reasoned explanation. The Board of Directors manages corporate assets. There are 9 members elected every year. They allocate corporate resources to various projects and cannot make any technical decisions. Out of 835 committers only 16 are Indians (~ 2%).

4.3. *Communication*

All communications are kept archived and done over asynchronous media like mailing lists etc. This allows others to join the communication or have a reference. This promotes geographic distributions.

4.4. *The Apache security team*

It forms the nuclei of all issues involving potential security vulnerabilities in Apache software. This team navigates through vulnerability reports and also acts as coordinator between different projects. Bug reports are duly forwarded and handled by project specific security teams (e.g., HTTP server and Tomcat). Table 1 lists the cumulative number of bugs that are reported and resolved between 2004 to the present date (6.03.2015). We have taken the data of bugs' transactions from ASF Bugzilla [11]. We have taken a total of 3564 bugs that were reported by 2472 different people/users and were resolved in the above mentioned period. Out of these 231 resolutions caused a change in Apache code or documentation. It clearly brings out the fact that a collaborative and open approach may lead to faster and efficient debugging. Figure 2 provides a column diagram to illustrate the efficacy of the open software in debugging, thus leading to the development of robust and secure FOSS products. The column diagram shows that 22% of the bugs are resolved on the same day they have been reported. Around 35% of the bugs are resolved in the next 7 days. It also shows that around half of the bugs are resolved within a month.

Table 1. Cumulative number of bugs resolved over the number of days.

No. of days	No. of bugs resolved
0	777
1	1010
2	1091
3	1140
4	1178
5	1197
6	1234
7	1264
30	1488
365	2293
730	2726
1000	2964

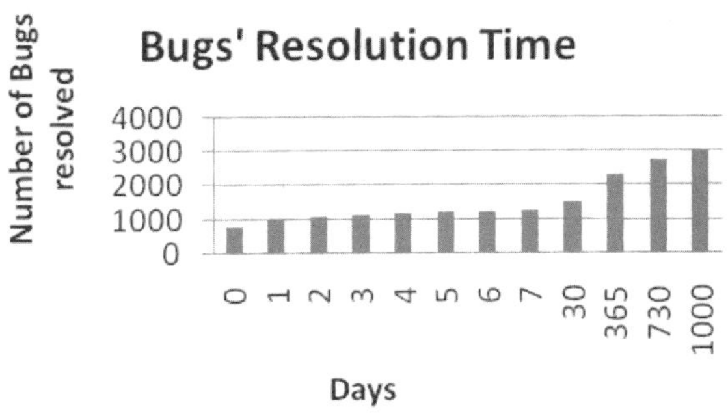

Fig. 2. Column graph of cumulative no. of bugs resolved in days.

4.5. *Performance*

A large variety of Multi Processing Modules (MPMs) are implemented at Apache instead of adopting a unitary architecture. Thus Apache operates in a process-based, hybrid (process and thread) or event-hybrid mode. This facilitates better matching of the demands of every infrastructure. This in turns means that the right choice of MPM and the configuration is very important. In cases where compromises in performance becomes necessary, the policy of Apache involves adoption of such practices which reduces latency and increase the output, relative to simply handling more requests. This ensures consistent as well as reliable processing of requests while respecting time-frames at the same time.

The ASF is only a virtual institution with no building which may house its offices. It maintains only a modest physical existence in the form of the necessary technical infrastructure which is required to runs its services viz. source code repositories, web services, issue/bug tracking services, distribution mirroring and mail services.

5. Concluding Remarks

A large scale acceptance of FOSS products is in the interest of the industry, government, academia and research organizations due to its intrinsic nature of openness and collaboration. It's very democratic nature dissuades vested interests thus FOSS products are generally secure and robust. The availability of source code promotes innovation and can be a companion on the learning curve of the software/IT industry. Useful lessons can be learnt from successful models of the FOSS. The Apache Software Foundation is an illustrative example of nurturing successful FOSS products which can be replicated for other OSS products because of its unique organizational hierarchy and a rigorous developmental matrix.

References

1. S. Williams, *Free as in freedom: Richard Stallman's crusade for free software.* Sebastopol, CA: O' Reilly & Associates, (2002).
2. P. Vixiel, Software engineering, In *C. DiBona, S. Ockman and M. Stone (Eds.), Open Sources: Voices from the open source revolution*, Sebastopol, CA: O' Reilly & Associates, 91 (1999).
3. R. M. Stallman, L. Lessig and J. Gay. Free Software, free society. Cambridge, MA: Free Software Foundation (2002).
4. M. O'Sullivam, Making copyright ambidextrous: An expose of copyleft, *The Journal of Information, Law and Technology*, **3** (2002).
5. E. S. Raymond, The Cathedral and the Bazaar. Sebastopol, CA: O' Reilly & Associates., 1999.
6. Mockus, R. T. Fielding and J. D. Herbsleb, Two case study of open source software development, Apache and Mozilla, *ACM Transactions on Software Engineering and methodology*, **11**, 309 (2002).
7. S. K. Saini, Use of Free/Open Source Software by Indian Organizations-A Preliminary Study and its Findings, *Doctoral thesis*, Anna University, Chennai (2013).
8. R. Ghosh and R. Glott, Results and policy paper from surveys of Government authorities, *Technical report*, MERIT, University of Maastricht (2005).
9. S. Waali, D. Gynn and B. Von Rotz, The growth of open source software in organizations, *Report available at http://www.optaros.com,* (2005).
10. *http://apache.org/foundation/how-it-works.html#meritocracy*
11. ASF Bugzila. Available at *http://issues.apache.org/bugzilla*

Android/Tizen Based Artificial Intelligence Techniques for Prognosis and Diagnosis of Electrical Machines

K. V. Satya Bharath[1] and Sheikh Suhail Muhammad[2]

M. Tech (Power Systems), Amity School of Engineering and Technology
Amity University, Noida, Uttar Pradesh-201313, India
E-mail: [1]kurukuru.bharath@student.amity.edu, [2]sheikh.mohammad@student.amity.edu

Prof. (Dr.) Priya Ranjan

Profesor-EEE, Amity School of Engineering and Technology
Amity University, Noida, Uttar Pradesh-201313, India
E-mail: pranjan@amity.edu

Electric machines are widely used in the present day industries which constitute about a 75–80% of the total deployed due to their user-friendly interface. Generally speaking machinery is prone to multiple types of faults and an intelligent monitoring layer can ensure smooth operation and inform as maintenance alerts mitigating any abrupt failures. In this work we present and ANDROID/TIZEN interface which integrates multiple AI and signal processing techniques to detect the different faults interrupting smooth performance of machinery in a very short span of time. We demonstrate the thought process, simulation, final software interface and test results to confirm its effectiveness.

Keywords: Prognosis and Diagnosis, AI, ANDROID, TIZEN.

1. Introduction

Induction motors are most commonly used electrical machines in industry because of their low cost, reasonably small size, ruggedness, low maintenance, and operation with an easily available power supply. Although these are very reliable, they are subjected to different modes of failures/faults. These faults may be inherent to the machine itself or due to operating conditions. The inherent faults may be due to the mechanical or electrical forces acting on the machine enclosure. If a fault is not detected or if it is allowed to develop further it may lead to a failure. A variety of machine faults have been studied in the literature such as winding faults, unbalanced stator and rotor parameters, broken rotor bars, eccentricity and bearing faults.

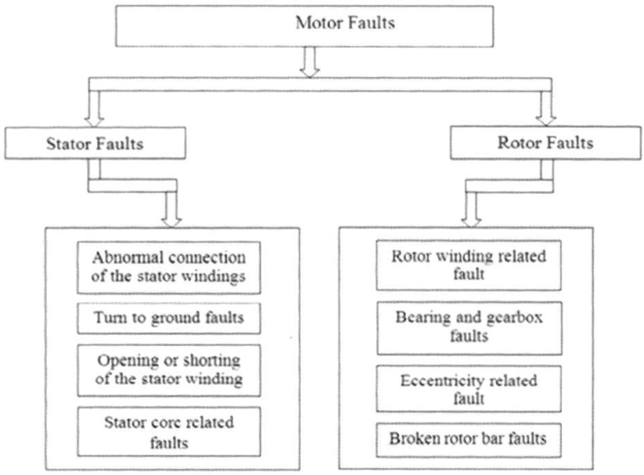

Fig. 1. Sources of motor faults.

Several fault identification method have been developed and been effectively applied to detect machine faults at different stages by using different machine variables, such as current, voltage, speed, efficiency, temperature and vibrations. Thus, for safety and economic considerations, it is essential to monitor the behavior of motors of different sizes, large and small. Traditionally, maintenance procedures in industry follow two approaches as follows: The first one is to perform fixed time interval maintenance, where the engineers take advantage of slower production cycles to fully inspect all aspects of the machinery. The second is to simply respond to the plant failure as and when it happens. However, making use of today's technology, new scientific approach was becoming possible for maintenance management which can transform abrupt disruptions to graceful degradation of service. One of the key elements to this new approach is predictive maintenance through condition monitoring, which depends upon the condition of the plant. Condition monitoring is used for upgrading performance of machinery, reducing consequential damage, increasing machine life, reducing spare parts inventories and reducing breakdown maintenance. An efficient condition-monitoring scheme is one that provides warning and predicts the faults at early stages. Monitoring system obtains information about the machine in the form of primary data and through the use of modern signal processing techniques it is possible to give vital diagnostic information to equipment operator before it catastrophically fails.

2. Literature Review

Condition monitoring is defined as the continuous evaluation of the health of the plant and equipment throughout its service life. It's important to detect incipient faults at early stages and fix them immediately. This is called incipient failure [1] detection. The incipient detection of motor failures also provides a safe operating environment. It is becoming increasingly important to use comprehensive condition monitoring schemes for continuous assessment of the electrical condition of electrical machine. By using the condition monitoring, it is possible to provide adequate warning of imminent failure.

2.1. *Existing condition monitoring techniques*

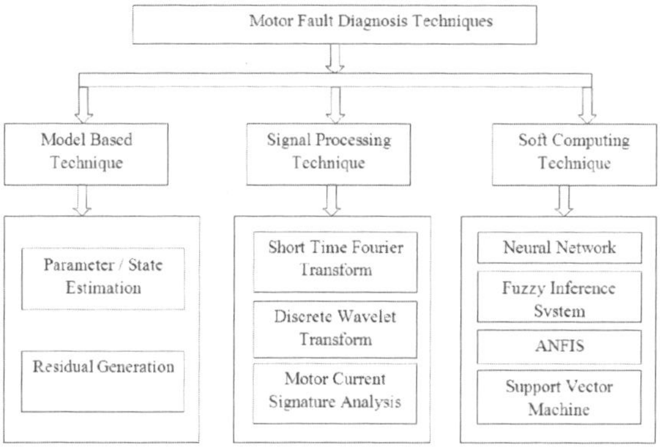

Fig. 2. Types of fault diagnosis techniques for electric motors.

Isermann [2] has presented a novel, unified model based fault detection and prediction (FDP) technique is developed for non linear multiple-input-multiple-output (MIMO) discrete time systems. The proposed scheme addresses both state and output faults by considering separate time profiles. The faults, which could be incipient or abrupt, are modeled using input and output signals of the system. The fault detection (FD) scheme comprises online approximation in discrete time (OLAD) with a robust adaptive term. An output residual is generated by comparing the fault detection (FD) estimator output with that of the measured system output. A fault is detected when this output residuals exceeds a predefined threshold. The asymptotic stability of the fault detection and prediction (FDP) scheme enhances the detection and time to failure accuracy. The effectiveness of the proposed approach is demonstrated by using a fourth-order MIMO satellite system.

2.2. *Soft Computing Technique*

Nejjari [8] and Benbouzid have used the Park's vector patterns for detecting different types of supply faults, such as voltage imbalance and single phasing. In addition a neural network based back propagation algorithm is used to obtain the machine condition by testing the shape of the Park's vector patterns. Two neural network based approach have been used, these are classical and decentralized. The generality of the proposed methodology has been experimentally tested and the authors claim that the results provide a satisfactory level of accuracy.

3. Motor Condition Monitoring Using Stator Current

Fuzzy subsets can be assigned to describe the stator current amplitudes by means of corresponding membership functions. A knowledge base pertaining to situations liable for faults in the motor comprising rule base and data base can be built to activate the fuzzy inference. The induction motor condition can thus be diagnosed using an apparatus of fuzzy inference. Induction motor under consideration is of *2-KW, 220/380 V, 15/8.6 A, 50-Hz, 4 pole, Δ connected squirrel cage* type. The judgment of induction motor condition made based on fuzzy inference is capable of better diagnosis as it shares the human's knowledge expressed in vague terms with computing abilities and knowledge based training etc. Such linguistic input prepositions can be directly expressed by Fuzzy System. The structure of Fuzzy Monitoring and Diagnosis is shown in Fig. 3.

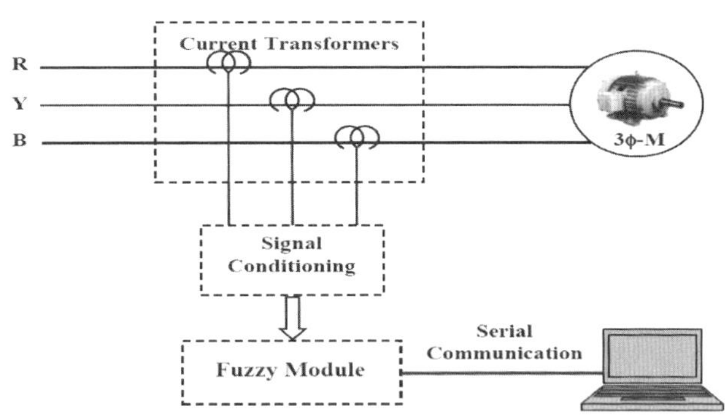

Fig. 3. Schematic representation of motor condition diagnosis layout.

The stator current signal reciprocates the promising information relating to motor fault. The Fuzzy systems rely on a set of rules. These rules allow the input to be fuzzy permitting the words to describe the fault conditions bringing a sort of robustness in information processing. Motor being an electro-mechanical system, the applied voltage, current drawn by the motor and internal temperature play a major role in generating the faults. In the present study the motor phase currents have been monitored and based on the subsequent trend in the current values diagnosis of motor status has been judged. For Fault Detection and Diagnosis of Induction Motor stator current amplitudes *Ia*, *Ib*, *Ic* have been considered as the input variables to the fuzzy systems. The motor condition, *Mc* has been chosen as the output variable. We have defined the fuzzy subsets $\tilde{I}_a$, $\tilde{I}_b$ and $\tilde{I}_c$ for input variables given to the fuzzy subsets and $\acute{M}_C$ for output variables as follows-

$$\tilde{I}a = {}^{\mu a(ia)}/_{ia'}\ \forall\ ia \in Ia$$

$$\tilde{I}b = {}^{\mu b(ib)}/_{ib'}\ \forall\ ib \in Ib$$

$$\tilde{I}c = {}^{\mu c(ic)}/_{ic'}\ \forall\ ic \in Ic$$

$$\acute{M}C = {}^{\mu mc(mc)}/_{mc'}\ \forall\ mc \in Mc$$

Where *ia*, *ib*, *ic* and are elements of the discrete Universe of Discourse (UoD) i.e., *Ia*, *Ib*, *Ic* and *Mc*. Fuzzy logic works with linguistic variables whose values are words or sentences in a natural or artificial language. This provides a means of systematic manipulation of vague and imprecise information. The term set (T) is a collection of linguistic values assigned to linguistic variable. The term sets used for stator current and motor condition are as follows-

T (*Ia*, *Ib*, *Ic*) = {Zero (*Z*), Small(*S*), Medium (*M*), Big (*B*), Very Big (*VB*)}

T (*Mc*) = {Open Phase, Damage, Critically Overloaded, Overloaded, Good}

The optimized rule base has been developed so as to encompass all possible healthy and faulty conditions of the motor.

Table 1. Rule base for motor health.

Rule	If I_a is	or I_b is	or I_c is		Motor Condition	Rule Weight
1	Z	Z	Z	THEN	*Open_phase*	(1)
2	B	B	B		*Critically_Overload*	(0.9)
3	M	M	M		*Over_Load*	(0.8)
4	S	S	S		*Good*	(0.7)
5	VB	VB	VB		*Damage*	(1)

4. Experiment, Simulation And Results

The complete set up of Fuzzy based Fault Detection and Diagnosis for Induction Motor has been built around the Fuzzy Inference System designed in Simulink for detection of motor health condition and it is shown in Fig. 4.

The built-in block set of Induction motor has been employed for the purpose of study. Simulink model is categorized into different categories like Induction Motor Drive with Power Supply, FIS, RMS to DC conversion and Fault creation. Three phase voltage supply is applied to induction motor. The RMS value is calculated over a running window of one cycle of the specified fundamental frequency. Fault creation block is for simulating different faults. Different faults like open phase, imbalance of input voltage are artificially generated in this block. Different blocks like Sum, Product, Step input and Constant are used in fault creation.

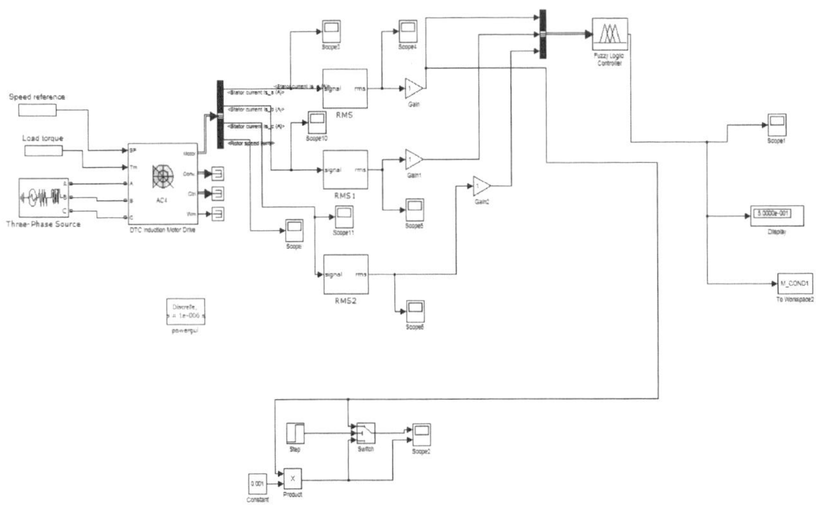

Fig. 4. Simulink diagram of fault detection technique.

The block FIS for Motor implements a fuzzy inference system (FIS) in Simulink. In our case, input is Stator current from each phase and output is condition of motor in linguistic form using fuzzy logic. The output is having range from 0 to 1 which is stored as variable M_COND in workspace and display on scope as well as its value on Display block.

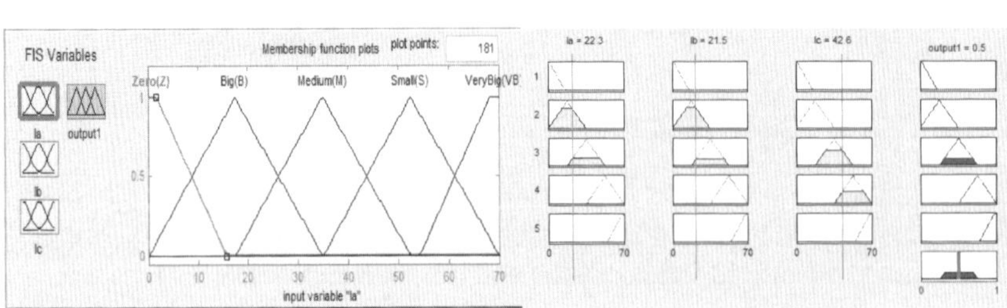

Fig. 5. Input membership functions and rule view of fuzzy.

5. Simulation Results

5.1. *Turn-Turn short in one phase winding*

After the simulation for normal operation of the induction motor model, simulation for the short circuit in the part of the winding in R phase has been carried out. At this condition the value of the stator resistance at short circuit fault is equal to Rstator, fault = 13.1Ω, we can find the value of the inductance at the fault state by using the ratio between the value of the resistance at both state (normal and fault).

5.1.1. *Open phase fault*

In this case after normal startup, at 0.5 second, R phase was open circuited and the corresponding results are shown in figures. The figure shows the condition monitoring of motor using MATLAB.

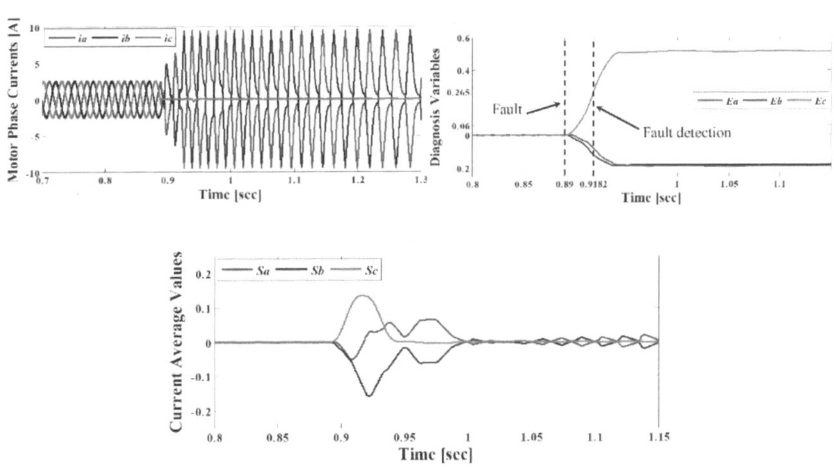

Fig. 6. Simulation results for a single phase open-circuit.

Fault in phase *c* of the (a) motor phase currents, (b) diagnosis variables and (c) current average values.

6. Integrating Android/Tizen With AI Techniques

Tizen is an open-source project that allows you to create feature-rich applications for multiple device categories, such as smart phones and tablets, using Web and native APIs. The Tizen platform allows you to create a wide range of applications for Tizen devices, customized with the common Tizen architecture as its basis. The Tizen platform allows you to specify the conditions for filtering out incapable devices based on the hardware and software components that your application requires to run correctly. Tizen also provides Web sites with tools for creating, distributing, and managing mobile applications efficiently. The Tizen platform provides 2 different types of frameworks for application development: the Web framework can be used to develop Web applications, while the native framework can be used to develop native applications. The Tizen platform also allows you to develop a hybrid application package where Web and native applications are packaged together to make more powerful applications. The Tizen platform ensures that all Tizen applications have consistent look and feel, regardless of whether you use the Web or native framework to create them.

6.1. *AI techniques with Android*

We present an open source Android/ Tizen library called jFuzzyLogicb which allows us to design and to develop FLCs following the standard for FCL. jFuzzyLogic offers a fully functional and complete implementation of a fuzzy inference system (FIS), and provides a programming interface (API) and an Eclipse plugin in order to make it easier to write and test FCL code. This library brings the benefits of open source software and standardization to the fuzzy systems community, which has several advantages:

- Standardization reduces programming work. This library contains the basic programming elements for the standard IEC 61131-7, removing the need for developers to attend to boiler plate programming tasks.
- This library extends the range of possible users applying FLCs. This provides a complete implementation of FIS following the standard for FCL, reducing the level of knowledge and experience in fuzzy logic control required of researchers. As a result researchers with less knowledge will be able to successfully apply FLCs to their problems when using this library.
- The strict object-oriented approach, together with the modular design used for this library, allows developers to extend it easily.
- jFuzzyLogic follows a platform-independent approach, which enables it to be developed and run on any hardware and operating system configuration that supports Java.

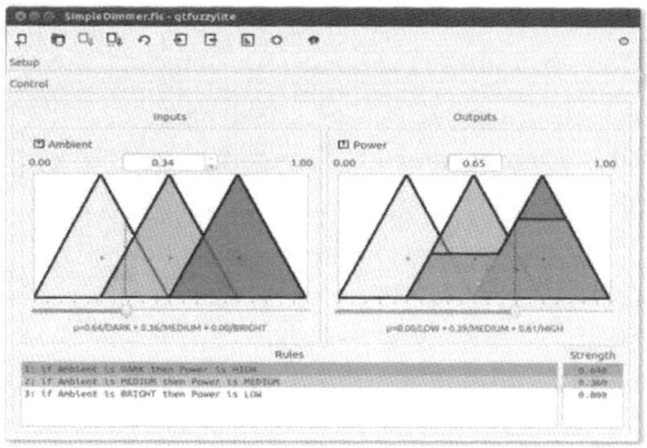

Fig. 7. Fuzzy Lite for ANDROID/Tizen through Ubuntu.

7. Conclusion

This paper discussed the use of conventional methods over modern mobile platforms like ANDORID/TIZEN to detect and diagnose faults in induction motors from electrical failures (inter-turn short circuit, phase unbalance and broken bars) beyond the normal operating condition. Prognosis and Diagnosis provides an optimal solution for detecting the various faults and hence improvising the performance of the machine. Its integration with ANDROID/TIZEN will provide access to the remote operation and detection which can be easily adapted in the industrial atmosphere. Further the technology is being developed for relay less operation in fault detection and diagnosis.

References

1. P. F. Albrecht, J. C. Appiarius, R. M. McCoy, E. L. Owen, D. K. Sharma, Assessment of the Reliability of Motors in Utility Applications – Updated, IEEE Trans. Energy Convers. vol.: EC-1, issue 1, pp. 39–46, 1986.
2. R. Isermann, "Model Based Fault Detection and Diagnosis—Status and Applications," Annu. Rev. Control, vol. 29, no. 1, pp. 71–85, 2005.
3. M. Arkan, D. Kostic-Perovic, P. J. Unsworth, "Modelling and Simulation of Induction Motors with Inter-Turn Faults for Diagnostics," Electric Power Systems Research, vol. 75, pp. 57–66, 2005.
4. F. L. Stanislaw, A. H. M. SadrulUla, A. M. Trzynadlowski, "Instantanious Power as a Medium for the Signature Analysis of Induction Motors," IEEE Trans. Ind. Appl., vol. 32, no. 4, pp. 904–909, 1996.
5. G. G. Yen, K. C. Lin, "Wavelet Packet Feature Extraction for Vibration Monitoring," IEEE Trans. Ind. Electron., vol. 47, no. 3, pp. 650–667, 2000.

6. H. Douglas, P. Pillay and A. K. Ziarani, “A New Algorithm for Transient Motor Current Signature Analysis Using Wavelets,” IEEE Trans. Ind. Appl., vol. 40, no. 5, pp. 1361–1368, 2004.
7. H. Nejjari, M. H. Benbouzid, “Monitoring and Diagnosis of Induction Motors Electrical Faults using a Current Park’s Vector Pattern Learning Approach,” IEEE Trans. Ind. Appl., vol. 36, no. 3, 2000.
8. Arfat Siddique, Member, IEEE, GXYadava and Bhim Singh. Review on Applications of Artificial Intelligence Techniques for Induction Machine Stator Fault Diagnostics.
9. Pablo Cingolani 1, Jes ´us Alcal´a-Fdez jFuzzyLogic: a Java Library to Design Fuzzy Logic Controllers According to the Standard for Fuzzy Control Programming In International Journal of Computational Intelligence Systems, Vol. 6, Supplement 1 (2013), 61−75.
10. Online References from: TIZEN https://developer.tizen.org/devguide/2.3.0 /org.tizen.gettingstarted/html/tizen_overview/tizen_overview.htm
11. TIZEN DESIGN: https://developer.tizen.org/devguide/2.2.1/org.tizen.web. appprogramming/htmltutorials/tizen_tutorial/creating_composite_filter.htm
12. Jover Rodríguez, P., Arkkio, A., 2008, Detection of Stator Winding Fault in Induction Motor Using Fuzzy Logic, Applied Soft Computing, Vol. 8, No. 2, pp. 1112−1120.
13. Aderiano m. Da silva, b.s. on Induction motor Fault diagnostic and monitoring methods, A thesis submitted to the faculty of the graduate school, Marquette university, In partial fulfillment of The requirements for The degree of Master of electrical and computer engineering Milwaukee, Wisconsin.
14. Aderiano M. Da Silva, Member, IEEE, Richard J. Povinelli, Senior Member, IEEE and Nabeel A. O. Demerdash, Fellow, IEEE on Induction Machine Broken Bar and Stator Short-Circuit Fault Diagnostics Based on Three-Phase Stator Current Envelopes in IEEE transactions on industrial electronics, vol. 55, no. 3, march 2008.
15. Fatiha Zidani, Demba Diallo, Senior Member, IEEE, Mohamed El HachemiBenbouzid, and RachidNaït-Saïd on A Fuzzy-Based Approach for the Diagnosis of Fault Modes in a Voltage-Fed PWM Inverter Induction Motor Drive in IEEE transactions on industrial electronics, vol. 55, no. 2, February 2008.
16. Millind S. Patil, Amir M. Shaikh, Ravindra R. Mudholkar on Motor Health Detection using Fuzzy Logic in International Journal of Engineering Research and Development e-ISSN: 2278-067X, p-ISSN: 2278-800X, www.ijerd.com Volume 3, Issue 9 (September 2012), PP. 73−83.

Performance Analysis of Quality of Service for Different Service Classes in WiMAX Network

Jokhu Lal
CSED, MNNIT
Allahabad, India
+91(532)2545171
pateljl@yahoo.co.in

Dr. Neeraj Tyagi
CSED, MNNIT
Allahabad, India
+91(532)2271367
neeraj@mnnit.ac.in

Broadband wireless access (BWA) is becoming quite popular today as it satisfy demand of users and also support different real time services and applications. WiMAX is Next Generation Networks which provides BWA solution for Wireless Metropolitan Area Networks (WMANs) with Point to Multipoint (PMP) support and high throughput with large covering distance. WiMAX supports Quality of Service (QoS) features. Quality of Service is the main consideration for different applications which are using network resources. These applications are voice over IP (VOIP), audio and video streaming, conferencing, gaming, web browsing etc.

In this paper performance of video traffic over WiMAX network using different service classes has been investigated. For this WiMAX module based on network simulator NS-2 is used. Various parameters that determine QoS of applications is analyzed. QoS parameters such as, throughput, packet loss, average delay and average jitter is compared for different services classes.

Keywords: WiMAX, QoS, BWA, VOIP, UGS.

1. Introduction

WiMAX (Worldwide Interoperability for Microwave Access) is a new Broadband Wireless Access Technology which provides high speed audio, video and data service over a large covering area. It is competing with wired broadband access like ADSL and cable modem as well as wireless broadband access like Wi-Fi and cellular networks. It is based on IEEE 802.16 standard [1] [2] which provides basic Internet Protocol (IP) connectivity to users. There is a greater demand for new services based on multimedia applications such as

VOIP, Audio and Video Streaming, Video Conferencing, File Transfer, web browsing, email, videogame etc. in recent times while user is on move. It is last mile solution which may be deployed rapidly with high scalability and lower investment and maintenance cost. Multimedia application sends large audio and video streams with variable bandwidth and delay requirement instead of these, there are other application such as remote monitoring for critical control, Banking and E-Commerce etc, which do not require strict bandwidth guarantees due to the bursty nature, but these require reliable and instant packet routing. Thus different types of traffic are present in network with different Quality of Service requirements.

WiMAX operates both in 10-66GHz (unlicensed frequency band) and in 2-11GHz (licensed frequency band) for Line of Sight (LOS) and Non-line of Sight (NLOS) operation respectively. This network uses orthogonal frequency division multiplexing which makes transmission resist to interference, fade and minimizes multipath effect. It can also work as a back haul trunk in point-to-point transmission with capability up to 72Mbps over a distance up to 30miles.

The rest of the paper is organized as follows: Section 2 gives overview of WiMAX network. Section 3 describes different QoS, Section 4 is for related works, Section 5 discussed QoS parameters, Section 6 covers simulation details, Section 7 discussed and analyze simulation results and Section 8 concludes the paper.

2. Overview of WiMAX Network

A Simple WiMAX network consists of a Base Station (BS) which is central controllers and number of Subscriber Stations (SSs) in its coverage area at varying distances. An SS can be an independent device equipped with wireless connectivity or building with wireless or wired local area network or a mobile station (MS)[15]. The SS has to scan downlink channel and perform initial ranging and registration process first to establish a wireless connection with BS. All the registered SSs and BS share the communication medium. The communication is in both ways. The downlink channel from BS to SSs is in broadcast mode where all SSs receive the information send by BS. The uplink channel from SSs to BS is shared by all the registered SSs. The BS allocates band width to each SS based on the request made by each SS. The frame size is fixed and divided in two parts: uplink and downlink sub frames. There are two modes of duplexing: Time Division Duplexing (TDD) and Frequency Division Duplexing (FDD). In TDD uplink and downlink sub frames are transmitted at different times, on FDD downlink and uplink sub frames can be transmitted simultaneously at different frequencies. A use case of WiMAX network is shown in Figure 1.

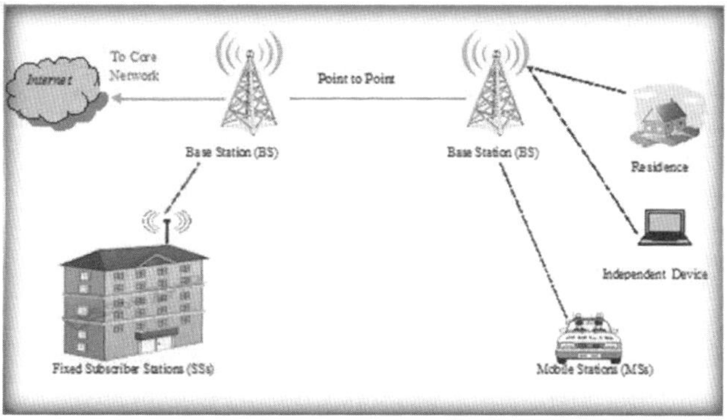

Figure 1. A simple WiMAX network use case.

3. Quality of Service in WiMAX Network

Quality of service is a broad concept which differs for Telephone and Data network. In case of packet-switched networks and computer networks, it means probability of successfully passing packets between two points in network. As the name suggest, it is qualitative measure of how much accurate and effective a network is, but there are number of parameters which can measure it quantitatively. Some of important are throughput, delay, jitter, packet loss etc. QoS is included at Mechanism Access Control (MAC) layer architecture of WiMAX. MAC layer also scheduled bandwidth for different users based on their requirements [3].WiMAX supports number of applications which require different levels of QoS [9][11]. For these applications WiMAX defines five services flow classes and each services flow is characterized by set of QoS parameters. These parameters describe the guarantees required by application to which it is designed for. For uplink connections there is also need to specific mechanism for bandwidth request this is necessary for end-to-end IP based QoS. Service flows can be created, changed or deleted by issuing Dynamic Service Addition (DSA), Dynamic Service Change (DSC) and Dynamic Service Deletion (DSD) messages. Either SSs or BS can initiate these messages by two or three - ways hand shake. The five QoS Service classes of WiMAX network are Unsolicited Grant Service (UGS), Real Time Polling Services (rtPS), Extended Real Time Polling Services (ertPS), Non Real Time Polling Services (nrtPS) and Best Effort (BS) services are summarized with their applications and delay constraints in Table 1.

Table 1. QoS services classes with applications and delay constraints in WiMAX.

QoS Service Class	Applications	Delay Constraints
UGS	VOIP, T1/E1	No Delay
rtPS	Streaming audio, video	High
ertPS	Voice with activity detection (VOIP)	Very High
nrtPS	FTP, Messaging, Online gaming	Moderate
BE	E-Mail, Web browsing (http)	Low

These QoS services classes are described as follows:

Unsolicited Grant Services (UGS): It supports service flows which generate fixed size data packets for real-time traffic such as Voice Over IP (VOIP), T1/E1 emulation at regular intervals. The BS provides fixed bandwidth allocations and SS cannot use contention or polling based request for bandwidth.
Real-Time Polling Services (rtPS): It supports service flows which generate variable size data packet for real-time traffic such as MPEG video transmission on a regular basis. The BS provides unicast poll request opportunities to SS at regular intervals, where SS specifies the amount of bandwidth it needs.
Extended Real-Time Polling Service (ertPS): It supports the service flows for applications based on real-time like voice over IP (VOIP) with silence suppression at regular intervals. Like in UGS here BS provides unicast grant without SS requests. But like rtPS, SS can send a request to change bandwidth size when polled by BS and bandwidths are not fixed here.
Non-Real-Time Polling Services (nrtPS): It supports the applications which generate variable size data packets like FTP traffic. A minimum amount of bandwidth is reserved for this type of connection. The bandwidth request is either by polling or contention based request (including piggybacking).
Best Effort (BE) Service: It supports the applications which do not require any guarantees on throughput or delay. Bandwidth request are made by SS based on contention or piggybacking and application supported are like E-Mail and http services List.

4. Related Works

Numbers of Author have worked on QoS parameters for different service classes in WiMAX networks. In [4], authors have analyzed quality of service parameters for FDD, but costs of equipments for FDD are high. In [6], authors have analyzed various QoS parameters for VOIP using NS-2 simulator. Their simulation was focused only for parameters of BE Service Class. In [7], simulation was done for UGS and ertPS service classes only and performance was compared. In [8], authors analyzed the QoS parameters but their performance analysis focused only on UGS service class. In [5], authors have simulated VOIP and multimedia traffic for analysis of QoS for three services classes (BE, UGS, rtPS) only. In [10] authors analyzed various QoS parameters for VOIP traffic using different codec for UGS, rtPS and BE service class only. Our work focuses on performance analysis of different service classes for video traffic. We have analyzed QoS parameters such as throughput, packet loss, average delay and average jitter based on simulation of WiMAX module in NS-2.

5. Quality of Service Parameters

QoS gauged the robustness and accuracy of the systems. It can be measured quantitatively in terms of parameters like throughput, packet loss, delay and jitter, which we considered in our analysis. These parameters are as follows:

Throughput: It is measured in terms of packets per second or per time slot. Throughput is data transfer rate or amount of packets effectively transferred in a network. In equation 1 for calculation of throughput TP, PacketSize*i* is the packet size of ith packet which reached the destination node, PacketSt*0* is the start time of first packet from the source node and Packetarr*n* is the last packet arrival time at destination node.

$$\text{TP}= \frac{\sum i \text{ PacketSize}i}{\text{Packetarr}n - \text{PacketSt0}} \qquad (1)$$

Equation 1. Calculation of throughput.

Here, in equation size of each packet was added for total data transferred. Total time is calculated as difference between time first packet started from source node and last packet reached at destination node.

Delay: Delay or Latency is the time taken by packet for reaching from source node to destination node. Sources of delay are: processing delay at source node, propagation delay, network delay and processing delay at destination node. In equation 2 average delay is calculated as packetarr*i* is time when packet reaches destination node and Packet St*i*, is the time when packet "i" leaves the source node and "n" is total number of packets.

$$\text{Average Delay} = \frac{\sum i \text{ Packetarr}i - \text{PacketSt}i}{\text{n}} \qquad (2)$$

Equation 2. Calculation of average delay.

Jitter: It is the end-to-end delay variation between two consecutive packets. Jitter reveals the variation in latency in network due to congestion, route

followed, queuing at nodes etc. Jitter is an indicator for stability and effectiveness of a network.
In equation 3 average jitter calculation step is shown. It is the average of absolute difference in time it takes for consecutive packets to reach the destination node.

$$\text{Average Jitter} = \frac{\sum i \left| (\text{Packet arr}ie - \text{Packet St}iet) - (\text{Packet arr}i - \text{Packet St}i) \right|}{\text{n}-1} \tag{3}$$

Equation 3. Calculation of average jitter.

Jitter determines the performance and QoS a network can provide.

Packet Loss: It affects the perceived quality of application. Several reasons are for packet loss such as bit error, buffer overflow, congestion in networks etc. Packet loss is the sum of all packets which are not reaching to destination node over the sum of packets which leaves the source node as shown in equation 4.

$$Packet\ Loss = \frac{\sum \text{Lost Packet Size}i}{\sum \text{Packet Size}j} \times 100 \tag{4}$$

Equation 4. Calculation of packet loss.

6. Simulation Details

Network Simulator (NS-2) [13] was used to simulate WiMAX network. WiMAX module patch release version 2.6 [14] was integrated with NS-2. Network Simulator is a simple event driven simulator which has proved useful in the dynamic nature of networks. It provides extensive support for simulator of TCP, routing and multicast protocols over wired and wireless networks. Simulator core is written in C++ and it has an OTcl script interpreter. Topology of network and traffic agents etc is specified in TCL file. Fig. 2 shows Different Simulation steps of WiMAX on NS-2.

First TCL Scripts were written to specify different scenarios and that was run on NS-2 which produces trace files. Perl scripts were used to find throughput, packet losses, delay and jitter from these trace files. The results are plotted on graph with gnu plot.

The traffic is started after some time to allow mobile node to complete registration because after simulation starts each node goes through basic registration procedure to get associated with base station.

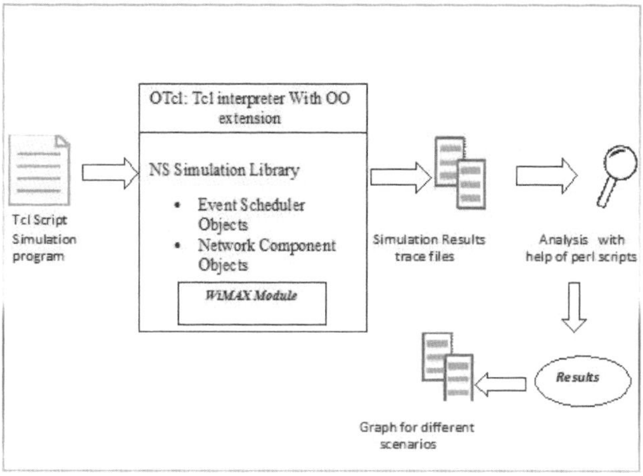

Figure 2. Different simulation steps of WiMAX on NS-2.

7. Simulation Results and Analysis

To analyze QoS in WiMAX network, a real life scenario is considered. As there is available large data pipe, today viewing videos on internet is a common application. So video streaming is analyzed for different service classes.

A number of subscriber stations generating video traffic are set up. Video streaming is a Variable Bit Rate (VBR) traffic, unlike CBR here packet size varied based on each frame type for example in case of MPEG traffic, 'I' and 'B' frames are smaller in size than a 'P' frame, Similarly for H.263 traffic there are 'I' frames, 'P' frames and 'PB' frames. As MPEG-4 is a simple video coding with enhanced compression performance and to provide a network friendly video representation. Here for simulation a high quality MPEG-4 data stream was used. Packets of Video stream is used for transportation over networks. Real-time Transport Protocol (RTP) is used as the transport protocol for this.

The VBR frame stream was obtained from the website in [12] from high quality MPEG-4 encoded Mr. Bean movie. Using this a trace file is generated.

As a traffic source, this trace file is attached to the UDP agent. The trace file contains information of time, frame number, frame type and packet size which is in binary format. For the video traffic, analysis is done using BE, nrtPS, rtPS and UGS services flows. The parameters for analysis are throughput, packet loss, delay and jitter. These are observed for each service flow as number of nodes with video traffic increases.

Figure 3 shows the variation of throughput for video traffic in case of increasing number of nodes.

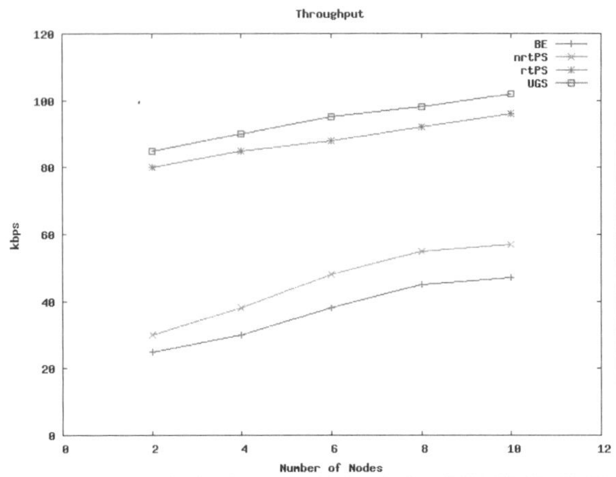

Figure 3. Throughput for video traffic for increasing number of nodes.

The rtPS and UGS service flows have higher throughput than BE and nrtPS service flows. The throughput for BE service flow is lowest. In comparison to UGS service flow with other three service flows, much higher throughput was observed in case of UGS than BE and nrtPS service flow and it was nearly equal to the rtPS service flow for same number of nodes. This is due to the fact that for UGS grants are fixed on periodic basis in each round of scheduling and there is no packet loss. For rtPS, BS polls each SS for grant of bandwidths in each frame, due to number of control packets for this some packets get loosed. In case of nrtPS there is contention based request so there is no guarantees of bandwidth allocation in each frame. For BE service bandwidth is allotted only when it is remaining after serving three types of service flows.

Figure 4 depicts packet loss for four service flows. There is constant packet loss in case of UGS service flow. The reason is that for UGS bandwidth grant is fixed in each round on regular basis, thus packets are sent in each round. Packet loss increases gradually as the number of nodes increases in case of BE and nrtPS service flows. Value of packet loss is less if number of nodes is less for BE and nrtPS service flow but it is higher for increasing number of nodes. As in case of BE and nrtPS service flows when number of nodes are less then much resources are available for these services but when node increases there are chances for collisions of packets and thus more packet loss. For rtPS service flow slight variation is in packet loss as number of nodes increases and it is lowest for all flows. The reason is that each SS is polled by BS to take request on regular basis and thus packets are sent without collisions.

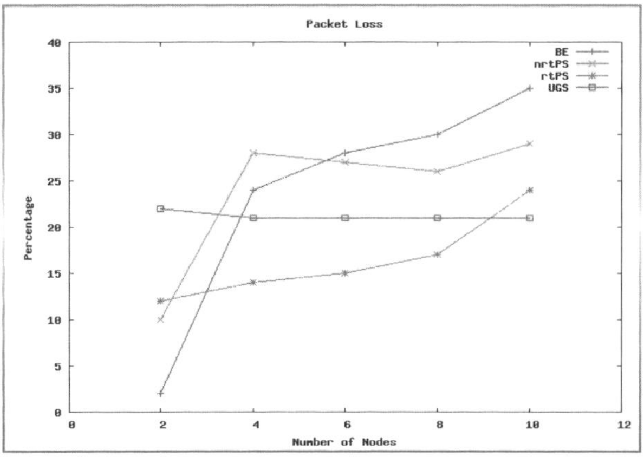

Figure 4. Packet loss for video traffic for four types of flows.

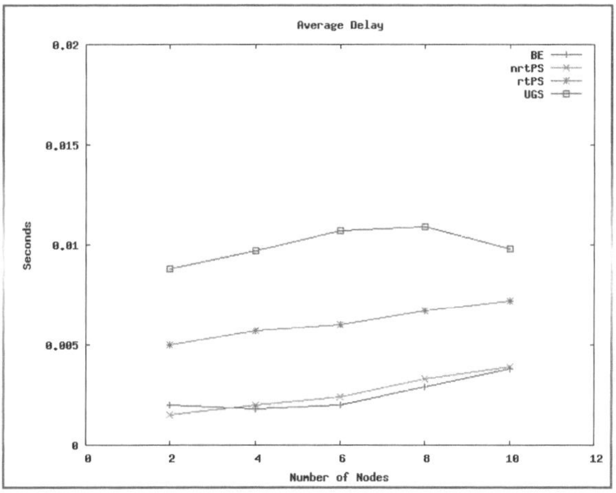

Figure 5. Average delay for four service flow.

Figure 5 shows the average delay for four service flows as number of nodes increases. In comparisons to BE, nrtPS and rtPS flows UGS has higher delay but delay variation is less for this. This is due to the fact that in case of UGS there is fixed grants on a regular interval. In case of BE service flow bandwidth is allotted only when there is remaining resources after serving all service classes. As number of nodes increases, for BE service flow average delay increases rapidly in comparison with rtPS service flow due to collision of packets. For rtPS service flow although value of average delay is higher than nrtPS service flow but variation in delay is less than nrtPS service flow. The reason is that for rtPS bandwidths are allotted on regular basis and for nrtPS it is allotted on contention basis in which there is collision of bandwidth request packets. Similar to BE service flow the value of delay increases in nrtPS service flow as number of nodes increases.

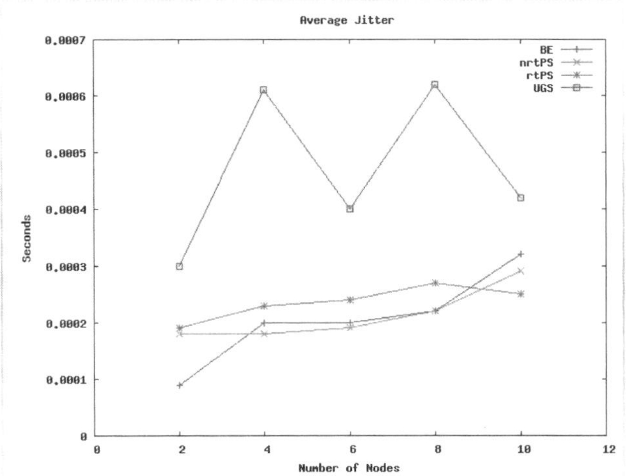

Figure 6. Changes in average jitter for all four service flows.

Figure 6 shows the changes in average jitter for BE, nrtPS, rtPS and UGS service flows. From above it seems that there is no any relation with number of nodes for UGS service flow and jitter value, because there is fixed allocation in this case and video traffic is variable bit rate service thus delay variation in packets are not fixed. But rtPS service flow has lowest jitter among all service flows for maximum number of nodes and it is nearly constant, as for rtPS service flow bandwidth request is sent by each SS based on packets in queue on regular basis. For small number of nodes, BE and nrtPS service flows have lower jitter, and it increases rapidly as number nodes increases. That means at the time when bandwidth is available, video traffic for BE and nrtPS service flows introduced least jitter, but as the nodes increases, the network resources are divided between all nodes, and results into increased jitter. For rtPS service flows there is small changes in average jitter with the increasing nodes and remains nearly constant.

8. Conclusion

The performance analysis for BE, nrtPS, rtPS, and UGS service flows with different QoS parameters like throughput, packet loss, delay and jitter was analyzed and compared for video traffic which passed through WiMAX network with increasing number of nodes. In case of packet loss and average jitter, rtPS service flow is better than all three service flows. There is slight changes in jitter for rtPS service flow and it has least value when number of nodes are being increased. For average delay value is high for rtPS than BE and nrtPS service flows, but there is slight variation in value as number of nodes increases and in case of BE and nrtPS it increases rapidly. The delay is maximum for UGS service as it is VBR type of traffic. rtPS service flows show least packet loss as number of nodes increases and throughput is much higher than BE and nrtPS service flows and is equivalent to UGS flow for video traffic. In case of UGS service flow it does not utilize network resource as grant of bandwidth is on periodic basis and traffic is Variable Bit Rate (VBR) for video traffic. In case of rtPS service flow bandwidth are periodically requested based on queue size and are not fixed as in case of UGS, where bandwidth may be or not be used. It shows that rtPS service flow is best choice for streaming video traffic.

References

1. IEEE standard 802.16, IEEE standard for Local and Metropolitan Area Networks, Part 16: *Air interface for fixed Broadband Wireless Access Systems*, (2004).
2. IEEE standard 802.16e, IEEE Standard for Local and Metropolitan Area Networks, Part 16: *Air interface for fixed and Mobile Broadband Wireless Access Systems*, 2005.
3. C. Ciconetti, L. Lenzini and C. Ekluud, Quality of Service Support in IEEE 802.16 Network, *IEEE Networks*, **20**, (2006).
4. C. Ciconetti, A. Erta, L. Lenzini and E. Mingozzi, Performance Evaluation of the IEEE 802.16 MAC for QoS Support, *IEEE Transactions on Mobile Computing*, **6** (2007).
5. R. A. Talwalkar and M. Ilas, Analysis of Quality of service (QoS) in WiMAX networks, *IEEE International Conference on Networking* (2008).
6. D. Joshi and S. Jangale, Analysis of VOIP traffic on WiMAX using NS2 Simulator, *International Journal of Advanced Research in Computer Science and Electronics Engineering*, **1** (2012).
7. I. Adhicandra, Measuring data and VOIP traffic in WiMAX networks, *Arxiv Preprint arXiv*; *1004.4583* (2010).
8. M. Vikram and N. Gupta, Performance Analysis of QoS Parameters for WiMAX Networks, *International Journal of Engineering and Innovative Technology*, **1** (2012).

9. Kalpalatha S. and Srikanth Subramarian, A Performance Study of Uplink Scheduling in WiMAX Network, *ICRTIT* (2012).
10. Tarik Anouri and A. Haqiq, Performance Analysis of VOIP Traffic in WiMAX using various service Classes, *International Journal of Computer Application(0975-8887)*, **52** (2012).
11. Jokhu Lal and Neeraj Tyagi, Proposal of a Combined Downlink and Uplink Scheduling Architecture for Mobile WiMAX, *International Conference on Recent Trends in Computer Science and Engineering,* (Patna, 2014).
12. MPEG-4 and H.263 Video Traces for Network Performance Evaluation, (Master's Thesis) Athamnoh K. http://trace.eas.asu.edu/TRACE/pics/FrameTrace/mp4/Verbose_bean.dat
13. The Network Simulator ns-2 http://www.isi.edu/nsnam/ns/.
14. NIST, Seamless and Secure mobility, NS-2 NIST WiMAX Module, http://www.antd.nist.gov/seamlessandsecure (2009).
15. C. So-In, Raj Jain and A. K. Tamini, Scheduling in IEEE 802.16e Mobile WiMAX Networks: Key Issues and a Survey, *IEEE Journal on Selected areas in communications*, 27 (2009).

A Review of Application of Artificial Neural Network in Ground Water Modeling

Neeta Kumari[1], Gopal Pathak[2]

Civil and Environmental Engineering, BIT, Mesra
Ranchi, Jharkhand. PIN – 85215, India.
E-mail: [1]neetak@bitmesra.ac.in, [2]gopalpathak@bitmesra.ac.in
www.bitmesra.ac.in.

Om Prakash

Integrated M.Sc., Center for Environmental Sciences, CUJ,
Ranchi, Jharkhand., India
E-mail: om.op1525@gmail.com,
www.cuj.ac.in

Artificial Neural Networks (ANNs) are modelling tools having the ability to adapt to and learn complex topologies of inter-correlated multidimensional data. ANNs are inspired by biological neuron processing, have been widely used in different field of science and technology incorporating time series forecasting, pattern recognition and process control. ANN has been successfully used for forecasting of groundwater table and quality parameters like nitrate, total dissolved solids. In case of groundwater quality prediction, availability of good quality data of better precision is required. ANNs are classified as Feed-forward neural networks (FFNNs), Recurrent neural networks (RNNs), Elman Backpropagation Neural Networks, Input Delay feed-forward Backpropagation Neural Network, Hopfield Network. The artificial neural networks (ANNs) ability to extract significant information provides valuable framework for the representation of relationships present in the structure of the data. The evaluation of the output error after the retraining of an ANN shows us that this procedure can substantially improve the achieved results. Through this review work it is observed that in most hydrological modeling cases FFNN and LM algorithm performed well till today's published research work.

Keywords: Template; Proceedings; World Scientific Publishing; ANN; Groundwater; FFNN; LM algorithm; Groundwater Modelling.

1. Introduction

The groundwater models are extensive data driven, needs sophisticated programs for calibration using rigorous optimization techniques and a better understanding of the underlying physical processes (Tahmasebi and Zomorrodian, 2004). When data insufficient, empirical models are a better option as it can give useful results without calibration (Krishna *et al.*, 2008). In the last decade, artificial neural networks being capable of analyzing long series and large-scale data and has become increasingly popular in hydrology and water resources people (Maier & Dandy, 1999) and practicing engineers. within water resources related course ANNs have been used as data driven models in rainfall-runoff prediction (Hsu *et al.*, 1997), groundwater simulation (Amir Jalalkamali & Navid Jalalkamali, 2011), variation in groundwater quality (Fi-John Chang *et al.*, 2010), forecasting groundwater contamination and water demand forecasting, water table depth fluctuation (Paulin Coulibaly *et al.*, 2001), prediction of nitrate distribution in groundwater (Almasri & Kaluarachchi, 2004) and other applications. ANNs are very effective in modeling virtually any nonlinear function to an arbitrary degree of accuracy (Amir Jalalkamali & Navid Jalalkamali, 2011).

Hydrologic models are of three categories: Physical Distributed Models, Lumped Conceptual Models, and Black Box Models. Black box models are important in operational hydrology. Example is Neural Network Models. It requires less data and are easy to construct, that is why they are so much popular in the Water Resources field (Mohammad Ali Shamim *et al.*, 2004). As the water is a dynamic system, the conventional data processing techniques do not give good result (Wu *et al.*, 2000; Xiang *et al.*, 2006) if they are used for modeling because the factors are non-linearly related to each other that complicate the problem. Regression analysis is also one of the data processing method but is not as efficient in understanding the relation between variables as ANN is (P. Sirisha *et al.*, 2008). ANN is an error tolerant, self adaptable, self-organizing and widely adopted technology for forecasting in these systems (Niu *et al.*, 2006).

2. Artificial Neural Network

Artificial neural networks have been known to be a very important tool for non-parametric modeling techniques where output is non linearly related to the input (Mayilvaganan & Naidu, 2011). Artificial neural network (ANN) has been designed by copying the human brain artificially. The concept of Neural network was introduced through the work of McCulloch and Pitts in 1943. An ANN is an architecture consisting of a large number of neurons organized in different layers and the neurons of one layer are connected to those of another layer by means of weights. It can be trained to perform a particular task by making proper adjustment of its architecture, connecting weights and other parameters. The ANNs are inspired by biological neuron processing, have been widely used in different field of science and technology incorporating time series forecasting, pattern recognition and process control (P. C. Nayak *et al.*, 2006). Basic processing element of an ANN is neuron which performs its work by two processes: i) internal activation; and ii) transfer function (Singh & Datta, 2006).

2.1. *Advantages*

- No physics based algorithm is required for development of model so the modeling approach is flexible and faster.
- Can handle non-linear relationships easily and properly.
- The expertise and user experiences may be incorporated easily into the model structure.

Thus the essential features of Artificial neural networks are (1) They learn by example (2) They constitute a distributed, associative memory (3) They are fault tolerant (4) they are capable of pattern recognition.

2.2. *Groundwater quality modeling using artificial neural network (ANN)*

Table 1. Groundwater modeling using soft computing techniques mostly ANN.

Modelling Parameter	Technique used	Place/Data collection time period	References
Nitrate	ANN & GA	Iran (2008-2010)	Sayyed Ali Moasheri *et al.*, 2012.
Nitrate	ANFIS	Ishfahan Province, Iran (2000-May 2001)	Mousavi, S. F. & Amiri, Md. Javad; 2012
Groundwater table & Nitrate	ANN & GA	Tehran	M. Karamouz *et al.*, 2005
Nitrate	ANN & GA	Dry region of Iran	Saeid Eslamian & Niloufar Lavaei, 2009
Groundwater level	ANN & GA	Kerman plain Iran	Amir Jalalkamali & Navid Jalalkamali, 2011
Groundwater level fluctuation	ANN & GA	Konan groundwater basin of Japan	Madan K Jha & Sasmita Sahoo, 2014
Groundwater level simulation	ANN	Edward's aquifer in Texas, USA	Ioannis C *et al.*, 2011
Groundwater level forecasting	ANN	River Island of eastern India	Sheelabhadra Mohanty *et al.*, 2010
Groundwater level	ANN	Fractured Media	S. Lallahem *et al.*, 2005
Groundwater level forecasting	ANN	Shallow aquifer	Purna C. Nayak *et al.*, 2006
Aquifer water level elevation	ANN	Aquifer	Emery A Coppola *et al.*, 2003

Table 1. (*Continued*)

Modelling Parameter	Technique used	Place/Data collection time period	References
Groundwater level	ANN	Minqin Oasis,lower reach of Shiyang river basin in NW China	Shaoyuan Feng *et al.*, 2008
Parameter estimation in Groundwater	ANN		Luis A. Garcia & Abdalla Shigidi, 2006
Water table elevation fluctuation	Fuzzy logic and ANN	Budaun District, UP	Dinesh Bisht *et al.*, 2013
Aquifer parameter determination	ANN		Khaled S. Balkhair, 2002
Aquifer parameter determination	PCA & ANN	Non-leaky confined aquifer	N. Samani *et al.*, 2007
Intelligent characterization and diagnosis of the Groundwater quality	ANN	Urban fractured rock aquifer	Yoon-Seok Hong & Michael R. Rosen, 2001
Groundwater level in hard rock region	ANN	Hard rock terrain	Pallavi banerjee *et al.*, 2009
Groundwater quality	ANN	In the area of Blackfoot disease in Taiwan	Wang *et al.*, 2011
Groundwater level	ANN	Urban Coastal aquifer	B. Krishna *et al.*, 2008
Water table depth fluctuation	ANN		Paulin Coulibaly *et al.*, 2001
Groundwater level forecasting	ANN	Messara Valley in Crete (Greece)	Ioannis N. Daliakopoulos *et al.*, 2005
Groundwater table fluctuation	ANN & GA	Greater Bangkok Area	Uruya weesakul *et al.*, 2010
Total Dissolved solid variation in Groundwater	ANN	Tehran Plain, Iran	P. Abbasi Maedeh *et al.*, 2013

3. Types of ANN

A neural network can be of two types: Static and Dynamic neural networks. In Static neural network the output is not compared with the targeted value to determine the error. Therefore the error in the operation is neither calculated nor propagated back for updation of the network. On the contrary in dynamic the error is determined and is fed back to the network to modify the architecture and update the weights. Thus here is a chance of error compensation and thus it gives better prediction result then the static neural network.

ANNs are classified depending on the structure of the connections among the layers:

(1) Feed-Forward Neural Networks (FFNNs).

(2) Recurrent Neural Networks (RNNs).

(3) Elman Backpropagation Neural Networks.
(4) Input Delay Feed-Forward Backpropagation Neural Network.
(5) Hopfield Network.

4. Application Of ANN in Groundwater Modelling

The mechanism through which the network is modified by updation of weights, biases, and other parameter is known as training. During the training the parameters of the network are optimized and as a result of which it undergoes an internal process of curve-fitting. It is then said that the network has passed through the learning phase. Training algorithms are Bayesian regularization, Conjugate gradient backpropagation, Gradient descent backpropagation, Gradient descent with momentum backpropagation, Levenberg-Marquardt backpropagation, One step secant backpropagation, Resilient backpropagation, scaled conjugate gradient backpropagation. Learning rate is the rate at which network learns. If the rate is slow it means it takes more time in training the system and faster rate may cause more error in training. The value ranges between 0 to 1. If the learning rate is low then training takes longer time but result is stable. There are certain learning laws on the basis of which network learns like hebb's rule, hopfield Law, the delta rule, kohonen's Law. After training of the network, it is validated by the use of independent set of data using chosen criteria. This set should not be used as training set at any stage. The error obtained during validation is much different then during training then it is possible that the data is not representative of the same population or the model is overfitted (Masters, 1993).

In an artificial neuron n number of inputs remain connected to a neuron through weights. The function of summing junction of an artificial neuron is to collect weighted inputs and sum them up. The activation function also known as transfer function. The output of summing junction may sometimes become equal to zero and to prevent such a situation a bias of fixed value is added to it. In ANN, the output of a neuron, largely depends on its transfer function. Transfer functions are Competitive transfer function, Hard-limit transfer function, symmetric hard-limit transfer function, log-sigmoid transfer function, inverse transfer function, positive linear transfer function, linear transfer function, radial basis transfer function, saturating linear transfer function, symmetric saturating linear transfer function, softmax transfer function, hyperbolic tangent sigmoid transfer function, triangular basis transfer function. Most research paper have used Levenberg-Marquardt backpropagation with log-sigmoid or tan- sigmoid or pure linear transfer function (Jha & Sahoo, 2014; Mayilvaganan & Naidu,

2011). The sigmoid functions adopted for the last layer may result in the limiting of output of the entire network in small range but if purelin a linear function is applied then the output can be any arbitrary value so the output neurons transfer function is preferable purelin (ZHAO, Ying *et al.*, 2007).

ANNs with large number of weights such as having too many degrees of freedom in relation to the amount of data available, can cause over-fitting that will generate poor predictive performance, because it would usually amplify minor fluctuations in the data. To avoid over-fitting it is necessary to use additional techniques like The leave-one-out (LOO) cross validation; Modified performance function (MPF); Principal component analysis (PCA) (Fi-John Chang *et al.*, 2010).

For the evaluation of model predictability six goodness of fit statistical measures or statistical indicators have been suggested like RMSE, R^2, bias, Coefficient of Variation (CV), Nash-Sutcliffe efficiency (NSE), Index of agreement (IOA) (Jha & Sahoo, 2014) among which many literatures have selected two or three (Mayilvaganan & Naidu, 2011; Amir Jalalkamali & Navid Jalalkamali, 2011) for the evaluation of the model predictability. Finally sensitivity analysis is done for the assessment of uncertainty.

Sensitivity analysis = Ratio = RMSE without one variable / RMSE with ANN complete model (Zalin Huo *et al.*, 2010).

5. Conclusion

Many past successful research works have proved that ANN is a very promising tool for the modeling of groundwater behavior. But the process is extensive data driven and needs suitable depth into the mechanism of quality fluctuations with time and predicting it by means of the most decisive pattern so as to properly delve into the situation of groundwater table and the quantity of accessible water.

References

1. Almasri, N. Mohammad & Kaluarachchi, Jagath J., Multi-criteria decision analysis for the optimal management of nitrate contamination of aquifers, *Journal of environmental management*, 20 (4) 1 (2004).
2. Balkhair, Khaled S., Aquifer parameters determination for large diameter wells using neural network approach, *Journal of Hydrology* 265(1), 118 (2002).

3. Banerjee, P., Prasad, R., Singh, V., Forecasting of groundwater level in hard rock region using artificial neural network, *Environmental Geology*, 58(6), 1239 (2009).
4. Bisht, D., Jain, S. & Mohan Raju, M., Prediction of water table elevation fluctuation through fuzzy logic and artificial neural networks, *Int. Jour. Adv. Sc. & Tech.*, 51, 107 (2013).
5. Coppola E., Anthony J. R., Poulton M., Szidarovszky F., Vincent W., A neural network model for predicting aquifer water levels elevations, *Groundwater*, 43(2) 231 (2005).
6. Coulibaly, P. Anctil, F., Aravena, R., & Bobee, Bernard, Artificial neural network modeling of water table depth fluctuations, *Water resources research*, 37 (4), 885 (2001).
7. Daliakopoulos I. N., Coulibalya P., Tsanis I. K., Groundwater level forecasting using artificial neural networks, *J. Hydrol.*, 309, 229 (2001).
8. Eslamian, S. & Lavaei, N., Modelling nitrate pollution of groundwater using artificial neural network and genetic algorithm in an arid zone, *International journal of water*, 5(2), 194 (2009).
9. Fi-john Chang, Li-shan Kao, Yi-Ming Kuo, Chen-Wuing Liu, artificial neural networks for estimating regional arsenic concentrations in a blackfoot disease area in Taiwan, *Journal of Hydrology*, 388 (2010).
10. Garcia, Luis A. & Shigidi, Abdalla, Using neural networks for parameter estimation in groundwater, *Journal of Hydrology*, 318, 215 (2006).
11. Hong, Yoon-Seok & Rosen, M. R., Interlligent characterization and diagnosis of the groundwater quality in an urban fractured – rock aquifer using an artificial neural network *Urban Water*, 3(3), 193–204 (2001).
12. Hsu, K., X. Gao, S. Sorooshian and H. V. Gupta, Precipitation estimation from remotely sensed information using artificial neural networks, *Journal of Applied Meteorology*, **36**, 1176 (1997).
13. Ioannis C. Trichakis, Ioannis K. Nikolos, G. P. Karatzas. Artificial Neural Network (ANN) Based Modeling for Karstic Groundwater Level Simulation, *Water Resources Management*, 25 (4), 1143 (2011).
14. Jalalkamali, A. & Jalalkamali, N., Groundwater modeling using hybrid of artificial neural network with genetic algorithm, *African journal of agricultural research*, **6** 5775 (2011).
15. Jha, Madan K. & Sahoo, Sasmita, Efficacy of neural network and genetic algorithm techniques in simulating spatio-temporal fluctuations of groundwater, *Hydrological processes*, DOI: 10.1002/hyp.10166. Wileyonlinelibrary.com (2014).

16. Karamouz, M., Mohd. Rezapour Tabari, M., Kerachian, R. & Zahraie, B. conjunctive use of surface and groundwater resources with emphasis on water quality, *EWRI (ASCE,2005).*
17. Krishna, B., Satyaji, Y. R., Vijaya, T., Modelling groundwater levels in an urban coastal aquifer using artificial neural networks, *Hydrol. Process*, 22, 1180 (2008).
18. Lallahem, S., J. Maniaa, A. Hani, On the use of neural networks to evaluate ground water levels in fractured media, *Journal of Hydrology*, 307(1-4) 92 (2005).
19. Maier, R. Holger & Dandy, Greame, C., Neural networks for the prediction and forecasting of water resources variables: a review of modeling issues and applications", *Environmental modeling and software,* 15 101 (1999).
20. Masters, T., Practical Neural Network Recipes in C 1 1. Academic Press, San Diego, CA (1993).
21. Mayilvaganan, M. Kavitha & Naidu, K. B. Application of artificial neural network for the prediction of groundwater level in hard rock region, *CCSEIT, CCIS* 204, 673 (2011).
22. McCulloch, W. S., Pitts, W., A logical calculus of the ideas imminent in nervous activity, *Bulletin and Mathematical Biophysic*s, 5, 115 (1943).
23. Moasheri, S. A., Tabatabaie, S. M., Razaghi, P., Sarani, N. & Abadi, S. H. E. M., Estimating the groundwater nitrate by using artificial neural network and optimizing it by genetic algorithm, *International conference on transport , environment and civil engineering (ICTECE' 2012) (Kuala Lumpur (Malaysia*), 2012).
24. Mousavi, Sayed Farhad & Amiri, Mohammad Javad, Modelling Nitrate concentration of Groundwater using adaptive neural- based fuzzy inference system, Soil & Water Research, 7(2), 73 (2012).
25. Nayak, P. C., Satyaji Rao, Y. R. & Sudheer, K. P., Groundwater level forecasting in a shallow aquifer using artificial neural network approach, *Water resources management*, 20, 77.
26. Nayak, P. C., Satyaji Rao, Y. R. & Sudheer, K. P., Groundwater level forecasting in a shallow aquifer using artificial neural network approach, *Water resources management*, 20, 77 (2006).
27. Niu, Z. G., Zhang, H. W., Liu, H. B., Application of neural network to prediction of coastal water quality, *Journal of Tianjin Polytechnic University*, **25**(2) 89 (2006).
28. Rajasekaran, S. & Vijayalakshami Pai, G. A., Neural networks, Fuzzy logic and Genetic algorithms, Prentice hall of India Pvt. Ltd. 2 1 (2004).

29. Samani, N., Gohari-Moghadam, M., Safavi, A. A., A simple neural network model for the determination of aquifer parameters, *Journal of Hydrology* 340, 1 (2007).
30. Sarle, Warren S., Neural Networks and Statistical Methods, *Proceedings of the Nineteenth Annual SAS Users Group International Conference* (2007).
31. Shamim, Mohammad Ali., Ghumman, A. R. & Ghani, U. Forecasting groundwater contamination using artificial neural networks, *International conferences on Water Resources & Arid Environment.*, 1 (2004).
32. Shaoyuan Feng, Shaozhong Kang, Zailin Huo, Shaojun Chen, and Xiaomin Mao. "Neural Networks to Simulate Regional Ground Water Levels Affected by Human Activities. *Groundwater*", 46 (1), 80 (2008).
33. Sheelabhadra Mohanty, Madan K. Jha, Ashwani Kumar, K. P. Sudheer. Artificial Neural Network Modeling for Groundwater Level Forecasting in a River Island of Eastern India, *Water Resources Management*, 24 (9), 1845 (2010).
34. Singh, R. M. & Datta, B. Artificial neural network modeling for identification of unknown pollutant sources in groundwater with partially missing concentration observation data *Water resource management*, 21, 557 (2006).
35. Sirisha, P., Sravanti, K. N., Ramakrishna, V., Application of artificial neural networks for water quality prediction, *International Journal of Syatems and Technology*, **1**, 115 (2008).
36. Tahmasebi, A. R., Zomorrodian, S. M. A., Estimation of soil liquefaction potential usingartificial neural network, *Second national Student Conference on Water and Soil Resources* (2004).
37. Wang, Sheng-Wei, Kuo, Yi-Ming, Kao, Yu-Hsuan, Jang, Cheng-Shin, Maji, Sanjoy Kumar, Chang, Fi-John, Liu, Chen-Wuing. Influence of hydrological and hydrogeochemical parameters on arsenic variation in shallow groundwater of southwestern Taiwan, *Journal of Hydrology*, 408 (3-4), 286 (2011).
38. Weesakula, U., Watanabe, K. and Sukasem, N., Application of Soft Computing Techniques for Analysis of Groundwater Table Fluctuation in Bangkok Area and Its Vicinity, *International Transaction Journal of Engineering, Management, & Applied Sciences & Technologies*, 1(1) 1906 (2010).
39. Wu, H. J., Lin, Z. Y., Guo, S. L., The application of artificial neural networks in the resources and environment. *Resources and Environment in the Yangtze Basin*, **9**(2), (in Chinese) 237 (2000).

40. Zalin Huo, Shaoyuan Feng, Shaozhong Kang, Xiaomin Mao and Fengxin Wang, Numerically modeling groundwater in an arid area with ANN – generated dynamic boundary conditions, Hydrological processes, DOI: 10.1002/hyp.7858, Wiley online library, 1 (2010).
41. Zhang, Y. H., Mastering MATLAB5. Tsinghua University Press, Beijing, 1 (in Chinese) (1999).
42. ZHAO, Ying, NAN Jun, CUI Fu-yi, GUO Liang. Water quality forecast through application of BP network at Yuqiao reservoir, *Journal of Zhejiang University SCIENCE A*, 8(9), 1482, (2007).

Density Based Outlier Detection (DBOD) in Data Mining: A Novel Approach

Govind Kumar Jha
Research Scholar
gvnd.jha@gmail.com

Neeraj Kumar
Head, Dept of Computer Science
Nalanda College of Engg Govt. of Bihar
javaneeraj@ gmail.com

Dr. Prabhat Ranjan
Asst. Professor, Central University of South Bihar
Patna, Bihar
prabhatranjan@cub.ac.in

K. G. Sharma
Assistant Professor
GLA University Mathura
hollyhoc@gmail.com

Outlier detection is a great area of interest in the field of data mining. It has been observed that there exist several application domains in which direct mapping is possible between outliers in data and real world anomalies. Outlier detection is an important research topic in various application domains and knowledge disciplines. This paper presents a new approach DBOD to overcome the disadvantage of well-known outlier detection algorithm LOF. The computation of LOF score is a tedious task because a large number of K nearest neighbor queries are handled. Our proposed algorithm is based upon a new definition of densities of data points known as point density. Point density is k-nearest neighbor divided by k-distance. We compared our results with LOF and found that our proposed DBOD detects outliers more accurately and effectively.

Compared the result on real world data set taken from UCI repository, kdd cup-99 and Wisconsin breast cancer dataset on simulator, and it has been found that our proposed DBOD detects outliers more accurately and effectively in comparison with LOF.

Keywords: outlier detection, point density, DBOD.

1. Introduction

In the field of data analysis a large number of observations or measurements are being sampled. Some measurement may be incorrectly observed, recorded or sampled. These incorrect observations are termed as outliers. The first steps of coherent analysis are the detection of these Outliers. Although outliers are often considered as an error or noise but they may carry important information. Because noise for one person may be signal to others [2]. Outliers may adversely affect the analysis and may lead to incorrect results. It is therefore important to identify them a priory before applying in modeling or data analysis (Williams *et al.*, 2002; Liu *et al.*, 2004).

Most commonly used definitions of outlier was given by Hawkins (Hawkins, 1980). It states "outlier is an observation that deviates so much from other observations as to arouse suspicion that it was generated by a different mechanism". According to Barnet and Lewis (Barnett and Lewis, 1994) "an outlying observation, or outlier, is one that appears to deviate markedly from other members of the sample in which it occurs", similarly, Johnson (Johnson, 1992) defines outlier is an observation in a data set which appears to be inconsistent with the remainder of the data sets. And Ramaswami's definition is "An outlier in a set of data is an observation or a point that is considerably dissimilar or inconsistent with the remainder of the data" (Ramasmawy *et al.*, 2000) [9]. All these definitions emphasize the fact that Outlier is different than other data in dataset they are exceptions if we will apply in a Model. As we mentioned above Exception can be error or information both. In both cases identification is necessary.

Outlier detection has been studied extensively in the past decades. It has been applied for numerous applications such as credit card fraud detection, clinical trials, voting irregularity analysis, data cleansing, network intrusion detection, financial applications, severe weather prediction, geographic information systems, athlete performance analysis, marketing and other data-mining tasks [1][2]. Basically it can be classified into two categories: *classic outlier approach* and *spatial outlier approach*. The classic outlier approach [5] analyzes transaction dataset, Spatial deals with spatial information. Classical approach can be grouped into statistical-based approach, distance-based approach [6], deviation-based approach, density-based approach. Density based approach is found to be proven approach. LOF methods is one of the most efficient and used algorithm.

In this work we proposed an efficient outlier detection algorithm based on the novel definition of density. This algorithm detects outlier more effectively while keeping time complexity lower than LOF.

This paper is organized as follows Section 2 discusses the related work on density based outlier detection method specially LOF. In Section 3 we introduce our proposed algorithm DBOD in details and in Section 4, experimental results has been discussed and finally Section 5 conclude the text.

2. Related Work

The traditional approaches to outlier detection can be classified as either distribution based, depth-based, clustering based and new approaches are distance-based and density-based.

2.1. *LOF (Local Outlier Factor) algorithm*

The concept of Local Outlier factor (LOF) was given by Markus M. Breunig *et al.* [3] in 2000 is used frequently. The Local Outlier Factor (LOF) is a very powerful density based outlier detection technique in data mining. An outlier as defined by the LOF quantifies how an object is outlying with respect to other objects in a given database. Necessary definitions to explain LOF are given below. For further details [1] can be referred. In all the definitions of LOF, k is a user supplied natural number and MinPts is minimum points. MinPts and k can be used interchangeably. p and o are object in database.

LOF algorithm can be illustrated within following three steps :In a first step, the k-nearest-neighbors are found according to the euclidean distance, which is the most intensive computational step. In general, all data instances have to be compared with all other remaining ones and computational complexity of this step is $O(N^2d)$, whereas d is the dimension of the data.

Definition 2.1: k-distance (p): This is the distance between p and its kth nearest neighbor.

Definition 2.2: $N_{k\text{-distance}(p)}(p)$: Given the k-distance(p), the $N_{k\text{-distance}(p)}(p)$ contains every object whose distance from p is not greater than the k-distance.

In second step, the Local Reachability Density (LRD) is computed for all data points p based on the set of k neighbors Nmin(p), such that the LRD can be defined as the inverse of the average reachability distance of the nearest neighbors. In this context, the reachability distance reach *distmin* is defined as the k-distance or the Euclidean distance of the two objects, depending on which value is larger [3].

Definition 2.3: reach-distk(p, o): This is the maximum distance out of k-distance of o and real distance between p and o. In LOF formulation MinPts is an indication of mass and the values $\text{reach-dist}_{\text{MinPts}}(p,o)$ for $o \in N_{\text{MinPts}}(p)$, is an indication of volume to determine the density in the neighborhood of p (lrd, as given in the next definition, is an indication of density).

Definition 2.4: $\text{lrd}_{\text{Minpts}}(p)$: The local reachability density of p can be calculated dividing one by the average reachability distance based on the MinPts-distance neighborhood of p.

In a third step, the LOF score can be computed using the LRDs from the k nearest-neighbors, such that From the above equation we see that the LOF is a ratio of the LRDs.

Definition 2.5: (local outlier factor of an object p).

The LOF of p is defined as

$$LOF_{MinPts}(p) = \frac{\sum_{o \in N_{\text{MinPts}}(p)} \frac{lrd_{Minpts}(o)}{lrd_{MinPts}(p)}}{|N_{MinPts}(p)|}$$

The local outlier factor of object p captures the degree to which we call p an outlier.

A large LOF will be assigned if the density of all neighbors is higher than the density of the data instance p itself, indicating a possible outlier. On the other hand, if the densities of the neighbors are approximately as high as of the instance itself, the resulting ratio will be close to one. A rule of thumb states, that outliers have larger LOF scores than a threshold in the range of 1.2 to 2.0, depending on the data. However, the weakness of the LOF algorithm is that it cannot detect the outliers existing in a subspace of full feature space.

3. Proposed Work Density Based Outlier Detection (DBOD)

From the above discussion, it is clear that in LOF algorithm local reachability density (lrd) is an indication of density of the region surrounding the object point while MinPts-dist is for MDV because small MinPts-dist maps to dense region (high density) and large MinPts-dist maps to sparse region (low density) in MDV algorithm, calculating MinPts-dist variance is simple and efficient because more no of outlier were found.

In our proposed DBOD algorithm we are moving towards more simplicity with still better result as we will see it in experiments. In our proposed Algorithm instead of calculating lrd or MinPts-dist variance, simply point density has been calculated. If this **point density** for a given data object is greater than some user defined threshold than this data object is outlier.

Now we formalize DBOD. Let p and q be data objects of a given dataset D. First we calculate k-distance and k-distance neighborhood of an object p denoted by k-distance(p) and $N_{k\text{-distance}(p)}(p)$ respectively. And then we calculate point density as defined below:

Definition 7: (Point Density of an object p): Let p be an object of a dataset D. Point density of an object p denoted as PD(p) is defined as:

$$PD(p) = \frac{N_{k\text{-distance}(p)}(p)}{k\text{-distance}(p)} \quad \text{Where } p \in D$$

Our next definition defines Outlier for our algorithm

Definition 3.1: Outlier or Noise: Let D be given data set and p be an object belonging to D and δ be some user defined threshold value. p will be outlier if PD(p) <δ.

Algorithm: The pseudo code of the proposed technique DBOD to find point density.
Input: Data set of size n, natural no k and threshold value δ.
Output: Density for each data points and outliers object.

Beauty of this method is that the density of a point is not fluctuating too much by changing the parameters. In contrast, the local reach ability density [3] that was used in LOF is affected by the parameter MinPts. This property allows us to pre-compute the densities of all data objects.

Pseudo code (DBOD)

```
Begin
//Initialize no of nearest neighbors k_n, k←0, PD={ }, Outliers={ },
    For i=1 to n
      For j=1 to n
D[i,j]= Distance between object i and j

      End For
    End For
          For p = 1 to n
          k-distance(p) = distance of k^th nearest data object from current p^th
          object.
          N_k-distance(p)(p) =no of data objects with distances less than or equal to
```

```
            k-distance(p)
```

$$PD(p) = \frac{N_{k\text{-distance}(p)}(p)}{k\text{ - distance }(p)}$$

```
If PD[p] <δ
Then p is outlier.
Else
P is normal data.
End if
            end_for
end
```

3.1.1. *Complexity analysis*

To explain time complexity of the DBOD algorithm, Let D be a data set of n d-dimensional data points and k is any positive integer. Our DBOD algorithm is designed according to incremental approach. Since in DBOD algorithm has three components described above and time complexity is based on the frequency of steps. First step of DBOD calculating the distances between data objects and the time complexity is $O(n^2)$. Second step finding the k-distance and k-distance neighborhood of each data objects and the time complexity is $O(nk^2)$. Third step of the DBOD is computing the point densities of each data object and the time complexity of the step is O(n). Hence the time complexity of DBOD algorithm is $O(n^2+nk^2)$ and this is not more than LOF.

4. Simulations and Results

We evaluated our proposed algorithms on real data sets. The following experiments were conducted on a Core 2 Duo2.10GHz computer with 2GB main memory running Windows XP. All algorithms were implemented in our own simulator. We used the data set KDD cup 1999 intrusion detection data set from the UCI repository [7] which contains a record of a military computer network where there have been multiple intrusions and attacks. This dataset contains 1000 records. Each record has 38 attributes. Out of these 1000 records in KDD dataset there are 43 outliers.

We had used Wisconsin breast cancer data sets taken from the UCI repository. For measuring the accuracy of our proposed algorithms, we count the outliers. We apply the DBOD algorithm and obtain outlier. This is due to the fact that the data points are indexed and initially calculated based on our definition of density and KNN connectivity. Thus we obtain a competent result.

Thus we can have some results to be close to our algorithm. So, the DBOD outperforms the LOF.

We also apply our DBOD algorithm on Wisconsin breast cancer data sets contain some of overlapping data points. Thus the use of path connectivity should be improved to include the similarity data points' distributions in which the overlapping clusters may be distinguishable from each other according to their distributions.

We apply another data set which is Wisconsin breast cancer data set from UCI repository to show the efficiency of our proposed algorithms.

Wisconsin breast cancer data sets taken from the UCI repository [7]. This dataset contains 500 records. Each record has 9 attributes. Out of 500 records 444 records are benign while 56 records are malignant. We considered benign records as normal data while malignant as outliers. Out of these 500 records in Wisconsin breast cancer dataset there are 56 outliers.

Since our proposed algorithm DBOD is based upon new definition of density, it is strongly felt that it should be compare with some exiting density based popular and efficient algorithm. Therefore we selected LOF algorithm for this purpose. DBOD gives better result also included in comparison.

For measuring the accuracy of our proposed algorithms, we count the outliers. Algorithm was executed for different input parameters and DBOD detects outliers more effectively in most of the cases.

Summarized results obtained after executing all density based algorithms i.e. LOF and DBOD are shown in Table 1. In this experiment, we have considered MinPts and k equal to 10.

Table 1. Comparison of LOF and DBOD.

No. of outliers reported	**No. of correct detection**	
	LOF	**DBOD**
10	6	10
20	9	20
30	13	22
40	16	26
50	18	30
60	20	30
70	22	32
80	22	34
90	25	36
100	25	37

And graphical representation of the above data is shown in Fig. 1.

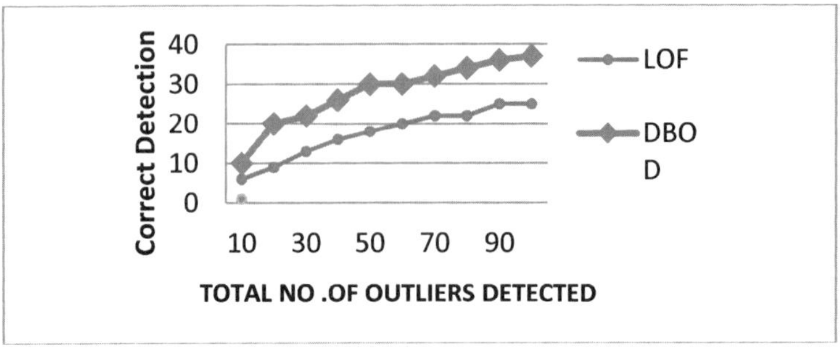

Fig. 1. Comparitive graph for LOF and DBOD for kdd cup 99.

Graph and table shown above makes it very clear that DBOD is working efficiently in comparison to LOF. Important point is that top twenty detction in case of DBOD is correct while in case of LOF only 15 detection are right and five detection are wrong and this fact has to be taken positively and seriously because exactly top twenty outliers detections are more important. DBOD gives 100% accuracy in the detection of top 20 outliers.

Table 2. Comparison of LOF and DBOD.

No. of outliers reported	**No. of correct detection**	
	LOF	**DBOD**
10	2	10
20	10	20
30	15	30
40	20	40
50	29	46
60	35	50
70	41	54
80	47	54
90	53	55
100	56	56

Graphical representation of the above data is shown in Fig. 2.

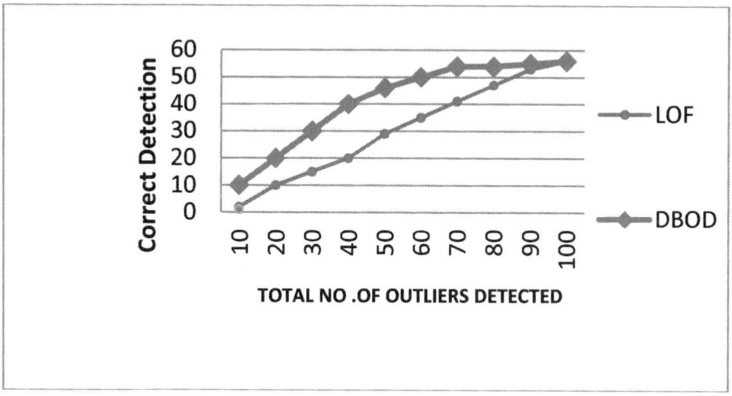

Fig. 2. Comparitive graph for LOF and DBOD for Wisconsin breast cancer data sets.

5. Conclusion

Outlier detection is an important task for many KDD applications. Proposed algorithm DBOD showing improvment in performance in comparison to LOF and DBOD in outlier detection. DBOD is computing point density of data points to find dense and sparse region in the given data sets. Sparcity belongs to outlierness. Thus DBOD detects outliers effectively. In this algorithm k is user defined parameter. Further enhancement could be in the direction that k is found automatically by algorithm itself.

References

1. D. Hawkins, Identification of Outliers, Chapman and Hall, London (1980).
2. J. Han and M. Kamber, Data Mining, Concepts and Techniques, Morgan Kaufmann, San Francisco (2001).
3. M. M. Breunig, H.-P. Kriegel, R. T. Ng and J. Sander. Lof: identifying density-based local outliers. SIGMODRec., **29**, 93 (2000).
4. K. G. Sharma, Efficient Denity Based Outlier Handling Technique in Data Mining, Proc. *1st international Conference on Computer science and Information Technology*, CCSIT, Part 1, 542 (2011).
5. V. Chandola, A. Banerjee and V. Kumar. Anomaly detection: A survey. *ACM Comput. Surv.*, **41**, 1.
6. Knorr, E. M., Ng, R. T., Tucakov, V., Distance-based outliers:algorithms and applications, *The VLDB Journal*, **8**, 237 (2000).
7. Frank, A. & Asuncion, A., UCI Machine Learning Repository [http://archive.ics.uci.edu/ml]. Irvine, CA: University of California, School of Information and Computer Science(2010).

8. Wikipedia: Variance at http://en.wikipedia.org/wiki/variance (accessed: 18 December 2013)
9. Ramaswamy, R. Rastogi and S. Kyuseok, Efficient algorithms for mining outliers from large data sets. In *Proc. of the ACM SIGMODInternational Conference on Management of Data*, 93 (2000).

Enhanced Velocity BPSO and Convergence Analysis on Dimensionality Reduction

Shikha Agarwal[1], R. Rajesh[2] and Prabhat Ranjan[3]

Dept. of Computer Science, Central University of South Bihar
Patna, India
E-mail: [1]shikhaagarwal@cub.ac.in, [2]rajeshr@cub.ac.in, [3]prabhatranjan@cub.ac.in
www.cub.ac.in

High Dimensional data spaces are very common in areas like medicine, biology, bio-informatics, web data, text documents. This high dimensionality brings different challenges when applied with algorithms such as slowness, sensitivity to initial values, either early or slow convergence etc. Various algorithm for the large data set has been proposed in literature. In this paper a literature survey on nature inspired particle swarm optimization (PSO) used for dimensionality reduction is presented and a new variant of PSO, Enhanced Velocity PSO (EVPSO) has been proposed. Convergence analysis of proposed method has been conducted. This work is an attempt of computational, mathematical and statistical analysis to provide direction to the researchers who are working in the field of dimensionality reduction with PSO.

Keywords: High Dimensionality; Feature Selection; Gene Expression Profile.

1. Introduction

High dimensional data is characterized with the very high number of features and relatively few samples. Identification of important features out of pool of features important for the analysis on the data set. Many statistical, mathematical and computational methods are reported by many research to for the feature selection but almost all methods gives comparable result while dealing with ultra-high dimensional data (like gene expression microarray data sets). Particle Swarm optimization (PSO) is a nature inspired computational methods is a population based evolutionary stochastic algorithm developed by Kennedy and Eberhart in 1995 [1]. The back bone of particle swarm optimization algorithm are its four characters exploration, exploitation capability, cognitive and social behavior. Due to its fast convergence and promise to get optimal solution it has gain popularity among researchers, although while dealing with high dimensional search space its convergence is slow and shows poor quality of result also in complex system. Many flavor of PSO has been given by the research to solve the problem of local optima and poor convergence in case of high dimensional data due to degeneracy of some feature. The main focus of this paper is to provide a survey of PSO for dimensionality reduction and to propose a new average velocity PSO (EVPSO) for the dimensionality reduction. In EVPSO the time step at which *pbest* and *gbest* have been calculated is taken as the distance in the velocity update equation to find the cognitive velocity and the social velocity component of the particle. Over the years researchers are working on two aspects of PSO; Either performance improvement or application of PSO in different field. Little attention has been paid to the convergence analysis of the PSO. In this work convergence analysis of proposed method has been conducted followed by the consistency analysis of the results. This paper is organized in a way that it first introduces the standard PSO, then a small survey of PSO for dimensionality reduction has been given, after that proposed modification in PSO velocity update equation and finally convergence and consistence analysis of EVPSO has been provided.

2. Particle Swarm Optimization

PSO is a population based stochastic search method. It is inspired by the flock of birds, fish and other animals searching for food. To understand the mathematical representation of let us design a ith particle $X_i = (x_1, x_2, x_3 \ldots x_d)$. Here d is the number of feature or dimension of the search space. The optimal particle of the swarm represents the global solution *gbest* and best position of any particle based on its fitness is represented as local best or *pbest*. Position of each particle is a candidate solution. Based on the fitness *pbest* and *gbest* are gets calculated. which is used to update velocity and position. Eq. (1) and Eq. (2) given by Kennedy and Eberhart [1] to update the velocity and position.

$$v(t+1) = v(t) + c_1 r_1 (pbest(t) - X(t)) + c_2 r_2 \big(gbest(t) - X(t)\big) \quad (1)$$

$$X(t+1) = x(t) + v(t+1) \quad (2)$$

In 1998 Shi and Eberhart [2] introduced the coefficient in equation of velocity (omega w) which is needed to control the velocity and balance between exploration and exploitation

$$v(t+1) = wv(t) + c_1 r_1 (pbest(t) - X(t)) + c_2 r_2 \big(gbest(t) - X(t)\big) \quad (3)$$

Kenndy and Eberhart has published two books on PSO [3][4]. PSO for the dimensionality reduction was first initiated in 1997 by Kennedy and Eberhart [5] when they proposed the binary PSO for the discrete optimization problem by changing in the probability of the fact that the value of position coordinate is zero or one. In 2009 L. Y. Chuang *et al.* [6] developed the improved the binary

PSO for the feature selection problem and to avoid trapping in local solution. M. Middendorf in 2005 proposed the hierarchy PSO, [7] in which a dynamic hierarchy is used to define the neighborhood in PSO. Pampara *et al.* 2005 [8] used the trigonometry function to generate a bit string. It reduce all of the high-dimensional problems to a four-dimensional problem. Felix *et al.* 2007 [9] in his work cooperative multiple PSO solve the multidimensional or large problems by using all conductivity and control properties of the PSO. In cooperative coevolving PSO of Yao 2008 [10] large scale problem is solved by breaking the problem into some smaller-scaled problem. Hybridization of PSO with harmony search by Li and Li 2007 [11] solve high dimensional problem. Simulated Annealing PSO of Wang and Li 2004 [12] is hybrid of SA and PSO to overcome the problems of both. Voss 2005 have developed principal component PSO (PCPSO) [13] to solve the problem in high dimensions, the particles are flown simultaneously in two separate spaces. In 2008 Cui *et al.* have proposed the LSI-HPSO, RPHPSO [14] for dimension reduction in document clustering. Few variants of PSO for dimensionality reduction. It cover brief description of few variants of PSO used for dimensionality reduction of PSO.

2.1. *Catfish BPSO*

Li-Yeh Chuang *et al.* 2011 [15] has proposed Catfish BPSO for the feature selection. Catfish BPSO m introduces catfish particles to BPSO to avoid the trapping in the local optimum solution. In this “sardine” particles are added to renew the search. In Catfish BPSO, the position and velocity of each particle are updated by Eqs. (4)–(5). If the distance between *gbest* and the surrounding particles is less than the threshold, each surrounding particle will move very less in the next generation. To avoid this situation, catfish particles are added and it replaces the ten percent of the particles with the worst fitness from the swarm. Catfish particles guarantees the success of a given optimization task.

2.2. *PSO-Adaboost*

H. Pan *et al.* 2013 [16] has proposed a reliable face detector system by fusing multi-feature descriptors and feature selection using PSO-Adaboost. Hybrid of PSO-Adaboost replaces exhaustive search used in original Adaboost framework with PSO technique for efficient feature selection. For a given set of training data (x_n, y_n), y_n->[1 0] & (n = 1, …., N) where y_n is the binary class label for data point x_n. In PSO-Adabosst algorithm during each iteration for each feature, a weak classifier is trained for the classification of positive patterns from negative patterns. In PSO-Ad boost particles of PSO are structured as X = [type;

x_s,y_s, width, height, sampling points, radius] where type denotes the feature descriptor type, (x_s, y_s) denotes the feature position in the detection window, denotes the orientation of GH and MBLBP features, (width, height) denotes the size of GH features in 36x36 cascade classifiers, sampling points, denote the number of neighboring sample points (P) and radius denotes the radius of sample circle (R). The search is done to find the optimum best feature descriptor set over entire feature space using PSO algorithm, according to the fitness function defined as the normalized classification error rate. According to H. Pan as compare to simple Adaboost algorithm for feature selection, PSO-Adaboost algorithm has showed 20 times less training time for classifier.

2.3. *Gini PSO-SVM*

N. Allias *et al.* 2014 [17] has proposed two stages of feature selection based on Hybrid of Gini index, PSO, SVM and Taguchi methods to reduce the high dimensionality of features and obtain a good classification result for spam email filtering. Firstly Preprocessed training data undergo Gini Index filtering to rank the feature in decreasing order of their importance. Attribute with the highest weighting score is selected varying from 25 features until 125 features. Selected features has been processed using Tagauchi method. In which an orthogonal array is used to reduce the number of features. Final selected attributes were passed to the PSO-SVM to optimize the selected features. This hybrid method has been trained and tested on Lingspam dataset. Testing done using naive bayes classifier and performance of this method showed a good precision and recall value with few number of feature as compare to traditional methods.

2.4. *PSO+SVM*

S. W. Lin *et al.* 2008 [18] has applied the PSO for the parameter tuning of SVM along with the feature selection for the classification task. SVM with kernel function helps the SVM in obtaining the optimal solution. The RBF kernel function is the most commonly used, because it can classify multi-dimensional data also. S. W. Lin has tuned the parameter of RBF-SVM using PSO. Two main parameters of RBFSVM are C and Ʈ. If C is too high, then the classification accuracy is very high during training, but very low during testing phase. If C is too small, then the model will be useless. Parameter Ʈ has a much greater importance than parameter C. High Value of Ʈ results in over-fitting and low value leads to the leads to under-fitting (Pardo & Sberveglieri, 2005). Without feature selection, two parameters are required; C and, are required. when dealing with feature selection, then 2 + n decision variables must be chosen with the value range of [0 - 1]. If the value of a parameter is less than or equal to 0.5, then its corresponding feature is rejected. Similarly if the value of a parameter is greater than 0.5, then its corresponding feature is accepted. The structure of particle of PSO is shown in Figure 1.

1	2	3	....	n+2
p_c	p_Γ	a_1	...	a_n

Figure 1. Structure of particle in SVM+PSO.

p_c = C, p_Υ =ϒ, a_n=Feature n is selected or not. Simulation result of author has revealed that PSO + SVM approach has a better classification accuracy than others methods. SVMPSO with feature selection applied to eliminate unnecessary features, and for the tuning of parameter values, resulting in improving the overall classification results.

2.5. *Improved binary PSO*

Kennedy and Eberhart (1997) [5] developed the binary PSO in which each particle is a binary bits string of 0 and 1. Each particle adjust its position according to personal learning and neighborhood learning. Li-Y. Chuang *et al.* 2009 [19] has been proposed an improved binary PSO [19] for the feature selection to avoid the local optimum. IBPSO keep a memory of all the *pbest* and *gbest*. Each particle's position and velocity is updated according the three equation of IBPSO.

$$v_{new}^{pd} = w.v_{old}^{pd} + c_1r_1\,(pbest^{pd} - x_{old}^{pd}) + c_2r_2\,(gbest^{pd} - x_{old}^{pd}) \quad (4)$$

$$S(v_{new}^{pd}) = \frac{1}{(1 + e^{v_{new}^{pd}})} \quad (5)$$

$$If rand < S(v_{new}^{pd})\ then\ x_{new}^{pd} = 1\,; else\ \ x_{old}^{pd} = 0 \quad (6)$$

Here in Eq. (4) p and d indicated the particle and dimension receptively means velocity of is vector of length equal to the dimension of dataset. In these equations, w is the inertia weight, c_1 and c_2 acceleration (learning) factors, and rand, r_1 and r_2 are random numbers. In order to avoid getting trapped in local optimum if *gbest* value does not change after three iterations, the *gbest* position is reset to zero. After that velocity is transformed using Eq. (5) and if transformed value is greater than the randomly generated value position bit at that place feature is selected otherwise rejected Eq. (6).

2.6. *Modified IBPSO*

In our previous work for dimensionality reduction [19] four modification in improved binary PSO has been suggested and simulated on high dimensional gene expression data set. Result showed that modified IBPSO-3 and modified IBPSO-4 has comparable results in term of fitness and in terms of feature selection modified IBPSO-3 is out performing than the other modifications. All four modification are proposed on the three Eqs. (4), (5) and (6) of IBPSO. In modified IBPSO-3 Eq. (6) of IBPSO rand number is replaced with the 0.5 and direct comparison of transformed velocity with the 0.5. Modified equation is Eq. (7). In modified IBPSO-4 no velocity transformation and direct checking of updated velocity whether it is negative or positive, modified equation is Eq. (8).

$$If\, S\left(v_{new}^{pd}\right) > 0.5\; then\; x_{new}^{pd} = 1\,; else\;\; x_{old}^{pd} = 0 \tag{7}$$

$$If\; V > 0\; then\; x_{new}^{pd} = 1\,; else\;\; x_{old}^{pd} = 0 \tag{8}$$

3. Proposed Method

The velocity update Eq. (4) contains three components namely; memory of previous velocity, cognitive component and social component. Cognitive component of velocity update equation represents nostalgia for previous best position of the particle. Social component quantifies the performance relative to the neighbors. The second component is the product of learning coefficient, a random number and difference between current position and previous best position. Similarly third component is the product of learning coefficient, a random number and difference between current position and global best position. The assumption taken by all the previous variants of PSO is that the difference between current position and previous best position is in unit time step therefore division of difference in position of *pbest* and *gbest* with respect to current position with in unit time step will be equal to the difference itself. Hence time step is not visible in the Eq. (4) because effect of unit time is negligible. In this work in place of unit time step, the difference between current time step and time step at which previous best has been found is taken as the length of time step and used in the second and third component of the velocity update equation to calculate the velocity. Since simply adding the three velocities does not justify the motion of particle therefore average of the three velocities has been taken. In cognitive part and social part of Eq. (4) is modified to $c_1 r_1\, (pbest^{pd} - x_{old}^{pd})/c\_timestp$ and $c_2 r_2\, (gbest^{pd} - x_{old}^{pd})/s\_timestp$ respectively. Here c_timestp and s_timesteps are respective length of time steps at which *pbest* and *gbest* has calculated. Final modified velocity update Eq. (9)

is the sum of previous velocity, new cognitive component, new social component divided by summation of w, c_1*r_1 and c_2*r_2.

$$v_{new}^{pd} = (w.v_{old}^{pd} + c_1r_1\,(pbest^{pd} - x_{old}^{pd})/c\_timestp + c_2r_2\,(gbest^{pd} - x_{old}^{pd})/s\_timestp)/(w + c_1r_1 + c_2r_2) \quad (9)$$

4. Result

IBPSO and proposed EVPSO have been applied for the dimension reduction of six gene expression data sets downloaded from http://www.gemssystem.org. Selected genes were evaluated using LOOCV method on 1-NN classifier. Interia Weight is set to 0.5, Learning rates are $c_1 = 2$ and $c_2 = 2$, velocity range [V_{min} V_{max}] = [-6 6], number of particles are 40, Number of maximum iteration are 150 and stopping criteria is max iteration = 150 or max accuracy = 100. IBPSO and EVPSO were applied ten times on each data set since both algorithms consist of the random number therefore relying on the finding of just one run do not justify the actual dimensionality reduction. The result in the Table 1 shows the average accuracy and average selected genes in both the cases.

Table 1. Average classification accuracy and selected genes.

	Improved BPSO	EVPSO
Datasets	Accuracy (Genes)	Accuracy(Genes)
SRBCT	95.59(1390)	95.66(1050)
DLBCL	79.51(3697)	78.868(1445)
Brain1	89.99(3924)	89.77(3354)
Brain2	78.2(5983)	78.7(3547)
Leukemia1	94.38(3643.8)	94.99(2233)
Leukemia2	87.22(4958)	89.44(3023)

The reported selected genes in case of SRBCT, DLBCL, Brain Tumor1, Brain Tumor2, Leukemia1 and Leukemia2 are 431, 754, 1197, 1034 and 1292 respectively reported by L. Y. Chuang [6]. In any algorithm which depends on the random number regeneration of the same result is not guaranteed. Table-1 shows that Average velocity PSO has better dimension reduction power with comparable accuracy as IBPSO. For the all data sets classification accuracy in EVPSO has comparable result with IBPSO classification accuracy. Number of selected genes in EVPSO are 1050, 1445, 3354, 3547, 2233 and 3023 for SRBCT, DLBCL, Brain Tumor1, Brain Tumor2, Leukemia1 and Leukemia2. For all data sets number of selected genes using EVPSO are lesser than the selected genes by IBPSO.

5. Discussion

Therefore in this section validation and convergence analysis of EVPSO has been discussed. Convergence-Divergence-Convergence indicates the way any algorithm approaches towards the global optimum and divergence indicated the attempt of coming out of the local optimum. Figure 2 clearly shows the convergence trajectory of EVPSO and divergence approach of EVPSO.

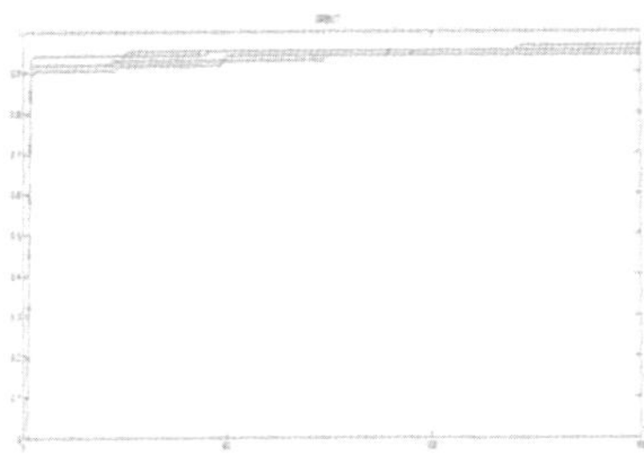

(a) Convergence of EVPSO on SRBCT

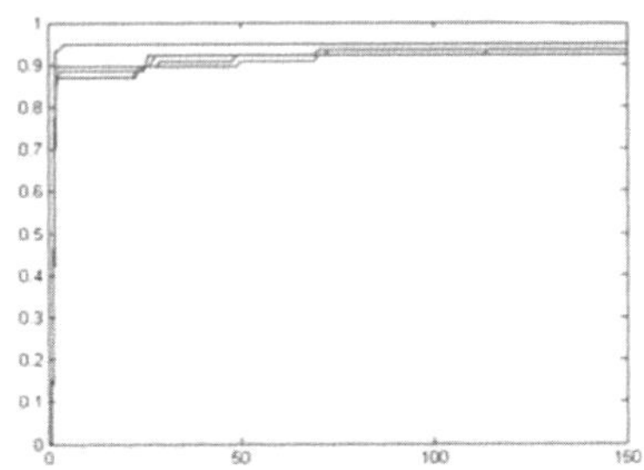

(b) Convergence of EVPSO on DLBCL

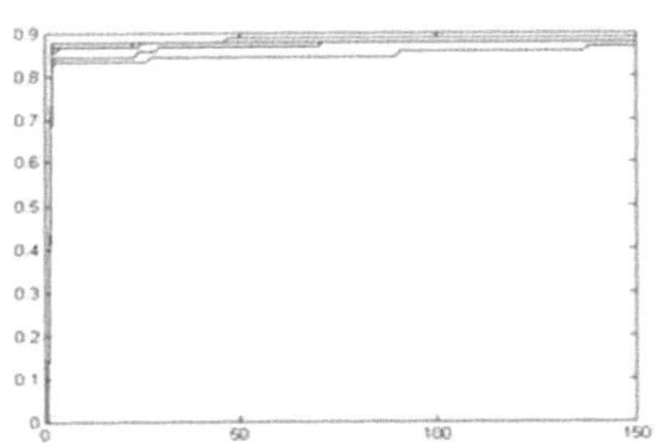

(c) Convergence of EVPSO on Brain Tumor1

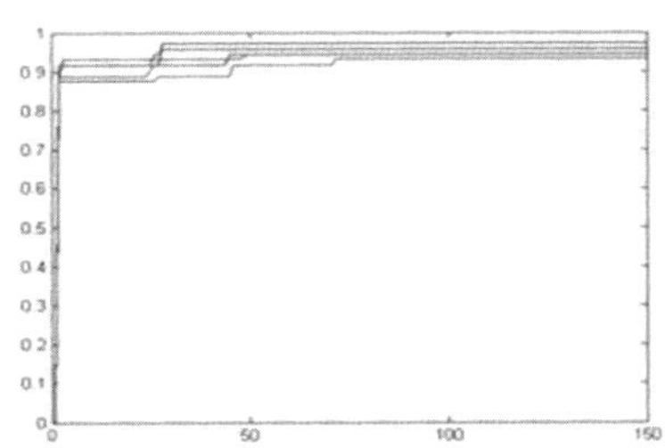

(d) Convergence of EVPSO on Leukemia1

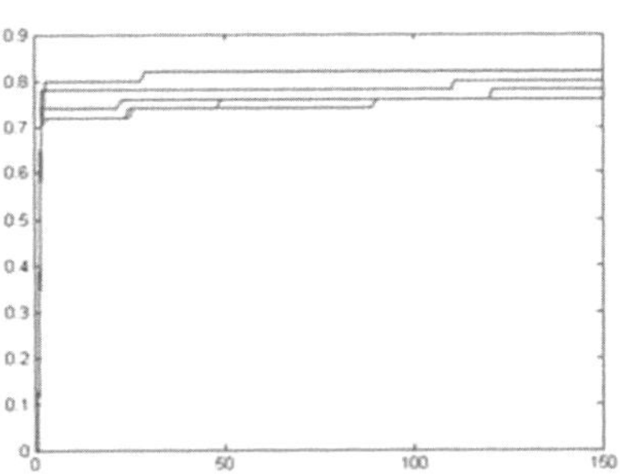

(e) Convergence of EVPSO on Lung cancer

Figure 2. Convergence of EVPSO on the six dataset. All the ten runs of EVPSO are following the same trajectory in all the cases. Overlapping of ten individual runs justifying the consistence nature of EVPSO in spite of having many random numbers.

5.1. *Convergence-Divergence*

Convergence indicates the way any algorithm approaches towards the global optimum and divergence indicated the attempt of coming out of the local optimum. Figure 2 clearly shows the convergence trajectory of EVPSO.

5.2. *Consistence analysis with standard deviation*

The convergence and fitness of an individual run is correlated. The definition of standard deviation shows that it represents the deviation degree of each run. Small standard deviation is then representing the less divergence nature among the each individual run where each run is consisting of many iterations. The Table 2 below describes the standard deviation of ten runs on each data set.

Table 2. Consistence table of EVPSO.

Dataset	Standard Deviation
SRBCT	0.0142
DLBCL	1.1703
Brain1	0.4321
Brain2	0.4559
Leukemia1	0.132
Leukemia2	0.213

6. Conclusion

In this study, a small survey of PSO variants used for dimensionality reduction has been presented and new average velocity PSO (EVPSO) have been implemented on six gene expression data sets for the better dimensionality reduction and classification of gene expression data using K-nearest neighbor (K-NN) evaluator. Experimental results shows that performance of EVPSO is comparable with the IBPSO along with satisfactory dimension reduction in all the six gene expression data sets. Convergence and standard deviation analysis represents the consistence of each run and reject the possibility of randomness of each individual run. Convergence analysis justify the use of EVPSO for the high dimensional analysis to avoid the local optimum results. Hence proposed method could help the researchers of for the feature selection of high dimensional datasets.

References

1. J. Kennedy and R. C. Eberhart, Particle swarm optimization *In: IEEE International Conference on Neural Networks, Perth, Australia*, 1942-1948, (1995).

2. Y. Shi and R. C. Eberhart, A modified particle swam optimizer, *IEEE Word Congress on Computational Intelligence*, 69–73, (1998).
3. J. Kennedy, R. C. Eberhart and Y. Shi, *Swarm Intelligence*, (Morgan Kaufman, San Mateo, CA, 2001).
4. J. Kennedy, R. C. Eberhart and Y. Shi, *Computational Intelligence*, (Morgan Kaufman, San Mateo, CA. 2007).
5. J. Kennedy and R. C. Eberhart, A discrete binary version of the particle swarm algorithm, *In: IEEE conference on systems, man, and cyber*, 5, 4104–4108, (1997).
6. L. Y. Chuang, H. W. Chang, C. J. Tu and C. H. Yang, *Improved binary PSO for feature selection using gene expression data,* Computational Biology and Chemistry, 32, 29–38,(2008).
7. S. Janson and M. Middendorf, A hierarchical Particle Swarm Optimization and Its Adaptive Variants, *IEEE Trans. on system, Man and Cybernetics-Part B: Cybernetics*, **35**, 6, December, (2005).
8. G. Pampara, N. Franken and A. P. Engelbrecht, Combining particle swarm optimization with angle modulation to solve binary problems, *IEEE Cong. Evol. Comput., 1*, 89–96, (2005).
9. T. S. C. Felix, V. Kumar and N. Mishra, A CMPSO algorithm based approach to solve the multi-plant supply chain problem. *Swarm intelligence, focus on ant and particle swarm optimization*, 447–474, (2007).
10. X. Yao, Cooperatively coevolving particle swarm for large scale optimization, *In: Conference of EPSRC, artificial intell technologies new and emerging computer paradigms* (2008).
11. H. Q. Li, L. Li, *A novel hybrid particle swarm optimization algorithm combined with harmony search for high dimensional optimization problems*, *In: Proceedings of the IEEE/IPC*, 94–97 (2007).
12. X.-H. Wang and J.-J. Li, Hybrid particle swarm optimization with simulated annealing, *In: Proceedings of the IEEE international conference on machine learning and cyber*, **4**, 2402–2405 (2004).
13. M. S. Voss, Principal component particle swarm optimization (PCPSO), *In: Proceedings of the IEEE symposium on swarm Intelligence*, 401–404 (2005).
14. X. Cui, J. M. Beaver, J. S. Charles and T. K. Potok, Dimensionality reduction particle swarm algorithm for high dimensional clustering, *In: IEEE swarm intelligence symposium*, 1–6 (2008).

15. L. Y. Chuang, S. W. Tsai and C. H. Yang, Improved binary particle swarm optimization using catfish effect for feature selection, *Expert Syst. Appl.,* **38**, 10, 12699–12707, Sep. (2011).
16. H. Pan, Y. P. Zhu and L. Xia, Fusing Multi-Feature Representation and PSO-Adaboost based feature Selection for Reliable Frontal Face Detection, *In: Proceedings of the IEEE symposium on swarm Intelligence*, 2998–3002 (2013).
17. N. Allias, M. N. Megat and M. N. Ismail, A Hybrid Gini PSO-SVM Feature Selection Based on Taguchi Method: An Evaluation on Email Filtering, *IMCOM, ACM,* January, 2014.
18. S. W. Lin, K. C. Ying, S. C. Chen and Z. J. Lee, Particle swarm optimization for parameter determination and feature selection of support vector machines, *Expert Syst. Appl.*, **35**, 4, 1817–1824, Nov. (2008).
19. R. Rajesh and S. Agarwal, Some Modification in Particle Swarm Optimization, *The 18th Online World Conference on Soft Computing in Industrial Applications* (2014).

Modification of the Android Operating System to Predict the Human Body Temperature Using Capacitive Touch

Shubhnkar Upadhyay[1] and Avadhesh Singh[2]
Dept. of CSE, National Institute of Technology
Patna, Bihar
E-mail: [1]shubhnkar123988@nitp.ac.in, [2]avadhesh123979@nitp.ac.in

Kumar Abhishek[3] and M. P. Singh[4]
Dept. of CSE, National Institute of Technology
Patna, Bihar
E-mail: [3]kumar.abhishek@nitp.ac.in, [4]mps@nitp.ac.in

In this paper, the authors have proposed a way to predict the core human body temperature through Android smartphones. The touch screen in any Android smartphone works on self-capacitance. The capacitance value is then converted into the corresponding frequency value. By slightly changing the source code of the Android operating system this frequency value can be used to predict the body temperature. The temperature change values corresponding to their frequency change values are stored in a database beforehand.

Keywords: Capacitive Touch; Frequency; Frequency Measurement.

1. Introduction

The capacitive touch technology can be used to identify the human body temperature. In this paper the authors have shown the inner workings of the Android operating system and how it can be used to detect human body temperature. The Android operating system is chosen because it is open source and the most widely used operating system in the market. The changes will be made in the source code of the operating system.

The paper describes the calculation of body temperature with respect to the normal body temperature previously stored. The relative values of the temperature change with respect to different touch gestures are required. Temperature change will be predicted through these values only. The database consisting of these values of temperature change has to be stored beforehand through crowd sourcing [6]. Currently no such API has been developed nor is in use.

1.1. *Motivation*

In the present time, quite a few numbers of products are available in the market, which detect the human body temperature. These include the devices which make use of microcontrollers and Arduino [3] LM35 temperature sensor [10]. A product called iCelsius [2] has also been developed. It is an external device which gets attached to the micro USB port of an iPhone or an Android phone. The external device has a stainless steel tip which is used to detect the human body temperature. The results are shown on the screen of the mobile phone used.

Though all these devices show accurate results, they have one demerit. All these devices are external, this means that apart from a mobile phone we need an external device which detects the body temperature [10]. In this paper, the use of this external device is uninvolved. This is done by changing the source code of the Android operating system and using the values detected by touch to predict the human body temperature. This would ensure more portability of a thermometer as no external device would have to be carried.

2. Related Work

Although not much work has been done in this field however, the temperature detecting applications so far proposed have all been dealing with the notion of calculating temperature with capacitance. Contrary to this idea, in this paper the temperature is predicted with the help of frequency. Frequency is used in place of capacitance because there is no direct relation by which temperature can be calculated using capacitance. But the temperature values corresponding to the frequency values can be stored beforehand in the database, which can be used for temperature detection.

As mentioned in the previous section there are a lot of devices available which detect temperature through Android devices. But all of them are external devices, this means that they have to be connected to the Android phone every time the temperature has to be calculated.

3. Capacitive Touch Screen

Many of the electronic appliances today, like smartphones, laptop touchpads and many more, use capacitive sensing for the touch screen. The capacitive touch screen is sensitive to the human touch or any other conducting material [1]. The basic principle behind the working of these touch screens is the continuously flowing bioelectricity through human body combined with the constant electric current that flows through screens' lower layer.

There are different types of capacitive touch screens present, what is used in this paper is self-capacitance [11]. Android smartphones make use of the self-capacitance touch screens.

3.1. *Layered architecture*

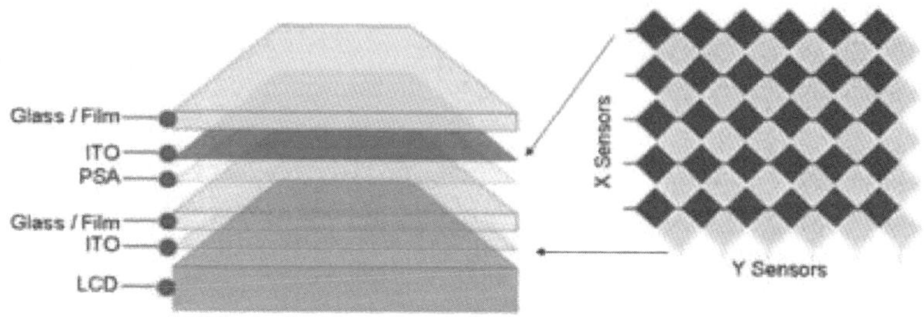

Fig. 1. Layered architecture of capacitive touch screen [4].

The capacitive touch screen is made up of different layers stacked on top of each other. These layers are arranged in a way such that the touch screen behaves as a capacitor.

At the very top is the glass film which acts as a substrate. Glass is used here because it is nonconductive and also prevents moisture. Two layers of Indium Tin Oxide are glued together and a constant voltage is supplied to them. These layers are glued using Pressure Sensor Adhesive which ensures complete vacuum between them. However, some air gap is usually formed between these layers, because of this touch sensitivity decreases. These two layers form the X-Y coordinate plane for the screen [11].

3.2. *Basic working principle*

When the screen is touched, additional capacitance from finger is detected. This additional capacitance is in parallel to the already present capacitance. And since capacitances in parallel are added to each other, hence the total capacitance is the sum of these two values. This total capacitance is then fed into the oscillator, which measures the frequency based on the RC time constant. The net capacitance is shown by the following equation.

$$C_s = C_p + C_f$$

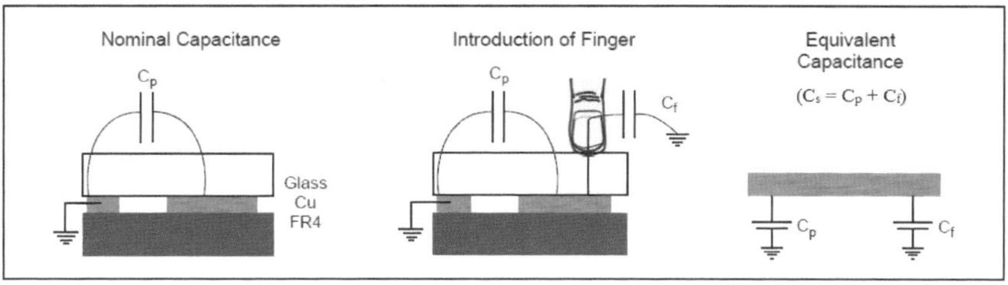

Fig. 2. Capacitance change due to touch [5].

Now, since the net capacitance has increased, this changes the RC time constant. And now, new frequency is calculated based on this new capacitance. The increased capacitance results in the decreased frequency of the oscillator.

4. Oscillator And Frequency

The oscillator mentioned in the previous section is responsible for the calculation of frequency.

4.1. *Oscillator circuit and timers*

The oscillator circuit comprises of two comparators and an SR latch. The oscillator functions according to the RC time constant. The capacitance in this time constant is the net capacitance calculated in the previous section. And the resistance in the time constant is the resistance of the oscillator.

In addition to this there are two timers that operate on the cycles of the oscillator. Although, both the timers are used for counting, they have different purposes. The output of the latch is fed into Timer 1 which increases accordingly. And Timer0 regulates the count of Timer1 [7].

4.2. *Frequency measurement*

The measurement of frequency is achieved with the help of oscillator circuit and timers. The positive inputs of the comparators are the upper and lower limits. The input of comparator1 is internally set while that of comparator2 is supplied externally to set the lower limit [5]. Their negative input is the net capacitance C_s, which is the sum of internal capacitance and the capacitance induced by touch.

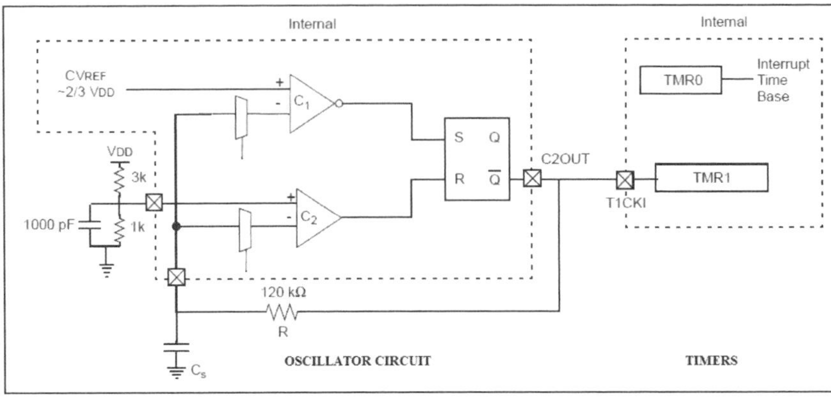

Fig. 3. Frequency measurement with oscillator and timers [5].

The work of the comparators is to compare the value of C_s with the upper and lower limits. And then accordingly output 1 or 0. For example, if C_s is less than both, the upper limit and the lower limit, then the value of both comparators will be 1, but since C_1 has NOT gate, hence S will be 0 and R will be 1. Now in this case, as shown in the truth table, the latch would output 1, and this means that charging of the capacitor will start.

Similarly, when the value of C_s lies between the upper limit and lower limit then C_2 will output 0 and C_1 will output 0, and the value of S and R will be 0. Now according to the truth table the capacitor will function according to the previous state, this means it will continue charging.

And the same can be shown when C_s exceeds both the upper limit and lower limit. In this case the values of S and R will be 1 and 0 respectively. And according to the truth table output of the latch will be 0. This means that at this point discharging of the capacitor will start.

Table 1. Truth table.

Truth Table of SR Latch		
S	R	$\overline{Q_{n+1}}$
0	0	$\overline{Q_n}$
0	1	1
1	0	0
1	1	N.D.

As shown in the truth table of the SR latch, the timer, functions according to the output of the latch. Every time the output of the latch changes from 0 to 1,

Timer1 is incremented. In this case, Timer1 will increase continuously and will eventually overflow.

But to detect the frequency, the value of Timer1 must be recorded before it overflows. For this reason Timer0 is used. It provides a fixed time over which the frequency has to be calculated [7]. Every time before the measurement of frequency Timer0 is cleared. Timer0 is set such that it also increments every time Timer1 increments. As soon as Timer0 overflows, an Interrupt Service Routine is initiated and the value of Timer1 is recorded. Timer0 must be set in a way that it overflows before Timer1 [5]. At the end of this Interrupt Service Routine both the timers are cleared for the next reading. Hence, every time a touch is detected we have a corresponding value of the frequency.

5. Predicting Body Temperature

The frequency obtained in the previous section is used to predict the human body temperature. This frequency is received by the Android Operating System which takes necessary steps to check whether a touch is made or not.

5.1. *Frequency-temperature relation*

$$f \propto \frac{1}{RC}$$

Frequency is inversely proportional to the capacitance. This relation between frequency and capacitance can be used to predict the body temperature. Since with the rise in temperature the capacitance rises, although some capacitors decrease their value with rise in temperature, but they are not considered here. Hence, the change of frequency can be related with change in temperature.

With the rise in body temperature the electrical activity also increases, this means that the value of C_f increases. This is in comparison with the normal body temperature. With the rise in C_f the net capacitance increases, and hence the frequency decreases. For example, as shown in Figure 4, a comparison is done of the required values of two different body conditions. The capacitance generated due to touch is more of the body having fever, hence the net capacitance is also more. Now, since frequency is inversely proportional to capacitance, therefore the body having fever generates less frequency as compared to the normal body. Conclusively, the change in frequency will be more for the body having fever. This frequency change can now be used to store the corresponding temperature change values.

	NORMAL BODY TOUCH	TOUCH OF BODY WITH FEVER	RELATION
CAPACITANCE DUE TO TOUCH	C_{f1}	C_{f2}	$C_{f1} < C_{f2}$
NET CAPACITANCE	C_{s1}	C_{s2}	$C_{s1} < C_{s2}$
FREQUENCY (f_i)	f_1	f_2	$f_1 > f_2$
CHANGE IN FREQUENCY ($\Delta f_i = f - f_i$)	Δf_1	Δf_2	$\Delta f_1 < \Delta f_2$

* f = average frequency when screen is not touched.

Fig. 4. Comparative study of required values at different body conditions.

5.2. *Creating the database*

In the source code of the Android Operating System, this frequency change can be associated with the temperature change already stored in the database. The values of the temperature change relative to the frequency change have to be stored beforehand. This has to be done practically through crowd sourcing. For example, people with different body temperature states are input as subjects to create the database. The subjects are required to use this application in order to store the temperature change with corresponding frequency change [6]. This method of collecting the temperature change values through crowd sourcing is essential because, there is no direct way of calculating temperature with capacitance value or frequency value.

5.3. *Temperature predicting algorithm*

The changes to be made in the source code are shown below. These changes are made in the ISR, which is invoked whenever a touch is detected. The algorithm checks for the conditions required for a touch to be detected. Once the touch is detected, then the ISR accesses the database and checks for the corresponding value of the temperature

```
unsigned int average;     //average of the previous readings
unsigned int raw;         //current reading from   TIMER1
unsigned int trip;        //change in frequency for normal body temperature
unsigned int t;           //minimum amount of time for touch
ISR
{
  . . .
  If(touch_time>= t)
```

```
    {
        If(raw < (average – trip))
        {
            match_with_database(raw);
            // show output temperature
        }
        else
            // body temperature is normal
    }
    else
    // touch should be made properly
...
}
```

The above algorithm shows how the temperature is calculated with the help of the Interrupt Service Routine (ISR). The basic working of the ISR has been modified for the calculation purposes. The average variable stores the average touch reading from the previous cycles. The raw variable stores the current reading from TIMER1. The trip is the change in frequency for the normal body temperature. The variable t is used for the minimum amount of time the screen should be touched.

In the ISR, firstly it is checked whether the touch has been made long enough to record the current frequency value. After this, it is checked that the frequency value detected is well below that of the normal body temperature. If this condition is met then this raw value is sent to the database through a method for predicting the temperature. The main function of this method is to access the database and check for the value of the temperature stored corresponding to the frequency value detected. If this temperature value is found then the result is displayed, however in the scenario where a value could not be matched up corresponding to the frequency then this frequency value is noted. The temperature value can then be stored corresponding to this frequency value.

5.4. *Implementing the API*

The method mentioned in this paper for predicting the core body temperature can be implemented in any Android phone. The temperature values for the corresponding frequency values have to be stored only once in the database. This is because, the temperature value will not change for a particular frequency value. Once the database is created then it can be extended further to create a well-stocked application with the help of this API. The flowchart below shows the step by step process for predicting the body temperature through the API mentioned.

6. Performance Analysis

The API mentioned here should be implemented at the industry sector and not on the small scale sector. This is because for changing the source code of the Android OS, machines with higher level of performance are required. Also, some of the hardware configuration should be known for effectively implementing the API, which is not possible at the small scale sector. The main idea behind this API is creating the database for the temperature values for their corresponding frequency values. Once this database is created then the performance can be analyzed for this API.

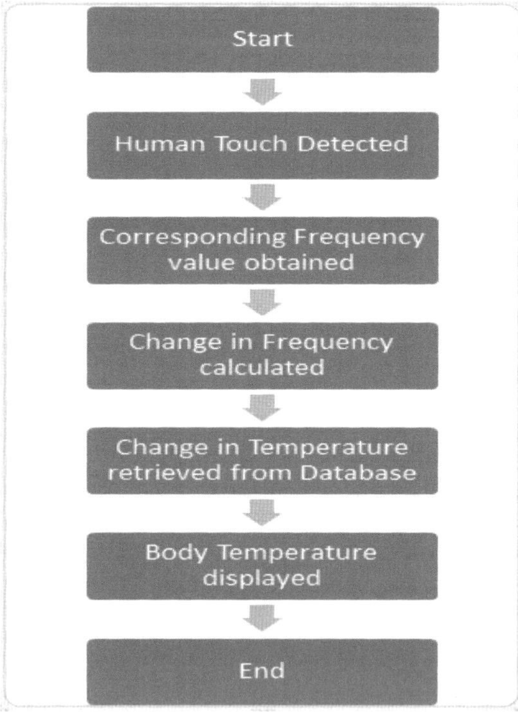

Fig. 5. Flowchart for the entire process.

7. Future Prospects

The temperature thus calculated can be further used to add various functions. Since, body temperature is the most essential part in figuring out the illness of a person hence, it can be applied in many fields for different outcomes. For example, an application can be designed in a way that it checks for the body temperature on a daily basis and shows this record graphically [8]. This temperature can further be combined with manually entered symptoms to predict the nature of illness.

8. Conclusion

Presently, calculating human body temperature requires the use of external hardware. This requirement may not always be met, because to carry an external kit is not possible everywhere.

In this paper, a technological enhanced way is proposed to predict the core human body temperature. This is done with the help of the Android smartphone, by changing the source code of the operating system.

References

1. Capacitance - Is a human (or other) electrical signal absolutely necessary to activate a capacitive touchscreen? - Electrical Engineering Stack Exchange.
2. iCelsius_http://www.icelsius.com/
3. Arduino http://blog.arduino.cc/category/languages/android/
4. CapacitiveTouch http://www.eizoglobal.com/library/basics/basic_understanding_of_touch_panel/
5. Tom Perme, AN1101 – Introduction to Capacitive Sensing, *Microchip Technology Inc.,* 1 (2007).
6. Crowdsourcing http://en.wikipedia.org/wiki/Crowdsourcing
7. Enrique Aleman, AN1171 – How to use Capacitive Sensing Module, *Microchip Technology Inc.,* 1 (2010)
8. B.R. Nandkishor, Aprna Shinde, Android Smartphone based Body Area Network for the evaluation of Medical Parameters in real life, *International Journal of Electrical, Electronics and Data Communication*, ISSN: 2320-2084, 3 (2014).
9. Sherin Sebastian, Neethu Rachel Jacob, Remote Patient Monitoring System Using Android Technology, *International Journal of Computer Science and Mobile Computing*, 191 (2013).

10. Uthayakumar, S. G., A. Sivasubramanian and A. Gratus Iruthaya Kisho. A novel technique to monitor human body vital signs. *International Journal of Biomedical and Advance Research,* **4**, 642 (2013).
11. Barrett, Gary and Ryomei Omote. Projected-capacitive touch technology. *Information Display,* **26**, 16 (2010).

Context-Aware Based Clustering in Wireless Sensor Networks — A Survey

Santu Paul[1], M. P. Singh[2], J. P. Singh[3] and Prabhat Kumar[4]

Computer Science & Engg., National Institute of Technology Patna
Patna, Bihar-800005, India
E-mail: [1]santupaul2807@gmail.com, [2]mps@nitp.ac.in,
[3]jps@nitp.ac.in, [4]prabhat@nitp.ac.in
www.nitp.ac.in

Wireless Sensor Network (WSN) is a popular research topic because of its multi-dimensional applications. Recent technological advancements in computation, communication, hardware and software play the most important role in the development of low cost, low power, small in size multifunctional sensor nodes. Since, radio transmission consumes lot of energy, limited battery power becomes an important issue in WSNs. Hence, battery power is the most critical parameter to decide the lifetime of sensor nodes in WSNs. WSN requires an energy efficient routing protocol for achieving longer life of sensor nodes. Different applications of WSNs (disaster management, border protection, security surveillance) need to deploy sensor nodes in large numbers. So, the main requirement of an algorithm is to form clusters by grouping sensor nodes in disjoint and non-overlapping manner. This paper presents a comparative study between energy efficient hierarchical cluster-based routing and context-aware cluster-based routing in Wireless Sensor Networks (WSNs).

Keywords: Wireless Sensor Networks (WSNs); Cluster-Head (CH); Hierarchical Cluster-Based Routing; Context-Awareness; Context-Aware Cluster-Based Routing.

1. Introduction

Continuous advancement in Micro Electronic Mechanical Systems (MEMS) and Wireless Communication technologies results in the development of tiny, low-cost, low-power multifunctional sensor nodes. Hence, WSNs become the most emerging technology in Twenty-First century and play an important role in our daily lives. The application of WSNs is in variety of civilian and military applications, for example, environmental monitoring, battle field surveillance and industry process control. Although the first application of WSN is battlefield surveillance, but now different civilian application areas, like environment and habitat monitoring, health-care applications, home automation, and traffic control use WSNs.

This paper presents the comparative study between energy efficient hierarchical cluster-based routing and context aware cluster-based routing. The organization of the paper is as follows: Section 2 introduces the main design of clustering. Section 3 discusses energy efficient hierarchical cluster-based routing protocols. Section 4 introduces different context aware cluster-based routing. Section 5 draws the comparison between energy efficient hierarchical cluster-based routing and context-aware cluster-based routing. And finally this paper concludes with Section 6, in which it examines some future research problems and draws conclusion on the current state of sensor network clustering.

2. Design Philosophy

There are several challenges (like limited energy, network lifetime) in implementing WSNs. To achieve the targeted design goals for a given implementation, there should be an efficient clustering algorithm. There are several key features that designers must carefully follow, which have some important role to play in WSNs.

- **Cost of Clustering**
- **Selection of Clusters and Cluster Heads**
- **Real time operation**
- **Synchronization**
- **Data Aggregation**
- **Repair Mechanisms**
- **Quality of Service (QoS)** So, it is important to consider all these metrics while choosing a clustering scheme.

3. Energy Efficient Hierarchical Cluster-Based Routing

Clustering approach may keep the transmission power level and power consumption in communication low, where it connects all members of cluster directly to the CH. So, sensor nodes in the same cluster can communicate with the CH without communicating any intermediate sensor nodes. CHs can transmit gathered information back to the base station through multi-hop communication among CHs. Designing an optimal mechanism considers various factors:

- **The Data Packet size**
- **Frequency of transmissions**
- **Maximum allowable latency**
- **Local computational processes**
- **Maintenance of partial database information.**

There are many research projects in WSNs implementing hierarchical clustering from different perspectives for last few years. Several papers [3][4][5][6] propose algorithm for prolonging the life of WSN and for routing the correct data to the BS. Each protocol has some advantages and disadvantages. Some of the hierarchical protocols are LEACH [5][6], EEHC [4], PEGASIS [7], HEED [8][9], DWEHC [10], TEEN [11][12], APTEEN [13].

Low Energy Adaptive Clustering Hierarchy (LEACH): LEACH [5][6][14] is one of the most and first popular hierarchical cluster-based routing protocol for WSNs, which reduces the power consumption, hence, increases the lifetime of a network.

In every cluster, one sensor node acts as CH and others act as member nodes. As CH does the data collection, aggregation and routing, it dissipates more energy and if any sensor node remains CH permanently, it dies quickly due to lack of energy. A sensor node becomes a CH for the current rotation round if the number is less than following threshold:

$$T(n) = \frac{p}{1-p*(r \bmod \frac{1}{p})};\ n \in G \qquad (1)$$

$$= 0;\ otherwise$$

Where, p is the desired percentage of CHs in the sensor population, r is the current round number and G is the set of sensor nodes that have not been CHs in the last $1/p$ rounds.

LEACH reduces energy dissipation by following features:

- Reducing the number of transmission to sink using CH.
- Reducing the amount of transmitted data through compression technique.
- Increases the life time of all sensor nodes through randomized rotation being as CH [5][6].

Some of the improvements of LEACH are as follows:

- **Extended Low-Energy Adaptive Clustering Hierarchy (E-LEACH)**
- **Multi-hop Low-Energy Adaptive Clustering Hierarchy (M-LEACH)**
- **Low-Energy Adaptive Clustering Hierarchy Centralized (LEACH-C)**
- **Low-Energy Adaptive Clustering Hierarchy with Fixed Cluster (LEACH-F)**
- **Solar-aware Low-Energy Adaptive Clustering Hierarchy (S-LEACH).**

Energy Efficient Hierarchical Clustering (EEHC): To maximize the network lifetime, Bandyopadhyay and Coyle [4] propose a distributed randomized clustering algorithm for WSNs, known as Energy Efficient Hierarchical Clustering (EEHC).

Power-Efficient Gathering in Sensor Information Systems (PEGASIS): PEGASIS [7] is an extension of LEACH [5], where sensor nodes form chains rather than forming multiple clusters. Here, each sensor node transmits and receives data from neighbor nodes.

Hybrid, Energy-Efficient Distributed Clustering (HEED): HEED [8][9] extends the basic scheme of LEACH, where the algorithm uses residual energy and node degree or density as a metric for cluster selection to achieve power balancing.

Distributed Weight-Based Energy-Efficient Hierarchical Clustering (DWEHC): Ding *et al.* [10] propose DWEHC to achieve more efficiency than HEED [9], where the algorithm optimizes cluster size and intra-cluster topology.

Threshold Sensitive Energy Efficient Sensor Network Protocol (TEEN): TEEN [11][12] is a hierarchical clustering protocol, which groups sensor nodes into clusters where each cluster has a CH.

Adaptive Periodic Threshold Sensitive Energy Efficient Sensor Network Protocol (APTEEN): The advancement of TEEN is APTEEN [13] to overcome some drawbacks of TEEN, which results in both capturing periodic data collections (LEACH) and reacting to time-critical events (TEEN).

Table 1. Comparison of different hierarchical energy-efficient routing protocols.

Protocol	Node Mobility	Cluster Overlapping	Energy Efficient	CH Selection	Algorithm Complexity	Methodology
LEACH	Fixed BS	No	High	Random	Constant	Distributed
E-LEACH	Fixed BS	No	Very High	Random	Constant	Distributed
M-LEACH	Fixed BS	No	Very High	Random	Variable	Distributed
LEACH-C	Fixed BS	No	Very High	Pre-assigned	Constant	Centralized
LEACH-F	Fixed BS	No	Very High	Pre-assigned	Constant	Centralized
S-LEACH	Fixed BS	No	Very High	Random	Variable	Distributed/ Centralized
EEHC	No	Yes	Very High	Random	Variable	Distributed
PEGASIS	Fixed BS	No	High	Random	Constant	Distributed
H-PEGASIS	Fixed BS	No	High	Random	Variable	Distributed
EB-PEGASIS	Fixed BS	No	Very High	Random	Constant	Distributed
HEED	Stationary	No	Very High	Random	Constant	Distributed
DWEHC	Stationary	Yes	Very High	Random	Constant	Distributed

Table 1. (*Continued*)

Protocol	Node Mobility	Cluster Overlapping	Energy Efficient	CH Selection	Algorithm Complexity	Methodology
TEEN	Fixed BS	Yes	High	Random	Constant	Centralized
APTEEN	Fixed BS	Yes	Very High	Random	Constant	Centralized

Table 1 shows the comparison of different hierarchical energy-efficient routing protocols.

4. Context-Aware Cluster-Based Routing

So, there exist several numbers of hierarchical protocols [4][7][8][11] depending on different aspects of hierarchical routing technique.
Context-Aware Computing
The core feature of ubiquitous and pervasive computing systems is context-awareness. Context-aware computing consists of five mechanisms:

i) Context-gathering or Context-acquisition
ii) Context-modeling
iii) Context-reasoning
iv) Context-conflict solving
v) Context-storage and management.

4.1. *Context-aware based clustering algorithm*

Some of the context-aware cluster-based algorithms are as follows:
Context-Aware Cluster-Based Hierarchy (CACH): The primary objective of CACH [23] protocol is to achieve maximum energy efficiency. Here, cluster formation is completely context-aware and distributed as it is quite evident that distributed cluster formation is one of the important features of WSNs. The CACH algorithm consists of two phases namely, i) Set up phase and ii) Steady-operation phase.

Context-Aware Unequal-Clustering Routing Algorithm (CAUCR): Energy hole is one of the major problems in WSNs which reduces the lifetime of WSNs due to the characteristics of the multi-hop forwarding model. CAUCR [24] proposes a partial solution to resolve the problem of energy hole, where it uses an Optimized Weighted Unequal-Clustering Algorithm (OWUCA) and a Reverse Minimum Energy (RME) multi-hop routing algorithm. In previously discussed algorithms [4][5][6][8][9][10][11][13], many algorithms [5][6][8][11] focus on the energy consumption in the routing phase and it ignores the energy efficiency of clustering phase.

Context-Awareness in Sea Computing Model (CASCR): The authors propose CASCR [25] for Internet of Things (IoT), which synchronizes the concept of Sea Computing model [25] and the routing mechanism of WSN.

Context-Aware Affinity Propagation Clustering Protocol (CAAPCP): Context-Aware Affinity Propagation Clustering Protocol (CAAPCP) [26] is a distributed, low-overhead context-aware clustering protocol for WSNs.

Table 2. Comparison of different context aware based clustering protocols.

Protocol	Context Type	Efficiency	CH Selection	Node Mobility	Cluster Overlapping	Failure Recovery
CACH	Sensed Data	Less efficient	Random	Yes	No	Yes
CAUCR	Minimum ho to base distance and residual energy of neighbor CHs	Efficient	Random	Yes	No	No
CASCR	IoT and Sea computing model	Efficient	Constant	No	No	No
CAAPCP	Location, Residual energy and contextual data	Efficient	Random	Yes	No	NO

5. Comparison of Hierarchical Cluster-Based Routing and Context-Aware Cluster-Based Routing

Table 3 shows the comparison of Hierarchical energy-efficient cluster-based routing protocols and Context-aware cluster-based routing protocols.

Table 3. Comparison of hierarchical energy-efficient cluster-based routing protocols and context-aware cluster-based routing.

Protocol	Clustering Attributes	Energy Efficiency	Cluster Overlapping	Clusterhead Selection	Problem of similar data traffic	Utilization of Energy
Hierarchical energy efficient protocol	Sensor attributes (Hardware approach)	Less efficient	Possible	Less efficient	Possible	Not properly utilized
Context aware cluster based protocol	Sensed data as context	Efficient	Not Possible	Efficient	Not possible	Properly utilized

Thus, for several reasons, Context-Aware Cluster-Based Routing reduces data traffic compare to Hierarchical Cluster-Based Routing. It results in more idle time and longevity of the WSNs.

6. Conclusion

WSN attracts significant attention due to its multidimensional applications. The application includes disaster management, border protection, combat field surveillance, where the application deploys large number of sensor nodes. To manage the sensor nodes effectively, grouping the sensor nodes into clusters plays an important role in WSNs. As a result, clustering mechanism becomes one of the important research domains in WSNs. This paper gives a survey on several important Hierarchical Energy Efficient Cluster Based Routing protocols as well as several Context-Aware Cluster Based Routing protocols. Each protocol has its advantages and drawbacks. Finally, this paper carries out an effective comparison between Hierarchical Energy Efficient Cluster Based Routing and Context Aware Cluster Based Routing.

The parameters affecting the cluster formation and CH communication are open issues for future research. On the other hand, the process of data aggregation and fusion among clusters is also an important domain to work on.

References

1. S. K. Singh, M. P. Singh and D. K. Singh, A survey of Energy-Efficient Hierarchical Cluster-Based Routing in Wireless Sensor Networks, in *International Journal of Advanced Networking and Applications*, **2**, 570 (2010).
2. Ameer Ahmed Abbasi and Mohamed Younis, A survey on clustering algorithms for wireless sensor networks, in *Computer Communications 2007*, **30**, 2826 (2007).
3. Qing Bian, Yan Zhang and Yanjuan Zhao, Research on Clustering Routing Algorithms n Wireless Sensor Networks, in *International Conference on Intelligent Computation Technology and Automation*, 1110 (2010).
4. S. Bandyopadhyay and E. Coyle, An energy efficient hierarchical clustering algorithm for wireless sensor networks, in *Proceedings of the 22nd Annual Joint Conference of the IEEE Computer and Communications Societies (INFOCOM 2003)*, 1 (San Francisco, California, 2003).
5. W. R. Heinzelman, A. Chandrakaasan and H. Balakrishnan, An Aplication-Specific Protocol Architecture for Wireless Microsensor Networks, in *IEEE transactions on Wireless Communications,* **1**, 660 (2002).
6. W. R. Heinzelman, A. Chandrakaasan and H. Balakrishnan, Energy-Efficient Communication Protocol for Wireless Microsensor Networks, in

IEEE Computer Society proceedings of the Thirty Third Hawali International Conference on System Sciences (HICSS '00) **8**, 1 (Washington, DC, USA, 2000).

7. S. Lindsey and C. S. Raghavendra, PEGASIS: Power-efficient Gathering in Sensor Information System, in *Proceedings IEEE Aerospace Conference, MT***3**, 1125 (2002).
8. Ossama Younis and Sonia Fahmy, Distributed Clustering in Ad-hoc Sensor Networks: A Hybrid, Energy-efficient Approach, in *INFOCOM,* 629 (2004).
9. Ossama Younis and Sonia Fahmy, Heed: A hybrid, Energy-efficient, Distributed Clustering Approach for Ad-hoc Networks, in *IEEE Transactions on Mobile Computing*, **3**, 366 (Oct.-Dec. 2004).
10. P. Ding, J. Holiday and A. Celik, Distributed energy efficient hierarchical clustering, in *Proceedings of the IEEE International Conference on Distributed Computing in Sensor Systems (DCOSS'05)*, 322 (Marina Del Rey, CA, 2005).
11. A. Manjeshwar and D. P. Agrawal, TEEN: A Protocol for Enhanced Efficiency in Wireless Sensor Networks,in Proceedings of the 1st International Workshop on Par- allel and Distributed Computing Issues in Wireless Networks and Mobile Computing, 1 (San Francisco, CA, 2001).
12. W. Lou, An Efficient N-to-1 Multipath Routing Protocol in Wireless Sensor Networks, *in Proceedings of IEEE MASS05*, 1 (Washington DC, 2005).
13. A. Manjeshwar and D. P. Agrawal, APTEEN: A Hybrid Protocol for Efficient Routing and Comprehensive Information Retrieval in Wireless Sensor Networks," in *Proceedings of the 2nd International Workshop on Parallel and Distributed Computing Issues in Wireless Networks and Mobile computing*, pp. 1–8 (San Francisco CA, 2002).
14. W. R. Heinzelman, Application-Specific Protocol Architecture for Wireless Networks, *PhD thesis, Massachusetts Institute of Technology*, 1 (2000).
15. Thiemo Voigt, Hartmut. Ritter, Jochen Schiller, Adam Dunkels and Juan Alonso, Solar-Aware Clustering in Wireless Sensor Networks, in *Proceedings of the Ninth IEEE Symposium on Computers and Communications*, 238 (2004).
16. S. Banerjee and S. Khuller, A clustering scheme for hierarchical control in multi-hop wireless networks, in *Proceedings of 20th Joint Conference of the IEEE Computer and Communications Societies (INFOCOM'01)*, 1028 (Anchorage, AK, 2001).

17. S. Lindsey, C.S. Raghavendra and K. Sivalingam, Data Gathering Algorithms in Sensor Networks using Energy Metrics, in *IEEE transactions on parallel and distributed systems*, **13**, 924 (September 2002).
18. Liu Yueyang, Ji Hong and Yue Guangxin, An Energy-Efficient PEGASIS-Based Enhanced Algorithm in Wireless Sensor Networks, in *China Communications Technology Forum*, 91 (2006).
19. W. R. Heinzelman, A. Chandrakasan and H. Balakrishnan, Energy-Efcient Communication Protocols for Wireless Microsensor Networks, in *Proceedings of Hawaiian International Conference on Systems Science*, 1 (2000).
20. C. Perera, A. Zaslavsky, P. Christen and D. Georgakopoulos, Context Aware Computing for The Internet of Things: A Survey, in *IEEE Communications Surveys and Tutorials*, **16**, 414 (2014).
21. G. Chong, L. Zhihao and Y. Yifeng, The research and implement of smart home system based on internet of things, in *Electronics, Communications and Control (ICECC), 2011 International Conference*, 430 (2011).
22. Y. Chen, J. Duo and X. Hu, The research of internet of things supporting technologies which face the logistics industry, in *Computational Intelligence and Security (CIS), 2010 International Conference*, pp. 659–663 (2010).
23. Md Enamul Haque, Noriko Matsumoto, and Norikho Yoshida, Context-aware cluster-based hierarchical protocol for Wireless Sensor Networks, in *Int. J. Ad Hoc and Ubiquitous Computing*, **4**, 379 (2009).
24. Longpeng Zheng, Zhiping Jia, Ruihua Zhang, Hui Xia, Lei Ju and Chuanhao Qu, Context-Aware Routing for WSNs Based on Unequal Clustering, in *2013 12th IEEE International Conference on Trust, Security and Privacy in Computing and Communications*, 1307 (2013).
25. Zhikui Chen, Haozhe Wang, Yang Liu, Fanyu Bu and Hen Wei, A Context-Aware Routing Protocol on Internet of Things Based on Sea Computing Model, in *Journal of Computers*, 7, 96 (2012).
26. Mahmoud ElGammal and Mohamed ElToweissy, Distributed Context-Aware Affinity Propagation Clustering in Wireless Sensor Networks, 1 (2010).

Speech Emotion Recognition Using Vowel Onset and Offset Points

Manish Kumar[1] and Jainath Yadav[2]
Dept. of Computer Science, Central University of South Bihar
Patna, India 800014
E-mail: [1]manishkumar08353@cub.ac.in, [2]jainath@cub.ac.in

Emotion plays an important role in speech processing. An experiment has been performed for speech emotion recognition using vowel region. Starting and ending points of vowel regions are identified using vowel onset and vowel offset points. In most of the previous work prosodic features have been extracted at the utterance level. This approach extracts Mel frequency cepstral coefficients (MFCCs) spectral features, which is used to classify eight different emotional classes such as Anger, Disgust, Fear, Happy, Sadness, Neutral, Sarcastic and Surprise. This experiment has been done on IITKGP: SEHSC database. Gaussian mixture model (GMM) is used as a classifier. Recognition rate of the model using utterance level speech signal for male and female speakers are 65.47% and 65%, while improved recognition rate using vowel region for male and female speakers are 75.59% and 69.58% respectively.

Keywords: Emotion Recognition; Mel frequency cepstral coefficients; Gaussian mixture model; Vowel Onset Point; Vowel Offset Point.

1. Introduction

Emotion recognition from speech has gained increasing attention in the recent years. Human beings extensively use emotions to convey the intended information along with the textual message. Majority of the today's speech systems are developed using neutral speech. Component of emotion is essential to make them more natural for practical applications. Speech emotion recognition has some of the important applications such as: call center conversation analysis that helps to improve quality of service of a call attendant [3], interactive movie [4], storytelling and E-tutoring [1] applications would be more practical if they can adapt themselves to listeners or students emotional states. The automatic way to analyze the emotions in speech is useful for indexing and retrieving the audio/video files based on emotions [5]. Medical doctors may use the emotional contents of the patients speech as a diagnosing tool for various ailments. Emotion analysis of telephone conversation between criminals would help crime investigation department. Conversation with robotic pets and humanoid partners would be more realistic and enjoyable, if they are able to express and understand emotions like humans. Emotion recognition based on speech is one of intensively studied research topics in the domains of human-computer interaction and affective computing. Speech emotion recognition can also be used to augment automated medical or forensic data analysis systems.

Emotions are often portrayed differently in different cultures and languages. For example, a specific type of intonation which indicates admiration in Japanese, indicates disbelief in English [6]. A method of translating emotions, in addition to text, between the languages may improve multi/cross lingual communication. The present study is aimed at recognizing the text-dependent and text-independent emotions from speech utterances using spectral features.

Gaussian mixture models are used to develop emotion recognition models. Text-dependent emotion recognition is performed by using the same text prompts for both training and testing of the models. However the utterances are chosen from different recording sessions. In text-independent case, the text prompts used for training and testing the emotion recognition models are different.

For the literature, it is observed that, majority of the studies on emotion recognition have used algorithms based on prosodic features [5] [7]. In the present work, the focus has been given to the spectral features. In particular the effectiveness of Mel frequency cepstral coefficients (MFCCs) is explored for speech emotion recognition.

Rest of the paper is organized in 5 sections. Section 2 contains the description of speech database, Section 3 describes feature extraction and development of emotion recognition models, Section 4 contains the steps of vowel onset and offset point detection. In Section 5, results are discussed. Paper is concluded with summary and some important citations.

2. IITKGP: SEHSC (Indian Institute of Technology Kharagpur: Simulated Emotion Hindi Speech Corpus)

The proposed database is recorded using 10 (5 male and 5 female) professional artists from Gyanavani FM radio station, Varanasi, India. The artists have sufficient experience in expressing the desired emotions from the neutral sentences. The male artists are in the age group of 28–48 years with varied experience of 5–20 years. Similarly female artists are from the age group of 20–30 years with 3–10 years of experience. For recording the emotions, 15 Hindi text prompts are used. All the sentences are emotionally neutral in meaning. Each of the artists has to speak the 15 sentences in 8 basic emotions in one session. The number of sessions recorded for preparing the database is 10. The total number of utterances in the database is 12000 (15 text prompts * 8 emotions * 10 speakers * 10 sessions). Each emotion has 1500 utterances. The number of words and syllables in the sentences vary from 4–7 and 9–17 respectively. The total duration of the database is around 9 hours. The eight emotions considered for collecting the proposed speech corpus are: anger, disgust, fear, happy, neutral, sad, sarcastic and surprise. The speech samples are recorded using SHURE dynamic cardioids microphone C660N. The distance between the speaker and the microphone is maintained to be around 1 ft. Speech signal was sampled at 16 kHz and each sample is represented as 16 bit number. The sessions are recorded on alternate days to capture the variability in human speech production mechanism. In each session, all the artists have given the recordings of 15 sentences in 8 emotions. The recording has done in such a way that each artist has to speak all the sentences at a stretch in a particular emotion. This provides the coherence among the sentences for each emotion category. Since, all the artists are from the same organization, it ensures the coherence in the quality of the collected speech data. The entire speech database is recorded using single microphone and at the same location. The recording was done in a quiet room, without any obstacles in the recording path [8].

3. Development of Emotion Recognition Models

Emotion recognition using pattern classifiers is basically a two stage process as shown in Fig. 1. In the first stage emotion recognition models are developed by training the models using the feature vectors extracted from speech utterances of known emotions. It is known as supervised learning. In second stage testing (evaluation) of the trained models is performed by using the speech utterances of unknown emotions.

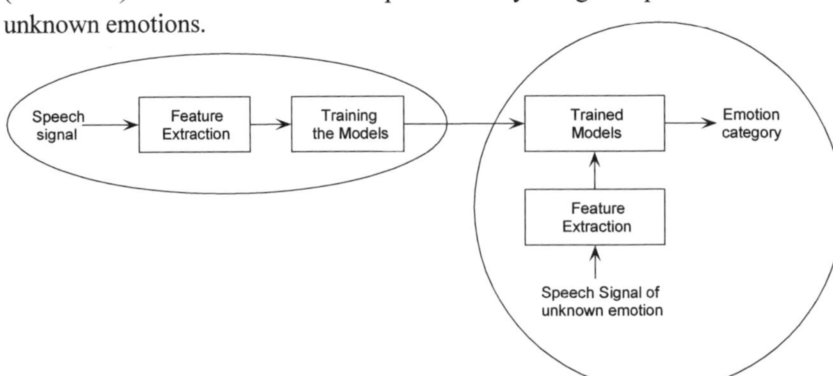

Fig.1. Block diagram of emotion recognition system.

3.1. *Feature extraction*

Mel frequency cepstral coefficients (MFCCs) have been used as features for recognizing the speech emotions. Thirteen mel frequency cepstral coefficients are extracted from speech signal to represent gradual spectral variations. The speech frames of size of 20 ms, each time overlapped by 10 ms are used to extract these 13 features. Hamming window is used while framing the speech signal. MFCCs features are extracted from these frames using the MFCC algorithm given in [12]. Cepstral mean subtraction and variance normalization are carried out in order to compensate recording variations. The normalization is done at the sentence level.

3.2. *Gaussian mixture models (GMM)*

In this work, GMMs are used to develop emotion recognition systems using spectral features. GMMs are known to capture distribution of data points from the input feature space. Therefore, GMMs are suitable for developing emotion recognition models using spectral features, as the decision regarding the emotion category of the feature vector is taken based on its probability of coming from the feature vectors of the specific model.

Gaussian mixture models (GMMs) are among the most statistically matured methods for clustering and for density estimation. They model the probability density function of observed data points using a multivariate Gaussian mixture density. Given a set of inputs, GMM refines the weights of each distribution through expectation-maximization algorithm. Once a model is generated, conditional probabilities can be computed for test patterns (unknown data points). Number of Gausses in the mixture model is known as number of components. They indicate the number of clusters in which data points are to be classified. In this work, one GMM is developed to capture the information about one emotion. The components within each GMM capture finer level details among the feature vectors of each emotion. Depending on the number of data points, number of components may be varied in each GMM. Presence of few components in GMM and trained using large number of data points may lead to more generalized clusters, failing to capture specific details related to each class. On the other hand over fitting of the data points may happen, if too many components represent few data points. Obviously the complexity of the models increases, if they contain higher number of components. In this work, GMM's are designed with 16 components and iterated for 100 times to attain convergence of weights. The emotion recognition systems developed using spectral features are shown in Figure 2.

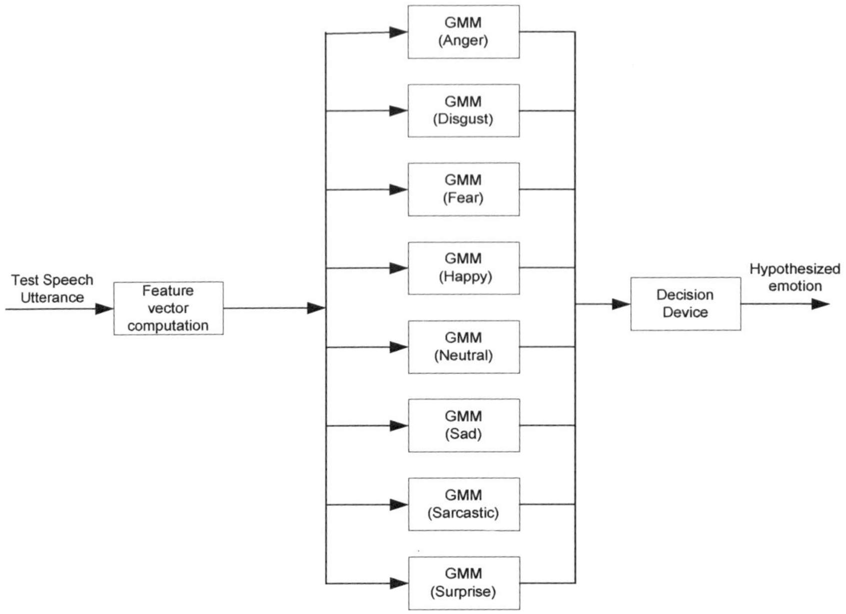

Fig. 2. Block diagram of emotion recognition system using Gaussian mixture models.

4. Detection of Vowel Onset and Offset Points [1] [2]

In Indian languages, more than 90% of the syllables are of consonant-vowel (CV) type [10]. In a CV unit, speech segment before the vowel onset point is the consonant region, and after the vowel onset point is the vowel region. These details are shown in Fig. 3.

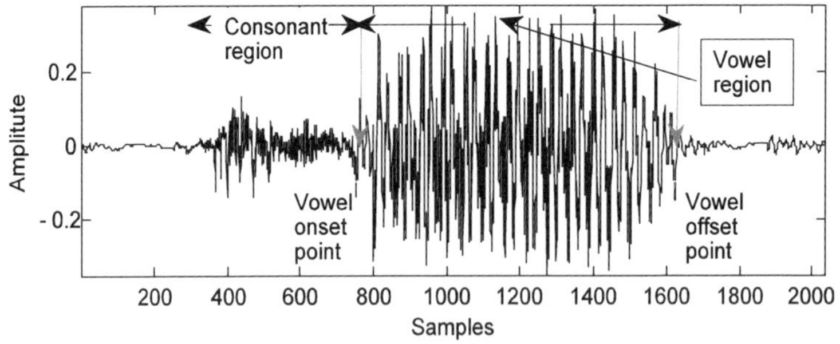

Fig. 3. Vowel onset and offset points for an utterance /chi/.

Vowel region starts with vowel onset point and ends with vowel offset point. In [10] [11], 80 and 100 ms regions respectively, after the vowel onset point are considered as vowel regions. However, vowel regions may not exactly represent 80 or 100 ms regions after each vowel onset point. There exists a lot of variability in durations of vowels within and across the vowel categories. Considering fixed duration of the vowel leads to inaccurate estimation of vowel regions, and applications based on these vowel regions may be severely affected. Following steps are used to find vowel onset and offset points:

(1) Determine the epoch locations or glottal closure instants (GCI) using zero frequency filtering (ZFF) method [9].
(2) Compute discrete Fourier transform for the speech samples present in 30% of glottal cycle starting from the GCI.
(3) Determine the spectral energy within the frequency band of 500–2500 Hz.
(4) Fluctuations in the spectral energy contour are smoothed by using mean smoothing with 50 ms window.
(5) The smoothed spectral energy is further enhanced by computing its slope using a first-order difference.
(6) The significant changes in the vowel evidence signal are detected by convolving the smoothed spectral energy contour with first order Gaussian difference operator of length 100 ms.
(7) Peaks and valleys in the vowel evidence plot represents the locations of vowel onset and offset points, respectively.

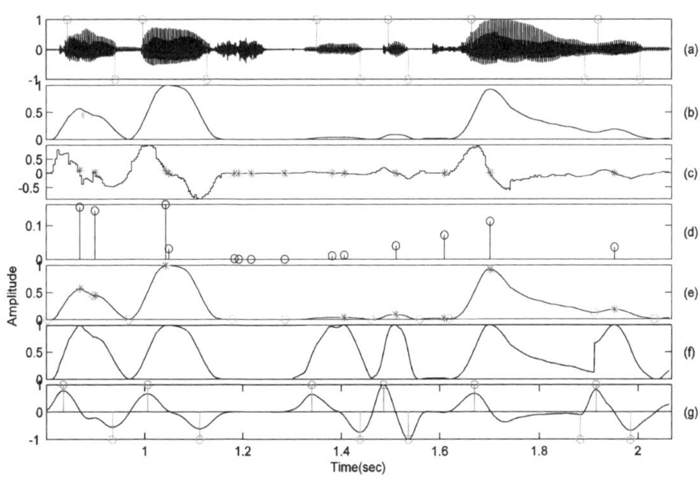

Fig. 4. Vowel offset point detection using SE-GC method for a speech utterance /Don't ask me to carry an/. (a) Speech signal with manually marked vowel onset and offset points. (b) Spectral energy in 500–2500 Hz band around each epoch. (c) Mean smoothed spectral energy. (d) Enhanced spectral energy signal. (e) Locations of vowel onset and offset points in the vowel evidence plot obtained using SE-GC method.

The outputs of each of the steps are shown in Fig. 4. Fig. 4(a) shows the speech signal /"Don't ask me to carry an"/ with manually marked vowel onset and offset points. Spectral energy in 500–2500 Hz band around each epoch is shown in Fig. 4(b) (step 3). Smoothed plot of spectral energy is shown in Fig. 4(c) (step 4). Fig. 4(d) shows the enhanced plot of Fig. 4(c) (step 5). Vowel evidence plot is shown in 4(e) (step 6).

5. Experimental Results and Discussion

Experiment is performed on IITKGP:SEHSC for male and female speakers. 70% of the database is used for the training purpose and remaining 30% of the database is used for testing the model. Firstly 13 MFCCs are extracted from the uttered speech. Eight different GMM models corresponding to each emotion class are build using the feature vector. The recognition rate for the male and female artists are 65.47% and 65% respectively. The result of the performed experiment is shown in Table 1 and Table 2 respectively. The diagonal elements of the confusion matrix shows the recognition performance for each emotion class.

Table 1. Emotion recognition percentage of male speaker's speech on IITKGP-SEHSC database, based on MFCC feature vector of uttered speech.

	Male Artist (Average: 65.47)							
	Anger	Disgust	Fear	Happy	Neutral	Sadness	Sarcastic	Surprise
Anger	**80.95**	9.52	0	4.76	0	0	0	4.76
Disgust	9.52	**71.43**	0	0	0	0	9.52	9.52
Fear	4.76	0	**85.71**	0	0	9.52	0	0
Happy	0	33.33	0	**57.14**	0	0	9.52	0
Neutral	0	9.52	0	0	**80.95**	9.52	0	0
Sadness	4.76	23.81	14.29	0	0	**47.62**	0	9.52
Sarcastic	0	42.86	0	0	0	0	**47.62**	9.52
Surprise	0	19.05	0	23.81	0	0	4.76	**52.38**

Table 2. Emotion recognition percentage of female speaker's speech on IITKGP-SEHSC database, based on MFCC feature vector of uttered speech.

	Female Artist (Average: 65)							
	Anger	Disgust	Fear	Happy	Neutral	Sadness	Sarcastic	Surprise
Anger	**53.33**	3.33	3.33	36.67	0	0	0	3.33
Disgust	3.33	**13.33**	6.67	20	0	0	3.33	53.33
Fear	0	3.33	**76.67**	0	0	0	0	20
Happy	10	0	0	**60**	0	0	20	10
Neutral	0	0	0	10	**83.33**	6.67	0	0
Sadness	0	0	3.33	3.33	20	**73.33**	0	0
Sarcastic	0	0	0	3.33	0	0	**90**	6.67
Surprise	0	0	0	0	0	0	30	**70**

In this approach, at first vowel onset and offset points are identified to find out the vowel region. The 13 Mel frequency cepstral coefficients (MFCCs) of vowel regions are extracted to form the feature vector. Feature vector is used to build 8 different GMM model. The recognition rate for the male and female artists are 75.59% and 69.58% shown in Table 3 and Table 4 respectively. The diagonal elements of the confusion matrix shows the recognition performance for each emotion class.

Table 3. Emotion recognition percentage of male speaker's speech on IITKGP-SEHSC database, based on MFCC feature vector of vowel region.

	Male Artist (Average: 75.59)							
	Anger	Disgust	Fear	Happy	Neutral	Sadness	Sarcastic	Surprise
Anger	**61.90**	14.29	4.76	19.05	0	0	0	0
Disgust	0	**76.19**	0	4.76	0	0	14.29	4.76
Fear	0	0	**95.24**	0	0	4.76	0	0
Happy	0	28.57	0	**38.10**	0	4.76	28.57	0
Neutral	0	0	0	0	**85.71**	14.29	0	0
Sadness	4.76	0	0	4.76	0	**95.24**	0	0
Sarcastic	0	9.52	0	4.76	0	0	**80.95**	0
Surprise	0	0	0	14.29	0	0	14.29	**71.43**

Table 4. Emotion recognition percentage of female speaker's speech on IITKGP-SEHSC database, based on MFCC feature vector of vowel region.

	Female Artist (Average: 69.58)							
	Anger	Disgust	Fear	Happy	Neutral	Sadness	Sarcastic	Surprise
Anger	**86.67**	0	10	3.33	0	0	0	0
Disgust	16.67	**40**	6.67	0	3.33	0	10	23.33
Fear	0	3.33	**73.33**	3.33	6.67	0	0	13.33
Happy	10	0	0	**60**	0	6.67	20	3.33
Neutral	0	0	0	0	**86.67**	13.33	0	0
Sadness	0	0	3.33	3.33	46.67	**46.67**	0	0
Sarcastic	0	0	0	6.67	0	0	**90**	3.33
Surprise	3.33	10	0	0	0	0	13.33	**73.33**

The recognition performance of both male and female speakers are shown in the histogram. Comparison between the percentage recognition rate of male speaker's uttered speech and vowel region of male speaker's uttered speech has been shown in Figure 5.

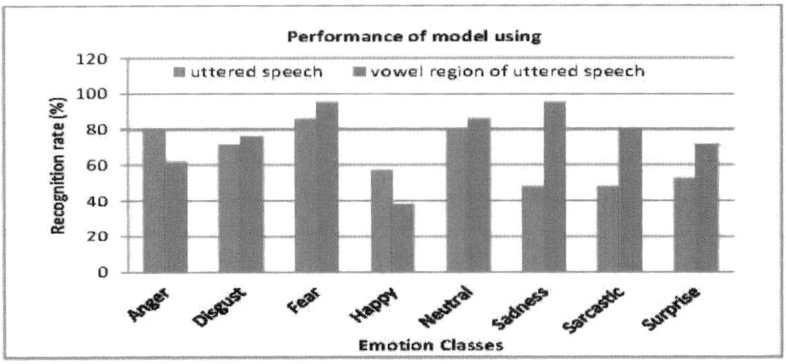

Fig. 5. Shows the percentage recognition rate comparison between male speaker's uttered speech and vowel region of mal e speaker's uttered speech.

Comparison between the percentage recognition rate of female speaker's uttered speech and vowel region of female speaker's uttered speech has been shown in Figure 6.

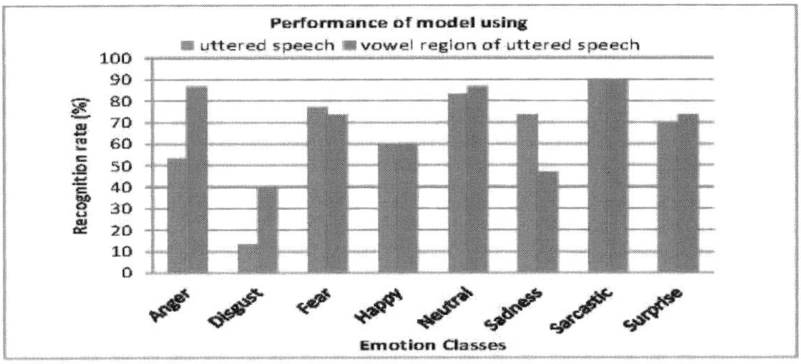

Fig. 6. Shows the percentage recognition rate comparison between female speaker's between uttered speech and vowel region of female speaker's uttered speech.

6. Summary and Conclusion

In this paper, a method of emotion recognition using vowel region of uttered speech signal has been proposed. To find the exact location of starting and ending point of vowel, vowel onset and offset points have been detected using SE-GC method. To construct the feature vector, Mel frequency cepstral coefficients (MFCCs) have been extracted. One GMM corresponding to each emotion class has been built for training purpose. All the vowel regions are concatenated after identifying the vowel regions for a test speech utterance. MFCC of concatenated vowel regions are extracted. Extracted MFCC of each test speech utterance is compared with the all trained model. On the basis of maximum likelihood, decision device classify speech utterance into its appropriate emotion class. IITKGP:SEHSC database has been used as a speech database. The recognition rate of the proposed model using utterance level speech signal is 65.47% for male speakers and 65% for female speakers. The improved recognition rate using vowel region of the uttered speech signal is 75.59% for male speakers and 69.58% for female speakers. With the use of vowel onset and offset points detection method, performance of the model has been increased.

References

1. J. Yadav and K. S. Rao, Detection of Vowel Offset Point From Speech Signal, *Signal Processing Letters, IEEE*, **20**, 299 (2013).
2. A. K. Vuppala, J. Yadav, S. Chakrabarti and K. S. Rao, Vowel Onset Point Detection for Low Bit Rate Coded Speech, *Audio, Speech and Language Processing, IEEE Transactions on*, **20**, 1894 (2012).
3. K. S. R. Murty and B. Yegnanarayana, Epoch Extraction From Speech Signals, *Audio, Speech and Language Processing, IEEE Transactions on*, **16**, 1602 (2008).
4. R. Nakatsu, J. Nicholson and N. Tosa, Emotion recognition and its application to computer agents with spontaneous interactive capabilities, *Knowledge-Based Systems*, **13**, 497 (2000).
5. T. V. Sagar, Characterization and synthesis of emotions in speech using prosodic features, Master's Thesis, Dept. of Electronics and communications Engineering, Indian Institute of Technology Guwahati.
6. J. Nicholson, K. Takahashi, R. Nakatsu, Emotion Recognition in Speech Using Neural Networks, *Neural Computing and Applications*, **9**, 290, (2000).
7. I. Luengo, E. Navas, I. Hernaez and J. Sanchez, Automatic Emotion Recognition using Prosodic Parameters, in *Proc. of INTERSPEECH*, (Lisbon, Portugal, 2005).
8. S. G. Koolagudi, R. Reddy, J. Yadav and K. S. Rao, IITKGP-SEHSC: Hindi Speech Corpus for Emotion Analysis, *Devices and Communications (ICDeCom), 2011 Internationqal Conference on*, **1**, 24 (2011).
9. K. T. Deepak, K. Ramesh, S. R. M. Prasanna, Extraction of glottal closure and opening instants using zero frequency filtering, *India Conference (INDICON), 2014 Annual IEEE*, **1**, 11 (2014).

10. C. C. Sekhar and B. Yegnanarayana, Neural network models for spotting stop consonant-vowel (SCV) segments in continuous speech, *Neural Networks, 1996, IEEE International Conference on*, **4**, 2003 (1996).
11. S. R. M. Prasanna and G. Pradhan, Significance of Vowel-Like Regions for Speaker Verification Under Degraded Conditions, *Audio, Speech and Language Processing, IEEE Transactions on*, **19**, 2552 (2011).
12. L. R. Rabiner and B. H. Juang, *Fundamentals of Speech Recognition*, (Prentice-Hall, 1993).

A Novel Algorithm for Magic Squares

Govind Kumar Jha
Research Scholar
gvnd.jha@gmail.com

Neeraj Kumar
Head, Dept of Computer Science
Nalanda College of Engg Govt. of Bihar
javaneeraj@gmail.com

Dr. Prabhat Ranjan
Assistant Professor
Central University of South Bihar
prabhatranjan@cub.ac.in

A. P. Shakya
Project Engineer
Wipro Technologies, Pune
amanshakya02@gmail.com

The arrangement of number in several ways gives us amazing & surprising objects, objects that we try to define historically mathematically, or relate them to religious context. This paper also presents such a magical arrangement of numbers called Magic Square, this paper is mainly concerned with the algorithm for finding the magic squares. In this text we will discuss two algorithms to find the magic square. The first one finds the magic square of odd order while the second one can be used to find the magic square of doubly even order (i.e. in multiple of 4 not 2). The algorithm presented in this paper will be explained by sample runs for magic square of small order i.e. 3 for odd order magic square while 4 for doubly even order magic square, but we have successfully tested these algorithms for higher order magic square too.

Keywords: Magic square, magic square of odd order, magic square algorithm, lo-shu tortoise.

1. Introduction

You A magic square of order $n \geq 3$ is the square matrix that is filled with distinct positive integers from 1 to n^2 in such a way that the sum of the elements of any row, any column or any main diagonal is same [1]. Magic squares have turned up throughout history, some in a mathematical context and others in philosophical or religious contexts [2]. According to legend, the first magic square was discovered in China. The story is told that around 2200 B. C. the emperor Yu observed a divine tortoise crawling out of the Yellow River. On the turtle's back, there was a three-by-three array of numbers. Of course, we do not have any eye-witness accounts, so we can only imagine that the turtle looked like Figure 2, here arranged in matrix form.

4	9	2
3	5	7
8	1	6

Fig. 1.

This square became known as the *lo-shu magic square.* One can see that the sum of the numbers in any row or column, or the diagonals formed by {4, 5, 6} or {8, 5, 2}, is 15. This story is recorded in Book of Permutations [3]

Later, magic squares appeared in India, after that they were known to the Arabs, who introduced them to the West. More research on the topic was done during the Renaissance by the mathematician Cornelius Agrippa (1486–1535), who constructed magic squares of orders 3 through 9 to represent various planets, the sun, and the moon [4].

2. Previous Work

2.1. *Arithmetic of magic squares*

a. The sum of all the elements of magic square will be
   Sum = $n^2 (n^2+1)/2$ [6]
   e.g. for n = 3, sum = 45

b. Magic sum is the number that comes out by adding all the elements of any row, any column or any main diagonal of a magic square and it can be calculated as follows
   Magic Sum = $n (n^2+1)/2$ [7]
   e.g. for n = 3, Magic Sum = 15

c. Central number = (Total Number +1)/2 [8]
   e.g. n = 3, central number = 5

2.2. *Algorithm*

2.2.1. *Algorithm for odd order magic square*

The magic square of order n (where n is a odd positive integer greater that 1) can be formed by running the following algorithm

1) *Data Structures*
   n is the order of magic square,
   S is the starting digit for columns,
   N is next digit &
   C is a constant.

2) *Sample run of the algorithm for order 3×3*
   We will start the algorithm by initializing the following variables
   a. Initialization
      n = 3, S = 1, N = 1, C = n+1, i.e. C=4
   b. Place the S in the middle column of the first row, as shown below

	1	

Fig. 3.

   c. Update N as N = N+C, i.e. N=5 and then move to the lower cell and place N there, as shown below

	1	
	5	

Fig. 4.

   d. Update N as N=N+C, i.e. N=9 then move to the lower cell and place N there, as shown in Fig. 5.

	1	
	5	
	9	

Fig. 5.

e. If the column gets completed then update the variables as follow: S=S+1, i.e. S=2, N=S, i.e. N=2. Now move to the left cell and place the N there, as shown below

	1	
	5	
2	9	

Fig. 6.

f. Update N as N=N+C, i.e. N=6 then move to the lower cell, since there is no lower cell so move to the first cell of the same column and place N there, as shown below

6	1	
	5	
2	9	

Fig. 7.

g. If we update N as N=N+C, then we get N=10.
Since maximum number allowed is n^2 (i.e. 9 in this example). So, condition breach occurs hence instead of N=N+C update the variables as follow

N=N+1, i.e. N=7, now move to the lower cell and place N there, as shown below

6	1	
7	5	
2	9	

Fig. 8.

h. Since the column gets completed, so taking reference to the step e, update the variables as follow
S=S+1, i.e. S=3, N=S, i.e. N=3
Now move to the left cell, since there is no left cell so move to the rightmost cell place the N there, as shown below

6	1	
7	5	3
2	9	

Fig. 9.

If we update N as N=N+C i.e. N=7. It already exists in the magic square, so again this is the case of condition breach. Hence update the variables as follow N=N+1

i. i.e. N=4. Now move to the lower cell and place N there, as shown below

6	1	
7	5	3
2	9	4

Fig. 10.

j. Update N as N=N+C, i.e. N=8. Now move lower cell but there is no lower cell so move to the first cell of the same column and place N there, as shown below

6	1	8
7	5	3
2	9	4

Fig. 11.

This is final magic square of order 3×3 with magic sum of 15.

We have developed a computer program in C to test the above algorithm. A snapshot of the output window is showing magic square of order 5 is listed below

```
Enter the order of magic square
5

15      8       1       24      17
16      14      7       5       23
22      20      13      6       4
3       21      19      12      10
9       2       25      18      11
```

Fig. 12.

2.2.2. *Algorithm for doubly even order magic square*

The magic square of order n (where n is an even positive integer divisible by 4) can be formed by running the following algorithm

We will fill the matrix by working in parallel on the rows from top side as well as from the bottom side.

a. Make a matrix of n×n and put 1 in the first row of first column and n^2 (i.e. 16) in the last row of last column as shown below

1			4
13			16

'Fig. 13.

b. Count the number of empty columns to the right hand side of 1 that will be 3 and add this count to 1, i.e. 1+3=4.
Then place that number in the rightmost empty cell in first row.
In parallel subtract that count from n^2 (i.e. 16-3=13) and place this number (i.e. 13) in the leftmost empty cell in the last row as shown in figure below

1			**4**
13			**16**

Fig. 14.

c. Now increase the 1 by 1 (i.e. 1+1=2) and put it just left to n^2. And subtract n^2 by 1 (i.e. 16-1=15) and put it just right to 1.

1	**15**		**4**
13		**2**	**16**

Fig. 15.

d. Follow the step b to find the corresponding element, i.e. Count the number of empty columns to the right of 15 (or to the left of 2), that will be 1 and add this count to 2 (i.e. 2+1=3). Then place this number (i.e. 3) in the leftmost empty in the last row.
In parallel, subtract that count from 15 (i.e. 15-1=14) and place that number (i.e. 14) in the rightmost empty cell in the first row.

1	15	14	4
13	3	2	16

Fig. 16.

e. Since the first & last rows have been filled so move on to the interior rows (i.e. row 2 and row n-1). Since we have used the number from 1 to 4 and 16 down to 13. So start the next rows with 5 and 12.
For better understanding remove the top and bottom row as shown below

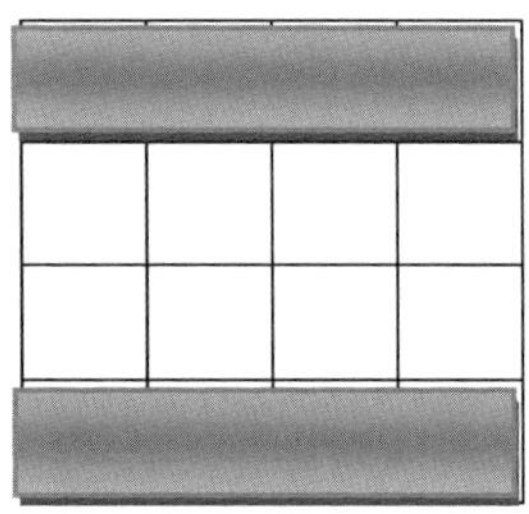

Fig. 17.

Note: After masking the first & last rows, we will refer those empty rows as first & last rows until the mask is removed.

Now put 12 in the first column of first row and 5 in the last column of last row as shown below

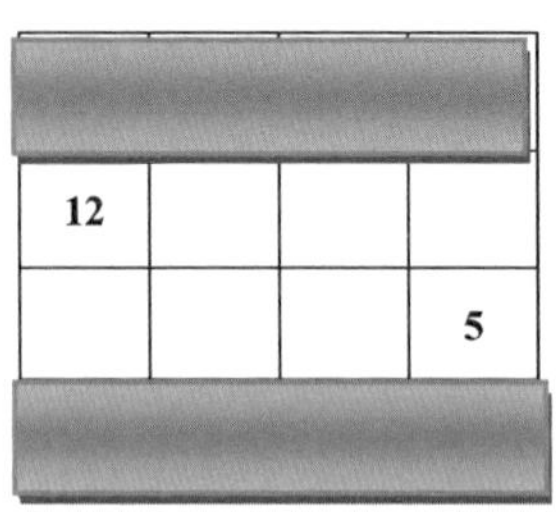

Fig. 18.

f. Now repeat steps 2, 3 & 4 intuitively to fill both of these rows. To avoid confusion whole procedure is shown in the subsequent steps

g. Count the number of empty columns to the right of 12 (or to the left of 5) that will be 3. Now subtract 3 from 12 (12-3=9) and place it in the rightmost empty cell in the first column.

In parallel, add 3 to 5 (i.e. 5+3=8) and place it in the leftmost cell of the last column.

12			9
8			5

Fig. 19.

h. Now increase the 5 by 1 (i.e. 5+1=6) and place it just right to 12.
In parallel, decrease the 12 by 1 (i.e. 12-1=11) and place it just left to 5.

12	6		9
8		11	5

Fig. 20.

i. Count the empty columns to the right of 6 (or to the left of 11) that will be 1. Now add this 1 to 6 (i.e. 6+11=7) and place it in the rightmost cell of the first row.

In parallel, subtract this 1 from 11 (i.e. 11-1=10) and place this in the leftmost cell in the last row.

12	6	7	9
8	10	11	5

Fig. 21.

Now both the rows have been filled so unmask the masked rows to get the final 4×4 magic square

1	15	14	4
12	6	7	9
8	10	11	5
13	3	2	16

Fig. 22.

This is the final magic square of order 4×4 with the magic sum of 34. We have also tested the doubly even order magic square algorithm for higher orders and here we have shown a magic square of order 8×8.

1	63	3	61	60	6	58	8
56	10	54	12	13	51	15	49
17	47	19	45	44	22	42	24
40	26	38	28	29	35	31	33
32	34	30	36	37	27	39	25
41	23	43	21	20	46	18	48
16	50	14	52	53	11	55	9
57	7	59	5	4	62	2	64

Fig. 23.

3. Conclusion

Magic squares have been studied from the ancient time and many attempts have been made to find the algorithms for generating the magic squares, we have also developed some of the algorithms that can find magic squares of odd orders and doubly even orders. Both the algorithms have been shown by a sample run for some small order. We have also tested the algorithm for magic square of odd order by making a computer program in C-language. The algorithm for doubly even order has been also tested for some higher orders and we have shown a magic square of order 8×8 that is formed by algorithm explained in this text.

References

1. S. Onkar, Magic Square, Wolfram demonstration project.
2. B. Matthias, C. Moshe, C Jessica and G. Paul, The number of magic, squares, cubes and hypercubes.
3. Eves, Howard, An Introduction to the History of Mathematics, 5th ed. (Philadelphia: Saunders College Publishing 1983).
4. Daryl Lynn Stephens, Matrix Properties of Magic Squares, (Denton Texas, April 1993).
5. C. Molar, Magic Square, chapter 10, p. 2 MathWorks Inc., (2011).
6. A. R. Ibrahim, Magic squares, (2006).
7. D Emanuele, Construction of magic squares.
8. D. Jonathan, Magic squares.

Magic Square and Nature

The Loh-Shu Tortoise

A Note on Intelligent Street Light System

Dr. J. Satheesh Kumar[1] and C. G. Sreekaviya
[1]*Department of Computer Applications, Bharathiar University, Coimbatore, India*
E-mail: jsathee@rediffmail.com
www.b-u.ac.in

Street light system is a significant challenge for researchers to save electricity. The power consumptions and control of street light system is not economically feasible in most of the existing systems due to its resource and maintenance. Wireless is an efficient and compatible environment for most of the applications due to higher utilization of Bluetooth, wifi, and global system for mobiles (GSM) or describes as Global packet radio service (GPRS) where the data pass through as a packet. Smart city people are moving towards modern communication as wireless sensor network provide comfort and satisfaction towards minimum power utilization and other resources without affecting city nature. Efficient control of street lights with minimum power consumption and optimum usage of light is a significant research problem in wireless sensor network. This paper focuses on various methods used for street light control system. It also discusses placement of existing lamps with Light Emitting diode (LED).

Keywords: Template; Proceedings; World Scientific Publishing.

1. Introduction

India is the largest producer and consumer of electricity. The main key role is utilization of energy around the world, which replicates the financial system of the country and can increase the demand of energy conservation in future. Each person speaks about energy preservation but each one of them should put an effort to reduce the energy usage by adapting smart technologies and can replace led lamp instead of olden lamps in indoor and outdoor lightings. People in today's contemporary world can adapt modern technologies of smart street light system without affecting the environment and free from pollution. Led light innovate a smooth way for renovation of lighting ability. The use of led light has longer life span and power usage is minimal while comparing with other olden lamps. Led light is not only used in home and other general purpose but also now a days in street lights. To conserve power solar is used along with led lights in each street lamp pole which helps to work independently. There are some cheap sensors which exploit low amount of current. It can attach with street lamp for automatic working and transferring of data's. If any fault occurs in the street light framework can be diagnosed easily by the host controller through the pc. Wireless sensor network technology can be used instead of wired street light system. This technological development helps in powering the modern street light system and save up to 90% of electricity. In Fig. 1 shows the architecture of olden street light system and shows how the current is distributed in the existing system.

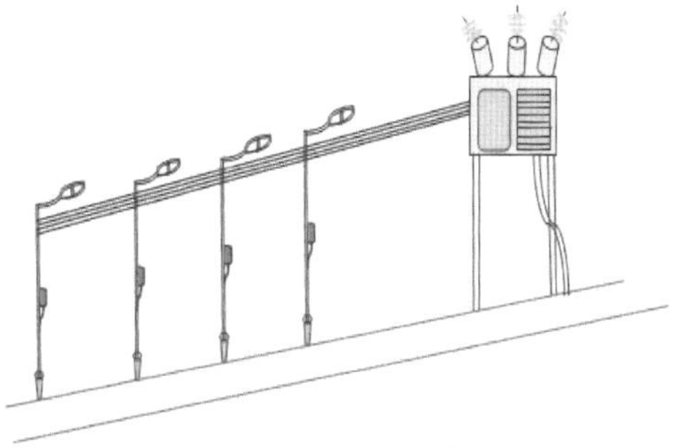

Fig. 1. Basic street light architecture.

2. Types of Lamps

Open arc street lamp was invented and used as flood lamp in the year 1880's. Incandescent lamp used for its colour rendering index (CRI) around the world in 1900's. Mercury vapour lamp was used due to its high brightness of light intensity in 1948. Metal halide has used for its long lifespan of street lamps in 1960's. Later in the year 1970's sodium vapour lamp was used a longer time due to its life time than other lamps and good colour rendering index. Led light has been used in many applications both in urban and rural from the 21st century. It vastly spread worldwide and it helps to save power usage.

3. Existing System

3.1. *Dynamic street light system*

Alexandru Lavric *et al.*, speaks about the decrease of electrical liveliness deployment has much dispute matter in an European level. Road lighting helps in various forms such as it expands comfort and user approval, expanded security and secure person on foot and street activity. The established road lighting control frameworks make use of photocells. The street light works with various conditions as 100%, 75% and 50% through programmed coding of the system [1]. Antal Haans *et al.*, tell about the forceful street light works with safety and secure condition during night. However it is modified by the people for its clever operational and existence of light in road. In olden lamps light pollution and inconvenience are found by the users. To avoid these conditions this work took an experiment in two ways, one is stationary and other one is walker. This has been done under the rule of safety, vision, disguise and escape [3]. Cleber costa Da Fonseca *et al.*, proposed the path light systems consist of wireless topology which drives using graphical user interface. It can be operated remotely so that working and managing of street light can be done by the user by saving time, human power and cost of energy. When "turn on/ off" command is passed by the user the street light works according to the instructions given through the controller. The street light architecture was designed by adding other street light nodes through GUI in an effortless way [8]. Fig. 2 shows the working status and condition of lighting level below.

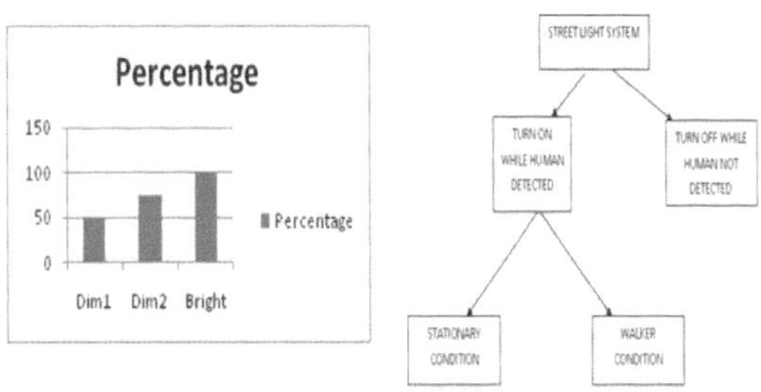

Fig. 2. Condition of light level and working status of dynamic street light system.

3.2. *Zigbee using street light system and concepts*

B. E. Biligin *et al.*, tells about the diverse applications progress based on zigbee such as load controller, power usage detection, distribution automation, etc. The act of zigbee has allowed the reduction of power and cost, end to end hold-up and smart framework development. From the results of experiments performed it is clear that zigbee is an ideal network for configuration and is reliable for real time applications [4].

C. Lozano *et al.*, works explains about the usage of wireless sensor network of zigbee without an uninterrupted flow of data. Zigbee is attached to the sensors like temperature, humidity, light, etc., which is used to communicate with the base station before the problem is going to happen in the forest. This study has made a research in the forest fire detection using zigbee plays a vital role [7].

Francesca Cuomo *et al.*, discuss that wireless sensor network, topology development is an essential problem. There are various criteria found in network topology formation such as delivery of information, life span, range of sensor and energy usage. This work categorizes the topology as network depth and node distributions in various levels; it investigates some performance of network and node formation of network topology. This work has approved to construct zigbee network standard [10].

Hung-cheng Chen *et al.*, developed a mechanical meter analysis based on the high performance, level of integration, least current consumption and cheap cost of zigbee network. This expertise has a strong constant growth due to its characteristics. It is a small range communication network but it study and develops data between meter and monitoring pc station repeatedly. This work not only provides less human resource but also makes the work easier. It also provides accurate power consumption rate without any interruption in flow of information [12].

J. Han *et al.*, tells about the standard of zigbee provides secured services to supervise the protocol defined by zigbee alliance. A number of energy nodes are produced by zigbee which can be expanded an integrated with other networks without any trouble. The outcome from this system deals with zigbee and its standards [17]. Fig. 3 shows the concept of zigbee.

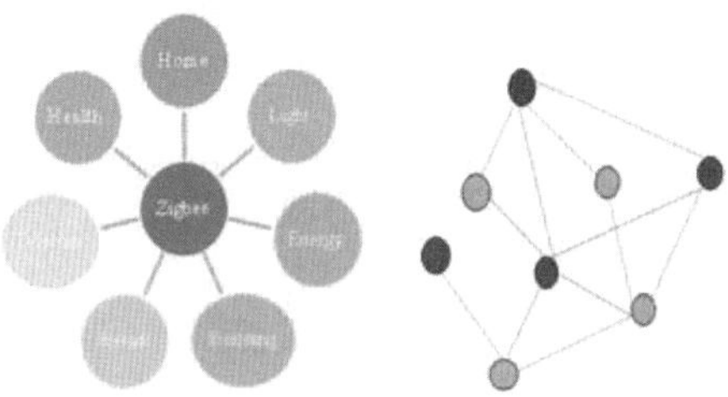

Fig. 3. Zigbee supplier and mesh network topology.

3.3. *Wireless sensor network*

Chaitanya Amin *et al.*, explains that the world current economy plays a key role in street light organization. This work explains about controlling and saving power using wireless sensor of gsm. This elegant street light structure execute mechanical turn on/ off/ dimming when essential by the atmosphere condition. The efficiency of the control station is designed and used by the user with a pc. This system can also be used with photocell [5].

Chakkor Saad *et al.*, discusses about the expansion in street light systems which uses sensors for quick presentation with better functionality. Wireless sensors such as wifi, wimax, uwb, Bluetooth, zigbee, zigbee pro, gsm and gprs are compared and assists to choose the better wireless sensor for uninterruptable flow of data from the street light system to the monitoring station [6].

F. Domingo-Perez *et al.*, explains the system works for providing savvy street light system in an efficient way using low rate wireless networks and digital addressable lighting interface protocol to get the bidirectional status for street lamp. This protocol can make up to utmost numbers of weight are controlled by the protocol in the system [9].

Giuseppe Anastasi *et al.*, deals with a remote sensor of zigbee framework embodies sensor centre points passed on over a geological domain for checking physical phenomena like temperature, clamminess, vibrations, other events, etc. Ordinarily, a sensor centre is a little contraption that fuses three fundamental portions: a sensing subsystem for data securing from the physical nature, a changing subsystem for adjacent data planning and stockpiling, and a remote correspondence subsystem for data transmission. This works survey the key method for power preservation using wireless system like zigbee. It helps in continuous data passing to the required area with no interruption and helps in saving current [11]. Jennifer Yick *et al.*, says that in present years zigbee has become cheap, tiny and smart in the sensor world. It has a high growth in industrial mutiny. This sensor is capable of conveying each other in remote areas. This work classifies three various issues such as domestic proposal and fundamental functioning system, communication procedure heap and network employment. The above said issues are considered and solved while developing a new lighting system [15].

Jin-Shyan Lee *et al.*, (2007) explains about the architecture of control systems such as correspondence, data, machine, control, and feeler which are built in a coordinated path to form a complete framework. This architecture was capable to have a better execution than the earlier framework. One of the inescapable important parts in production line computerization is the modern correspondence. Numerous diverse system sorts have been advertised for utilization on a shop floor, including control zone system, Process field bus, Mod bus, etc. The basic issue faced by the modern designer is the problem of choosing a system standard for a specific function [19]. Fig. 4 shows the comparative structure of WSN (uwb, Bluetooth, wifi, zigbee, wimax, gsm, gprs, etc.).

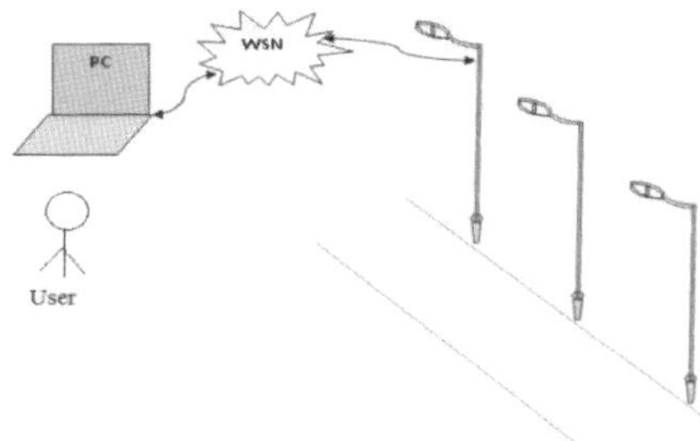

Fig. 4. Comparison and study of wireless sensor network using olden street lamp.

3.4. *Intellectual road light system*

Huaguo Zhou *et al.*, (2009) has reviewed the various types of road light safety system during night times and proposed a new innovative technology for light meter measurement. Measurement techniques for road lighting frameworks such as illuminance, luminance and minimum goal visibility were discussed. In order to avoid the difficulty of using regular light meters a new innovative technology for light meter measurement was proposed. This system had a modern instrument which is capable to collect huge amount of information about the luminance level of street light such as resourceful, safe and secured modes. It uses distance measurement instrument, a pc which connects with an interface device and hardware components, software which develops a link between computer and light meter to record the illumination information. After these steps are processed the system data is evaluated and validated with the site information. The proposed methodology minimised the collection cost of information and provides a secured data information [13].

Jennic Technology *et al.*, provides an advanced expansion in zigbee wireless network and outlay resourceful approach for street lighting system. This helps in reducing power cost, gas emissions, maintenance cost, fast payback and information oriented encoding. Jennic Technology produces wireless organizer and expands jennet networking with properties such as controlling, managing and self remedial properties. It works in short range for street light system for transmitting information to a particular terminal of control centre using wireless networks of zigbee, gprs, gsm, etc. [14].

Jun Zhang *et al.*, in this paper tells the self-governing control system of the street light uses group decision making. In this the street lights are combined and connected to a main node of street light through which all the communication between the user and the street lamps will take place. This can manage or supervise the street light system using tree topology format. As a result smart level of controlling street lamps is attained [20]. Fig. 5 shows the working status of intelligent street light system.

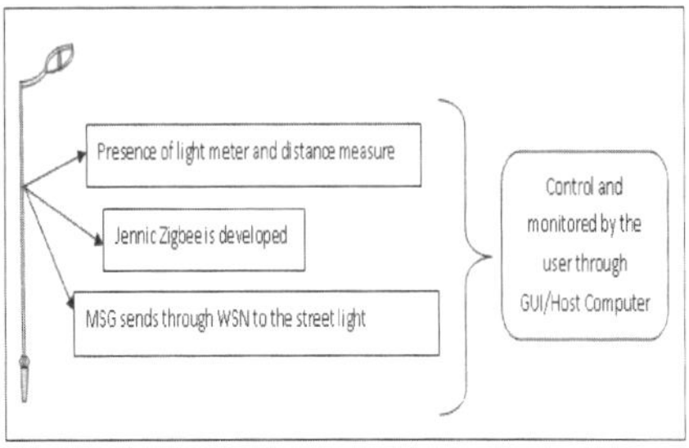

Fig. 5. Intelligent street light working conditions.

4. Proposed Prototype

In the existing system they describe that street light system combine with smart technologies to save current and cost. It uses solar panel to supply the sustainable energy for led street light, in some other cases it uses dynamic /olden street lamps. Thus the street light monitored and controlled by the base station via graphical user interface (GUI). The main drawbacks in the existing system are: Initially the intensity of the light cannot be decreased gradually so that power consumption will be high. Secondly when zigbee get repaired the fault message will not received by the site engineer via GUI. To avoid these drawbacks the proposed system can design the smart street light system by applying pulse width modulation (PWM) method, the values are passed to microcontroller so that street light intensity can be operated as dim or degradation or bright or off automatically or manually this helps to reduce the power consumption of street light lamps till 90%. Instead of olden lamp system led lamps can be used this helps to decrease the pollution in our country this has been shown in Fig. 6.

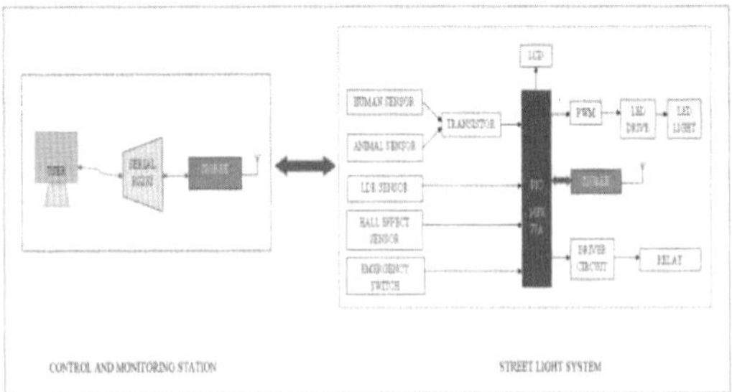

Fig. 6. Proposed system.

5. Conclusion

This paper discusses various conventional methods used for street light system. New technologies reaches a person soon all over the worldwide and it takes a turn to change our country by producing and saving current, Solar LED street light initialization cost is higher at once but it is used for long life time. Solar led efficient street light system with smart technology and adding new sensors gives a sustainable environment friendly surrounding due to the power saving and by providing proper light visibility, security and safety for the people and driving vehicle. This method helps us to save power up to 90% and make the country as green home.

References

1. Alexandru Lavric, Valentin Popa, Ilie Finis, Daniel Simion, The Design and Implementation Of An Energy Efficient Street Lighting Monitoring And Control System, *Proceedings Of The 2012. International Conference on Electrical and Power Engineering*, 312 (Romania, Iasi, 2012).
2. Amr Aminhafez, Mohamedamindessouky, Hanifikriragai, Design of A Low-Power Zigbee Receiver Front-End For Wireless Sensors, *Microelectronics Journal* **40**, 1561 (2009).
3. Antal Haans, Yvonne A. W. De Kort, Light Distribution in Dynamic Street Lighting: Two Experimental Studies On Its Effects On Perceived Safety, Prospect, Concealment, And Escape, *Journal Of Environmental Psychology* **32**, 342 (2012).
4. B. E. Bilgin, V. C. Gungor, Performance Evaluations Of Zigbee In Different Smart Grid Environments, Computer Networks **56**, 2196 (2012).
5. Chaitanya Amin, Ashutosh Nerkar, Paridhi Halani, Rahul Kaul, Gsm Based Autonomous Street Illumination System for Efficient Power Management, *Ijett*, **4**, 54 (2013).
6. Chakkor Saad, Baghouri Mostafa, El Ahmadi Cheikh, Hajraoui Abderrahmane, Comparative Performance Analysis Of Wireless Communication Protocols For Intelligent Sensors And Their Applications, *International Journal Of Advance Computer Science And Application*, **5**, 77 (2014).
7. C. Lozano, O. Rodriguez, Design Of Forest Fire Early Detection System Using Wireless Sensor Networks, *The Online Journal On Electronics And Electrical Engineering*, **3**, 402 (2011).
8. Cleber Costa Da Fonseca, Rodrigo Palucci Pantoni,Dennis Brandao. Public Street Lighting Remote Operation and Supervision System,

Computer Standards & Interfaces, Doi: 10.1016/J.Csi.2014.08.003. (2014).

9. F. Domingo-Perez, A. Gil-De-Castro, J. M. Flores-Arias *et al.*, Low-Rate Wireless Personal Area Networks And Dali Protocol Applied To Street Lighting, *Light Research And Technology*, **45** 90 (2013).
10. Francesca Cuomo, Emanuele Cipollone, Anna Abbagnale, Performance Analysis of IEEE 802.15.4 Wireless Sensor Networks: An Insight into the Topology Formation Process, *Computer Networks*, **53**, 3067 (2009).
11. Giuseppe Anastasi, Marco Conti, Mario Di Francesco, Andrea Passarella. Energy Conserving In Wireless Sensor Networks: A Survey, *Adhoc Networks*, 537 (2009).
12. Hung-Cheng Chen, Long-Yi Chang, Design And Implementation Of A Zigbee-Based Wireless Automatic Meter Reading System, *Przegląd Elektrotechniczny (Electrical Review)*, ISSN 0033-2097, R. 88 Nr 1b/2012, 64 (2012).
13. Huaguo Zhou, Fatih Pirinccioglu, Peter Hsu, A New Roadway Lighting Measurement System, *Transportation Research Part* C. **17**, 274 (2009).
14. Jennic Technology. White Paper. Intelligent Street Lighting. (2009).
15. Jennifer Yick, Biswanath Mukherjee, Dipak Ghosal, Wireless Sensor Network Survey, *Computer Networks*, **52**, 2292 (2008).
16. Jelena Misic Zigbee: A Long Way To Go? R&D Technology 32 (2007).
17. J. Han, H. Lee and K. Park. Remote-Controllable and Energy-Saving Room Architecture based on ZigBee Communication, *IEEE Trans. on Consumer Electronics*, **55**, 264 (2009).
18. J. L. Hill, System Architecture for Wireless Sensor Networks, Ph.D. dissertation, *Dept. Elect. Eng., University of California, Berkeley*, CA, (2003).
19. Jin-Shyan Lee, Yu-Wei Su and Chung-Chou Shen, A Comparative Study of Wireless Protocols: Bluetooth, Uwb, Zigbee, and Wi-Fi, *the 33rd Annual Conference of the IEEE Industrial Electronics Society (Iecon)*, 46 (Taipei, Taiwan, 2007).
20. Jun Zhang, Guifang Qiao, Guangming Song *et al.*, Group Decision Making Based Autonomous Control System for Street Lighting Measurement **46** 108 (2013).

An Overview of Test Case Optimization Using Meta-Heuristic Approach

Sushant Kumar[1], Prabhat Ranjan[2] and R. Rajesh[3]
Department of Computer Science, Central University of South Bihar, Patna, Bihar, 800014, India
E-mail: [1]sushantkumar@cub.ac.in, [2]prabhatranjan@cub.ac.in, [3]rajeshr@cub.ac.in, www.cub.ac.in

Software testing is one of the important and crucial phases of software development life cycle. In context of time, cost and effective testing the prime need is test case optimization. In present testing scenario the meta-heuristic methods are used for optimization problem and also provide good optimized result. The most popular algorithm used meta-heuristic software optimizations are genetic algorithm, particle swarm optimization, artificial bee colony, ant colony, memetic algorithm, harmony search and many more. In this paper we provide the comparative study of all meta-heuristic approaches and find which is more towards the solutions i.e. coverage of requirement (fault coverage) and also take less time (cost) for test case execution and support software quality.

Keywords: Search based software testing; Test case optimization; Test case generation; Test case selection; Test case prioritization; Meta-heuristic.

1. Introduction

In a software development life cycle [15] the software testing phase [16] is the most important and crucial step to ensure software quality. Increasing demand for software and more dependent on software it is highly desirable for software to meet customer requirements. Software may be changed several times as per need. Changes may modify the old or add new requirements, for this testing play crucial role. In the new scenario the role of regression testing is very important. Regression testing is an important strategy used in the software maintenance phase to make a high quality software product [22]. Regression testing is performed to make sure that modification is correct and have not adversely affected the unchanged portion of the software. The test suites that are previously available from earlier versions of software can be expensive to execute completely. Different techniques are proposed by researchers for reducing cost related to regression testing. They include regression test selection [25], regression test prioritization [27], and hybrid approach (selection and prioritization). In prioritization test cases are ordered such that highest priority tests case is executed earlier than lower priority test case. The priority is set according to chosen criteria to increase rate of fault detection or to achieve maximum coverage or to cover important features and so on. When execution time is short, we need to have time constraint prioritization techniques because it is proved that the test case optimization is NP complete problem [30]. Test case optimization is already starting to use for regression testing for different applications like embedded system [28]. The goal of software testing is to design a set of minimal number of test cases and test data such that it reveals as many faults as possible. In software testing process each test case has an identity and is associated with a set of inputs and a list of expected outputs. Structural testing is most frequently used in software testing. It should follow some standard coverage method. Path testing is one of the most widely used for coverage of software testing. In this paper, second part is about search based meta-heuristic approach that is applied in software testing. The comparison of meta-heuristic algorithm is also shown in second part. The third part is conclusion part.

2. Meta-Heuristic Techniques for Test Case optimization

Algorithms has classified as deterministic or meta-heuristic. If an algorithm works for mechanically deterministic manner, it is called deterministic. Generally it is provide same solution if our initial point is same. So it will desirable to provide in randomness in solution and it will give better result. Meta-heuristic nature-inspired algorithm is most widely used algorithms for optimization in many engineering field of application. Here meta-heuristic has applied for test case optimization problem. Testing optimization may be performed in black box testing and white box testing. Tester work to optimize the test case for both the testing type and found the fault as early stage [21]. The different algorithm is used for test case optimization like genetic algorithm, particle swarm optimization, Artificial bee colony, Ant colony and many more.

2.1. *Genetic algorithm*

Genetic algorithms are inspired by Darwin's theory about evolution. Genetic Algorithms (GAs) are adaptive Meta-heuristic search techniques premised on the evolutionary ideas of natural and genetic selection. Algorithm is started with a set of solutions (represented by chromosomes) called population. Solutions from one population are taken and used to form a new population. This is motivated by a hope, that the new population will be better than the old one.

Solutions which are selected to form new solutions (offspring) are selected according to their fitness. Crossover and mutation operator used in genetic algorithm for new population and diversity. Ahmed S. Ghiduk [5] has given tool for automatic generation of basis test paths using Genetic algorithm. Path testing is the strongest coverage criteria in structure testing.

2.2. *Particle swarm optimization*

Particle swarm optimization is compare to genetic is new technique to use for optimization problem. A complex problem are solved using PSO for optimization like traveling sales man problem, Job sequencing problem. In software testing PSO is also used because testing is complex problem and generation of test case is NP-complete problem. PSO can be used for solving either single objective and multiple objective [24]. Shujum Jiang *et al.* [17] has modified particle swarm optimization with reduced adaptive particle swarm optimization. This paper proposes a novel approach for generating the test data automatically. This method reduced the PSO equation and provides an evolution equation without velocity. This algorithm further modified with to achieve convergence rate of the algorithm.

2.3. *Ant colony optimization*

Ant colony optimization (ACO) is a meta-heuristic technique for solving computational problems. It is best effective approach for reduced reduce the path and find optimal path. Marco Dorigo has proposed ant colony optimization. The real power of ACO is the behavior of ants to find a path between their colony and source of food. In ACO ants are blind and communicate with their colony using chemical pheromone. Chengying Mao *et al.* [13] Ant Colony Optimization is used automatic test data generation. Author has applied ant colony optimization on structural testing. Search based Meta-heuristic technique is explored with ACO approach. Author has shown with result that Ant Colony method is better than SA And GA for coverage capability, convergence speed and for stability.

2.4. *Artificial bee colony*

Artificial bee colony (ABC) has proposed by karaboga in 2005. It is an optimization algorithm based on the intelligent behavior of honey bee. In ABC, the colony consists of three groups of bees (1) Employed bees (2) Onlooker bees 3) scouts bees. One artificial bee for each food source i.e. the number of employed bees equal to the number of food sources near hive. Ruchika Malhotra *et al.* [3] has purposed a new and implemented a new approach for test suite optimization using Mutated Artificial Bee Colony. It used both local search and global search. Author also compared different Meta-Heuristic algorithm for test case optimization. Author verified and validate the algorithm by running 10 benchmark C++ programs.

2.5. *Memetic algorithm*

Memetic algorithm is advanced hybrid approach for genetic algorithm. It supports local and global search. Gordon Fraser *et al.* [11] has extended the work of previous paper "Whole test generation". Author has used Memetic algorithm for branch coverage. The Evosuite tool is used for implementation of paper. This paper has more focused on local search operators.

2.6. *Cuckoo search*

Cuckoo search is comparatively new search technique in nature inspired computing. It has local search ability and also global search ability. But the main advantages of cuckoo search is for global optimization problem. Test case prioritization is solved by the cuckoo search and author shows the result that it also provide good result for global optimization problem [31] [42].

2.7. *Harmony search*

Harmonic search is a music inspired algorithm. It is given Zong Woo Geem in 2001 on first time, after he was inspired by improvisation of musicians. Harmony search also uses gradient search. Many engineering problem has been solved using harmony search. It is also used in test case optimization. Ming Huang *et al.* [43] has improved the HS algorithm for test case selection.

2.8. *Other meta-heuristic algorithm*

Apart from genetic algorithm, particle swarm optimization, Ant colony optimization, Artificial bee colony optimization and Memetic algorithm the other algorithm that applied on test case optimization are tabu search, hill climbing, simulated annealing, Artificial Immune system and more, some of the algorithm defined in Table 1. Meta-heuristic algorithm have applied in test case optimization in 3 category like generation, selection and prioritization, Table 2 shows the different category. Fuzzy logic and artificial neural network is also used for test case optimization. Ali M. Alakeel (2014) [18] has used Fuzzy logic for Test case Prioritization for regression Testing. Assertion have imposed using fuzzy logic for test case prioritization. Abraham *et al.* [45] has used hybrid approach ABC, FCM (Fuzzy C Means) and PSO for test case optimization to minimize the cost and time in regression testing. The experiment is done for test case prioritization.

2.9. *Analysis of different testing techniques using meta-heuristic approach*

In Table 1, different testing techniques have shown with using of meta-heuristic search based techniques. We have taken powerful coverage technique like path testing, mutation testing and branch coverage for optimization of test case. For the test case optimization we have to select the testing criteria the that has support maximum coverage and check the quality of the test case. From analysis, here analyzing that path testing is still most powerful coverage of program and validation of test case. Other approach branch coverage, mutation testing is also used for check the quality of test case. Functional coverage is also important because we check the function without uses the internal detail of code structure. Other method is also used for test case coverage like slice based testing. For complex program slice based has produce good result to check the coverage of program through testing.

Table 1. Software testing problem and meta-heuristic approach.

Sr. No.	Testing approaches	Meta-heuristic algorithm	Author
1	Path testing	Artificial Bee Colony	Soma Sekharababu *et al.* [7]
		Genetic Algorithm	Ahmed S. Ghiduk [5]
		Genetic Algorithm	P. R. Srivastava *et al.* [12]
		Firefly Algorithm	Praveen R. S. *et al.* [14]
		Genetic Algorithm	Lodha G. M. [21]
		Firefly Algorithm	P. R. Srivastava *et al.* [38]
2	Mutation testing	Genetic Programming	William B. Langdon et.al.[6]
3	Functional testing	Genetic Algorithm, ACO	Javier Ferrer *et al.* [10]
4	Branch coverage	Genetic Algorithm	Gordan Fraser *et al.* [9]
		Memetic Algorithm	Gordon Fraser *et al.* [11]
5	Test case optimization	Ant Colony Optimization	Chengying Mao *et al.* [13]
		RAPSO	Shujaun Jiang *et al.* [17]
		Fuzzy Logic	Ali M. Alakeel [18]
		ACO	Elanthiraiyan, N. [22]

2.10. *Analysis of different test case optimization techniques using meta-heuristic approach*

In Table 2, test case optimization has shown in broadways. Three techniques are used like test case generation, test case selection and test case prioritization. The analysis shows that test case generation has covered by many authors different approach is used to generation of the test case and test data. Here we cover only those papers that used meta-heuristic approach to solve the optimization problem. The other approaches are test case selection and test case prioritization. Test case selection is also effective method for test case optimization because in large dimensional test data the selection of good reliable test data is very important. Test case is selected according to data driven and knowledge driven for test optimization. And the last approach I have taken is test case prioritization. For optimization of test case prioritization is one of the important and crucial task, because it decide the time bound for testing. In the given time slot all the critical test case must be execute to find the fault. So need the good prioritization technique to search the all test suite (collection of test case) to find the critical test case, to produce new optimized test suite with prioritized result.

Table 2. Test case optimization techniques.

Sr. No.	Test case optimization	Author
1	Test case generation	Bestoum *et al.* [2], Ahmed S. Ghiduk *et al.* [5], Soma Sekhara *et al.* [7], Gordon Fraser *et al.* [9], Javier Fraser *et al.* [10], Gordon Fraser *et al.* [11], C. Mao *et al.* [13], P. R. Srivastava *et al.* [14], Shujuan Jiang *et al.* [17], Shujuan Jiang *et al.* [31], P. R. Srivastava *et al.* [38], Gordan Fraser *et al.* [39]
2	Test case selection	Luciano *et al.* [23], Luciano *et al.* [24], Luciano *et al.* [25], Swarendubiswas *et al.* [26]
3	Test case prioritization	Yogeshsingh [1], Md. Saeed Siddik *et al.* [8], Ali M. Alakeel [18], Elanthiraiyan *et al.* [22], G. Rothermal *et al.* [27], Khinhla *et al.* [28]

2.11. *Comparison of different meta-heuristic algorithms*

In Table 3, based on the analysis of different meta-heuristic algorithm apply in software testing we have done the comparison of algorithms. The pros and cons is provided in the table form. This analysis help to select which algorithm is more reliable and suitable for test case optimization problem. Meta-heuristic algorithm which has easy to implement, less parameters to adjust and have local and global search ability is more reliable and near to solutions. The complexity of algorithm is also decide that the algorithm is effective for both type of testing i.e. black box testing and white box testing. Since test case optimization is NP-Complete problem so select the best algorithm that can solve NP problem to solve optimization problem for software testing.

Table 3. Comparison of meta-heuristic algorithm.

Sr. No.	Algorithm	Advantages	Disadvantages
1	Genetic Algorithm	• Answer gets better with time • Do not break easily	• Give slow results • Weak local search ability
2	Artificial Bee colony	• Good quality solutions of difficult optimization problem can be discovered	• Extra overhead
3	Hill climbing	• Simple • Give Fat results	• Leads to local optima • Highly dependent on stating solutions
4	Simulated annealing	• Less dependent on the starting solutions • Less restrictions	• Stuck in local optima • Cooling problem
5	Memetic algorithm	• Strong local search ability • Hybrid approach	• Gives slow result • Hard to implement
6	Ant colony optimization	• Record memory to entire colony • Less affected by initial poor solutions	• Convergence is guaranteed • Time of convergence uncertain
7	Particle swarm optimization	• Easy to perform • Efficient in global search • Few parameters to adjust	• Weak local search ability • In refined stage convergence is slow

3. Conclusion

The application of meta-heuristic for solving software engineering problems is part of a relatively new area called search based software engineering (SBSE). Search based software testing is using meta-heuristic approach for solving test case optimization problem. The test optimization techniques have been applied by many researchers for automatic test case generation, selection and prioritization but not any one could achieve the best performance for every piece of code to select the test case and also to find early fault in testing process. Due to large test case, the test case optimization has become an complex problem. So scope remains open to apply some new meta-heuristics techniques to achieve better results. We found that from literature genetic algorithm not provide better result. And most of the techniques is solving single objective Now we need to develop an optimization technique for test case optimization using meta-heuristic techniques. Therefore, meta-heuristic approach play key role for automatic selection, generation and prioritization for test case. Some swarm optimization like PSO, ACO and ABC are solving in better way for test case optimization.

References

1. Yogesh Singh Systematic Literature Review on Regression Test Prioritization techniques, *Informatica* **36** 379 (2012).
2. Bestoun S. Ahmed, Mouayad A. Sahib, Moyad Y. Potrus, Generating combinational test cases using Simplified Swarm Optimization (SSO) algorithm for automated GUI functional testing, *Engineering Science and Technology*, An International Journal (2014).
3. R. Malhotra and M. Khari, Test Suite Optimization using Mutated Artificial Bee Colony, *Proc. of Int. Conf. on advances in Communication, Network and Computing*, CNC, Elsevier, (2014).
4. Samatha K. Shreesha Chokkadi, Yogananda Jeppu, A Genetic Algorithm Approach for test case optimization of safety critical control, *International Conference on Modeling, Optimization and Computing (ICMOC 2012) Procedia Engineering* **38** 647 (2012).
5. Ahmed S. Ghiduk, Automatic Generation of basis test paths using variable length genetic algorithm, *Information Processing Letters* **114**, 304 (2014).
6. William B. Langdon, Mark Harman, Optimizing Existing Software with Genetic Programming, *IEEE Transaction On Evolutionary Computation*, **19**, (2015).
7. Soma Sekhara Baba Lam, M. L. Hari Prasad Raju, Uday Kiran M., Swarj Ch, Praveen Ranajn Srivastav, Automated Generation of Independent Paths and Test Suite Optimization Using Artificial Bee Colony, *Procedia engineering*, Elsevier **30**, 191 (2012).
8. Md. Saeed Siddik, Kazi Sakib, RDCC: An Effective Test Case Prioritization Framework using Software Requirements, *IEEE conference* on *Design and Source Code Collaboration* (2014).
9. Gordon Fraser, Andrea Arcuri, Whole Test Suite Generation, *IEEE Transaction On Software Engineering*, **39**, 2 (2013).
10. Javier Ferrer, Peter M. Kruse, Francisco Chicano, Enrique Alba, Search Based Algorithm For Test Sequence Generation In Functional Testing, *Information and Software Technology* (Elsevier) **58**, 419 (2015).
11. Gordon Fraser, Andrea Arcuri, Phil Mcminn, A Memetic Algorithm for hole test suite Generation, *The Journal of System and software*, Elsevier (2014).
12. Praveen Ranjan Srivastava, Tai-hoon Kim, Application of Genetic Algorithm in Software Testing, *International Journal of Software Engineering and its application* **3**, 4 (2009).

13. Chengying Mao, Lichuan Xio, Xixin, Jinfu Chen, Adapting Ant Colony Optimization to generate test data for software structural testing, *Swarm and Evolutionary Computation*, Elsevier (2015).
14. Praveen Ranjan Sriatsava, B. Mallikarjun, Xin-She Yang, Optimal Test Sequence Generation using firefly algorithm, *Swarm and Evolutionary Computation*, Elsevier, **8**, 44 (2013).
15. Ian Somerville, Software Engineering Eight edition, (Pearson Edition Boston, 2009).
16. Aditya P. Mathur, Foundation of software Testing, (Pearson Education, India, 2007).
17. Shujuan Jiang, Jiaojiao Shi, Yanmei Zhang, Han, automatic Test Data generation based on reduced adaptive particle swarm optimization, *Neurocomputing*, Elsevier (2015).
18. Ali M. Alakeel, Using Fuzzy Logic in Test case prioritization, The Scientific World Journal, Hindwai Publishing Crop. (2014).
19. Mark Harman, S. Afshin Mansouri, Yuanyuan Zhang, Search Based Software Engineering: Trend, Techniques and Applications, *ACM Computing Surveys* **45**, 1 (2012).
20. McMinn P., Search-Based Software Testing: Past, Present and Future, *IEEE Fourth International Conference on Software Testing, Verification and Validation Workshops* (ICSTW), (2011).
21. Lodha, G. M., Gaikward, R. S., Search Based Software Testing with Genetic using fitness function, *IEEE conference on Computational Intelligence on Power, Energy and Controls with their impact on Humanity* CIPECH) (2014).
22. Elanthiraiyan, N., Arumugam, C., Parallelized ACO algorithm for regression testing prioritization in hadoop framework, *IEEE International Conference on Advanced Communication Control and Computing Technologies* (ICACCCT), (2014).
23. Luciano S. de Souza, Ricardo B. C. Prudencio, Flavia de A. Barros, Multi-Objective Test Case Selection: A study of the influence of the Catfish effect on PSO based strategies, *Anaios do XV Workshop de Testes e Tolerancia a Falh*a –WTF (2014).
24. Luciano S. de Souza, Ricardo B. C. Prudencio, Flavia de A. Barros, A Multi objective Particle Swarm Optimization for Test Case Selection Based on Functional Requirements Coverage and Execution Effort, *IEEE International Conference on Tools with Artificial Intelligence*, (2011).

25. Luciano S. de Souza, Ricardo B. C. Prudencio, Flavia de A. Barros, Search Based constrained test case selection using execution effort, *Expert Systems with Applications* **40** (2013).
26. Swarendu Biswar and Rajib Mall, Regression test selection Techniques: a Survey. *Informatica* **35** 289 (2011).
27. Gregg Rothermel, Roland H. Untch, Mary Jean Harrold, Prioritizing Test Cases For Regression Testing, *IEEE TRANSACTIONS ON SOFTWARE ENGINEERING*, **27** (2001).
28. Khinhla, Young Sikchoi, Jong Sou Park, Applying Particle Swarm Optimization to Prioritizing Test Cases for Embedded Real Time Software Retesting, *IEEE International Computer and Information Technology* Workshops, (2008).
29. Rajib Mall, Fundamentals of software engineering *PHI*, (2009).
30. Manojkumar, Arun Sharma, Rajesh Kumar, Towards Multi-Faceted Test Cases Optimization, *Journal of Software Engineering and Applications*, **4**, 550 (2011).
31. Shujuan Jiang, Yanmei Zhang, Dandan Yi, Test, Data Generation Approach for Basis Path Coverage, *ACM SIGSOFT Software Engineering Notes* **37**, 3 (2012).
32. Mala D. J., Kamalapriya M., Mohan V., Automated Software Test Optimisation Framework an Artificial Bee Colony Optimization-Based Approach, *IET Software* **4**, 334 (2010).
33. Ruchika Malhotra, Manju Khari, Heuristic search-based approach for automated test data generation: a survey, *Int. J. Bio-Inspired Computation*, **5**, 1, (2013).
34. Bojana Koteska, Bisera Dugalic, Recent and future trends and challenges of software testing, 9th CIIT, (2012).
35. Anusharma, Arpita Jadhay, Parveen Ranjan Srivastava, Renu Goyal, *J. software Engineering and applications* (scientific research) **3**, 477 (2010).
36. B. Beizer, *software testing Techniques*, 7th edition (van Nostrand Reinhold, New York, 1990).
37. I. Sommerville, Software Engineering, (Pearson Education, TMH, India, 2005).
38. Parveen Ranjan Srivatsava, B. Malikarjun, Xin-She Yang, Optimal test sequence generation using firefly algorithm, *Swarm and Evolutionary computation* **8** 44 (2013).
39. Gordon Fraser, Andrea Arcuri, Phil McMinn, A. Memetic Algorithm for whole test suite generation, *Journal of System and software*, Elsevier (2014).

40. B Suri, Shweta Singhal, Analyzing test case selection and prioritization using ACO, *ACM SIGSOFT Software Engineering Notes*, (2011).
41. P. R. Srivastava, D. V. P. kumarreddy, M. Srikanthreddy, Ch. V. B. RamarajuIch. Manikanthnath, Test case prioritization using cuckoo search, *Advanced automated software testing*, 2012.
42. Xin-She Yang, Suash Deb, Cuckoo search:Recent advances and Applications, *Neural Computer Application (springer)*, **24**, 169, 2014.
43. Ming Huang, Shujieguo, XuLiang, Xuanjiao, Application of Improved harmony search algorithm in test case selection, *journal of software*, **9** 5 (2014).
44. Chintala Abhisek, Veginati Pavankumar, Harish Vitta, Parveen Ranjan Srivastava, *Software Engineering and application*, **3**, 331 (2010).
45. Abraham Kiran Joseph and G. Radhamani, Fuzzy C means Clustering Based hybrid Swarm Intelligence algorithm for Test Case optimization, *Research journal of Applied sciences, engineering and Technology* **8** 76 (2014).

Smart City Traffic Management and Surveillance System for Indian Scenario

Tarun Kumar[1], Rohit Kumar Sachan[2] and Dharmender Singh Kushwaha[3]

Department of Computer Science and Engineering
MNNIT, Allahabad
Allahabad, India
E-mail: [1]ertarun123@gmail.com, [2]sachan.rohit100@gmail.com, [3]dsk@mnnit.ac.in

In India, speed and red light violation are the main cause of accidents. Over speed reduces our reaction time and it makes difficult to control a vehicle during obstacle moves into our path. Violating the red light increases our chances of crashing. At present in India to detect the over speed vehicles and red light violation there should be an efficient and automated system. In this paper we proposed the framework of smart city traffic management and surveillance system for automatic over speeding vehicle detection and red light violation detection system. The proposed model will help to detect the vehicles having over speed and red light violation. This proposed system is based on pattern recognition and image processing techniques.

Keywords: License Plate Detection Character Segmentation; Character Recognition; Projection; Template matching.

1. Introduction

Road accident occurrences have increased recently so there needs to be a system that allows detecting over speeding cars. Unfortunately the number of road accidents in India is among the highest in the world. According to Indian Road Accident Data (IRAD), 4,90,000 road accidents were reported during the year 2012 and in those 1,38,000 persons were killed. This represents every hour, 56 accidents were reported and in those 16 persons dead in India. Analysis reports reveals that, causes of road accidents are mainly due to fault of drivers (over speed and red light violation) followed by fault of pedestrians, defect in condition of the vehicle, defect in road condition, fault of the cyclist and weather conditions. In India for speed detection there are police personnel who check car speed by using handheld gun and then manually inform authorities about the vehicle. So there is a need for an automated system for the same. In many other countries, there is a traffic surveillance camera on traffic light, which records the all the activities occurs on signals. This will reduce the manpower cost on the traffic surveillance. In India, CCTV cameras are already there on mostly traffic signals. These cameras can be used for video surveillance. We proposed the system which is based on video recorded by these cameras in real time. Our system is responsible for the real time data collection, processing, information retrieval, and analysis and for report generation. Our system includes a database to store the vehicle details likes- vehicle registration number, speed, time, etc.

2. Proposed System

Our proposed system covers the three scenarios for traffic surveillance in India. In first scenario, the system will consist of a camera with a processing device (can be PC or small PCB). The camera in our system will be on traffic light as shown in Fig. 1.

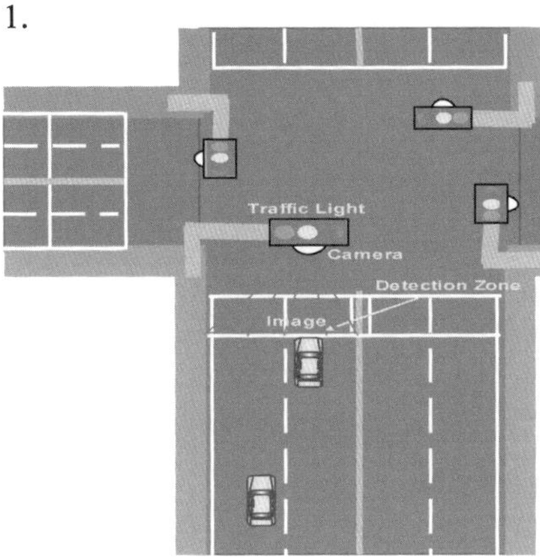

Fig. 1. Traffic surveillance system, scenario 1.

In second scenario where only speed limit violation detection needed, the system will consist of two cameras with a processing device (can be PC or small PCB). The cameras in our system will be at some distance on the road as shown in Fig. 2.

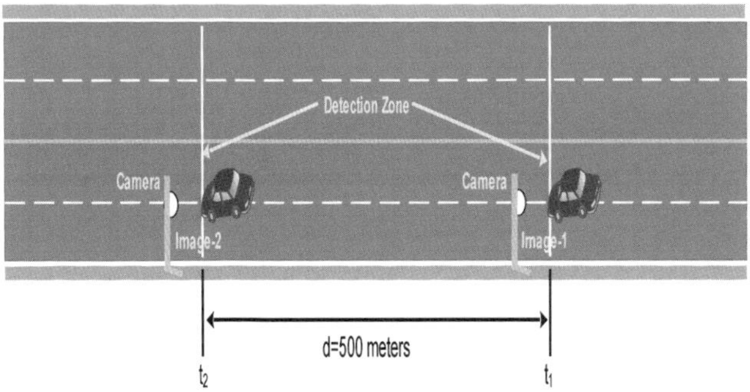

Fig. 2. Traffic surveillance system, scenario 2.

For this scenario we put two cameras at roadside at detection zone. Distance between that detection zones is d (500 meters). We extract vehicle details at time t1 and at time t2. Time t1 is at detection zone-1 and t2 is at detection zone-2. Now we do calculation the speed of vehicle via formula.

$$Speed = Distance / Time = d / (t2\text{-}t1)$$

The registration number of vehicle and speed is stored in our database. If speed of vehicle is crossed the limit then we sends the vehicle details to traffic police. In third scenario where stop look and go rule violation detection needed, the system will consist of a camera on the T-point of the road and processing unit as shown in Fig. 3.

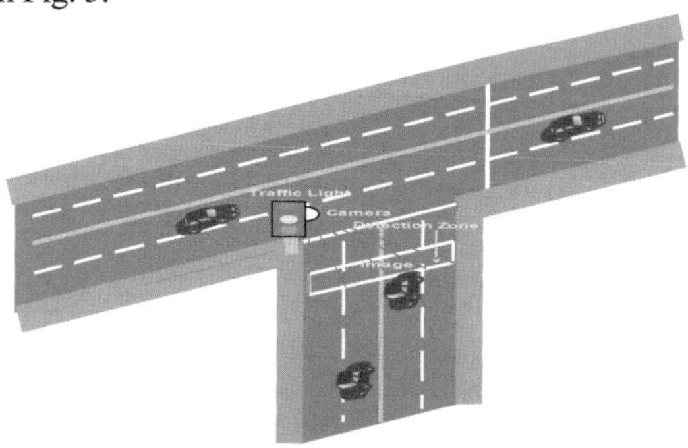

Fig. 3. Traffic surveillance system, scenario 3.

The Camera in this system will capture the video and will send the digital video to the processing unit. After getting video stream from camera now processing unit will begin analysis. For video capturing IR Bullet Camera (Day/Night, Resolution: 600 TVL, Lens: 6mm, IR Range: 20mtrs) can be used. The frame rates of these cameras vary from 15 frames per second to 30 frames per second. The proposed system mainly consists of three units.

- Video Capturing Unit
- Processing Unit
- Storage Unit

The working model of the system is shown in Fig. 4. In this model the video capturing module will capture the video and give the video stream to the frame extraction module of processing unit. Frame extraction module from the video streaming extracts the video frames and store these frames into frame buffer. The Frame extraction [1] module also stores the status of the traffic light like RED/GREEN/YELLOW. Now this module sends the frame to the next module. At the next level there are two modules. One module is speed limit violation detection module and another the module is Red light violation detection module. One module according to traffic light status selected. If the at the time at which frame is capture the light status is Red then this frame will passed to the Red light violation detection module of the light status is Green or Yellow then this frame will processed by the speed limit detection module. Both modules identify the object that violate the speed limit or red light and passes the key frames to the License number extraction module.

Now license number extraction module extract the license plate and recognize the number by template matching and send the numbers to the storage and notify unit of the system. Now the storage unit keeps the records of defaulter objects with their numbers and notify to the traffic control unit.

2.1. *Video capturing unit*

This unit contains a camera that will capture the traffic and generate the video stream.

2.2. *Processing unit*

Processing unit of the system will contain the mainly four modules.

2.2.1. *Frame extraction module*

Frame extraction module extracts the frames from the video stream and stores the frames into the frame buffer with the adding some information about the traffic light status. Also this module makes the decision of selection next module to process the frame. This module takes decisions on the basis of the status of traffic light, if it is GREEN or YELLOW then it selects the speed limit violation detection module otherwise it select the red light violation module and pass the frame to the next module.

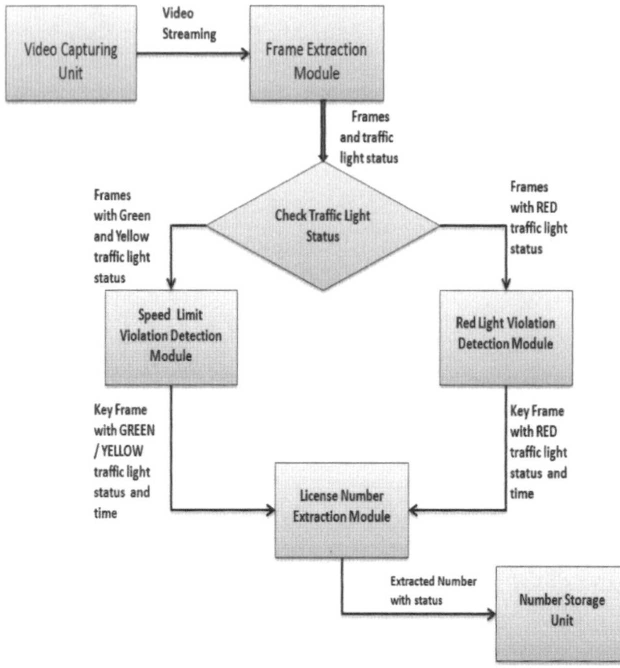

Fig. 4. Working model of automatic traffic surveillance system.

2.2.2. *Speed limit violation detection module*

This module takes input a frame and stores the frame into a buffer and process the frame. In this module, first step is to identify the objects on the frame by using background subtraction method [3] [4] and finding the edge of the identified objects. Now generate the id for each object present in the frame. Now this module compares the object of this frame from the previous identified and stored objects. In this module any new object is stored with some properties like *frame_no*, time, *frame_count* and *object_shape*, *object_color*, *object_contour*. Now the objects of the current frame are matched [6] [8] [11] with each of the previous object. During this matchin process, three cases can occurs. In first case case if one of the objects is matched completely to previous object then this module update the *frame_count* of previous match object and update information like *shape*, *object_contour*. In the second case object of the current frame does not match with any other previous object then it will consider as new object and so stores it. In third case if any one of the previous object does not match with any new object then it will consider as the previous object crossed camera range. In this case the speed of the object is measured on the basis of *frame_count*. There is a camera which captures the video at f frame/second and range of the camera is d meters then time of one frame is $1/f$ seconds. Now if an object appears in n frames then it means this object cross the d meters distance in n/f seconds. Now we can measure the speed of object is as s.

$$s= \frac{d \, X \, f}{n} \quad (1)$$

If the speed of the object is greater than the predefined speed limit then it will beconsidered as defaulter object and the frame on which this object appears is passed to the next module which is license number extraction module.

2.2.3. *Red light violation module*

This module takes input a frame from frame extraction module. This module detects the object into the range by using the background subtraction method. If it identifies any object in the range then it will consider asa red light violation case. Now this module passes the frame to the next module which is license number extraction module otherwise it discard the frame.

2.2.4. *License number extraction module*

Now this module gets the key frame input either from speed limit violation detection module or from red light violation detection module. This module extracts the license plate [8] [10] of the object. To detect the licenses plate it uses the sobel operator, edge detection [10] [11] [12] [13], Morphological operation methods on the gray scale frame. Now the second step after license plate recognition is the identifying the character segment from the extracted license plate region of the key frame. In this character segment is obtained by inversing the binary image of the plate after skew correction [5]. Now the third step is to recognition of the characters from the license plate. To recognize the characters from the plate template matching is used. In template, there is dataset which contains the different images of the English characters and digits of size M x N. The recognized character [10] sends to the storage unit.

2.3. *Storage unit*

Storage unit will store the license number of the vehicle who violates the speed limit or red light signal. Also this unit notifies the traffic control unit.

3. Conclusion

In this paper we proposed the framework of speed limit and red light violation detection system for Indian traffic control system. As this system will contains the template of different license number styles used in India. Also this system is easy to install as it uses the single camera to detect the both speed limit and red light violation. This framework is compiled of different independent modules. If we want to use single module then it can be possible from this module. So if it is needed to detect red light violation only we can easily remove the speed limit violation module or vice versa.

References

1. R. R. Schultz and R. L. Stevenson, Extraction of high-resolution frames from video sequences, IEEE Trans. on Image Processing, **5**, 1011 (1996).
2. Mohammad Salahshoor, Ali Broumandnia and Maryam Rastgarpour, Application of Intelligent Systems for Iranian License Plate Recognition, IEEE (2014).
3. M. Piccardi, Background subtraction techniques: A review, in Proc. IEEE Conf. Syst (Man Cybern) **4**, 3099 (2004).
4. A. Elgammal, D. Harwood and L. Davis, Non-parametric model for background subtraction, *in Proc. Eur. Conf. Computer. Vis.—Part II*, 751 (2000).
5. Ms. Sonali Sonavane, Prof. Alka Khade and Prof. V. B. Gaikwad, Localization & Segmentation of Indian Car Number Plate system: A Projection based Multistage Approach, *International Journal of Scientific & Engineering Research*, **4**, 2229 (2013).
6. Shan Du, Member, IEEE, Mahmoud Ibrahim, Mohamed Shehata, and WaelBadawy, IEEE, Automatic License Plate Recognition (ALPR): A State-of-the-Art Review, *IEEE Transactions On Circuits And Systems For Video Technology*, **23**, 2, (2013).
7. Nuzulha Khilwani Ibrahim, Emaliana Kasmuri, Norazira A Jalil, Mohd Adili Norasikin, Sazilah Salam, License Plate Recognition (LPR): A Review with Experiments for Malaysia Case Study, in *Proceedings International Conference on Soft Computing and Software Engineering* 2013 [SCSE'13], (San Francisco State University, CA, U.S.A. 2013).
8. J. Jagannathan, A. Sherajdheen, R. Muthuvijaydeepak, N. Krishnan, License Plate Character Segmentation Using Horizontal And Vertical Projection with Dynamic Thresholding, *IEEE International Conference on*

Emerging Trends in Computing, Communication and Nanotechnology (ICECCN 2013).

9. N. Vishwanath, S. Somasundaram, Nishad A., N. Krishnan Nallaperumal, Indian License Plate Character Recognition using Kohonen Neural Network.
10. Kumar Parasuraman and P. Vasantha Kumar, An Efficient Method for Indian Vehicle License Plate Extraction and Character Segmentation, *IEEE International Conference on Computational Intelligence and Computing Research* (2010).
11. Mayank Garg & Sahil Goel, Real-Time License Plate Recognition and Speed Estimation from Video Sequences, ITSI Transactions on Electrical and Electronics Engineering (ITSI-TEEE) **1**, 5, 2013.
12. Dongwook Kim1, Liu Zheng, Car License Plate Detection Based on Line Segments, *Advanced Science and Technology Letters*, **58** (Signal Processing 2014).
13. Mostafa Ayoubi Mobarhan, Asadollah Shahbahrami, Saman Parva, Mina Naghash Asadi and Atefeh Ahmadnya Khajekini, A License Plate Detection Algorithm Using Edge Features, *Springer-Verlag Berlin Heidelberg* 2012.

Improving Attribute Inference Attack Using Link Prediction in Online Social Networks

Ashish Kumar[1] and N. C. Rathore[2]
Department of Computer Science, Central University of South Bihar, Patna, Bihar, India
E-mail: [1]ashishmzp06@gmail.com, [2]nemichandra@cub.ac.in
www.cub.ac.in

Online social networks (OSNs) act as a platform to build online community. Recent years show an unprecedented surge in the OSN users. It leads not only to interesting opportunities but even new challenges. In link-prediction problem, we are given a snapshot of an OSN at time t and aim to predict links (e.g., friendships) that may emerge in the network between t and a later time $t`$ or we can imagine the setting in which some links existed at time $t$ but are missing at $t`$. Link prediction problems can also be viewed as predicting hidden or missing links in OSNs. In attribute inference problem our aim is to infer the missing or partial attribute data for network nodes. This situation is usual, when users in OSNs set their profiles to be publicly invisible or create an account without providing any attribute information. In this paper we show that accuracy of attribute inference attack can be improved significantly by first inferring missing links.

Keywords: OSN; Attribute Inference; Link Prediction; Attack.

1. Introduction

OSNs are web based user's networks. It is a medium that enables people to stay in touch with their contacts and provides an interactive means of online social interaction. The popularity of OSN is increasing day-by-day. This popularity attracts both faithful users and adversaries. We can represent an OSN as a graph consisting of nodes and edges. Node represents an actor varying from profile of an individual to a group. Edges are social links among these actors (nodes).

Web Technology plays a vital role in OSNs. It's evolution has enhanced inter-connectivity, self-expression and information sharing. These reasons are responsible for emergence of online social networking services. In OSNs, users are focal objects and virtually all communication happens among users and applications triggered by them. The set of participants in OSNs grow and fall over time. Users can solicit others to join and create online communities. Inside an OSN, a user is capable of using instant messaging and bulletin board similar to physical world.

One of the most common features of the OSNs is that each user has his/her own profile, which includes his/her personal information. This personal information consist of user attributes like age, gender, friends list, list of social groups or relations that users have joined.

Large volume of personal information on OSN attract a lot of malicious users. These information includes demographic information, contact information, comments, images, videos etc. Most users publish their information publicly without any careful consideration. These careless activities of users make available a large amount of sensitive information, for attacker to exploit.

The rest of the paper is organized as follows. In Section 2, we briefly introduce the background and related work. In Section 3, we explain the problem formulation and attacker model. In Section 4, we propose various methods for link prediction and attribute inference. In Section 5, we discuss the experiment and result.

2. Related Work

Till date, a wide range of attribute inference methods have already been proposed. He *et al.* proposed the approach of inferring privacy information from social networks [1]. They use Bayesian Network Model to infer one's private information. The proposed model exploits the social relationship among people. Lindamood *et al.* [2], presented a modified version of naive Bayes classification that takes into account, both the node and link structure details. It is more effective for sensitive attribute inference such as political affiliation. Li *et al.* propose the approach of inferring privacy information via social relations [3]. They aim at determining the missing gender of users, provided genders of all users are given, in some social network. They consider three methods for nodes label classification i.e. local, global and Bayesian method. Kozikowski *et al.* [4] used large scale data mining based approach and show a significant increase in accuracy in predicting attribute like age, gender, location etc.

Lei *et al.* [5] proposed a mutual-friend based attacks in social network systems. They showed that mutual-friend based attacks help adversary in locating all friends of target user. Tang *et al.* [6] propose two inference attacks and also discuss the impact of a fraction of hidden friends in those two attacks. They adopt machine learning algorithm to classify user based on user's high school names. To build the classifier, they use Weka toolkit and in classification algorithm they use a Bayes network classifier. Gergely *et al.* [7] proposed machine learning based approach to infer hidden value from public information available in Social Network (friends, group membership). In his work, Gergery carries out attribute inference attack based on the concept of multiple layer perceptron (MLP). Ratan Dey *et al.* [8] propose estimating age privacy leakage in Online Social Networks. They focus on estimating birth year, that is private for many individual, using Reverse Lookup Technique.

Ahmadinejad, Anwar and Fong address two classes of attacks (i.e. SNS API inference attack and authentication attack) [9]. They argue that the success of one will increase the likelihood chances in occurring of the other. Their work presents, how a subscriber of third party application, who is very choosy regarding disclosure of his information, may inadvertently give access to his/her data stream like wall post, photo containing words like Happy Birthday wishes etc. to a third party application. Using several data analysis tools or parser, adversary can obtain answer to several private attributes.

3. Problem Formulation

In our problem setting, we use an undirected, simple graph $G = (V, E)$ to represent an OSN, where set of edges, E represents connections between $N = |V|$ nodes. For instance, in the Facebook social network, nodes are users, edges represent friendship (or some other relationship) among users, and node attributes are derived from user profile information which include fields such as gender, employer, school, hometown, etc.

3.1. *Attacker model*

We use α and t to represent an attacker and a target user, respectively. $F(t)$ represents the friend list of the target user. Given a social network $G = (V, E)$, and an attacker $\alpha \in V$, and target $t \in V$, we assume the following-privacy settings of t does not allow α to view its profile information. But privacy settings of t allows α to view $F(t)$ i.e. users in $F(t)$ are known to α. Users in $F(t)$ also allow α to view their profile information. Attacker α has access to local snapshot of target user social network.

In our problem setting, attacker (α) aims to infer hidden attribute of user u ($u \in V$) and also tries to improve inference attack using link prediction in online social networks. In next section, we propose methods to infer user attributes.

4. Methods

Since we considered only the topological property of a graph for attribute inference, our proposed method uses the centrality measure (closeness and betweenness) for attribute inference. The proposed methods are as follows:

4.1. *Closeness measure*

Closeness can be regarded as a measure of how long it takes to spread information from a node to all other nodes in the network, sequentially [10]. The Closeness centrality of a node n and the distance between user n and i are denoted as $C(n)$ and $d(n,i)$ respectively. Algorithm 1 uses Closeness centrality for inferring missing/hidden attribute (gender) of user.

Algorithm 1 Closeness algorithm

```
procedure CLOSENESS(G)
    for each t ∈ V do
        list1 ← F(t) + t
        m ← |list1|
        Draw a sub-graph taking all nodes ∈ list1
        for each node n ∈ list1 do
            C(n) = (m−1) / Σ_{i=1}^{m−1} d(n,i)
            list2 ← [n, C(n)]
        end for
        Select [max[C(n)], n] from list2
        if n.gender = true then
            gender(t) = gender(n)
        else
            list2 ← list2 − [n, C(n)]
            Go to 10
        end if
    end for
end procedure
```

4.2. *Betweenness measure*

Betweenness calculates the number of times a node lies along the shortest path between a pair of nodes in the network. Linton Freeman [11] firstly introduce it as a measure for quantifying the control of a human on the communication between other humans in a social network. Algorithm 2 uses the Betweenness centrality for inferring missing/hidden attribute (gender) of user. $B(n)$ denotes the Betweenness centrality of node n. $\sigma st(n)$ is the number of shortest paths from s to t that n lies on. σst denotes the number of shortest paths from s to t. s and t are nodes in the network different from n.

Algorithm 2 Betweenness algorithm

```
1: procedure BETWEENESS(G)
2:     for each t ∈ V do
3:         list1 ← F(t) + t
4:         m ← |list1|
5:         Draw a sub-graph taking all nodes ∈ list1
6:         for each node n ∈ list1 do
7:             B(n) = (Σ_{s≠n≠t} σst(n)/σst) / (((m−1)(m−2))/2)
8:             list2 ← [n, B(n)]
9:         end for
10:        Select [max[B(n)], n] from list2
11:        if n.gender = true then
12:            gender(t) = gender(n)
13:        else
14:            list2 ← list2 − [n, B(n)]
15:            Go to 10
16:        end if
17:    end for
18: end procedure
```

We propose the following method, for link prediction:

4.3. *Common neighbors*

Algorithm 3 is used to predict missing links. $F(x)$ and $F(y)$ denote the friends of user x and y respectively. Here, unlikely to common neighbors [12], we have included new parameter i.e. common groups for link prediction. $GR(x)$ and $GR(y)$ denotes the groups the users x and y joined respectively.

Algorithm 3 The Common Neighbor Algorithm

procedure COMMONNEIGHBOR(G)
    for each $node(x, y) \in V(G)$ and $edge(x, y) \notin E(G)$ **do**
        $f \leftarrow F(x) \cap F(y)$
        $g \leftarrow GR(x) \cap GR(y)$
        $w \leftarrow |f| + |g|$
        $list \leftarrow [(x, y), w]$
    end for
    $mw \leftarrow max(w) | \forall w \in list$
    for each $[(x, y), mw] \in list$ **do**
        $E(G) \leftarrow E(G) \cup edge(x, y)$
    end for
end procedure

5. Experiment and Result

Here, our objective is to improve attribute inference attack using link prediction in OSNs.

5.1. *Experimental setup*

Our algorithm is implemented on an Intel Core i7 @3.40 GHz CPU with 4 GB RAM. For our experimental purpose, we needed a large and real data-set of an OSN. The dataset was created by crawling profiles of several connected users, using a Facebook application named Netvizz. The characteristics of our data-set have been shown in Table 1.

Table 1. Properties of dataset.

Total Number of Nodes	5618
Total Number of Edges	102578
%age of Male Users	49.89
%age of Male Users	50.11

5.2. *Obtained result*

The performance of attribute inference before link prediction and after link prediction has been compared, by creating two copies of the same data-set. Then fifteen different data groups of each have been devised by randomly hiding gender attribute values of 10%, 15%, 20%, ..., 70%, 75%, 80% users respectively. For each such data group of the hidden gender attribute values (10%, 15%, ..., 75%, 80%); on one hand, we executed all the three attribute inference methods and noted the results, whereas on the other hand, we at first applied link prediction method before executing all the three attribute inference methods. Since, all the experiments are performed by randomly hiding the different percentages of gender, hence all the executions are carried by iterating it ten times and then averaging the result. The final results have been shown in Table 2.

Table 2. Result of attribute inference before link prediction and after link prediction.

% of Hidden Attribute	**Attribute Inference before link prediction**			**Attribute inference after link prediction**		
	Closeness	Betweenness	Bayesian	Closeness	Betweenness	Bayesian
10	46.25	46.29	47.07	52.43	52.63	54.16
15	45.34	45.52	47.15	51.96	51.82	53.85
20	45.85	46.36	47.17	51.77	51.33	53.04
25	46.13	46.88	46.99	51.90	51.42	53.20
30	46.36	45.95	47.73	49.66	49.33	52.97
35	45.92	45.22	48.81	49.22	49.39	53.21
40	45.85	45.45	48.03	48.53	48.48	53.04
45	45.07	44.62	47.64	48.76	48.67	51.73
50	44.57	44.25	47.94	47.09	46.76	51.94
55	44.05	44.19	48.44	46.89	46.60	51.08
60	42.20	42.07	48.45	44.94	44.94	51.99
65	42.50	42.53	48.27	43.88	43.63	51.75
70	39.24	39.36	48.29	43.12	43.01	51.70
75	37.71	37.68	48.15	40.43	40.65	51.74
80	36.54	36.66	48.46	38.41	38.23	51.59

Since the difference between attribute inference before link prediction and after link prediction are relatively small. Hence in order to check the significance of the difference, we have applied paired t-test [13]. For Closeness method, the calculated value of t is -8.6864. Similarly for Betweenness method and Bayesian method, the obtained value of t is -9.1460 and -13.1112 respectively. The value of t in t-distribution table, at 5% level of significance, for 14 degrees of freedom, is -1.860. The paired t-test result shows that there is a significant difference between attribute inference before link prediction and after link prediction.

On the basis of observation of Table 2, one can infer that link prediction improves attribute inference. It is also inferred that, more the gender information is hidden, more the accuracy of all algorithm decreases.

The Impact of link prediction method on each of the attribute inference methods can be clearly visualized by Figures 4, 5 and 6.

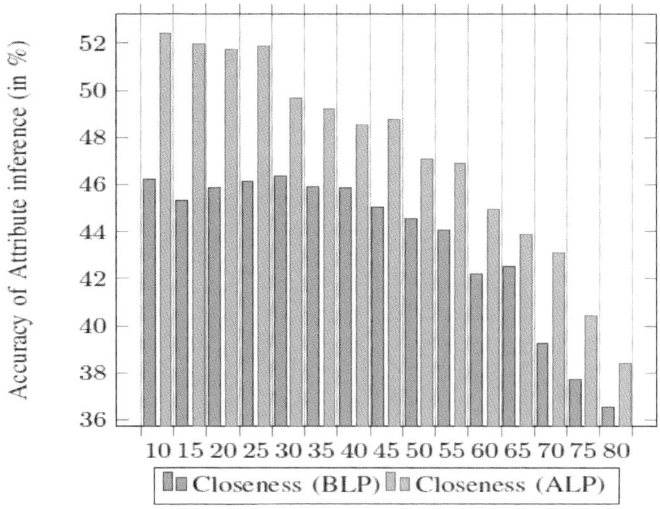

Figure: 4

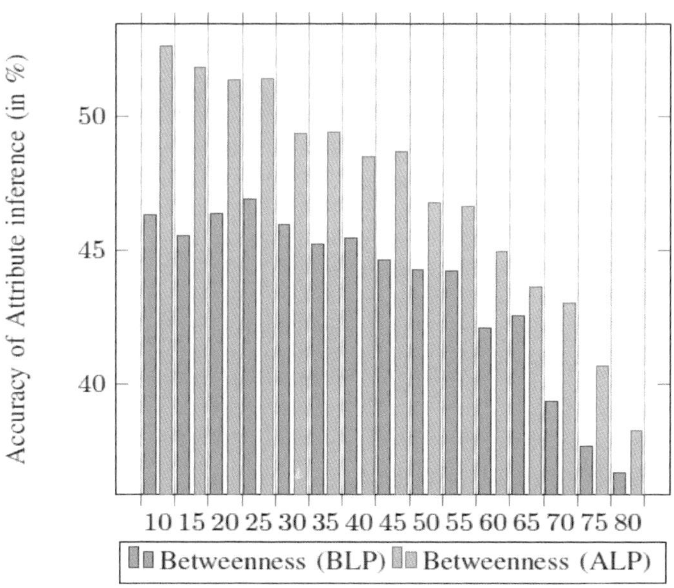

Figure: 5

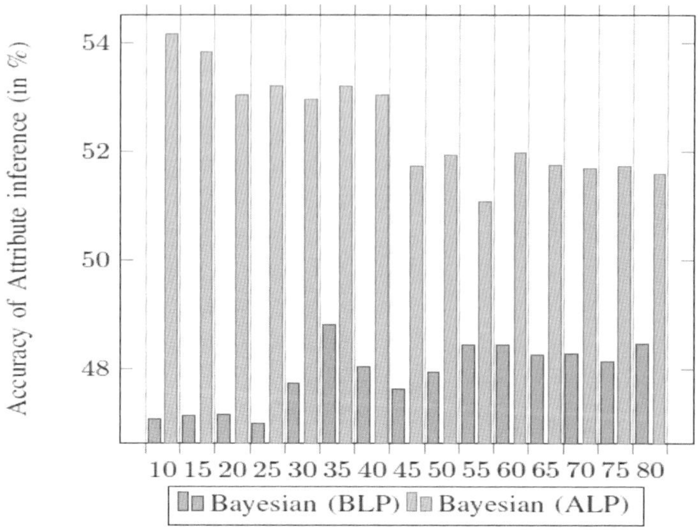

Figure: 6

6. Conclusion

In this paper we made a hypothesis that accuracy of attribute inference attack is improved by first inferring missing links, and we have proved the same. This paper considers the topological property of online social network for link prediction and attribute inference. As a future work we plan to use node similarity measure for attribute inference.

References

1. J. He, W. W. Chu and Z. V. Liu, Inferring privacy information from social networks, *Intelligence and Security Informatics*, 154 (2006).
2. J. Lindamood, R. Heatherly, M. Kantarcioglu and B. Thuraisingham, Inferring private information using social network data, in *Proceedings of the 18th international conference on World wide web*, 1145 (2009).
3. W. Xu, X. Zhou and L. Li, Inferring privacy information via social relations, *Data Engineering Workshop, 2008. ICDEW 2008. IEEE 24th International Conference on*, 525 (April 2008).
4. P. Kozikowski and G. Groh, Inferring profile elements from publicly available social network data, in *Privacy, Security, Risk and Trust (PASSAT) and 2011 IEEE Third Inernational Conference on Social Computing (SocialCom), 2011 IEEE Third International Conference on*, 876 (2011).

5. L. Jin, J. B. Joshi and M. Anwar, Mutual-friend based attacks in social network systems, *Computers & security*, **37**, 15 (2013).
6. C. Tang, Y. Wang, H. Xiong, T. Yang, J. Hu, Q. Shen and Z. Chen, Need for symmetry: Addressing privacy risks in online social networks, in *Advanced Information Networking and Applications (AINA), 2011 IEEE International Conference on*, 534 (2011).
7. G. K´otyuk and L. Butty´an, A machine learning based approach for predicting undisclosed attributes in social networks, in *Pervasive Computing and Communications Workshops (PERCOM Workshops), 2012 IEEE International Conference on*, 361 (2012).
8. R. Dey, C. Tang, K. Ross and N. Saxena, Estimating age privacy leakage in online social networks, in *INFOCOM, 2012 Proceedings IEEE*, 2836 (2012).
9. S. Ahmadinejad, M. Anwar and P. Fong, Inference attacks by third party extensions to social network systems, in *Pervasive Computing and Communications Workshops (PERCOM Workshops), 2011 IEEE International Conference on*, 282 (2011).
10. M. E. Newman, A measure of betweenness centrality based on random walks, *Social networks*, **27**, 39 (2005).
11. L. C. Freeman, A set of measures of centrality based on betweenness, *Sociometry*, 35 (1977).
12. D. Liben-Nowell and J. Kleinberg, The link-prediction problem for social networks, *J. Am. Soc. Inf. Sci. Technol.*, **58**, 1019 (2007).
13. C. Kothari, *Research Methodology: Methods and Techniques*. New Age International (P) Limited, (2004).

A Dynamic Model on Computer Virus

Upendra Kumar
Dept. of Computer Science and Engg, BIT
Patna-14, Bihar, India

An attempt has been made to develop SIRA model to balance the number of nodes in computer network. The SIRA model has Susceptible(S)-Infectious(I)-Recovered(R)-Appended(A) compartments. It has been assumed that an updated antivirus runs on computer network in continuous interval of time to avoid attacks of computer virus. The concept of natural death has been has been implemented. The threshold number and stability of the system have been established. Numerical methods are implemented to solve and simulate the system of equation developed under the real parametric values.

Keywords: Epidemic; Antivirus; Natural death; Eigen value; Stability.

1. Introduction

Computer viruses are software program arose in 1980. It can self-replicate and damaged executable files. The major source of computer viruses are electronic mail and secondary storage devices. The viruses damage computer system, erasing data, stealing information as well as modify normal operations. Their action throughout the network can be studied by using classical epidemiological model for diseases propagation [1–5]. The Kermack and McKendrick proposed SIR classical epidemic models for malicious objects propagation [8–9]. The Estimations for temporal evaluations of infected nodes depending on the network parameters of topological aspects of the network are implemented [10–14]. These kind of approaches are applied on email propagation schemes [6–7]. The modification of SIR model generates guide for infection prevention by using the concept of epidemiological threshold [25]. Richard and Mark proposed an improved SEI model to simulate virus propagation [23–24]. The new compartment has been introduced and its effect has been analyzed in this paper.

One way to control the propagation of viruses in network is implementation for updated antivirus software in continuous interval of time and finally maximum nodes of the network will be recovered from infected node [20–22]. In the SIRA model for infection control that do not confer immunity, susceptible nodes first go through the latent period and becoming infectious, therefore some infected nodes stay in infectious class(I) and then move into recover class(R). The computer nodes are temporally recovered after running of an antivirus. Since in the cyber world, acquired immunity is not permanent, so the recovered nodes are again return back into susceptible class nodes [15−19]. To balance total number of nodes in network due to attack of other than viruses, new nodes are recruit i.e. called Append class which is numerically equals to the natural death. The schematic diagram to the flow of viruses in the computer network has been depicted in Figure 1.

An assumption has been made that the population is a homogenous spatial distribution and mixing of host follow the law of Mass Action. More specifically I assumed that local density of the total population is constant. The total population size N(t)=S(t)+I(t)+R(t)+A(t) may vary with time in their compartments. Nodes of computer network are S(t), I(t), R(t) denote the size of S, I, R classes at the time t respectively. The per capita contact rate is β which is the average number of effective contact with other nodes per unit time. The total number of new susceptible nodes at time t is given by βAS where $\beta$is Mass Action coefficient. Further nodes are infected by the rate γ using law of Mass Action. Those nodes which are in infectious class will be recovered with δ per unit time and finally recovered nodes will be changed into Append class with the rate α per unit time. The natural death of nodes has been defined as a parameter X for the network system.

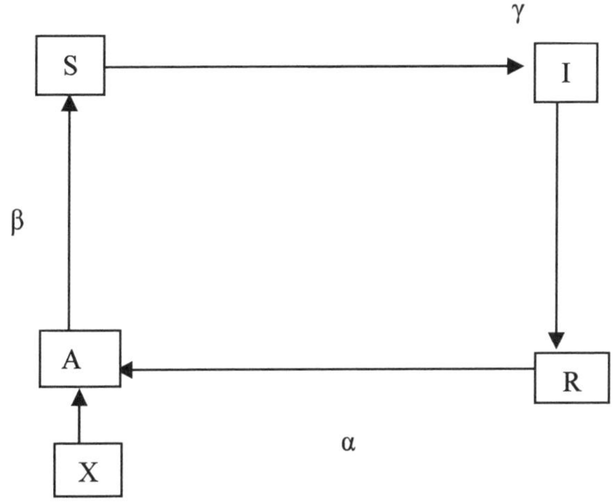

Fig. 1. Schematic diagram for the flow of computer viruses.

2. Mathematical Model Formulation

The number of nodes N(t) are partition into sub classes of nodes which are susceptible, infectious, recovered and appended with size S(t), I(t), R(t) and A(t) respectively. Our assumptions on propagation of the viruses are depicted in Figure 1. The SIRA model with law of mass action is formulated as

$$\begin{aligned} S' &= \beta AS - \gamma SI - \mu S \\ I' &= \gamma SI - \delta I - \mu I \\ R' &= \delta I - \alpha R - \mu R \\ A' &= \alpha R + X - \mu A - \beta AS \end{aligned} \tag{1}$$

Where parameters β, γ, δ, α and μ are non-negative constant. The constant X is the recruitment of new nodes in the network. μ is the per capita natural mortality rate (that is the crashing of nodes due to the reason other than attacks of malicious objects), δ is the rate at which nodes recovered temporarily after the running of updated antivirus software and recruitment into recovered class R from compartment I. α is the rate of recruitment from existing recovered class.

In the this model the nodes changes its classes from the S class to the I class, I class to the class R, R class to class A and again from class A to S class, because there is no permanent recovery in cyber world.

2.1. *Virus free conditions for computer network*

Using equation (1) we get

$$N' = X\mu N$$

Let
$$R \to \frac{A}{\mu} - S - I$$

Then
$$A' = \alpha\left(\frac{A}{\mu} - S - I\right) + X - \mu A - \beta AS$$

And N1=S+I+A, so $N1' = -\mu N1 - \alpha(S+I) - \delta I + X + \frac{\alpha A}{\mu}$

In disease free state I_0, therefore, $N1 \to \frac{A}{\mu}$

$$\Rightarrow N \to \frac{A}{\mu}$$

Diseases free equilibrium is a state corresponding to the absence of infected individuals (I=0). Apply this condition to the equilibrium of the model represented by equation (1), the equilibrium point is calculated as;

p1=(S, I, R, A)=(N,0,0,0)

Considering the equilibrium of p1, corresponding to linear system around it has a Jacobian J_{p1} [12], is given by

$$J_{p1}=\begin{bmatrix} -\mu & -\gamma\mu & 0 & \beta N \\ 0 & \gamma N-\delta-\mu & 0 & 0 \\ 0 & \delta & -\alpha-\mu-\lambda & 0 \\ 0 & 0 & \alpha & -\beta N \end{bmatrix}=0$$

Corresponding characteristic equation is given by

$$\begin{bmatrix} -\mu-\lambda & -\gamma\mu & 0 & \beta N \\ 0 & \gamma N-\delta-\mu-\lambda & 0 & 0 \\ 0 & \delta & -\alpha-\mu-\lambda & 0 \\ 0 & 0 & \alpha & -\beta N-\lambda \end{bmatrix}=0$$

An Eigen value of J_{p1} $\left[-\mu, \quad \gamma N-\delta-\mu, \quad -\alpha-\mu, \quad -\beta N\right]$.

Since maximum possible parametric values are negative, which shows that p1 is stable.

2.2. *Stability of network system*

To find stability, we have to equate value of S and A from equation 2(a) & 2(b) and have to put these values in equation (2) of (c).

The equation (2) is defined as

$$\begin{aligned} &\beta SI-\gamma SI-\mu S=0\cdots\cdots\cdots(a)\\ &\gamma SI-\delta I-\mu I=0\cdots\cdots\cdots\cdots(b)\\ &\alpha R+X-\mu A-\beta AS=0\cdots(c) \end{aligned} \qquad (2)$$

With the help of equation (1)(a) as well as (b) of equation (2) we get,

$$A=\frac{\gamma I+\mu}{\beta}$$

$$S=\frac{\delta+\mu}{\gamma}$$

after putting these values in equation (2) of (C), we have

$$\begin{aligned} &\alpha\left\{\left(\frac{\gamma I+\mu}{\beta\mu}\right)-\left(\frac{\delta+\mu}{\gamma}\right)-I\right\}+X\\ &-\mu\left(\frac{\gamma I+\mu}{\beta\mu}\right)-(\gamma I+\mu)*\left(\frac{\delta+\mu}{\gamma}\right)=0 \end{aligned} \qquad (3)$$

thus f(I) is clearly negative for large positive value of I, and also we have found in maximum cases eigen value is negative, thus local stability is found in the system [25].

3. Simulation and Discussion

Model has been simulated using N-R method of order four using real parametric values. The number of quarantined nodes are equals to recovered class. By considering Fig. 2 and Fig. 3, we can say as rate of recovery is increased, then the no of quarantined nodes will be decreases.

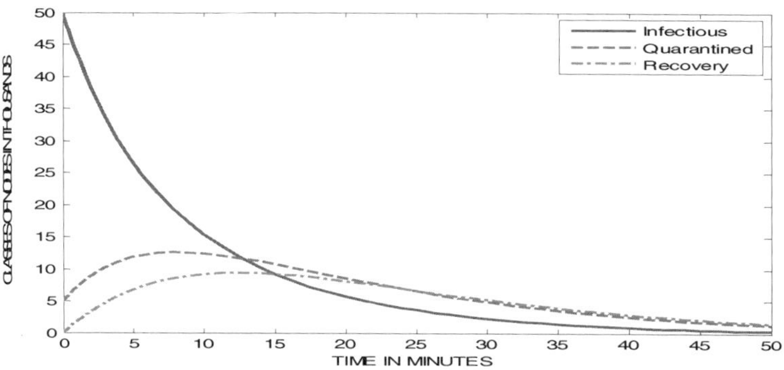

Fig. 2. Schematic diagram for infectious class and quarantined class to recovered class.

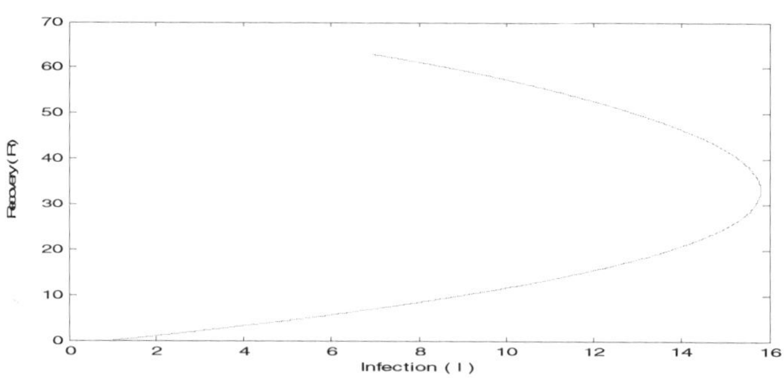

Fig. 3. Schematic diagram for infectious class to recovered class.

4. Conclusion

The Nodes of Computer network are categorized after attack of computer viruses. Updated antivirus is run to recovered nodes from infected class. Natural death of nodes is used to balance number of nodes in network. A Mathematical model is developed. Eigen value is calculated for local stability of the model which is negative. Reproductive number is defined having value less than one.

Stability of the model is discussed using Jacobian method. For authentication model is simulated.

References

1. M. E. J. Newman, S. Forrest, J. Balthrop, *Email networks and the spread of computer virus*, (E 66: 035101-1–035101-4 Phys. Rev, 2002).
2. W. O. Kermack, A. G. McKendrick, *Contributions of mathematical theory epidemics*, (Proc. Royal Soc. London – Series, A 115, 700 1927).
3. E. Gelenbe, Dealing with software viruses: a biological paradigm, Inform. SecurityTechnical (Rep, 12 (4) 2422007).
4. J. R. C. Piqueira, F. B. Cesar, Dynamical models for computer virus propagation, (doi:10.1155:/940526. Math. Prob. Eng, 2008).
5. S. Forest, S. Hofmeyr, A. Somayaji, T. Longstaff, Self–nonself discrimination in a computer, Proceedings of IEEE Symposium on Computer Security and Privacy, 202 (1994).
6. Y. Wang, C. X. Wang, *Modeling the effects of timing parameters on virus propagation*, ACM Workshop on Rapid Malcod ACM, 61 2003.
7. Allen, L. G. S., *some discrete time SI, SIR and SIS epidemic model*, **124**, Math Biol, 83 (1994).
8. W. O. Kermack, A. G. McKendrick, *Contributions of mathematical theory to epidemics* (A138, Proc. Roy. Soc. London – Series, 551932).
9. W. O. Kermack, A. G. McKendrick, *Contributions of mathematical theory to epidemics,* (A 141, Proc. Roy. Soc. London – Series, 94 1933).
10. C. C. Zou, W. B. Gong, D. Towsley, L. X. Gao, *The monitoring and early detection ofinternet worms*, **13** (5) (IEEE/ACM Trans. Network, 122, 2005).
11. J. O. Kephart, S. R. White, D. M. Chess, *Computers and epidemiology*, **26**, IEEE (1993).
12. M. J. Keeling, Eames, *Networks and epidemic models*, (J. Roy. Soc. Interf. 2005).
13. Ma. M. Williamson, J. Leill, An Epidemiological Model of Virus Spread and Cleanup, (2003).<http://www.hpl.hp.com/techreports/>.
14. W. T. Richard, J. C. Mark, Modeling virus propagation in peer to peer network, *IEEE International Conference on communication and signal, processing*. 981 (2005).
15. Ping Yan, Shengqiang Liu, *SEIR epidemic model with delay*, (J. Aust.; 48(1), Math. Soc. series B-Apple. Math, 119 2006).

16. W. H. Murrary, The application of epidemiological to computer viruses, *Computer& Security*, **7**, 130 (1988).
17. Alan Solomon, *Epidemiology and Computer Virus*, (unpublished, 1990).
18. Cliff Stoll, An epidemiology of viruses and network worms, *12th National Computer Security Conference*, 369 (1998).
19. Cohen F., Computer viruses theory and experiments, *computer and security*, **6** (1987) 22.
20. Kephart. J.O. Biological inspired defense against computer virus, *Proceeding of IJCA*, 19 (1995).
21. Williamson, M. M., Leveille, J., *An epidemiological model of virus spread and cleanup* (Hewlett-Packard, Feb 2003).
22. W. O. Kermack, A. G. McKendrick, *Contributions of mathematical theory to epidemics*, (Proc. Royal Soc. London – Series A, pp. 700–721, 1927).
23. Geoffrey Canright and Keneth Engoe-Monsen, *Spreading on networks, Atopographics view*, (2001).
24. F. Cohen, *On the implication of computer viruses and method of defense Computer & Security*, (V7 #2, April, 1998).
25. Paul Walkman, *Deterministic Threshold models in the Theory of Epidemics*, (Springer Leverage, New York, 1974).

State of the Art In-Service Condition Monitoring Techniques of Rotary Machines

Krishna Kant Agrawal
Ph.D. Research Scholar,
Indian Institute of Information Technology, Allahabad
Allahabad, India
E-mail: rs92@iiita.ac.in

Dr. Shekhar Verma
Associate Professor and H.O.D. (I.T.),
Indian Institute of Information Technology, Allahabad
Allahabad, India.

Prof. G. N. Pandey
Visiting Professor,
Indian Institute of Information Technology, Allahabad,
Allahabad, India.

In-service condition monitoring signifies a continuous assessment of health of industries and its equipment continuously throughout its serviceable life without any shutdown or hampering its normal working. Various condition based maintenance techniques comprises of detection of parameters like vibration, machine temperature, acoustic, current monitoring, etc are getting growing interests and showing considerable development for a current scenario of health monitoring of industrial machines that advice maintenance decisions based on different parameters, by investigating these parameters it enables us achieve the condition of the machine under consideration. The major purpose of implementing condition based monitoring is not only for planned maintenance rather to increase productivity, decline overall maintenance cost and enhance safety as well. These techniques substantially lowers the under lying operation cost by able to select the most suitable technique for a specific industrial scenario. Different literature are serene over the years on a variety of parameters of CBM techniques of machines and an exhaustive survey is sensed essential with aim to reproduce the current state of the art developments emphasizing on in-service machinery without having a shutdown.

Keywords: Condition based Monitoring (CBM); Sensors; Rotary Equipments; Maintenance Strategies; Vibration Monitoring.

1. Introduction

Condition based In order to maintain continuous production, sales and stay into market, it is becoming necessary for companies to increase the efficiency of their production processes continuously. Various production policies like Justin time and on-demand Production demand high availability of equipment under production in order to timely delivery of goods, as a result, maintenance strategies have gained in importance for ensuring availability, quality, on-time deliveries and overall safety of industries. Major constrains hindering the usage of maintenance functions are its initial costs which increased mainly due to improper planning, extra human efforts, and wrong choice of preventive maintenance on specific areas.

The necessity of condition based monitoring can be viewed from two angles. First, the health and safety in industries are so important that an unexpected fault may result in a great accident and high loss in output cost, particularly under current scenarios of continuous production. It is important that companies have to implement suitable maintenance techniques to avoid sudden breakdown, reduce underlying maintenance cost since machines themselves are very expensive, and subsequently extend the lifetime of machines. Condition based Monitoring is the only solution to all these problems to utilizing machines in a best possible way. Second, since the technology is growing at a fast speed, various sensors, signal acquiring and analysis together various techniques like Support Vector Machine (SVM) and Artificial Neural Network (ANN) has made it possible for implementing of condition monitoring techniques more effectively, which are making these techniques more reliable, more intelligent and easily to implement for workers into industries.

Different challenges of implementing Wireless Sensor Networks (WSN) for monitoring and protection of oil and gas operations are also addresses in this paper.

2. Progress in Condition Based Maintenance Technologies

Condition Based Maintenance Technologies are being driven by demands for up-gradation in: productivity, quality, and expenditure on machinery. Recent technological advancement includes improvement in knowledge of material failure, advancements in failure predicting techniques, advancements in various monitoring and sensory devices, advancements in diagnostic and prognostic software, and developments in maintenance software and networking technologies etc.

G. Vachtsevanos [1] state that neural networks and fuzzy logic are technologies designed to aid the performance of condition based maintenance systems. These system capabilities are judged on their ability to diagnose failure conditions and further prognosis of remaining working life before machine failure, as shown in Figure 1.

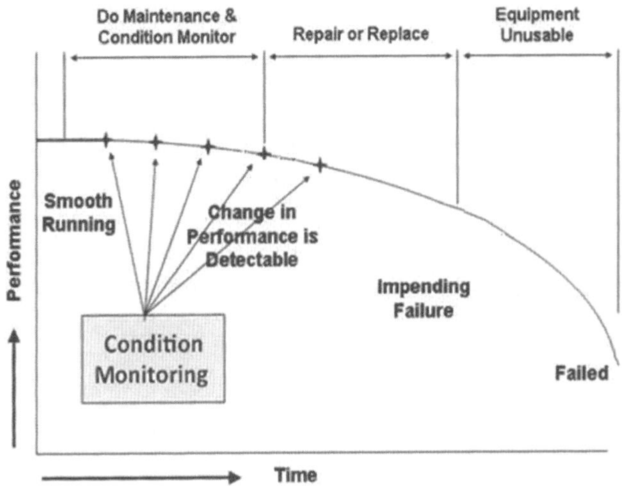

Fig. 1. Progress in machine performance reliability vs time [10].

2.1. *Intelligent condition based maintenance*

The term intelligent implies a system which is able to understand and make decisions itself without any external human intervention, technologies underlying include: sensors with built in intelligence, which are able to sense and transmit high grade information [2]; on-line sensors [4], designed to be reconfigured that detect patterns change; designed for analyzing trends and patterns within recorded sensory data, and decide the probability of failure of machines under monitoring [4]; various AI algorithms are capable of providing substitute data for failing the gapped data patters within recorded data [5].

Further theses intelligence could be integrated with automating parts ordering in case of prediction of any future failure into a specific part of machineries [2].

2.2. *Localized condition based maintenance*

Localized condition based maintenance is an independent prediction based maintenance technique, implemented right over the components under monitoring, by a specific maintenance personals. This technique involves recording maintenance data at given intervals in order to determine the condition of the machine under monitoring, and further analysis is performed on the received data whether the condition of machine under monitoring is acceptable or not.

2.3. *Remote condition based maintenance*

Remote condition based maintenance can be classified into standalone system or connected to another systems under monitoring. It involves monitoring the condition of machine away from physical machine. Theses monitoring could be performed automatically depending on the systems needs and at the frequency the data of machine is required and in the same way diagnosing of the machine again could be performed automatically or could be manual on time-to time basis or as per need of the industry. These all could be made possible through wireless networks and sensors present which could be mounted on various location of machines to receive continuous data of machines. These automated collections of data plays a vital role in collecting data from places where humans can't record data by itself like into high temperature zones or heavy noisy environment etc by F. Zorriassatine [5].

Currently monitoring through online- web based mechanism accessible remotely via internet are termed as 'Internet CBM', or 'E-Monitoring'. V. R. Kennedy [6] gives details of how Active Server Pages (ASPs) are used for interaction between users and the system is performed. ASPs, offer new areas to system designers for interactive pages. Lloyd Dewey Lee [7] gives details of vibration levels could easily be transmitted over the internet using remote CBM, by which analysis could be performed over the machines in actual time intervals by displaying graphs, tables and trends of vibrations or any other parameters involved.

3. Analysis of Various Monitoring Technique

3.1. *Motor current signature monitoring (MCSA) [3]*

In this technique of motor current signature analysis, machine active condition is monitored through analyzing the different spectrum generated from machine's current signature. Various differences into electric background signature are directly associated with the any defect or alteration in the specific mechanical part or components of the machine by which, fault signature can be identified by motor current signature analysis technique [9]. In this technique there is a limitation that in most of the cases minor bearing damage or initial defects is not much significant to produce a noticeable current fluctuations, due to which it becomes quite difficult to discriminate the signatures from current fluctuations due to voltage imbalances or stator faults should be centered and numbered consecutively.

3.2. *Oil analysis [3]*

One of the most important aspects to make ensure that the machine is in healthy condition is by ensuring the oil examination for maintaining the condition and reliability of that machine. Many factors involved such as machine operating environment are important factors in the selection of a specific lubricant for a specific type of machines. The application of Tribology department is frequently referred to as an Oil Analysis in heavy oil and gas industry. This technique of regular Oil Analysis of different parts of a machine is a common Condition-Based Maintenance practice within industry, particularly in oil and gas industries where machines are a critical element. By quick oil analysis by an expert or maintenance engineers on the Condition of their machinery and can make useful outcome for corrective/maintenance actions, such as adding oil/grease or to completely replace the component.

3.3. *Temperature monitoring [3]*

As the name indicates in temperature monitoring technique which is mostly used in different high temperature and high speed environment, typically suitable for heavy industries and industrial purpose tools. Temperature sensor consists of two metals which are joined together, and when exposed to some temperature difference generates a thermoelectric voltage which is dependent on the specific material properties and temperature difference [9]. Specially bearing related defects generate some excessive heat in the rotating apparatus of machines. Continuous monitoring of temperature difference of a housing, casing or lubricant is one of the effortless methods for fault identification into machines by temperature analysis. Machinery defects generate excessive heat which results in deviation from the allowable temperature range into rotating components [9]. Monitoring temperature of bearing housing is not very efficient for huge industrial machinery scenario as small deviation into a temperature becomes quite hard to detect.

3.4. *Neural network based monitoring [3]*

Artificial neural networks (ANN) are combination of different mathematical models inspired by the brain nerves. This technique of fault identification has been used in fault diagnoses, pattern matching and classification of defects. Neural Network is capable to model complex processes with multiple inputs and output as required. The neural network model can learn by observing the input and outputs, and adjusting internal weights in between the stages. Neural Network can identify different patterns that can be later used for fault classification. Physical modeling and current knowledge base for fault analysis is not sufficient enough to identify a specific defect into a machine rather using a proactive modeling using a neural network could predict the state of machine into different stages.

3.5. *Acoustic monitoring [3]*

One of the useful techniques in condition based monitoring is acoustic based monitoring of machines. It was the first to be developed as a Non-Destructive Testing (NDT) technique to detect different faults into a real industrial scenario. Specific acoustic sensors are deployed, which are able to detect acoustic signals generated by the vibration waves in wear or crack growth which is a transient impulse generated in any solid matter which is under mechanical or thermal stress [7]. Detection of fracture in parts of machines is a major application of acoustic monitoring technique and consequently, the technique is be used as a means for a condition based monitoring of damages into a bearing and shaft cracks. The technique of acoustic monitoring can be a measurement of machine sound also is engaged for detecting the defects in bearings. In general, the accuracy varies with sound pressure and intensity [7]–[9].

3.6. *Loose parts monitoring [3]*

This technique is performed in many industries on a regular basis. This work engage different sensors installed at several locations over machines. Both audio signals and noise are used for loose parts monitoring. The audio signals enables alarms if any part of the system is found significantly loose. The different alarm set points are selected on the basis of specific plant and the sensitivity of the loose parts monitoring machine. If an alarm is activated, sensor output data are analyzed to identify its size and location. Localization of a loose part is can be achieved by comparing signals from different sensor in various locations by identifying signal transmission times.

3.7. *Vibration monitoring [3]*

In recent years, predictive maintenance through vibration analysis has become one of the most significant practices in industrial processes. By means of accelerometers or vibration sensors and many other similar sensors, the vibration of machinery is measured and further to identify any deviations from the expected or normal performance. By this technique it is simple to successfully identify the current status of machines and many upcoming troubles by predicting the condition of machines with industrial equipment and especially with rotating machinery.

Condition / Parameters	MCSA	Temperature	Acoustic	Oil Analysis	Loose Parts	Vibration
Misalignment		✓	✓		✓	✓
Bearing Damage	✓	✓		✓		✓
Out-of-Balance		✓	✓		✓	✓
Excess Noise			✓			✓
Gear Damage	✓	✓		✓		✓
Looseness			✓		✓	✓

Table 1. Comparative analysis of different monitoring techniques [3].

4. Overall Maintenance Strategies

Through exhaustive literature survey it's now becomes clear that maintenance strategies can be performed broadly in two major types: preventive maintenance and corrective maintenance as shown in chart below. Both of the maintenance techniques are widely used in all industrial scenarios.

Corrective maintenance, as the name suggests, is undertaken after a break-down or when specific faults are located which is quite similar to traditional repairing process, on the other hand Corrective maintenance is "Maintenance carried out after fault recognition and intended to put an item into a state in which it can perform a required function." In repairing process, the failed part is restored to its normal state and it continues to its normal working and in some cases some modeling techniques are also available to use, but the part just repaired could again fail to work as soon as repaired as the conditions for the part or machinery are as still the same.

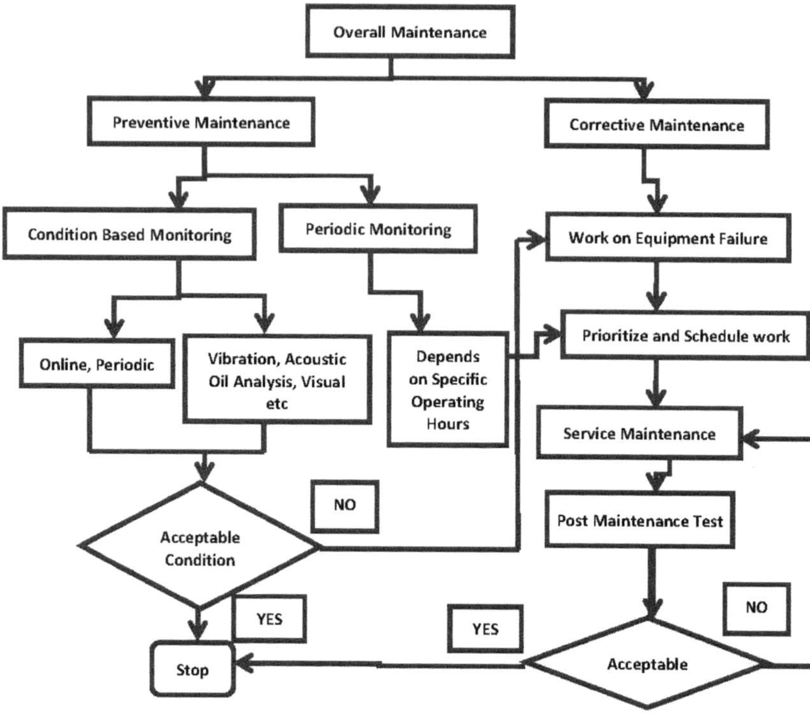

Fig. 2. Overall maintenance strategies.

These types of maintenance usually being performed in case of urgency such as lack of time or availability of spare parts. The other technique, as shown in Figure 2, condition based maintenance or CBM, does not work on pre-determined time intervals like that of preventive maintenance, and instead it continuously monitors the condition of machine since many components do not have a clear wear-out region and thus are not suitable for preventive maintenance. Condition based maintenance, with condition monitoring in particular, are introduced as one solution for these potential failures.

The other technique, condition based maintenance or CBM, does not work on pre-determined time intervals like that of preventive maintenance, and instead it continuously monitors the condition of machine since many components do not have a clear wear-out region and thus are not suitable for preventive maintenance. Condition based maintenance, with condition monitoring in particular, are introduced as one solution for these potential failures.

Condition based maintenance is mainly performed for the following two purposes [3]:

- Determine if a problem exists, severity of problem, and the remaining lifetime of the machine before its complete failure, and
- Delectation and identification of specific parts on the machine under monitoring which are degrading and further diagnose the exact problem.

5. Conclusion and Future Work

This paper attempts to summarize the various techniques that appeared in recent research literature involved in machine diagnostics and prognostics for implementing CBM. Various techniques of condition based maintenance have been categorized depending on different areas of implementation. Condition monitoring has become a very important technology in the field of mechanical equipment maintenance, and is attracting more and more attention worldwide.

The potential functions of failure prediction, identification, and overall life estimation of a machine carry a lot of advantage for different companies, benefits includes: reduced maintenance cost, extended equipment's overall life, enhanced safety, minimizing accident as well as improving power quality.

Due to above said benefits and best utilization of existing resources under current environment of huge competition, Condition based monitoring is currently a hot topic to managers and engineers in industries as well as among researchers too. Maintenance strategies which help to achieve precise prediction of status are receiving a great deal of attention recently due to massive maintenance cost. Many technologies are being adopted from other domains such as Artificial Intelligence and patter recognition and modified accordingly for precise predication of current status and predicting lifetime of machines.

Condition based monitoring is most widely used into Petroleum refinery, chemicals, petrochemical plants etc, among various identified techniques of condition monitoring, Vibration Analysis is the most widely used technique of CBM and NDT. Adopting predictive maintenance, reduced machine breakdowns are the basic motivations behind implementing CBM systems. Improved product quality, to adopt change or improve existing maintenance strategies are additional benefits of implementing CBM techniques.

Technological integration of various other fields into CBM techniques develops a variety of different mediums of representation of status of machines like discrete values, different charts, graphics, setting up of automatic alarms values etc. More than 80% of reliability of implementing different CBM systems exists with respects to meeting user's expectations, introducing predictive maintenance, creating future maintenance initiatives and increasing awareness of maintenance issues.

Various experiments are being carried out and analysis of recorded data of acoustic and vibration of bearing at different states of faults including various race faults and subsequent validation of analysis results with the available set of standard bearing faults at CWRU are under process.

Acknowledgments

This research is supported by Research Grants provided by Indian Oil Corporation, Research & Development Center, Faridabad, India and Indian Institute of Information Technology, Allahabad, India.

References

1. G. Vachtsevanos and P. Wang, An intelligent approach to fault diagnosis and prognosis, *Georgia Institute of Technology* (2001).
2. J. Coy *et al.*, Proceedings, International Conference Mail Technology; Evolution to e-Revolution, UK, ISBN 186058 327X, 99 383 (2001).
3. Agrawal, K. K., Pandey, G. N., Chandrasekaran, K., Analysis of the condition based monitoring system for heavy industrial machineries, *Computational Intelligence and Computing Research (ICCIC), 2013 IEEE International Conference*, 1 doi: 10.1109/ICCIC.2013.6724183 (2013).
4. Philip Jacob, Ideas for the exploration of distributed intelligence in condition monitor systems, The University of Manchester.
5. Rolf Orsagh *et al.*, An internet-based machinery health monitoring system, *MFPT Committee Meeting*, Virginia Beach, VA, May (2000).
6. F. Zorriassatine *et al.*, Wireless sensors for intelligent condition monitoring – a survey, *Proceedings ICOM03, International Conference on Mechatronics*, ISBN 1 86058 420 9 427 (2003).
7. Andrew K. S. Jardine, Daming Lin, Dragan Banjevic, A review on machinery diagnostics and prognostics implementing condition-based maintenance, *Mechanical Systems and Signal Processing* 20 1483 (2006).
8. Ying Peng, Ming Dong, Ming JianZuo, Current status of machine prognostics in condition-based maintenance: a review, *International Journal of Advanced Manufacturing Technology* 297 (2010).
9. Y. Han and Y. H. Song, Condition Monitoring Techniques for Electrical Equipment—A Literature Survey, *IEEE TRANSACTIONS ON POWER DELIVERY*, **18**, (2003).
10. http://www.lifetime-reliability.com

Image Segmentation: A Survey

K. M. Pooja[1] and R. Rajesh[2]
Dept. of Computer Science, Central University of South Bihar,
Patna, Bihar, India
E-mail: [1]poojam102@gmail.com, [2]kollamrajeshr@ieee.org
www.cub.ac.in

Segmentation is needed for extracting the useful or concerned information. There are lots of segmentation algorithms available. Here, we have discussed various segmentation strategies based on edge, threshold, super pixel, texture and some preprocessing technique for dealing with complex structured images. Experimental are carried out for the segmentation of pedestrians on OSU thermal pedestrian dataset based on GBSF and ARMSF proposed by Z. Lieu *et al.* The results on pedestrian segmentation are promising.

Keywords: Pedestrian Segmentation; Global Background Subtraction Filter; Adaptive Row Mean Subtraction Filter; OSU thermal database.

1. Introduction

Image segmentation with 100% perfection is always a challenging task and is an important step for processing/analyzing an image in applications like object detection/tracking/recognition [5][6]. There is no single segmentation algorithm that work for all types of images/situations [7]. Most of the cases only certain portion of the image is important for specific purpose [1]. Image segmentation is a process of combining/partitioning the pixels that are of similar on color/texture/intensity [2][3][4].

This paper presents a quick survey on segmentation. Experimental are carried out for the segmentation of pedestrians on OSU thermal pedestrian dataset based on GBSF and ARMSF proposed by Z. Lieu *et al.* The results on pedestrian segmentation are promising.

Rest of the paper is organized as follows. Section 2 discuss about preprocessing techniques, Section 3 discuss different segmentation strategies, Section 4 discuss the experimental result of segmentation using GBSF and ARMSF [8] on OSU thermal pedestrian dataset and Section 6 discuss the conclusion.

2. Preprocessing Techniques

Image that are processed and analyzed are of different types. Some of the images are having complex background structures (like IR images captured during night) and complex texture structure (like hyper spectral images). Preprocessing techniques are used for suppressing the complex background or for smoothing the complex texture structure or for noise reduction or for making the segmentation task easier. The most popular techniques done before segmentation are background subtraction [8] and mean shift filtering [9].

3. Segmentation Strategies

This section provides the various strategies for image segmentation.

3.1. *Texture based*

Texture of an image gives a the feel of surface or describes spatial organization of color/intensity and is useful in segmentation or classification of an image. Texture is repetition of pattern or shape and texture based segmentation is useful in pattern recognition/machine learning/ground classification/medical imaging/aerial imagery.

Textured based segmentation algorithm are basically used for segmenting hyper spectral images. Some of the segmentation schemes are (i) extracting the complex shape information from textured VHR images [12], (ii) oriented edge based feature for drawing out local edge feature which are useful in dealing with complex background image and gray scale image [28], (iii) variable weighting parameter using local and global weight for extracting edge and texture information [3], (iv) textural spatial information along with novel tropic model to highlight texture in very high resolution remote sensing images [13].

3.2. *Threshold based*

Threshold based segmentation is simplest among the available segmentation methods. Basic idea is to select a threshold value (either local or global threshold). Global threshold based segmentation method does not work when there is large variation in illumination.

Some of the popular thresholding based methods are (i) Otsu thresholding technique [14], (ii) hierarchical segmentation based on multiple thresholding [15], (iii) iterative tri-class thresholding technique [16], (iv) fast two dimensional Otsu methods [17], (v) normalized graph cut as thresholding [18].

3.3. *Super-pixel based*

Super pixel are blocks of irregular shapes based on texture similarity, brightness similarity and contour continuity [19] [21]. The algorithms used for dividing the image into super pixel are watershed algorithm [20] and simple linear iterative clustering [22].

3.4. *Edge based*

Objects are separated from the background through edges or boundaries. There are several edge detection techniques some of them are based on first order derivative (eg. Sobel, Prewit and Canny methods), second order derivative (eg. Laplacian of Gradient operator) and on basic morphological operation (eg. erosion, dilation, opening, closing) [23][24][26][27].

Other hybrid methods are (i) histogram of normalized gradient with self-adaptive threshold [25], (ii) texture information combined with gradient information [25].

4. Pedestrian Segmentation using GBSF and ARMSF

IR images capture the heat emitted from the object. Temperature of objects vary from its surrounding background and this variation is captured in IR imaging.

In pedestrian images, captured in night by thermal camera, temperature difference of different part of pedestrian are significant due to clothing and hence the brightness varies within the pedestrian.

In [29] GBSF has been applied to suppress the complex background of ship target. Then ARMSF using saliency, gradient and foreground image has been applied to background suppressed (foreground image) to highlight the target. In ARMSF saliency has been used to focus on target and gradient using sobel operator has been used for highlighting the edge.

We have applied GBSF and ARMF [29] for pedestrian segmentation. Benchmarking OSU thermal pedestrian database [11] is used for the study which consists of 10 sequences of images of size 360×240 in 8 bit gray scale format (We have not used sequence 3).

The result of segmentation using iteration less GBSF and ARMF is shown for three images in Figure 1. Table 1 shows that number of pedestrians segmented is 767 out of 878 pedestrians and there are 213 hot objects which are wrongly segmented as pedestrians. Hence there is still scope in improving the results based on size based features in addition to segmentation.

Table 1. Segmentation using GBSF [8] and ARMSF [29] on OSU pedestrian database.

Pedestrian detected	**767**
Pedestrian not detected	116
Non pedestrian detected	213

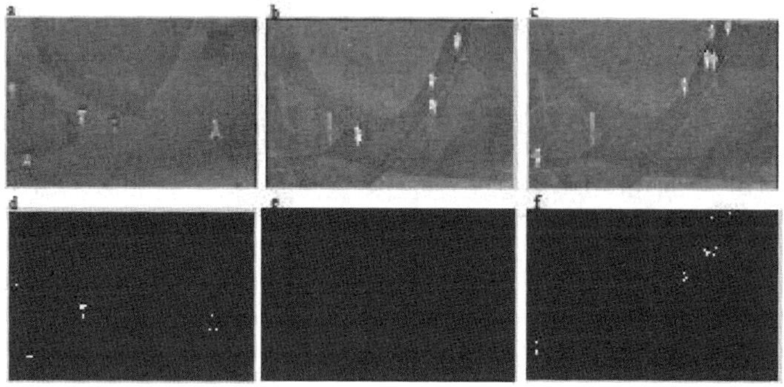

Fig. 1. Original image (a,b,c). Segmentation using GBSF and ARMSF (d,e, f).

5. Conclusion

A quick survey of image segmentation is presented in this paper. Experiments are conducted on pedestrian dataset using GBSF and ARMSF proposed by Z. Lieu *et al.* The results on pedestrian segmentation are promising. There are still scope in improving the results based on size based features in addition to segmentation.

References

1. B. Padmapriya, T. Kesavamurthi and H. W. Ferose, Edge based image segmentation technique for detection and estimation of the bladder wall thickness, *Procedia engineering*, **30**, 828 (2012).
2. N. M. Nimbarte and M. M. Mushrif, Multi-level thresholding algorithm for color image segmentation, *Int. Conference on Computer Engineering and Applications (ICCEA)*, **2**, 231 (2010, Bali Island).
3. C. Zheng, Image segmentation based on hierarchical belief propagation with variable weighting parameters, *Optik-International Journal for Light and Electron Optics*, **125**, 1158 (2014).
4. H. Fleyeh, D. Jomaa and M. Dougherty, Segmentation of low quality fingerprint images, *IEEE Int. Conference on Multimedia Computing and Information Technology (MCIT)*, (Sharjah, 2010).

5. A. Faruquzzaman, N. R. Paiker, J. Arafat, Z. Karim and M. A. Ali, Object segmentation based on split and merge algorithm, *IEEE Int. conference on TENCON*, 1 (Hyderabad, 2008).
6. M. A. Ali, L. S. Dooley and G. C. Karmakar, Object based image segmentation using fuzzy clustering, *in Proc. IEEE Int. Conference on Acoustics, Speech and Signal Processing (ICASSP)*, **2**, (Toulouse, 2006).
7. S. Ben, Chaabane, M. Sayadi, F. Fnaiech and E. Brassart, Colorimage segmentation using automatic thresholding and the fuzzy c-means techniques, in *IEEE conference on Electro technical Conference*, (Ajaccio, 2008).
8. J. W. Davis, Sharma and V. Sharma, Background-subtraction using contour-based fusion of thermal and visible imagery, *Computer Vision and Image Understanding*, **106**, 162 (2007).
9. L. Wang, G. Liu and Q. Dai, Optimization of segmentation algorithms through mean-shift filtering preprocessing, *Geoscience and Remote Sensing Letters, IEEE* **11**, 622 (2014).
10. D. Martin, C. Fowlkes, D. Tal and J. Malik, A database of human segmented natural images and its application to evaluating segmentation algorithms and measuring ecological statistics, in *Proc. IEEE Int. Conference on Computer Vision, (ICCV)*, **2**, 416 *(Vancouver, BC,* 2001*).*
11. J. W. Davis and M. A. Keck, A two-stage template approach to person detection in thermal imagery, *WACV/MOTION*, **5**, 364 (2005).
12. S. Balla-Arabe, X. Gao, B. Wang, F Yang and V Brost, Multi-Kernel Implicit Curve Evolution for Selected Texture Region Segmentation in VHR Satellite Images, *IEEE Transactions on Geoscience and Remote Sensing*, **52**, 5183 (2014).
13. H. Feng, Z. Jiang and X. Han, Texture segmentation for remote sensing image based on texture-topic model, in *IEEE Int. Symposium Geoscience and Remote Sensing (IGARSS)*, 2669 (Vancouver, BC, 2011).
14. M. Cheriet, J. N. Said and C. Y. Suen, A recursive thresholding technique for image segmentation, *IEEE Transactions on Image Processing* **7**, 918 (1998).
15. J. Ma, D. Wen, S. Yang, L. Wang and J. Zhan, Hierarchical segmentation based on a multilevel thresholding, *Int. Congress on Image and Signal Processing (CISP)*, **3**, 1396 (Yantai, 2010).
16. H. Cai, Z. Yang, X. Cao, W. Xia and X. Xu, A new iterative triclass thresholding technique in image segmentation, *IEEE Transactions on Image Processing*, **23**, 1038 (2014).

17. C. Yu, C. Dian-ren, Y. Xu and L. Yang, Fast two-dimensional otsus thresholding method based on integral image, *in Proc. IEEE Int. Conference on Multimedia Technology*, 1 (Ningbo, 2010).
18. J. Zhao, Y. Heng, X. Gu and S. Wang, The edge detection of river model based on self-adaptive canny algorithm and connected domain segmentation, in *World Congress IEEE on Intelligent Control and Automation (WCICA)*, 1333 (Jinan, 2010).
19. S. Liu, L. Zhang, Z. Zhang, C. Wang and B. Xiao, Automatic cloud detection for all-sky images using superpixel segmentation, *in Geoscience and Remote Sensing Letters*, **12**, 54 (2015).
20. C. Xu, J. Xie, G. Chen and W. Huang, An infrared thermal image processing framework based on superpixel algorithm to detect cracks on metal surface, *Infrared Physics & Technology*, **67**, 266 (2014).
21. L. Li, W. Feng, L. Wan and J. Zhang, Maximum cohesive grid of superpixels for fast object localization, in *IEEE Conference on Computer Vision and Pattern Recognition (CVPR)*, 3174 (Portland, OR, 2013).
22. R. Achanta, A. Shaji, K. Smith, A. Lucchi, P. Fua and S. Susstrunk, SLIC superpixels compared to state-of-the-art superpixel methods, *IEEE Transactions on Pattern Analysis and Machine Intelligence*, **34**, 2274 (2012).
23. M. Sridevi and C. Mala, A survey on monochrome image segmentation methods, *Procedia Technology*, **6**, 548 (Coimbatore, 2012).
24. Y.-T. Hsiao, C. L. Chuang, J.-A. Jiang and C. C. Chien, A contour based image segmentation algorithm using morphological edge detection, in *IEEE International Conference on Systems, Man and Cybernetics*, **3**, 2962 (2005).
25. J. Qin and G. He, A segmentation method for high resolution remote sensing image based on texture gradient, in *IEEE Int. Conference on Advanced Computer Theory and Engineering (ICACTE)*, **1**, 1 (Chengdu, 2010).
26. Z. Yu-qian, G. Wei-hua, C. Zhen-cheng, T. Jing-tian and L. Ling-yun, Medical images edge detection based on mathematical morphology, in *IEEE Int. Conference on Engineering in Medicine and Biology Society (EMBS)*, (Shanghai, 2006).
27. B.-G. Kim and D.-J. Park, Novel non contrast-based edge descriptor for image segmentation, in *IEEE Transactions on Circuits and Systems for Video Technology*, **16**, 1086 (2006).
28. L. Sun and T. Shibata, Unsupervised Object Extraction by Contour Delineation and Texture Discrimination Based on Oriented Edge Features,

IEEE Transactions on Circuits and Systems for Video Technology, ***24***, 780 (2014).

29. Z Liu, F Zhou, X Chen, Bai X, C Sun. Iterative infrared ship target segmentation based on multiple features. *Pattern Recognition*, **47**, 2839 (2014).

Empirical Reliability Modeling of Transaction Oriented Autonomic Grid Service

Dharmendra Prasad Mahato[1] and Ravi Shankar Singh[2]

Department of Computer Science and Engineering,
Indian Institute of Technology (BHU), Varanasi, India 221 005
E-mail: [1]dpmahato.rs.cse13@iitbhu.ac.in, [2]ravi.cse@iitbhu.ac.in

This paper presents reliability modeling of transaction oriented autonomic grid service. Transaction oriented autonomic grid technology is aimed at providing reliable services for users by hiding the complexity of the service and protecting the system from various failures. The reliability in these systems is greatly affected by the occurrence of failures. A coloured petri net model, CPN-TOGS (CPN-Transaction Oriented Grid Service), is presented in this paper for analyzing the empirical reliability in transaction oriented autonomic grid service. The model maintains the recovery of the failed processes by using both local level and replicated level recovery.

Keywords: Reliability; Transaction Oriented Grid Service.

1. Introduction

Grid computing provides a service oriented infrastructure with standardized protocols and services for enabling access to and for sharing of geographically distributed hardware, software and information resources [1], [2]. The grid technology is limited not only to scientific computing but also it is expanding in large scale to business applications by providing autonomous environment and reducing the users' interventions as much as possible [3]. In order to create reliable and real-time applications with multiple transparencies on location, replica, concurrency and failure, transaction processing and its management in grid technology are needed as effective means by sharing a large number of resources among different organizations [4], [5], [6], [7], [8], [9]. To make the system consistent and free from the various failures, today's grid technology provides autonomic environment with some recovery mechanisms in a self-configuring, self-optimizing, self-healing, and self-protecting fashion so that complexity associated with the grid can be reduced and the system can be made easy to use [10]. Transaction oriented autonomic grid service, which is a set of operations executing on heterogeneous grid sites, is responsible for ensuring the reliable and real-time execution of inherently business applications [11]. But the handling of real time transactions in autonomic grid is a challenging task, because it is an important means to protect the systems from various failures and to complete the task by the deadline [10]. Transaction oriented autonomic grid service reliability is the probability of all of the subtasks (short-lived or long-lived) involved in the considered service to be executed successfully [1], [12], [13]. This paper presents CPN-TOGS model for the empirical evaluation of reliability of transaction oriented autonomic grid service. The model also describes two types of recovery of failed processes: local level recovery and replicated level recovery.

The remainder of this paper is organized as follows. Section 2 reviews related research and discusses existing approaches to grid service and transaction oriented grid service reliability modeling and analysis. In Section 3 some background information is presented. Section 4 presents the proposed CPN-TOGS model in detail. Section 5 presents simulation setup and the results in different scenarios. Section 6 concludes the paper.

2. Related Work

Although the modeling and analysis of grid service reliability has attracted a lot of research attention, reliability modeling and analysis in transaction oriented autonomic grid service has not been sufficiently studied. Azgomi *et al.* have presented a model to analyze grid service reliability [14] where task scheduling by RMS and the task execution within grid resources were modeled using coloured Petri nets (CPNs). To improve the grid service reliability, Guo *et al.* presented a fault recovery technique in grid systems and conducted an empirical modeling and analysis of grid service reliability with fault recovery [2]. Later on, Guo *et al.* presented a multi-objective task scheduling model and proposed an ant colony optimization (ACO) algorithm to solve the fault recovery to enhance the grid service reliability [1]. For effective and accurate assessment of grid service reliability Doguc *et al.* presented a new method based on data-mining algorithm K2 to discover the grid system data using Bayesian networks (BN) [15]. Dai *et al.* again presented a virtual approach for modeling grid services and obtained grid service reliability using Morkov models, Queuing theory, and the graph theory [16]. Levitin and Dai introduced grid service reliability, performance indices and a fast numerical algorithm for the evaluation were presented [12]. In [17, 18], Dai *et al.* presented an algorithm to achieve greatest possible expected service performance or reliability, based on graph theory, Bayesian approach, and the evolutionary optimization approach and presented a virtual tree-structured model. In [19, 20], Dai *et al.* presented models for reliability and performance analysis of tree-structured grid service considering data dependence and failure correlation.

Recovery of failed process also plays an important role in analyzing the reliability of the grid systems and it is found that fault recovery studied in most related research is achieved by migration mechanism, i.e., when a failure occurs on a grid node, the state information is migrated to another node on which the subtask execution is resumed from the interrupted point [9]. Heddaya and Helal analyzed the impact of local fault recovery on the service reliability of distributed computing system where the service is restarted locally, thus, the global communication is minimized. However, the research on the impact of local fault and migration (replicated) fault recovery on the transaction oriented autonomic grid service reliability is very scarce [21].

3. Background

Grid service reliability R_{gs} is defined as the probability that a set of programs contained by a grid service can be completed successfully on multiple resources and links [2]. Considering the situations of node failure, communication link failure, and software failure in the reliability analysis, there are some key assumptions that must be considered such as,

1) The user jobs are generated with exponential distribution,
2) When a service request reaches the RMS, the RMS reply immediately,
3) Each node can execute more than one task at a time as a node consists of a cluster of homogeneous or heterogeneous processors,
4) The failures in different elements are independent,
5) The failure processes of nodes are modeled by Poisson's distribution.

When a service S is received by the RMS, the service S is divided into m subtasks and these subtasks are assigned to w nodes. The processing time of subtask i $(i = 1, 2, \ldots, m)$ on node k is

$$\tau_{ik} = \frac{c_i}{s_k}, \quad (1)$$

where c_i is the computational complexity of subtask i and s_k is the processing capability of node k $(k = 1, 2, \ldots, w)$ [2].

When λ_k be the failure intensity of node k, the probability that the node k functions without any failure during the processing time of subtask i is

$$p_{ik} = e^{-\lambda_k \tau_{ik}} \quad (2)$$

where τ_{ik} is given from equation (1).

When subtask i is being executed, the communication time between RMS and node k is as

$$l_{ik} = e^{a_{ik}/y_k}, \tag{3}$$

where a_{ik} be the amount of data exchanged between RMS and node k when subtask i is being executed and y_k be the mean speed of communication link between RMS and the node k [2].

When ϵ_k be the failure intensity of communication link between RMS and the node k, the probability that the communication link functions without any failure is

$$q_{ik} = e^{(-\epsilon_k l_{ik})}, \tag{4}$$

where l_{ik} is given by the equation (3) [2].

When λ_{prog} be the failure intensity of the program running on the node k where the subtask i is being executed, the probability that the program on node k functions without any failure is

$$r_{ik} = e^{(-\lambda_{prog}\tau_{ik})} . \tag{5}$$

Now the probability that the subtask i is successfully completed by node k or the reliability of subtask i is given by

$$R_{ik} = p_{ik}q_{ik}r_{ik} \cdot \tag{6}$$

For improvement in grid service reliability, a subtask is assigned to several nodes for parallel execution. When a node completes a subtask successfully, it returns the output to the RMS. If Ni be the node set to which subtask i is assigned, then the reliability of subtask i, R_{Sub_i} is given by

$$R_{Sub_i} = 1 - \prod_{k \in N_i}(1 - R_{ik}) , \tag{7}$$

where R_{ik} is from equation (6).

When output of all the subtasks are received by the RMS, the grid service is said to be completed successfully. Hence, the grid service reliability (Rgs) is given by

$$R_{gs} = \prod_{i=1}^{m}[1 - \prod_{k \in N_i}(1 - R_{ik})] . \tag{8}$$

But transaction processing in grid is not easy to achieve because, grid services are autonomous and loosely coupled and they do not allow to be locked by outside application [5].

A. Transaction oriented autonomic grid service availability

Transaction oriented autonomic grid service availability A_i can be defined as the probability that at any time a required minimum fraction of transactions are finishing within a given deadline [22]. It can be derived in terms of mean time to failure *(MTTF$_i$)* and mean time to repair *(MTTR$_i$)* for i^{th} component used in a transaction processing in grid environment. Therefore,

$$A_i = \frac{MTTF_i}{MTTF_i + MTTR_i}. \tag{9}$$

B. Mean transaction oriented autonomic grid service delay

Mean transaction oriented autonomic grid service delay *D* can be defined as the sum of all non-overlapped delays from service initiation time to service result time. If S_{it} is the service initiation time, S_{tt} is the service transmission time, S_d is the service propagation delay, S_{pt} is the service processing time, and S_{rdt} is the service result display time, then

$$D = S_{it} + S_{tt} + S_d + S_{pt} + S_{rdt}. \tag{10}$$

4. CPN-TOGS Model

4.1. *Reliability analysis of transaction oriented grid service*

The probability that there are exactly *f* failures in component *i* in time interval *t* and λ_{tg} being the mean number of failures per unit time is given by

$$R_{togs_i}(f, t) = \frac{(\lambda_{tg}t)^f * e^{-\lambda_{tg}t}}{f!} * A_i, \tag{11}$$

where A_i is the availability of component *i* from equation (9).

When *t* is replaced by mean transaction oriented grid service Delay D_i of component *i*, then the equation (11) reduces to as

$$R_{togs_i}(f, D_i) = \frac{(\lambda_{tg_i}D_i)^f * e^{-\lambda_{tg_i}D_i}}{f!} * A_i, \tag{12}$$

where λ_{tg_i} is the expected number of failures per unit time and it is equal to $\frac{1}{MTTF_i}$. In equation (11), if $f = 0$, then

$$R_{togs_i}(0, D_i) = e^{-\lambda_{tg_i}D_i} * A_i. \tag{13}$$

For improvement in transaction oriented grid service reliability, a subtask is assigned to several nodes for parallel execution. When a node completes a subtask successfully, it returns the output to the RMS. If N_i be the node set to which subtask *i* is assigned, then the reliability of subtask *i*, R_{togs_i} is given by

$$R_{togs_i}(Sub_i) = 1 - \prod_{k \in N_i}(1 - R_{togs_i}(f, D_i)). \tag{14}$$

When output of all the subtasks are received by the RMS, the grid service is said to be completed successfully. Hence, the transaction oriented grid service reliability *(R_{togs})* is given as

$$R_{togs} = \prod_{i=1}^{m}[1 - \prod_{k\in N_i}(1 - R_{togs_i}(f, D_i))] \tag{15}$$

5. Simulation and Results

5.1. *Simulation set up*

In simulation set up, the CPN-TOGS model is categorized into four categories using CPN tool.

- Category 1: In this category, the model has no transaction management and the recovery of failed processes are done on local node.
- Category 2: In this category, the model has no transaction management and the recovery of failed processes are done on replicated nodes.
- Category 3: In this category, the model has transaction management and the recovery of failed processes are done on local node.
- Category 4: In this category, the model has transaction management and the recovery of failed processes are done on replicated nodes.

5.2. *Results*

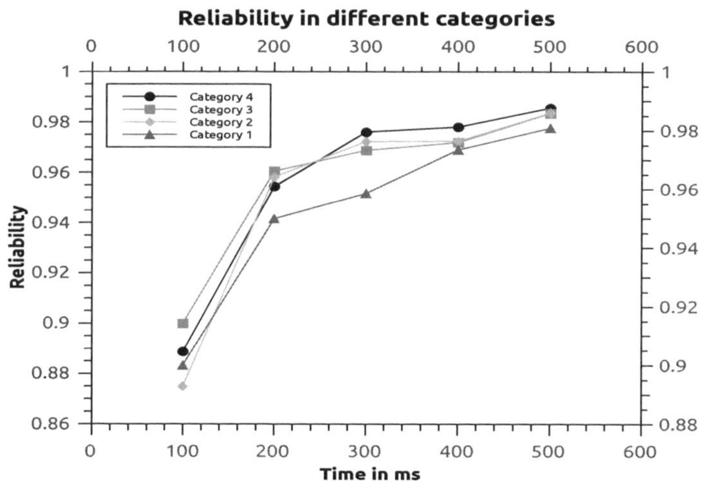

Fig. 1. Reliability in different categories.

1) *Reliability in different categories:*

In Fig. 1, the different values of reliability with respect to time (in milli seconds) has been shown. when the range of deadline be from 150 to 500 milli seconds, we got that the grid service with transaction management having local recovery (Category 3) had the highest value of reliability and the grid service without transaction management having local recovery (Category 1) had the lowest value of reliability.When the range of deadline be from 150 to 500 milli seconds, the grid service with transaction management having replicated recovery (Category 4) had the highest value of reliability and the Category 1 had the lowest value of reliability. In the table I, the reliability of different categories are given, when deadline ranges from 150 to 500 ms.

2) *Reliability analysis with local recovery and replicated recovery*:
Here we conclude that Category 3 is better option than Category 1. In Fig. 2, the analysis is shown to be done having local level recovery while Fig. 3 demonstrates the analysis having replicated level recovery. Here it is concluded that Category 4 is better than Category 2, provided that the deadline be within 300 ms and if the deadline is above this 300 mark, both of the categories are better options to use.

3) *Reliability analysis with and without transaction management*:
When transaction management is used, Category 4 is better than Category 3 (as we see in Fig. 4). According to the Fig. 5, it is found that at all possible values of deadline, the reliability in grid service (without transaction management) which uses replicated recovery (Category 2) is more than that in Category 1.

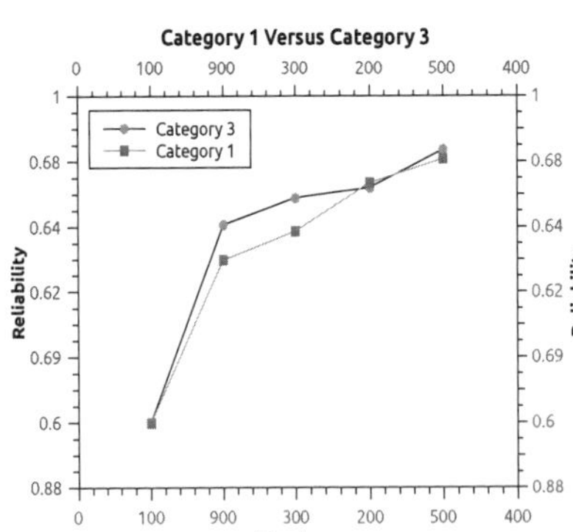

Fig. 2. Reliability with local recovery.

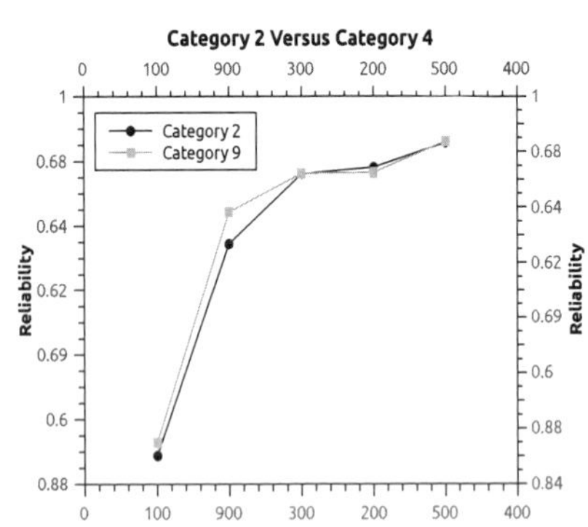

Fig. 3. Reliability with replicated recovery.

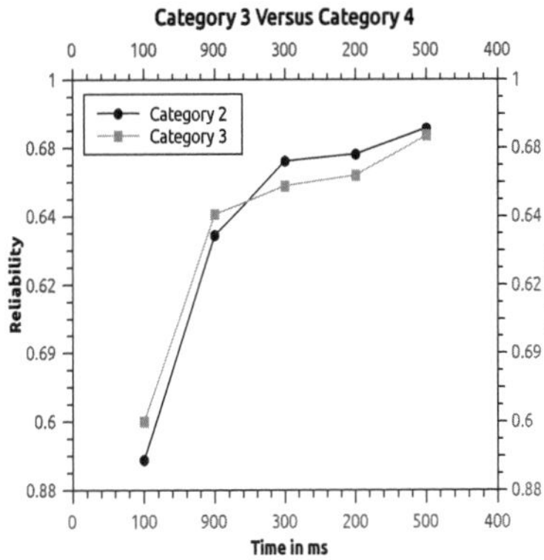

Fig. 4. Reliability with TM.

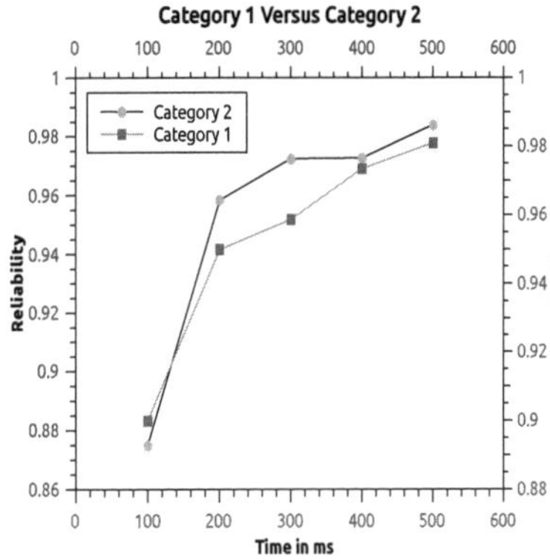

Fig. 5. Reliability without TM.

Table 1. Reliability analysis in different categories.

deadline (ms)	Category 1	Category 2	Category 3	Category 4
150	0.901	0.901	0.875	0.888
200	0.955	0.960	0.958	0.954
300	0.958	0.968	0.972	0.976
400	0.973	0.972	0.972	0.978
500	0.980	0.983	0.983	0.985

6. Conclusions and Future works

The proposed model CPN-TOGS is based on the reliability definition and the service provided by a transaction oriented grid environment has been modeled. From the experimental results, it was analyzed that the short-lived transactions and long-lived transactions have different values of reliability. The grid service having local recovery also had the lowest value of reliability. The transaction oriented grid service having replicated recovery had the highest value of reliability. The future work will be on the empirical analysis of dependability in transaction oriented grid service where not only reliability term will be concerned but also availability, safety, maintainability, and integrity terms will have to be focused.

References

1. S. Guo, H.-Z. Huang, Z. Wang and M. Xie, Grid service reliability modeling and optimal task scheduling considering fault recovery, *Reliability, IEEE Transactions on*, **60**, 263 (2011).
2. S. Guo, H.-Z. Huang and Y. Liu, Modeling and analysis ofgrid service reliability considering fault recovery, *New Generation Computing*, **29**,

345–364 (2011). [Online]. Available: http://dx.doi.org/10.1007/s00354-009-0114-8.

3. I. Foster, The grid: A new infrastructure for 21st century science, *Grid Computing: Making the Global Infrastructure a Reality*, 51 (2003).
4. P. Bernstein and E. Newcomer, *Principles of Transaction Processing: For the Systems Professional*. San Francisco, CA, USA: Morgan Kaufmann Publishers Inc., 1997.
5. Y. Cardinale, J. E. Haddad, M. Manouvrier and M. Rukoz, CPN-TWS; a coloured petri-net approach for transactional-QoS driven web service composition, *Int. J. Web Grid Serv.,* vol. 7, no. 1, pp. 91–115, Feb. 2011. [Online]. Available: http://dx.doi.org/10.1504/IJWGS.2011.038389.
6. J. Gray, The transaction concept: Virtues and limitations (invited paper), in Proceedings of the Seventh International Conference on Very Large Data Bases - **7**, ser. VLDB '81. VLDB Endowment, 144 (1981). [Online]. Available: http://dl.acm.org/citation.cfm? id=1286831.1286846
7. J. Gray and A. Reuter, *Transaction Processing: Concepts and Techniques,* 1st ed. San Francisco, CA, USA: Morgan Kaufmann Publishers Inc., 1992.
8. M. Husemann, M. von Riegen and N. Ritter, Transactional coordination of dynamic processes in service-oriented environments, *in Web Services, 2007. ICWS 2007. IEEE International Conference on*, July 2007, pp. 1024–1031.
9. D. P. Mahato, L. S. Umrao and R. S. Singh, Adaptability in transaction oriented grid service, *in Parallel, Distributed and Grid Computing (PDGC), 2014 International Conference on*, Dec 2014, pp. 239–244.
10. F. Tang, M. Li and J. Z. Huang, Real-time transaction processing for autonomic grid applications, *Engineering Applications of Artificial Intelligence,* **17**, 799 (2004), autonomic Computing Systems. [Online]. Available: http://www.sciencedirect.com/science/article/pii/S0952197604001228.
11. F. Tang, M. Guo, M. Li and L. Li, Transaction management for reliable grid applications, *in Advanced Information Networking and Applications,* 2009. AINA '09. International Conference on, 427 (2009).
12. G. Levitin, Y.-S. Dai and H. Ben-Haim, “Reliability and performance of star topology grid service with precedence constraints on subtask execution, *Reliability, IEEE Transactions on*, **55**, 507 (2006).
13. G. Levitin and Y.-S. Dai, “Service reliability and performance in grid system with star topology,” *Reliability Engineering & System Safety*, **92**, 40 (2007).

14. M. A. Azgomi and R. Entezari-Maleki, "Task scheduling modeling and reliability evaluation of grid services using coloured petri nets," *Future Generation Computer Systems*, **26**, 1141 (2010). [Online]. Available: http://www.sciencedirect.com/science/article/ pii/S0167739X10001093.
15. O. Doguc and J. E. Ramirez-Marquez, An automated method for estimating reliability of grid systems using bayesian networks, *Reliability Engineering & System Safety*, **104**, 96 (2012).
16. Y.-S. Dai, Y. Pan and X. Zou, A hierarchical modeling and analysis for grid service reliability, *Computers, IEEE Transactions on*, **56**, 681 (2007).
17. Y.-S. Dai, G. Levitin and X. Wang, "Optimal task partition and distribution in grid service system with common cause failures," *Future Generation Computer Systems*, **23**, 209 (2007). [Online]. Available: http://www.sciencedirect.com/science/article/ pii/S0167739X06001038.
18. Y.-S. Dai and G. Levitin, Optimal resource allocation for maximizing performance and reliability in tree-structured grid services, *Reliability, IEEE Transactions on*, **56**, 444 (2007).
19. Y.-S. Dai, G. Levitin and K. S. Trivedi, Performance and reliability of tree-structured grid services considering data dependence and failure correlation, *Computers, IEEE Transactions on*, **56**, 925 (2007).
20. Y.-S. Dai and G. Levitin, Reliability and performance of tree-structured grid services, *Reliability, IEEE Transactions on*, **55**, 337 (2006).
21. A. Heddaya and A. Helal, Reliability, availability, dependability and performability: A user-centered view, (1997).
22. V. Mainkar, Availability analysis of transaction processing systems based on user-perceived performance, *in Reliable Distributed Systems, 1997. Proceedings., The Sixteenth Symposium on*, 10 (1997).

Performance Degradation of Language Identification System in Noisy Environment

Randheer Bagi[1] and Jainath Yadav[2]

Department of Computer Science, Central University of South Bihar, Patna, Bihar 800014, India

E-mail: [1]randheer.bagi11@gmail.com, [2]jaibhu38@gmail.com

www.cub.ac.in

Spoken language identification refers to the set of algorithms and procedures to identify language with high accuracy of recognition rate. Accuracy and speed are two major concerns in language identification for real time applications. It becomes more challenging in the noisy environment due to the addition of different types and levels of noise from the environment. This paper covers language identification performance in a noisy environment in the context of Indian languages by using speaker independent Indian speech corpus (IITKGPMLILSC). The experiment is carried out by extracting acoustic features (13 MFCCs, 1 Delta and 1 Delta-Delta) from the raw speech signal. Gaussian mixture models (GMMs) are trained for each language with varying number of mixture components. For the noisy environment, white noise with zero mean is added at different SNR (dB) values. The performance of the proposed LID system is evaluated for clean and noisy speech signal. The average recognition rate of clean speech is 56.48%, while of noisy speech signal is 14.84%. The recognition rate of the proposed LID system varies according to different noise level, duration of speech utterance and mixtures of GMMs.

Keywords: Acoustic features; White noise; Gaussian Mixture Model (GMMs); Indian speech corpus (IITKGP-MLILSC).

1. Introduction

Language Identification (LID) is one of the most recent advance research in speech signal processing. Its sole aim is to replicate human capability of language identification through computational means without the interference of human. To reduce the cost and training periods, a LID system can be trained once and can be run simultaneously on multiple machines. It has a wide range of application domain like spoken language translation, multilingual speech recognition, spoken document retrieval, online banking, checking into the hotel and arranging a meeting or making travel arrangement for non-native speakers [1] [2]. An ideal LID system should exploit each aspect of speech information in detail with accuracy. It must be capable to distinguish different languages from given a huge amount of the target language and flexible enough to adapt the variation of different speakers. Modelling of speech signal depends on different types of features such as acoustic, phonetic, phonotactic, syllable and prosodic information that can be directly obtained from raw speech [1] [3]. It is observed that for spoken language identification, acoustic and phonotactic features are the most effective language cues.

Acoustic refers to the physical sound pattern, whereas phonotactic refers to constraints that determine permissible syllable structures in a language [3]. A LID system has been broadly divided into two sections, the first one is the front end, which extract a sequence of features for observation and another is the back end for language modelling. It is presumed that each language model is equally likely. Observation of results shows that low-level acoustic features are easier to obtain, but they are volatile due to the presence of the speaker or channel variation [4]. The LID system should not be biased towards any language or a group of languages. It must take shorter computation time while increasing the number of target languages or decreasing the duration of the test utterance respectively, provided the performance of the system does not degrade [5]. It must be robust to variation of speakers or channels and low signal-to-noise ratio (SNR) [6][7][8].

This paper is organized as follows: Section 2 covers simple review of background work, Section 3 deals with short description of Indian speech corpus (IITKGP-MLILSC), Section 4 consists a note on acoustic feature extraction, Section 5 represents Gaussian mixture model (GMM) classifier, Section 6 shows the effect of noise on LID performance, Section 7 presents the proposed model, Section 8 explains the experimental setup, Section 9 shows results and discussion of the proposed LID system. Finally, Section 10 covers the conclusion part with some future directions.

2. Prior Work

In literature, a number of significant results are obtained using different types of front end combinations, i.e. Mel Frequency Cepstral Coefficients (MFCCs), Delta, Delta-Delta, Linear Prediction Cepstral Coefficients (LPCCs) and back end combinations i.e. GMM, SVM and DNN in LID system [1] [3]. Sadanandam Manchala *et al.* (2014) have proposed a combination of MFCC and formant frequency [9] at the front end. The significance of feature vectors is evaluated by measuring the usefulness between the pair of languages. In [10] language branch discriminative is used by Xianliang Wang *et al.* (2014) which are based on factor analysis technique. It is obtained by concatenating low dimensional factors into the language branch variability space. In [5] [11] pitch synchronous, formant frequency and glottal closure (GC) [12] [13] based spectral features are used to extract specific information present in the speech at multi-level is proposed by V. Ramu Reddy *et al.* & K. Sreenivasa Rao *et al.* (2013). David Martinz and Eduardo Lleida *et al.* (2013) presents prosodic feature, formant frequency based LID system using i-vector followed by GMM [14].

In [15] Pinki Roy and Pradip K. Das (2013) proposed, a hybrid VQ-GMM approach for identifying 4-Indian languages, which is carried out over recorded data of 50 native speakers. S. Jothilakshmi *et al.* (2012) reported a two level language identification system for Indian language with database of 9-Indian languages [2]. In the first step, it identifies the family of the spoken language where as in second step particular language is identified in the corresponding family. In [16] Marc A. Zissman, IEEE Member (1996) presents GMM and SVM based LID system, which do not require phonetically labelled training of speech and runs faster for real-time applications, while PRLM and PPRLM require phonetically labelled training and take long computation time. D. A. Reynolds, W. M. Campbell *et al.* (2008) represents a spectral based system, where SVM is 3 times faster than GMM in recognizing speed, but in terms of performance GMM have 3% less EER as compare to SVM [17] in real environment. The performance of the LID system heavily depends on the selection of preprocessing techniques, features and classifiers. Unfortunately, the LID system reviewed above all have one criteria common i.e. they were designed to work in a controlled environment.

3. Speech Corpus

3.1. *Indian language speech corpus (IITKGP-MLILSC)*

IITKGP-MLILSC consist 27 Indian languages for the extraction and analysis of valuable information present in the speech signal [18]. These 27 languages are named as, Arunachali, Assamese, Bengali, Bhojpuri, Chhattisgarhi, Dogri, Gojri, Gujrati, Hindi, Indian English, Kannada, Kashmiri, Konkani, Manipuri, Mizo, Malayalam, Marathi, Nagamese, Neplai, Oriya, Punjabi, Rajasthani, Sanskrit, Sindhi, Tamil, Telugu and Urdu. Speech data have been collected from TV and radio channels and have been recorded using the audacity software. Sixteen languages have been recorded from news bulletin which are broadcasted over a radio channel and the remaining are from TV talk/live shows and interviews. The average numbers of speaker for each language are 10, including both male and female with average recording duration of 1 hour. The collected speech corpus has similar channel and language characteristics, for each language. The environment of recording is a studio, which ensures that the recorded speech corpus is free from background noise. For each speaker, data are collected within 5–10 minute, with the sampling rate of 16 kHz. Also the original speech has been down-sampled at 8 kHz [11] for my experiment.

4. Feature Extraction

4.1. *Mel frequency Cepstral coefficient, delta and delta-delta*

Mostly used acoustic feature, which uses filter-bank based parameterization method is Mel Frequency Cepstral Coefficients (MFCCs) [1] [4] [9]. It is best suited for several applications of speech processing, such as language identification/verification, speech recognition and speaker verification/identification. It approximates the non-linear frequency resolution with respect to the human ear and is computed to capture short term speech dynamics. Extraction of MFCCs is a 8 step process which is detailed in Figure 1 MFCCs fails to capture the dynamic aspects of human speech which are very important for distinguishing different languages. Thus, in order to overcome the deficiency of MFCCs, delta and delta-delta cepstra are used. It provides an estimation of the local temporal derivatives of the speech cepstrum. It has been implemented as a least square estimation of the local slope, which is calculated over multiple frames. Delta and Delta-Delta are more robust to channel artifacts and include temporal information [12].

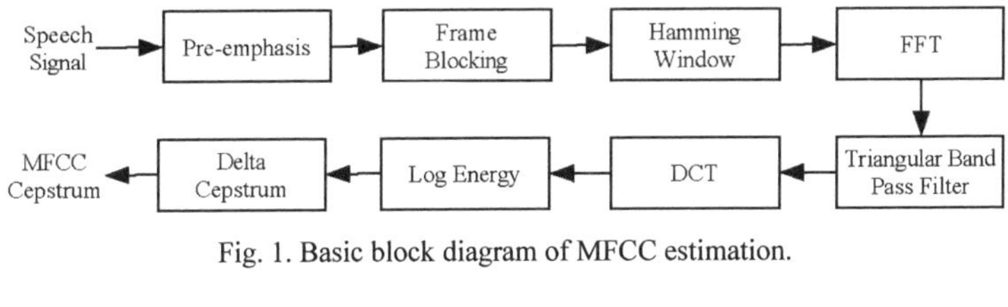

Fig. 1. Basic block diagram of MFCC estimation.

5. Gaussian Mixture Model

GMM is a parametric representation of a probability density function based on a weighted sum of multi-variant Gaussian distribution [1] [3] [19]. It uses a generative approach for classification which follows normal distribution. It tries to model probability density individually for each class. Training of a GMM involves forming an estimate of the probability distribution that best characterizes the set of training data. Use of k-mean vectors, k-covariance and weighted sum matrices completely define the GMM. In the proposed method, training of 27 languages from the IITKGP-MLILSC database indicates that 27 different GMMs are trained using extracted acoustic features (13 MFCCs and 1 Delta and 1 Delta-Delta). The basic motive to use GMM is its ability to smoothly approximate the arbitrarily shaped distributions and to model the complicated or widely varying distribution of speech.

6. Effect of Noise on LID Performance

Noise is an unavoidable part of our life. It is scientifically studied and analysed in many applications. Addition of different type and level of noise causes serious adversary impact on the performance of the LID system. In the experiment, average recognition of clean speech is 56.48%, which degrades rapidly and becomes 14.84% when noise is added into it. In the proposed LID system, white noise with zero mean is added to the speaker independent speech corpus of IITKGPMLILSC. The addition of noise is done at different SNR (dB) values (0dB, 5dB, 10dB and 20dB) and it is observed that at low SNR (dB) value the performance of the proposed LID system falls down drastically. The Figure 2 clearly represents the impact of noise over clean speech. Figure 2(a) shows a waveform of the clean speech signal of 10 sec utterance duration. The waveform of noisy speech signal at 0 dB SNR for the same 10 sec utterance duration is represented in Figure 2(b).

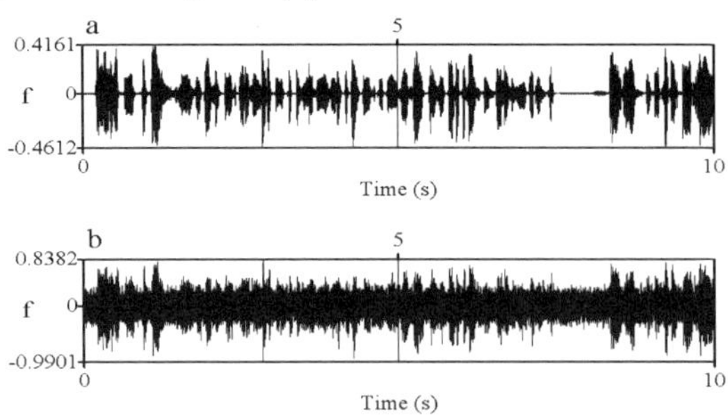

Fig. 2. Waveform of speech utterance from Hindi news channel in (a) clean speech (b) noisy speech.

7. The Proposed Model

7.1. *Training phase*

In the proposed model, for the training phase acoustic features are extracted from the clean speech signal $s(n)$ over speaker independent database of IITKGP-MLILSC in n discrete time index. Primarily, clean speech signals are transformed into the stream of feature vectors $X = [x_1, x_2, x_{3,..} x_{p....}]$, where p is the frame index and x_p is N dimensional feature vector. Let λ is the language model and L is the total number of languages (i.e. 27) in the training phase, then for each language a language model λ_l is built according to extracted feature vectors.

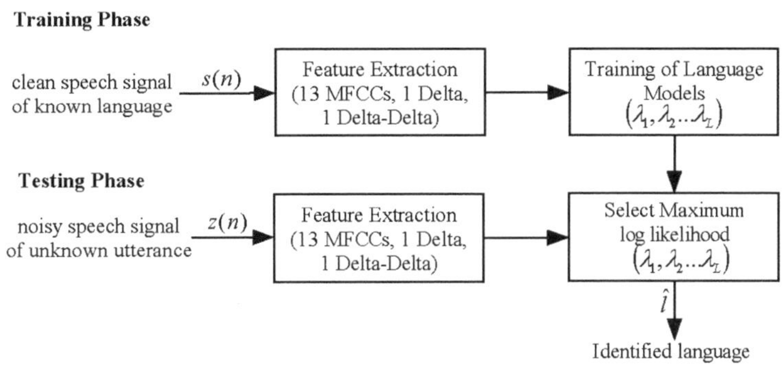

Fig. 3. Basic block diagram of proposed LID system in noisy environment.

7.2. *Testing phase*

Let $d(n)$ is the noise signal with zero mean added to clean speech $s(n)$ then the output of noise added speech signal is represented as: $z(n) = s(n) + d(n)$. After addition of the noise, the noisy speech signal $z(n)$ transformed into a stream of feature vectors $Y = [y_1, y_2, y_{3,..}y_p ....y_p]$, where p represents the frame index and y_p is a N dimensional feature vector. Then same acoustic features are extracted from Y and are compared with the training language model set $\{\lambda_l | 1,2,3 \ldots . L\}$. Finally the identification task has been executed using log likelihood.

$$P(Y|\lambda_l) = \prod_{i=1}^{N} P(Y^{(i)}|\lambda_l) \tag{1}$$

The degree of confidence is measured by Bayes theorem to ensure the goodness of the results.

$$\hat{l} = \arg max_{1 \le l \le L} \frac{P(Y|\lambda_l)P(\lambda_l)}{P(Y)} \tag{2}$$

It is assumed that each model is equally likely i.e. $P(Y)$ will be same and irrespective of training language model. Finally, the language model in which Y has the highest probability of occurrence is equivalent to the identified language.

8. Experimental Setup

The experiment has been carried out using MATLAB 13 on Windows 7 platform and speaker independent database of IITKGP-MLILSC. The acoustic features (13 MFCCs, 1 Delta and 1 Delta-Delta) are extracted from each short term window. Modelling of languages is performed by varying the number of Gaussian mixtures (32, 64, 128, 256 and 512) using extracted acoustic features. After building model for each language using clean speech, white noise at different dBs was added into the 27 languages of clean speech to make noisy speech corpus. Testing has been performed by using test utterance (of durations 2 sec, 5 sec and 10 sec) for clean speech as well as for noisy speech at different SNR values (0 dB, 5 dB, 10 dB, 15 dB and 20 dB).

9. Result and Discussion

The performance of the proposed LID system has been evaluated and depicted in Table 1 and Table 2 for the clean speech and for noisy speech respectively. It shows the recognition rate increases with increasing the mixture of GMMs. Table 2 shows the performance of noisy speech signal gradually increases with increasing the SNR (dB) values (from low level SNR to high level SNR value).

Table 1. LID performance rate (%) for clean speech signal using IITKGP-MLILSC over different mixtures of GMM.

Speech Signal Duration	LID Performance rate (%) with different mixture of GMM over clean speech				
	32	64	128	256	512
2 sec	49.6296	51.8519	52.2222	51.8519	50.9259
5 sec	54.4444	55.9259	56.6667	58.5185	59.2593
10 sec	59.0741	59.8148	60.9259	62.4074	63.7037
Average	54.3827	55.8642	56.6049	57.5926	57.9629

The average recognition rate of the proposed LID system using clean speech reaches maximum for 512 mixtures of GMMs i.e. 57.96%. From Table 2 and Figure 4 it is analysed that, the use of 128 mixture components of GMMs is better, instead of 256 and 512 mixture components in real time applications. It takes lesser time in training and testing phase as compared to 256 and 512 mixture components of GMMs. It is because, 256 and 512 mixture components of GMMs show identical average recognition rate, but 128 mixture components present significant result in average recognition rate i.e. 56.60%.

Table 3 and Figure 5 show the average recognition rate of clean and noisy speech (with varying mixtures of GMMs). Increasing the mixture components of GMMs, the recognition rate has been enhanced. The overall average performance of the proposed LID system was depicted in Table 4 which shows the impact of different test speech utterance duration over performance. The histogram in Figure 6 clearly represents the improved recognition rate with 10 sec test utterance duration over 2 sec and 5 sec for both (clean & noisy speech signal).

Table 2. LID performance rate (%) of the proposed LID system over speaker independent database of IITKGP-MLILSC with noisy speech using different mixtures of GMM.

SNR (dB)	Speech Signal Duration	LID Performance rate (%) with different mixtures of GMM				
		32	64	128	256	512
0	2 sec	4.2593	4.2593	4.0741	3.7037	3.7037
	5 sec	3.8889	3.7037	3.7037	3.7037	3.7037
	10 sec	4.8148	3.7037	3.7037	3.8889	3.8889
5	2 sec	4.2593	5	4.4444	4.6296	3.5185
	5 sec	4.0741	5.7407	4.2593	3.3333	3.1481
	10 sec	6.6667	6.2963	4.8148	4.4444	3.8889
10	2 sec	7.7778	9.4444	9.4444	9.4444	9.0741
	5 sec	8.1481	9.8148	9.2593	9.4444	9.8148
	10 sec	9.6296	10.9259	11.4815	10.9259	11.2963
15	2 sec	13.3333	16.1111	15.9259	17.7778	20
	5 sec	15.9259	18.7037	20	21.4815	21.6667
	10 sec	20.5556	22.2222	23.5185	25.5556	26.2963
20	2 sec	27.4074	31.2963	32.4074	32.037	33.1481
	5 sec	33.1481	36.8519	37.5926	32.037	40
	10 sec	37.5926	40	40.7407	42.4074	42.963
Average	2 sec	10.2160	13.2222	13.2592	13.5185	13.8888
	5 sec	13.0370	14.9629	14.9629	15.4814	15.6666
	10 sec	15.8518	16.6296	16.8518	17.4444	17.6666

Table 3. Average performance rate (%) of the proposed LID system over speaker independent database of IITKGP-MLILSC with different mixtures of GMM.

Test Speech Signal	Average LID performance rate (%) with different mixtures of GMM				
	32	64	128	256	512
Clean speech	54.3827	55.8642	56.6049	57.5926	57.5929
Noisy speech	13.0350	14.9383	15.0247	15.4815	15.7407

Table 4. The average performance rate (%) of the proposed LID system with test speech utterance duration of 2 sec, 5 sec and 10 sec.

Speech Signal Duration	Average LID performance rate (%)	
	Noisy speech	Clean speech
2 sec	12.8209	51.2963
5 sec	14.8222	56.9629
10 sec	16.8888	61.1851
Average	14.8439	56.4814

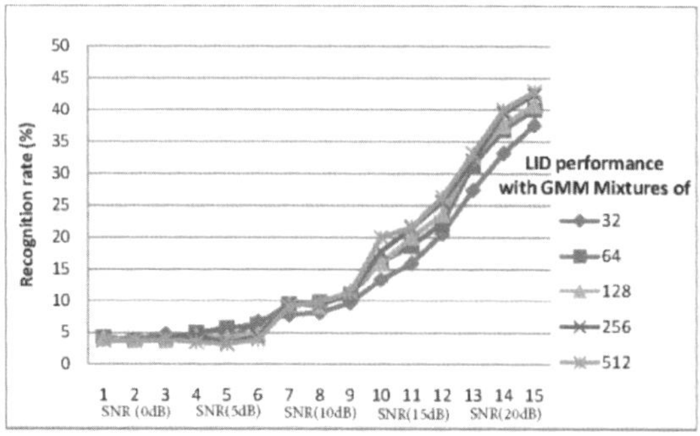

Fig. 4. The performance of noisy speech signal for different mixtures of GMM over speaker independent database of TTIKGP-MLILSC.

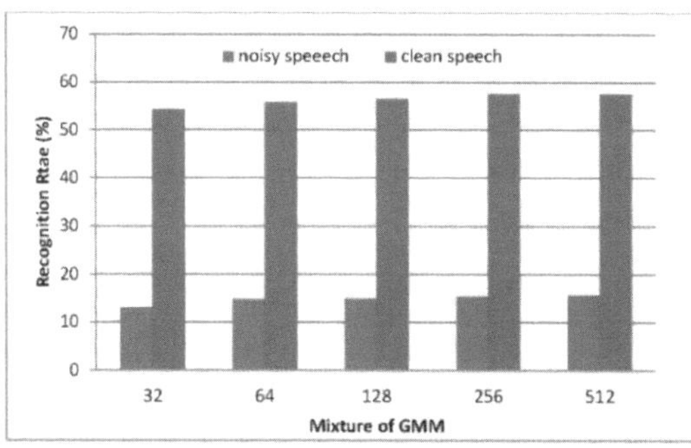

Fig. 5. Average performance of the proposed LID system for noisy and clean speech with different mixtures of GMM over speaker independent database of TTIKGP-MLILSC.

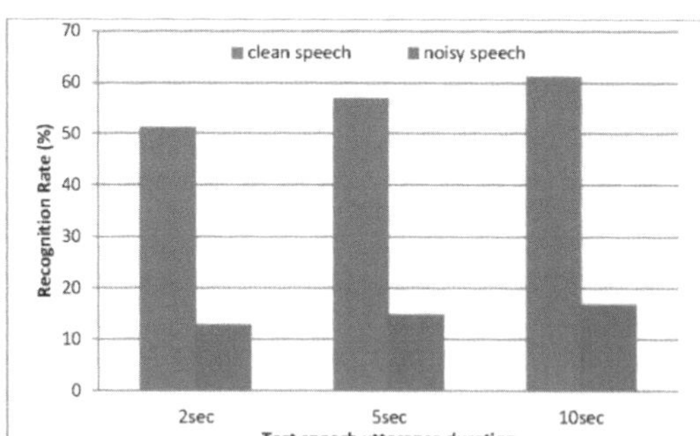

Fig. 6. Average performance of the proposed LID system for noisy and clean speech with different test speech utterance over speaker independent database of TTIKGP-MLILSC.

10. Conclusion

The proposed method shows the adverse role of noise in language identification, when LID system is exposed in noisy environment. A comparative study of clean and noisy speech signal suggests that the use of 128 mixture components of GMMs are better choice than 256 and 512 mixture components. Analysis shows, when LID system exposed in noisy environments, its performance degrades rapidly. It is due to the addition of noise in the clean speech signal from the environment. The key challenge faced in modelling the speech data is, mismatch of speech conditions in training and testing data. So, for language identification in noisy environment, the proposed system may further be analysed by selecting the features closer to excitation source. Fusion of suitable classifiers can be used to improve the results. Another approach suggests the use of the appropriate noise suppression technique before feature extraction from noisy speech. It should be noted that, it not only reduces noise from noisy signal, but also recovers the missing data, lost during the process of noise suppression.

References

1. E. Ambikairajah, H. Li, L. Wang, B. Yin and V. Sethu, Language identification: a tutorial, *Circuits and Systems Magazine, IEEE*, **11**, 82 (2011).
2. S. Jothilakshmi, V. Ramalingam and S. Palanivel, A hierarchical language identification system for indian languages, *Digital Signal Processing*, **22**, 544 (2012).
3. H. Li, B. Ma and K. A. Lee, Spoken language recognition: from fundamentals to practice, *Proceedings of the IEEE*, **101**, 1136 (2013).
4. M. A. Zissman and K. M. Berkling, Automatic language identification, *Speech Communication*, **35**, 115 (2001).
5. K. Rao, S. Maity and V. Reddy, Pitch synchronous and glottal closure based speech analysis for language recognition, *International Journal of Speech Technology*, **16**, 413 (2013).
6. S. Boll, Suppression of acoustic noise in speech using spectral subtraction, *Acoustics, Speech and Signal Processing, IEEE Transactions on*, **27**, 113 (1979).
7. P. Scalart *et al.*, Speech enhancement based on a priori signal to noise estimation, in *Acoustics, Speech, and Signal Processing. ICASSP - 96. Conference Proceedings., 1996 IEEE International Conference on*, **2**, 629 (1996).

8. M. Djendi and P. Scalart, Reducing over-and under-estimation of the a priori snr in speech enhancement techniques, *Digital Signal Processing*, (2014).
9. S. Manchala, V. K. Prasad and V. Janaki, Gmm based language identification system using robust features, *International Journal of Speech Technology*, **17**, 99 (2014).
10. X. Wang, Y. Wan, L. Yang, R. Zhou and Y. Yan, Language recognition system using language branch discriminative information, *IEEE International Conference on Acoustic, Speech and Signal Processing (ICASSP)*, (2014).
11. V. R. Reddy, S. Maity and K. S. Rao, Identification of indian languages using multi-level spectral and prosodic features, *International Journal of Speech Technology*, **16**, 489 (2013).
12. L. R. Rabiner and B.-H. Juang, *Fundamentals of speech recognition*. PTR Prentice Hall Englewood Cliffs, **14** (1993).
13. K. S. Rao and B. Yegnanarayana, Duration modification using glottal closure instants and vowel onset points, *Speech communication*, **51**, 1263 (2009).
14. D. Martinez, E. Lleida, A. Ortega and A. Miguel, Prosodic features and formant modelling for an i-vector based language recognition system, in *Acoustics, Speech and Signal Processing (ICASSP), 2013 IEEE International Conference on*, 6847 (2013).
15. P. Roy and P. K. Das, A hybrid vq-gmm approach for identifying indian languages, *International Journal of Speech Technology*, **16**, 33 (2013).
16. M. A. Zissman *et al.*, Comparison of four approaches to automatic language identification of telephone speech, *IEEE Transactions on Speech and Audio Processing*, **4**, 31 (1996).
17. W. Shen, Automatic language recognition via spectral and token based approaches, in *Springer Handbook of Speech Processing*, 811, (2008).
18. S. Maity, A. Vuppala, K. Rao and D. Nandi, Iitkgp-mlilsc speech database for language identification, in *Communications (NCC), 2012 National Conference on*, 1 (2012).
19. P. A. Torres-Carrasquillo, D. Reynolds and J. Deller, J. R., Language identification using gaussian mixture model tokenization, in *Acoustics, Speech, and Signal Processing (ICASSP), 2002 IEEE International Conference on*, **1**, 757 (2002).

Analysis of Software Fault Detection and Correction Processes with Log-Logistic Testing-Effort

Md. Zafar Imam[1], Ishrat Jahan Ara[2] and N. Ahmad[3]

Dept. of Statistics and Computer Applications, T. M. Bhagalpur University
Bhagalpur-812007, India
E-mail: [1]zaf.imam@gmail.com, [2]ish_num@rediffmail.com, [3]nesar_bgp@yahoo.co.in

Software reliability growth models are helping the software society in predicting and analyzing the software product in term of quality. During the software-testing phase, software reliability is highly related to the amount of development resources spent on detecting and correcting latent software errors, i.e. the amount of testing effort expenditures. Some researches proposed in the literature to study in the fault detection and fault correction processes. However software reliability can be enhanced considerably during testing with faults being detected and corrected by testers. The allocation of testing resources such as man power and CPU hours during testing phase can largely influence fault detection speed and the time to correct a detected fault. The testing resource allocation is usually depicted by testing effort function, which has been incorporated into software reliability model. Fault correction process is usually modelled as a delayed process of fault detection process. In this paper we proposed a model of fault detection and fault correction processes based on Log-logistic testing effort function. In addition the methods of data analysis and comparison criteria are presented. Models parameters are estimated by the least square estimation method, and computational experiments performed on actual software failure dataset. A comparative analysis to evaluate the effectiveness of the proposed model with the other existing model are also performed. Results show that the proposed model can give fairly better prediction.

Keywords: Fault detection process; Fault correction process; Testing effort function; Least square estimation.

1. Introduction

Reliability is one important element for software testing where it defines quality. During testing many errors are identified and corrected later. Testing consumes a large amount of testing resources, such as manpower and CPU hours, which are usually not constantly allocated during testing phase. The amount of testing resources consumed in a time interval largely influences the number of faults detected during the period, thus it is more reasonable to incorporate this information into modeling the fault detection process (FDP_s). The function that describes how testing resources are distributed is usually referred to as testing effort function (TEF) and it has been incorporated into software reliability studies by some researchers (Kapur *et al.*, 2008).

A detected faults generally cannot be corrected immediately and the time required to correct a detected fault is usually called debugging lag/delay. The idea of modeling fault correction process was first proposed in the so-called Schneidewind (1975) model. Schneidewind modeled the fault correction process (FCP_s) as a separates process following the FDP_s will a constant time lag. Some extensions are made in (Xie *et al.*, 2007) and (Wu *et al.*, 2007) by incorporating other assumptions of debugging delay into this framework.

In this paper we show how to integrate a log logistic testing-effort function into NHPP growth models (Ohba, 1984; Ahmad *et al.*, 2011) to get a better description of the software fault detection and fault correction processes. Moreover, we assume that fault correction commenced with fault detection i.e the time delay between fault detection and fault correction time. A framework is proposed to obtain testing effort dependent paired fault detection and fault correction process models with consideration of fault introduction. We used the real dataset to illustrate the application of proposed models which satisfying results.

2. Log-Logistic Testing-Efforts Function

The testing effort indicating how the errors are detected effectively in the software and can be model by different distributions (Musa *et al.*, 1987; Yamada *et al.*, 1986, 1993; Kapur *et al.*, 1999; Ahmad and Bokhari, 2014). Gokhale and Trivedi (1998) proposed the Log-logistic SRGM that can capture the increasing/decreasing nature of the failure occurrence rate per fault. Recently, Bokhari and Ahmad (2006), and Ahmad *et al.*, (2010, 2011) also presented how to use Log-logistic curve to describe the time dependent behaviour of testing-effort consumptions during testing.

The cumulative testing-effort expenditure consumed in (0,t] is depicted in the following:

$$W(t) = \alpha\left[1 - \left\{1 + (\beta t)^{\delta}\right\}^{-1}\right] = \alpha[(\beta t)^{\delta}/(1 + (\beta t)^{\delta})] \qquad (1)$$

Therefore, the current testing-effort expenditure at testing t is given by:

$$w(t) = \left[\alpha\beta\delta(\beta t)^{\delta-1}\right]/\left[1 + (\beta t)^{\delta}\right]^{2} \qquad t > 0, \qquad (2)$$

Where α is the total amount of testing-effort consumption required by software testing, β is the scale parameter, and δ is the space parameter. The testing effort $w(t)$ reaches its maximum value at time,

$$t_{max} = 1/\beta(\delta - 1\ /\ \delta + 1)^{\frac{1}{\delta}} \qquad (3)$$

The fault detection and fault correction processes with Log-logistic testing-effort function is formulated on the following assumptions (Ahmad *et al.*, 2010).

i. Testing-effort expenditures are described by the Log-logistic curve.
ii. The fault detection process follows a nonhomogeneous Poisson process with intensity function.
iii. The mean number of faults detected in the time interval $(t, t+\Delta t)$ to the current testing-effort expenditures to the mean number of remaining faults in the system.

3. Fault Detection Process Model

We use fault content rate function a(t) to denote the expected total number of faults at time t. which is the sum of the number of initial faults in the software and the number of faults introduced from the beginning of testing to testing time t. The expected initial number of faults in the software is denoted by a. We use the mean value function $m_d(t)$ to depict the expected number of faults detected till time *t* and fault intensity function $\lambda_d(t)$to denote $\frac{dm_d(t)}{dt}$. The number of faults detected during time interval $[t, t+\Delta t)$ by current testing effort expenditure is usually assumed to be proportional to the number of remaining faults at time *t* (Lin and Huang, 2008). Hence we have

$$\lambda_d(t) = \frac{dm_d(t)}{dt} = b(t).w(t).\big(a(t) - m_d(t)\big), \tag{4}$$

Where $b(t)$ is the current fault detection rate per unit testing effort (TE) at time *t*, and $w(t)$ is the current TE expenditure at time *t*. Furthermore we have

$$\frac{dm_d(t)}{dt} + b(t).w(t).m_d(t) = b(t).w(t).a(t). \tag{5}$$

That is

$$\frac{d(m_d(t)e^{\int_0^1 b(x).w(x)dx})}{dt} = b(t).w(t).a(t).e^{\int_0^1 b(x).w(x)dx} \tag{6}$$

Substituting the marginal condition $m_d(0) = 0$ into (6) gives

$$m_d(t) = a(t) - ae^{-\int_0^1 b(x).w(x)dx}$$
$$-e^{-\int_0^1 b(x).w(x)dx}.\int_0^t a'(x)e^{\int_0^x b(y).w(y)dy}\,dx \tag{7}$$

Where $a'(x) = \frac{da(x)}{dx}$. Various $m_d(t)$ can be derived based on different assumptions of a(t), b(t) and w(t). λ_d *(t)* can be obtained by combining (4) and (7) together as

$$\lambda_d(t) = ab(t)w(t)e^{-\int_0^1 b(x).w(x)dx}\left(1 + \int_0^t \frac{a'(x)}{a} e^{\int_0^x b(y).w(y)dy}\,dx\right). \tag{8}$$

4. Fault Correction Process Model

We use mean values function $m_c(t)$ to denote the expected number of faults removed till time t and the fault intensity function $\lambda_c(t)$ to denote $\frac{dm_c(t)}{dt}$. Since a removed fault must be first be detected, FCP can be modeled as a separate process following FDP with a debugging delay. We define correction effort to be the testing effort consumed during the period of detection of a fault and final removal of the fault. Furthermore we use $\Delta(t)$ and $\Delta W(t)$ to denote the debugging delay and the correction effort associated with a fault which is corrected at time t. Accordingly we have

$$W(t - \Delta(t)) = W(t) - \Delta W(t) \tag{9}$$

Generally correcting different faults requires different amounts of testing resources, hence we model correction effort $\Delta W(t)$ as a random variable. We use $f(x)$ to denote the probability density function of $\Delta W(t)$. Given that the correction effort associated with a fault which is finally corrected at time t equals to x, the mean number of faults corrected during time interval $[t, t + \Delta t)$ is equal to the mean number of faults detected during time interval $[W^{-1}(W(t) - x), W^{-1}(W(t + \Delta t) - x)]$. We use $X(t)$ to denote the number of faults corrected up to time t, thus

$$E[X(t + \Delta t - X(t)|\Delta W(t) = x] = \lambda_d\big(W^{-1}(W(t) - x)\big)$$
$$[W^{-1}(W(t + \Delta t) - x)W^{-1}(W(t) - x)]. \tag{10}$$

In addition we have

$$E(X(t + \Delta t) - X(t)] = \int_0^{W(t)} E(X(t + \Delta)$$
$$-X(t)|\Delta W(t) = x]f(x)dx. \tag{11}$$

Combining (10) and (11) gives

$$\lambda_c(t) = \int_0^{W(t)} \lambda_d\left(W^{-1}(W(t) - x)\right)f(x)\frac{w(t)}{w(W^{-1}(W(t)-x))}dx. \tag{12}$$

Furthermore $m_c(t)$ can be derived as

$$m_c(t) = \int_0^1 \lambda_c(x)dx. \tag{13}$$

Different $m_c(t)$ can be derived based on $m_d(t)$ and different $f(x)$.

Furthermore, Fault detection rate function $b(t)$ is usually assumed to be constant and it is denoted as b here (Lin and Huang, 2008). From (7) we have

$$m_d(t) = a(t) - ae^{-bW(t)} - e^{-bW(t)}.\int_0^t a'(X).e^{bW(x)}dx, \tag{14}$$

In this section we assume that the total number of faults increases exponentially with the increase of the testing effort consumed, which can be denoted as

$$a(t) = e^{\gamma W(t)}, \gamma \geq 0. \tag{15}$$

Combining (14) and (15) gives

$$m_d(t) = \frac{ab}{b+\gamma}(e^{\gamma W(t)} - e^{-bW(t)}). \tag{16}$$

Also

$$m_d(t) = \frac{ab}{b+\gamma}(e^{\gamma\alpha[(\beta t)^\delta/(1+(\beta t)^\delta)]} - e^{-b\alpha\left[\frac{(\beta t)^\delta}{1+(\beta t)^\delta}\right]}) \tag{17}$$

Combining (4) and (14) and (15) gives

$$\lambda_d(t) = b.w(t)(\frac{a\gamma}{b+\gamma}e^{\gamma W(t)} + \frac{ab}{b+\gamma}e^{-bW(t)}). \tag{18}$$

Combining (12) and (18) gives

$$\lambda_c(t) = \frac{a\gamma b W(t) e^{\gamma W(t)}}{b+\gamma}\int_0^{W(t)} e^{-a\gamma}\mathrm{f}(x)dx$$

$$+\frac{ab^2W(t)}{b+\gamma}e^{-bW(t)}\int_0^{W(t)} e^{bx}\mathrm{f}(x)dx. \tag{19}$$

5. Exponential Distributed Correction Effort

In this case we have

$$f(x) = ce^{-cx} \tag{20}$$

Where c is fault correction rate per detected fault.
Combining (19) and (20) together gives

$$\lambda_c(t) = \frac{a\gamma bcw(t)}{(b+\gamma)(c+\gamma)}\left(e^{\gamma W(t)} - e^{-cW(t)}\right)$$

$$+\frac{ab^2cw(t)}{(b+\alpha)}e^{-bW(t)}\int_0^{W(t)} e^{(b-c)x}dx. \tag{21}$$

If c=b, we have

$$\lambda_c(t) = \frac{a\gamma b^2 w(t)}{(b+\gamma)^2}\left(e^{\gamma W(t)} - e^{-bW(t)}\right) + \frac{ab^3w(t)}{(b+\gamma)}W(t)\,e^{-bW(t)} \tag{22}$$

Furthermore from (13) we have

$$m_c(t) = \frac{ab}{(b+\gamma)^2}\left(be^{\gamma W(t)} + \gamma e^{-bW(t)}\right) - \frac{ab}{(b+\gamma)}(1+bW(t))e^{-bW(t)} \tag{23}$$

If c≠b, we have

$$\lambda_c(t) = \frac{a\gamma bcw(t)}{(b+\gamma)(c+\gamma)}\left(e^{\gamma W(t)} - e^{-bW(t)}\right) + \frac{ab^2cw(t)}{(b+\gamma)(b-c)}W(t)\,e^{-bW(t)} \tag{24}$$

Furthermore from (13) we have

$$m_c(t) = \frac{a}{(1+\gamma/b)}\left(\frac{ce^{\gamma W(t)}+\gamma e^{-cW(t)}}{c+\gamma} + \frac{ce^{-bW(t)}-be^{-cW(t)}}{b-c}\right) \tag{25}$$

Also we have

$$m_c(t) = \frac{a}{(1+\gamma/b)}\left(\begin{array}{c}\frac{ce^{\gamma\alpha[(\beta t)^{\delta}/(1+(\beta t)^{\delta})]}+\gamma e^{-c\alpha[(\beta t)^{\delta}/(1+(\beta t)^{\delta})]}}{c+\gamma}\\ +\frac{ce^{-b\alpha[(\beta t)^{\delta}/(1+(\beta t)^{\delta})]}-be^{-c\alpha[(\beta t)^{\delta}/(1+(\beta t)^{\delta})]}}{b-c}\end{array}\right). \quad (26)$$

6. Least Square Estimation

Least square estimation is a fairly general technique which can be applied in more practical situations and it is better for small or medium sample sizes (Musa *et al.*, 1987). In order to estimate the parameters, the total squared difference between observed and predicted values is minimized.

The parameters α, β, δ in the Log-logistic testing-effort function (1) can be estimated by the method of least square estimation. These parameters are determined for n observed data pairs in the form (t_k, W_k) $(k = 1, 2, \ldots\ldots, n; 0 < t_1 < t_2 \ldots\ldots < t_n)$, where W_k is the cumulative testing-effort consumed in time $(0, t_k]$. The least square estimation $\hat{\alpha}, \hat{\beta},$ and $\hat{\delta}$ can be obtained by minimizing:

$$S(\alpha, \beta, \delta) = \sum_{t=1}^{n} [\ln W_k - \ln \alpha - \delta \ln(\beta t_k) + ln\left(1 + (\beta t_k)^{\delta}\right)]^2$$

Once the estimated testing-effort parameters $\hat{\alpha}, \hat{\beta},$ and $\hat{\delta}$ in Log-logistic testing-effort function have been obtained. Now we estimate the parameters of fault detection and fault correction models. The parameters of a, b and c in fault detection and fault correction models are obtained by minimizing the sum of squared residuals, which is the difference between mean value function and the observations for both detected and corrected faults as

$$S(a, b, c) = \sum_{i=1}^{n} [(m_d(t_k) - d_k)^2 + (m_c(t_k) - c_k)^2]$$

where d_k and c_k denote the cumulative number of detected and corrected faults collected until time t_k respectively, t_k, $k = 1,2,3 \ldots\ldots$, are the running times from the beginning of testing.

Due to the complexity of the models we used SPSS package for the model parameter estimation.

7. Numerical Examples

7.1. *Model comparison criteria*

The Mean Square Errors (MSE): MSE $=\frac{1}{k}\sum_{i=1}^{k}[m_d(t_i)-m_i]^2$

The Accuracy of Estimation (*AE*): $AE = \left|\frac{m_a - a}{m_a}\right|$

R-squared $R-squared = \frac{\sum_{i=1}^{n}(\hat{m}(t_i)-\bar{m})^2}{\sum_{i=1}^{n}(m_i-\bar{m})^2}$, $\quad \bar{m} = 1/n\sum_{i=1}^{n} m_i$

7.2. *Performance analysis*

This section evaluates the performance of proposed model and comparison some other existing models using System T1 data (Musa *et al.*, 1987). Here we estimate the parameter by using least square estimation. But due to the complexity of the models we used SPSS package for the model parameter estimation. The estimated parameters are obtained as;

$$\hat{\alpha} = 33.110, \qquad \hat{\beta} = .057, \qquad \hat{\delta} = 7.151$$

Using the estimated parameters $\hat{\alpha}, \hat{\beta}$ and $\hat{\delta}$, the other parameters *a*, *b* and *c* in (17) and (26) can be estimated by the method of least square estimation. These parameters are:

$$\hat{a} = 71.043, \qquad \hat{b} = 0.678 \qquad \hat{c} = 0.816$$

Compares the performance of various SRGMs for the actual fault data set and all estimated values for various models are listed in Table 1. From Figures 1 and 2 graphically illustrate the observed failure data and estimated Log-logistic testing effort data. Here fitted curve are shown as a dotted line and solid line represent the actual software data. Figure 3 plots the comparison between failure data and estimated by Eq. (17), Figure 4 plots the comparison between corrected data and estimated by Eq. (26). Following the work in Musa *et al.* (1987), we compute the relative error in prediction for this data set and result are plotted in Figure 5. We observed that relative error approaches zero as t_e approaches t_q and the error curve is usually within ±5 percent. Altogether, from Figures 1 to 5 and Table 1 we can see that proposed model has better performance and predicts the future behavior well.

Table 1. Comparison results of different SRGMs for data set with log-logistic.

Models	Estimated Parameters	AE(%)	MSE	R-square
$m_d(t)$ in Equation (17)	a = 71.043 b =0.678	62.211	MSE_d= 60.5356	0.975
$m_c(t)$ in Equation (26)	c =0.816		MSE_c= 41.4503	0.933
$m_d(t)$ in Nan Zhang *et al.*, [18]	a = 139.6 b = 0.1297	25.745	MSE_d = 42.94	0.9826
$m_c(t)$ in Nan Zhang *et al.*	c = 0.8794		MSE_c= 60.476	.9745
Huang *et al.*, [9]	a = 138.026 b = 0.1451	-	MSE = 62.41	0.9758
Lo [10]	a = 133.119 b = 0.1597 c = 0.7825	-	MSE = 58.70	0.9812

Graphs for system T1

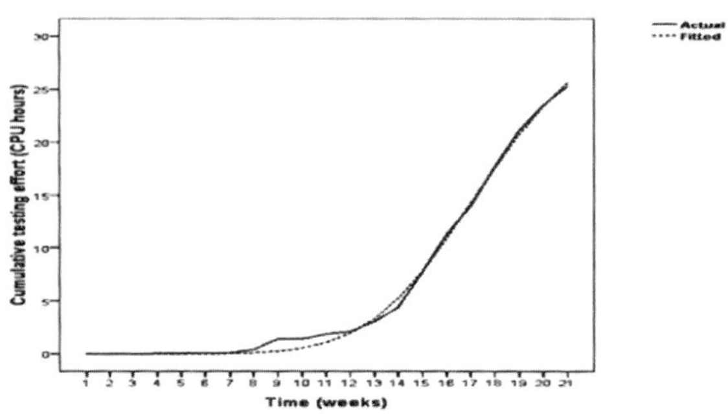

Fig. 1. Estimated cumulative testing effort function v. time.

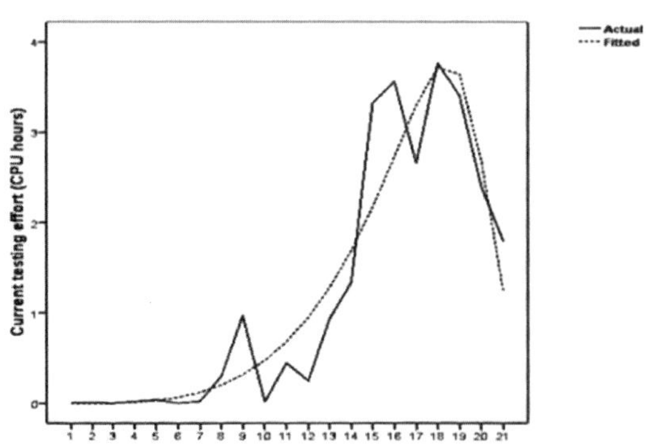

Fig. 2. Estimated current testing effort function v. time.

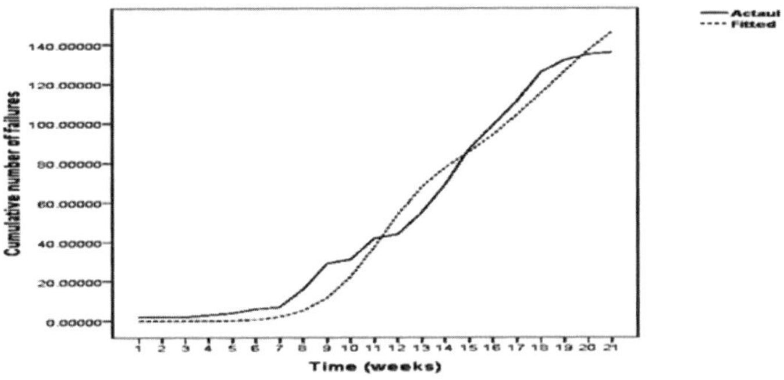

Fig. 3. Estimated cumulative number of failure v. time.

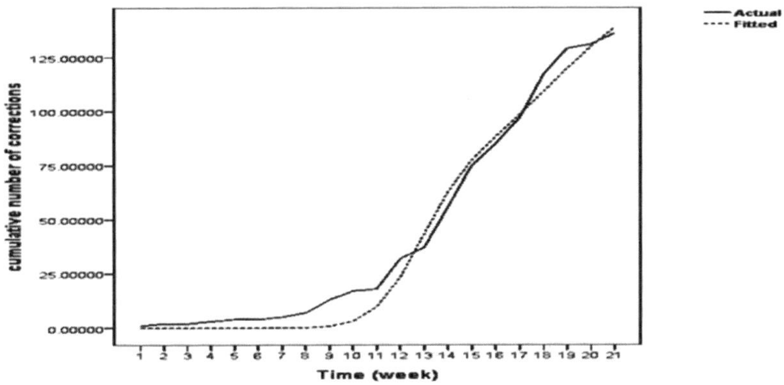

Fig. 4. Estimated cumulative number of correction v. time.

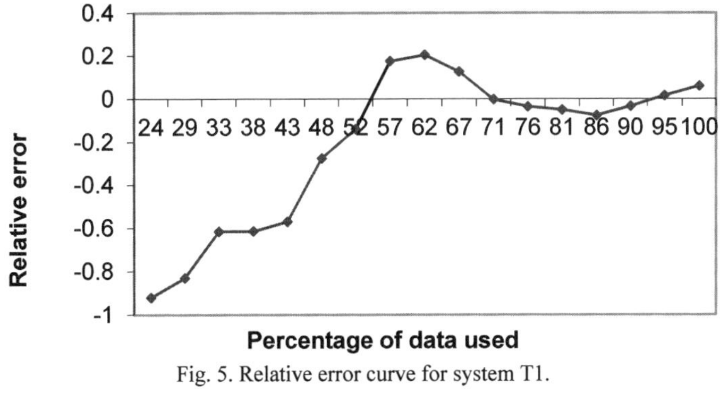

Fig. 5. Relative error curve for system T1.

8. Conclusion

In this paper we have proposed a flexible software reliability growth model based on NHPP model which incorporate Log-logistic testing effort function into software fault detection and fault correction process that can consider the influence of resources on software debugging phase, and enhance the prediction and assessment of software reliability. Moreover a framework of incorporating testing effort into fault detection and fault correction process with consideration of the effect of fault introduction. Several models of fault detection and fault correction process are derived on different assumption of correction effort. We discussed the model comparison criteria where this will judge the performance of the proposed model. Experiments are performed based on real software failure data. We conclude that incorporated Log-logistic testing-effort function is a flexible and can be used to describe the actual expenditure patterns more faithfully during software development. We also conclude that proposed model has better performance as compare to the other SRGM and experimental results show that the proposed models give a goodness of fit to the observed data.

References

1. Ahmad, N., Khan, M. G. M. and Rafi, L. S., A Study of Testing-Effort Dependent Inflection S-Shaped Software Reliability Growth Models with Imperfect Debugging, *International Journal of Quality and Reliability Management*, **27**, 89 (2010).
2. Ahmad, N., Khan, M. G. M. and Rafi, L. S., Analysis of an Inflection S-shaped Software Reliability Model Considering Log-logistic Testing-Effort and Imperfect Debugging, *International Journal of Computer Science and Network Security*, **11**, 161 (2011).
3. Bokhari, M. U. and Ahmad, N., Incorporating Burr Type XII Testing-efforts into Software Reliability Growth Modeling and Actual Data Analysis with Applications, *Journal of Software*, **9**, 1400 (2014).
4. Bokhari, M. U. and Ahmad, N., Analysis of a software reliability growth models: the case of log-logistic test-effort function, *In Proceedings of the 17th IASTED International Conference on Modeling and Simulation (MS'2006)*, Montreal, Canada, 540 (2006).
5. C. Y. Huang, S. Y. Kuo and I. Y. Chen, Analysis of software reliability growth models with Logistic testing effort function. *Proc. Of the 8th International Symposium of Software Reliability engineering, USA,* 378 (1997).

6. Goel, A. L., Okumoto, K., Time-dependent error-detection rate model for software reliability and other performance measures, IEEE Trans. Reliab. 206 (1979).
7. Gokhale, S. S., Lyu, M. R., Trivedi, K. S., Software reliability analysis incorporating fault detection and debugging activities, *Proceedings of the 9th International Symposium on software Reliability Engineering*, 64 (1998).
8. Huang, C. Y., Kuo, S. Y. and Chen, I. Y. Analysis of Software Reliability Growth Model with Logistic Testing-Effort Function, Proceeding of the 8th International Symposium on Software Reliability Engineering (ISSRE'97), Albuquerque, New Maxico, 378 (1997).
9. Huang, C. Y., Kuo, S. Y., Analysis and assessment of incorporating logistic testing effort function into software reliability modeling. IEEE Transactions on reliability 51 (3), 261 (2002).
10. Lo, J. H., An integrated framework of the modeling of fault detection and fault correction processes in software reliability analysis, *Proc of IEEE International Conference on Industrial Informatics, Korea*, 557 (2008).
11. Lo, J. H. and Huang, C. Y., Incorporating imperfect debugging into software fault correction process, in: *Proceeding of the IEEE Regional 10 Annual International Conference (TENCON'2004), Chiang Mai, Thailand*, 326 (2004).
12. Kapur, P. K. Garg, R. B. and Kumar, S., Contributions to Hardware and Software Reliability, *World Scientific Publications, Singapore* (1999).
13. Kapur, P. K., Goswani, D. N., Bardhan, A., Singh, O., Flexible software reliability growth models with testing effort dependent learning process. *Applied mathematical modeling* **32**, 1298 (2008).
14. Lin, C. T., Huang, C. Y., Enhancing and measuring the predictive capabilities of testing effort dependent software reliability models. *The journal of systems and software* **81**, 1025 (2008).
15. Lyu, M. R., Handbook of software reliability Engineering. McGraw-Hill, New York (1996).
16. Lyu, M. R. and Nikora, A., Applying Software Reliability Model More Effectively, *IEEE Software*, **43** (1992).
17. Musa, J. D., Lannino, A., Okumoto, K., Software Reliability Measurement, Prediction and Application. McGraw-Hill, New York (1987).
18. N. Zhang, G. Cui, H. W. Liu, Software Reliability Analysis using Queuing based Model with Testing Effort, in: *Proc. Journal of Software*, **8**, 1301 (2013).

19. Ohba, M., Software reliability analysis models, IBM Journal of Research and Development, **28**, 428 (1984).
20. Pham, H., An imperfect-debugging fault-detection dependent-parameter software, *International Journal of Automation and Computing*, **4**, 325 (2007).
21. Schneidewind, N. F., Analysis of error processes in computer software. Sigplan Notices **10**: 337 (1975).
22. Wu, Y. P., Hu, Q. P., Xie, M., Ng, S. H., Modeling and Analysis of Software Fault Detection and Correction Process by Considering Time Dependency, IEEE Trans on Reliab, **56**, (2007).
23. Xie, M., Hu, Q. P., Wu, Y. P., Ng, N. G., A study of the modeling and analysis of software fault detection and fault correction processes, *Qual. Reliab. Engng. Int.* **23**: 459 (2007).
24. Yamada, S., Hishitani, J. and Osaki, S., Software Reliability Growth Model with Weibull Testing-Effort: A model and Application, *IEEE Transactions on Reliability*, **42**,100 (1993).
25. Yamada, S., Ohtera, H. and Narihisa, H., Software Reliability Growth Models with Testing-Effort, *IEEE Transactions on Reliability*, **35**, 19 (1986).

Skewness Removal of LEACH Protocol for Wireless Sensor Networks

Vishal Gupta[1] and M. N. Doja[2]
Dept. of Computer Science, Jamia Millia Islamia,
Delhi, India
E-mail: [1]vishalg26@rediffmail.com, [2]mndoja@gmail.com
www.jmi.ac.in

Wireless Sensor Network is a wireless network that consists of independent sensors which communicate with each other in distributed fashion to sense and monitor the environment. The goal of Wireless sensor network is to have long life time and high reliability with maximum coverage. LEACH is one of the first and the most discussed hierarchical, cluster-based routing approaches for sensor networks. In WSNs, the LEACH protocol has got high attention due to its simplicity, energy-efficiency and load-balancing nature.

In this paper we have proposed an improvement over LEACH protocol for the Wireless Sensor Networks by putting a check on how the cluster-heads are chosen. For this, the clusters are sub-divided into groups. Choosing the location of cluster heads properly, can greatly improve the life-time of the network. Simulation results demonstrate that our proposed modification achieve our claim.

Keywords: Wireless Sensor Network/s (WSN/s), LEACH, cluster, life-time, skewness, Bayesian Network (BN), Base Station (BS), Cluster-Head (CH).

1. Introduction

In Wireless Sensor Network (WSN), the sensor nodes generally employ distributed algorithms to gather the environmental information and communicate this information wirelessly to a central gateway, from where this information may be processed and analyzed.

The routing protocols in WSN, that are based on energy-efficiency, can be classified into three categories: data-centric protocols [2][3], location-based protocols [3][4] and hierarchical protocols [1][3][5][6][7]. The hierarchical routing protocols are characterized by clustering the nodes to reduce transmissions wherein the nodes in the cluster talk to their respective cluster head only using short-range radio. Low-Energy Adaptive Clustering Hierarchy protocol (LEACH) [1][7] is a dominant representative protocol with clustering structure.

However, the number of cluster heads and the number of cluster members is a major performance issue that cannot be controlled by LEACH [8]. To add, there is also no check on how these chosen cluster-heads are positioned in the area under consideration. This problem may lead into a situation where the chosen cluster heads are skewed towards a particular area or direction. The result of this situation is the reduced life-time of the network as the nodes need to communicate the distant located cluster head.

This work aims at this issue and proposes a check on how the cluster-heads are chosen thereby prolonging the lifetime of the cluster-based wireless sensor network. The remainder of this paper is organized as follows: Section 2 gives the overview of LEACH protocol. Section 3 describes the related work. The basic idea of our proposed protocol is presented in Section 4. Section 5 presents the simulation environment and the performance evaluation of our protocol. The paper ends by the concluding remarks in Section 6.

2. LEACH Protocol

The LEACH [1] protocol forms the clusters in distributive fashion and works in rounds. There are two phases in each round:

2.1. *The set-up phase*

In this phase, the node picks a random number between 0 and 1. A node is picked as the cluster-head node for the current round if the chosen number is less than the following threshold:

$$T(n) = \begin{cases} \frac{P}{1 - P\left(r.\mathrm{mod}\frac{1}{P}\right)} & ,if\ n \in G \\ 0 & ,otherwise \end{cases} \qquad (1)$$

where,

P → the %age of the cluster-head nodes in the population, as desired.

r→ the current round number.

G → G is the set of nodes that have not acted as CH nodes in the last 1/P rounds.

The Non-CH nodes get themselves associated with the CH node that gives the minimum communication overhead and inform this to the selected CH node.

After this cluster-setup sub-phase, each of the CH knows the number of its member nodes. Then, the CH creates a TDMA schedule and broadcasts this TDMA schedule to its cluster members.

2.2. *The steady-state phase*

The member nodes communicate the collected data during their allocated TDMA slot to the CH. The CH aggregates this collected data and sends it to the BS for further processing and decision making. After this, it's the time for the next round and the whole process is repeated over again.

However, the LEACH protocol does not check the location of the cluster-head chosen. On certain occasions, this unchecked selection may lead to a situation that the CHs are not evenly distributed/located in the sensor nodes population. At times it may be case that the CHs are all located in the close vicinity and/or towards a particular direction the area. This is the skewness problem of CH that may result into higher energy consumption by the member nodes as the nodes may require to communicate with a far located cluster head node.

3. Related Work

Much work has been conducted on the performance and efficiency enhancement of LEACH protocol but almost all improvements provide no guarantee that the cluster heads are uniformly positioned in the population.

LEACH-C [7] considers the residual energy when the clusters are formed. WEEC [9] utilizes the location of each node for calculating the probability of Ch-selection. The main drawback of this protocol is that it does not take into account the residual energy during the formation of cluster-heads. In BN-LEACH [10], the author propose to select the Ch using the Bayesian Network (BN) model based on three factors-distance to base station (BS), remaining energy and density. Op-LEACH [11] presents a modification to TDMA based MAC protocol of steady-state of LEACH in order to use the slots corresponding to nodes that don't have data to send at its scheduled slot. In [12] the authors have proposed a new cluster-head probability function which is based on the residual and the initial energy of the candidate nodes, in addition to an optimum number of the cluster heads. In [13] the authors determine the optimal number of the CHs by dividing the region into grids according to the grid theory and also take the energy factor into account for CH selection. Apart from these, a number of other variants in [14][15][16] are also proposed for improving the efficiency of LEACH protocol.

4. Proposed Modification

In this paper we have proposed an improvement over LEACH protocol for the Wireless Sensor Networks by checking the location of CH node. For this, we take an additional phase in which we ensure that the CH nodes chosen by the protocol are homogeneously located in the sensor population i.e. these are not located only towards a particular side of the area.

The careful selection of the CH nodes can greatly minimize the energy consumption of the member nodes and improve the life-time of the network.

Algorithm:

```
begin
    Initialize ( )
    {// Zone the area, depending upon the number of CHs to be chosen
      // Randomly deploy the sensor nodes in the area
      // Save the location of every node deployed
    }
    CH_Select( )
    {// Pick only one sensor node as the CH node from each of the zone,
    depending upon the selection criteria
    }
end
```

5. Performance Evaluation

The performance of the proposed modification is analyzed with the MATLAB simulator. The proposed modification has been compared with the probabilistic LEACH protocol in terms of picking the cluster head nodes in different rounds.

5.1. *Simulation environment*

For simulation, we have picked a random, topologically fixed wireless micro-sensor network of rectangular shape (100 * 100). All nodes in the network are assumed to be homogeneous and of finite energy with no mobility since in most of the applications of wireless sensor networks, the sensor nodes are fixed.

The following table represents the parameters used by the simulation:

Table 1. Simulation parameters.

Parameter	**Values**
Network size	100*100
Location of sink node/base station	(50,50)
Total number of nodes	100
The number of clusters	4

In addition, all the sensor nodes know their deployment co-ordinates and the same information is also shared with the base station.

5.2. *Simulation result*

The snapshot of the random deployment of the sensor nodes and the position of the base station is shown in the Figure 1:

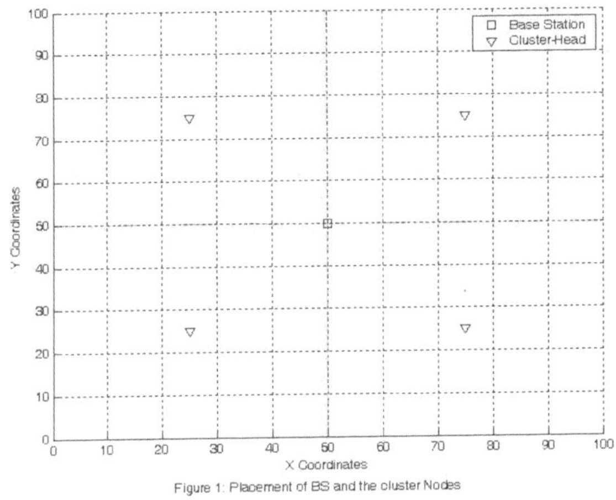

Figure 1. Simulation parameters.

The snapshot for some iteration for the probabilistic LEACH is shown in the Figure 2:

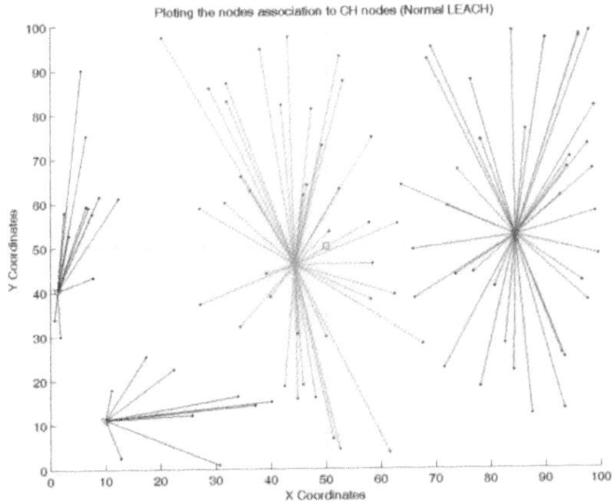

Figure 2. CH and the member nodes association in probabilistic LEACH.

The snapshot for some iteration for the modified LEACH is shown in the Figure 3:

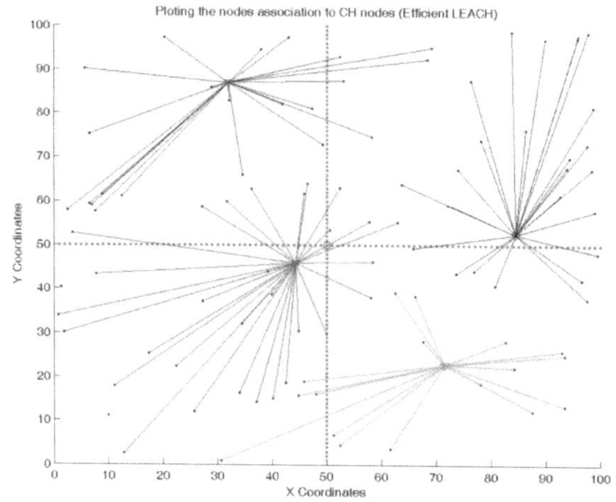

Figure 3. CH and the member nodes association in modified LEACH.

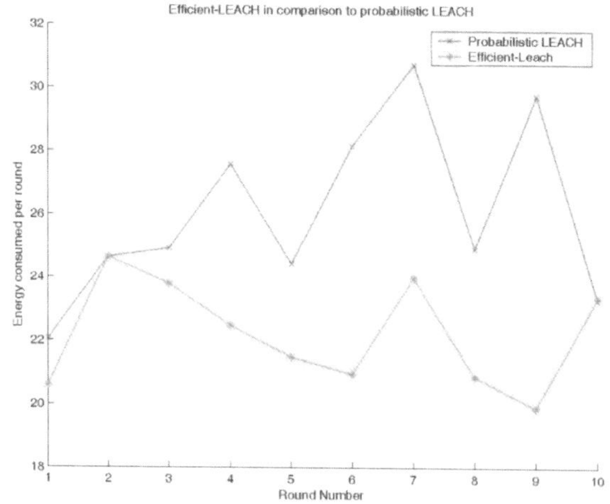

Figure 4. Probabilistic LEACH Vs G-LEACH.

Figure 4 demonstrates the comparison for the average distance of the member nodes from the cluster head nodes in the probabilistic LEACH and the modified LEACH proposed.

The result clearly indicates that in our proposed protocol, the member nodes of a cluster need to send the sensed data to a smaller distance as compared to the probabilistic LEACH protocol. This clearly proves our claim for better energy utilization & prolonging the life-time of the network.

Limitations

The limitation of our approach to LEACH protocol is that our approach requires that all the nodes know their deployed locations and we can virtually see the area clustered statically by the co-ordinates. This information is not required in LEACH protocol.

6. Conclusions

In this work, we have proposed that the carefully located cluster heads can extend the lifetime of the network. Based on the MATLAB simulation, our protocol confirms our claim of longer lifetime as compared to probabilistic LEACH.

References

1. W. R. Heinzelman, A. Chandrakasan and H. Balakrishnan, Energy-Efficient Communication Protocol for Wireless Microsensor Networks, in*proc. of the IEEE 33rd Hawaii International Conference on System Sciences (HICSS '2000)*, 1 (2000).
2. M. Younis, M. Youssef and K. Arisha, Energy-aware routing in cluster-based sensor networks, in proc. of the 10th International Symposium on Modeling, Analysis and Simulation of Computer and Telecommunications Systems (MASCOTS's 02), 129 (2002).
3. K. Akkaya and M. Younis, A Survey on Routing Protocols for Wireless Sensor Networks, in *Elsevier Ad Hoc Networks*, **3**, 325 (2005).
4. Y. Xu, J. Heidemarm and D. Estrin, Geography informed energy conservation for ad hoc routing, in *proc. of 7th annual IEEE/ACM International Conference on Mobile Computing and Networking (MobiCom '01)*, 70 (2001).
5. S. D. Muruganathan, D. C. Ma, RI. Bhasin and A. O. Fapojuwo, A Centralized Energy-Efficient Routing Protocol for Wireless Sensor Networks, in *IEEE Radio Communications*, (2005).
6. M. C. M. Thein and T. Thein, An Energy Efficient Cluster-Head Selection for Wireless Sensor Networks, in *proc. of International Conference on Intelligent Systems, Modelling and Simulation (ISMS'10)*, 287 (2010).

7. W. Heinzelman, A. Chandrakasan and A. Balakrishnan, An Application-specific Protocol Architecture for Wireless Microsensor Networks, in *IEEE Transactions on Wireless Communications*, **1**, 660 (2002).
8. Ying Wang, Mudi Xiong, Monte Carlo Simulation of LEACH Protocol for Wireless Sensor Networks, in IEEE Sixth International Conference on Parallel and Distributed Computing, Applications and Technologies, (PDCAT'05), 85 (2005).
9. N. Behboudi and A. Abhari, A Weighted Energy Efficient Clustering (WEEC) for Wireless Sensor Networks, in *proc. of IEEE Seventh International Conference on Mobile Ad-hoc And Sensor Networks (MSN)*, 146 (2011).
10. Heydar Ghasemzadeh, Mehdi Rezaeian, Fatemeh Dehghan and Mohammad Mohsen, BN-LEACH- An Improvement on LEACH protocol using Bayesian Networks for Energy Consumption Reduction in Wireless Sensor Networks, in *IEEE 7th International Symposium on Telecommunications (IST'2014)*, 1138 (2014).
11. Sapna Gambhir and Nida Fatima, Op-LEACH: An Optimized LEACH Method for busty traffic in WSNs, in *IEEE Fourth International Conference on Advanced Computing & Communication Technologies*, 22 (2014).
12. D. Q. Lu Cheng and W. Wu, An Energy Efficient Weight-Clustering Algorithm in Wireless Sensor Networks, in *proc. of the Japan-China Joint Workshop on Frontier of Computer Science and Technology, IEEE Computer Society*, Nagasaki, 30 (Japan, 27–28 Dec., 2008).
13. Linlin Wang, Jie Liu and Wei Wang, An Improvement and Simulation of LEACH Protocol for Wireless Sensor Network, in *proc. of the IEEE First International Conference on Pervasive Computing, Signal Processing and Applications*, Harbin, China, 444 (17–19 Sep., 2010).
14. A. Yektaparast, F. HodaNabavi, A. Sarmast, An Improvement on LEACH Protocol (Cell-LEACH), in*proc. of IEEE International Conference on Advanced Communications Technology (ICACT2012)*, 992 (2012).
15. J. Hu, Y. Jin and L. Dou, A Time-based Cluster-Head Selection Algorithm for LEACH, in *IEEE Symposium on Computers and Communications (ISCC'08)*, 1172 (2008).
16. H. Yang and B. Sikdar, Optimal Cluster Head Selection in the LEACH Architecture, in *IEEE International Conference on Performance, Computing, and Communications* (IPCCC'07), 93 (2007).

A Novel Approach for Fast Handoff in WLAN

Mithilesh Patel
Electrical Department, NIT Raipur,
Raipur, Chhattisgarh, India
E-mail: mr.mithilesh.patel@gmail.com

Bhavna Singh, Sonam Gupta, Anurag Jajoo and Pavan Kumar Mishra[†]
Information Technology Department, NIT Raipur,
Raipur, Chhattisgarh, India
[†]*E-mail: pavan_km.it@nitrr.ac.in*

In IEEE 802.11 based WLAN system; the mobile nodes (MN) are connected through access points (APs). During mobility a MN leaves one AP and is associated to new APs, A handoff process will occur. To provide a better seamless connectivity, Handoff process latency should be very small. Handoff latency is a combination of scanning, re authentication and reassociation latency. Reauthentication latency is major contributing factor that affects the performance of handoff and increase the handoff latency. In this paper we present a novel approach for reducing the Reauthentication latency, and network overhead. For reducing the re-authentication latency we apply pre-authentication mechanism which is preceded by the mobility prediction to consider the user mobility behavior as the contributing factor in the pre-authentication. With the help of mobility predication the central server sends a pre authentication key to the APs and also sends the Ids of AP to the MN and MN store into their buffer. The simulation results show a good factor of improvement over the latency values in WLAN environment.

Keywords: Seamless mobility, handover, authentication, latency, mobility prediction.

1. Introduction

Wireless network is the key technology which keeps connection with others during mobility. IEEE 802.11 based WLAN is a part of the wireless communication system which is still a challenging issue to provide a better seamless connectivity in mobility. In WLAN the MN is connected through the AP and all the APs are connected with a central server. When a MN enters into a radio coverage area of any AP it requires authentication password to establish connection with APs. In mobility condition when a MN leaves one AP and associates with others, handoff will occur.

A handoff process can be divided into three phase: scanning, re-authentication and re-association [2]. In the scanning phase the MN searches the free channel of new APs for the establishment of new connection. The scanning process can be categorized into two categories: passive scanning and active scanning [13]. In passive scanning APs broadcast beacon packet to MN after a regular interval while in active scanning MN broadest probe request message in a certain channel and collect the response message from other AP and APs also broadcast a beacon packet after regular interval. Active scanning is much more effective because the channel scanning time is 100 time less than passive scanning. After finding the free channel of APs, authentication process is done and after that association so that the MN connected to the new AP.

This paper focuses on reduction of the authentication latency in WLAN during handoff condition. We can pre-authenticate APs of a wireless region for the requesting mobile station. Pre-authentication process requires new message flow and management to be flowed and requires a decision process to select the neighbor cells which will be pre-authenticated [1]. As it is obvious to note the naïve method of pre-authentication can be performed by authenticating all the neighboring APs beforehand but this obviously will be too naïve to implement as in case of moving MN there is a mobility pattern involved with the MN depicting the usual behavior of the user in the network and also increases the network overload. Hence we can formulate a method to actively select few APs in accordance with the mobility patterns of the user in particular network region. This mobility pattern determination is a kind of prediction which is performed taking into consideration of the past history (recorded in some specific format) and hence formulating of probability of movement of the user (MN) in the pre-authenticated APs. Along with mobility prediction [10] we can also introduce the priority factor for the selection of pre-authentication APs. This priority factor is based on the user sensitiveness to the latency factor for its smooth working.

This paper has been organized into six sections. In Section 2 we have discussed about related work. In Section 3 we have proposed a novel method that reduce the authentication latency as well as reduce the overall network overhead. We have done simulation in Section 4 and Section 5 is used for performance analysis and discusses the result. In Section 6 discusses the conclusion and future work.

2. Related Work

In this section, an overview of the proposed handoff in WLAN will be given with the focus on fast and secure handover. A WLAN is a group of computer and peripheral devices that share a common medium. All the APs are connected with an AAA server [11] (Authentication, Authorization) through a LAN wire.

When a MN enters into WLAN area and tries to establish a communication, this server check their credential and after sending positive feedback corresponding AP so that the MN can able to re associates with AP and communicate into network.

To reduce the handoff latency several schemes has been proposed. In [2], the author has proposed an empirical study of handoff process at the link layer with the detailed breakup of the latency of various components and MAC layer function Probe is primary contributor for overall handoff latency which affect on quality on service. They also discuss with large variation in the latency with from one handoff to another and also among APs and STAs used from deferent vendors. In [3], the author has proposed important design considerations for mobility support in future IEEE 802.11 networks are suggested. Apart from that they introduce a fast-handoff framework which adaptively meets different application requirements via a cross-layer approach. In [4], the author has proposed an analysis of handover latency component model in IEEE 802.11b. They have performed their task on test bed which gives a significant result for reducing the handoff latency. In [4] the author proposed fast handover mechanism for real time multimedia application.

For identification of pre authentication APs, in [5, 12], the author has proposed use the mobility predication based on the behavior of the user. They categorize the behavior in short term and periodic behavior for prediction of next cell. In [6], the author proposed holistic approach to minimize handover latency in heterogeneous environment.

This paper is an extension to the paper [1, 7]. They proposed the term known as "Frequent Handoff Region (FHR)". The FHR is a subset of adjacent APs, which are likely to move to in the near future. In this prediction scheme, the Handoff probability between APs is estimated using the previous Handoff ratio and residence time collected and calculated by the central system. The method proposed by them involved the usage of log record for recording the handoff events. And using this weight value FHR set is defined, and the pre-authentication is performed on this set. They formulate a mathematical expression for the calculation of the weight for the selection of the FHR region. The weight is calculated using the following formula:

$$H(i, j) = \sum_{K=1}^{N(i,j)} \frac{1}{R(i, j)} \tag{1}$$

Where $H(i, j)$ and $N(i, j)$ denote the unit handoff ratio and the number of handoff events from $AP(i)$ to $AP(j)$, respectively. Let $R_K(i, j)$ be the residence time in the kth handoff event from $AP(i)$ to $AP(j)$. The residence time is calculated as follows:

$$R_{K(i,j)} = T_{out}\left(K\right) - T_{in}\left(K\right) \tag{2}$$

Where $T_{out}\left(k\right)$ and $T_{in}\left(k\right)$ are in-time and out-time of kth handoff event, respectively.

$$W(i,j) = \begin{cases} 0 & (if\ i == j) \\ \dfrac{1}{H(i,j)} & (i \neq j, AP(i)\ and\ AP(j)\ are\ adjacent) \\ \infty & \end{cases} \tag{3}$$

Equation (3), finally shows the weight function between calculated between $AP(i)$ and $AP(j)$. $w(i,j)$ denotes the link weight value.

3. Proposed Method

A MN enters into WLAN network; it searches for an AP with stronger RSS to associate with. This searching process can be done by MAC layer function knows as scan. The proposed method reduces the handoff latency. We have changes in the message sequences, changed few message structures and also added few buffers in MN order to reduce latency even more and increase the Handoff efficiency. The details of the proposed algorithm will be in this section.

3.1. *Message flow*

In the proposed method we are using the passive mode of scanning, i.e. the APs will be sending the beacon messages at the interval of 50 ms, containing the ID of the AP and other performance related parameters like RSS, current load etc.

In case of 1st time authentication: The beacon message if received by any active mobile station (user end entity referred as MS), then confirms the presence of the AP, the MS then response back with the access request message to access the corresponding AP. The AP responds back with the ID request message, requesting the ID and registration details of the MS, the MS responds back with the ID and password (the credential details). The AP check its local database for the credential record (assuming that in initial scenario this local database contains no record), it relays the message to AAA server, i.e. the central authentication server. After receiving the authentication request message, the central server checks for the verification of MS identity details, if authenticated then it checks for the weight table and calculate the FHR region. If authentication is successful then it sends pre-authentication messages to the AP selected in FHR (containing the credentials user ID and password of the so authenticated MS along with the current time stamp) and sends back to the requesting AP the success message along with the WEP keys, sessions keys types of details along with the IDs of the APs who were selected by the FHR (i.e. the ID of the APs who is now having pre-authentication details of the MS). On receiving the reply from the central server , the AP maintains the record for the requesting MS in its local database and open the port to the user (explained later) and also relays back to the MS , the success message along with AP IDs and time stamp and the control port number of the AP (explained in the later section). The MS so authenticated will store the IDs in the buffer provided to it (as proposed in the method), and it continues to exchange the user data on the port as specified by the AP. This whole message flow works in the case of authentication process is performed for first time for a MS in a particular network region.

In case of Handoff: Now we are going to examine the message flow so proposed in this project in case of handoff situation, when a user is pre-authenticated in a particular region and enters to the AP (consisting the pre-authentication details). In this situation also the MS senses and receive the same beacon message containing the AP ID as in the previous scenario. Now on reception of such message, the MS before replying back for association checks the buffer for the ID of the AP (whose beacon message is received). If the ID of the AP is present in the buffer that means, it is received while going through the 1st initial authentication process and hence its implied that the user is pre-authenticated and forced to consist the authentication record in its local database for this MS. Knowing this, MS now responds back with the special message containing the ID and password in advance (integrating the authentication and association steps of connection establishing) along with handoff info. Now on receiving such message, the AP check for the validity of the user using the corresponding user record in the local database and authorize the port and sends back the success message along with the control port information. The MS now can communicate with the AP using the control port.

Benefits through the proposed message passing:

a) Through the use of the pre-authentication in FHR region the handoff latency is reduced.
b) Further through the usage of buffer containing IDs of pre-authenticated APs we reduce the exchange of two messages between AP and MS, further reducing the time.

c) Through the concept of this we provide a way to more efficient record the Handoff event in the log table at central server for the calculation of weight and FHR region.

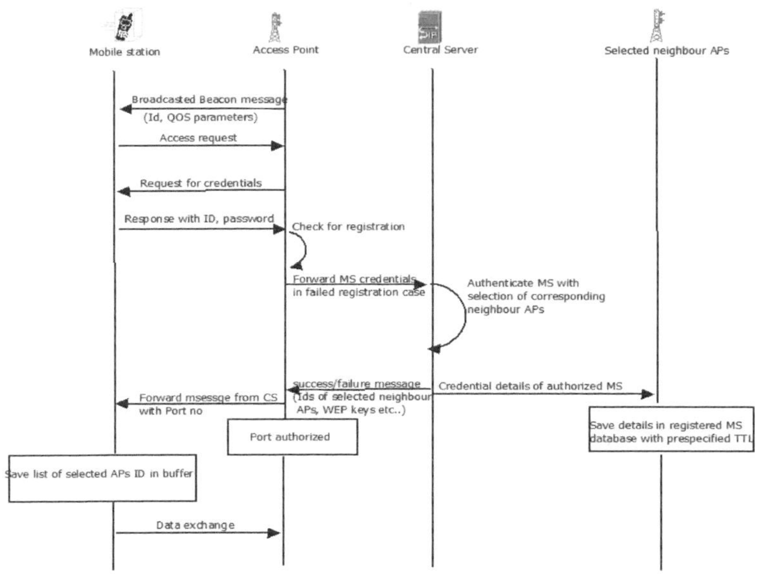

Figure 1. Proposed message passing (initial authentication).

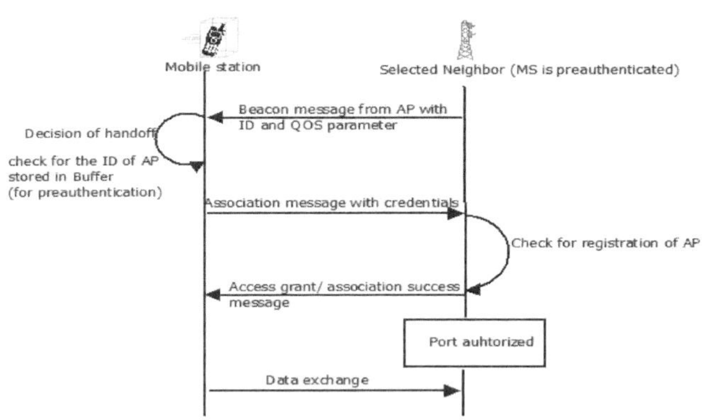

Figure 2. Proposed message passing (in pre-authenticated AP).

3.2. *Controlled port and uncontrolled port*

In the proposed method, the message exchange between the AP and the MS is done on two different ports, known as controlled port and uncontrolled port.

In the uncontrolled port MS can send messages to the AP, but the types of the messages that can be sent on this port are the authentication and association messages (i.e. only control and management types of message), no data can be sent through this port and hence its name uncontrolled port, as it can be accessed by any type of user whether authenticated or non authenticated. But to send the data through the AP it must be done through the controlled port (only then the data will be relayed to the destined user), and hence the controlled port can only be used by the authenticated user only. Also the information of the controlled port is only known to the corresponding AP, who informs about it only to the authenticated user and hence with the help of two different ports the security and feasibility of message passing is maintained.

4. Simulation Result

The simulation is performed in the NS-2 (Network Simulator version 2).

We have implemented a simulation environment for WLAN 802.11 consisting of 1 Central server and 10 APs covering the region. We have about 20 MS uniformly distributed in the region. Apart from these 20 MS, we have 1 moving MS, moving in accordance with the pre-specified destination coordinate, with pre-defined speed, to form 2 different cases (full authentication and pre-authentication). The simulation is depicted in the figure below:

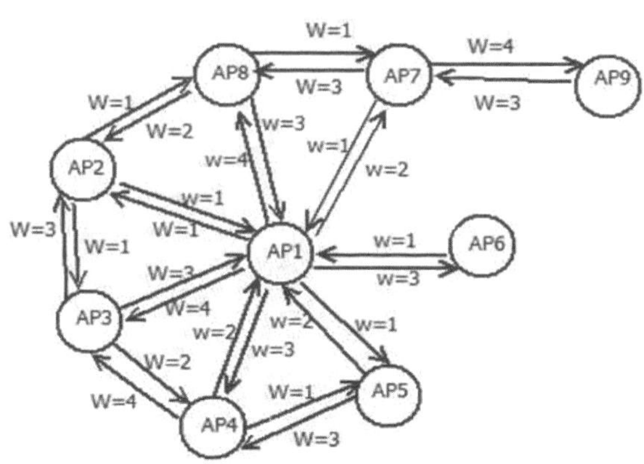

Figure 3. Implemented scenario (weight graph).

The corresponding weight matrix will be in the form of:

$$
W = \begin{bmatrix}
0 & 1 & 4 & 3 & 1 & 3 & 2 & 4 & \infty \\
1 & 0 & 1 & \infty & \infty & \infty & \infty & \infty & \infty \\
3 & 3 & 0 & 2 & 0 & \infty & \infty & \infty & \infty \\
2 & \infty & 4 & 0 & 1 & \infty & \infty & \infty & \infty \\
2 & \infty & \infty & 3 & 0 & \infty & \infty & \infty & \infty \\
1 & \infty & \infty & \infty & \infty & 0 & \infty & \infty & \infty \\
1 & \infty & \infty & \infty & \infty & \infty & 0 & 3 & 4 \\
3 & 2 & \infty & \infty & \infty & \infty & 1 & 0 & \infty \\
\infty & \infty & \infty & \infty & \infty & \infty & 3 & \infty & 0
\end{bmatrix}
$$

Here we have taken node AP1 as the initial node, where the MS will first connect to the WLAN region. This will perform whole steps for authentication first and pre-authentication step will be performed for the selected APs which will be done on basis of the purposed algorithm. Considering the scenario implemented for the node AP1, the FHR set which we will have is (calculated according to algorithm proposed in [1]). $M[1] = [110100000]$.

Now the AP1 (colored green) and other APs (colored red) will be having the authentication record for the current MS (as result of pre-authentication). As shown in the diagram below:

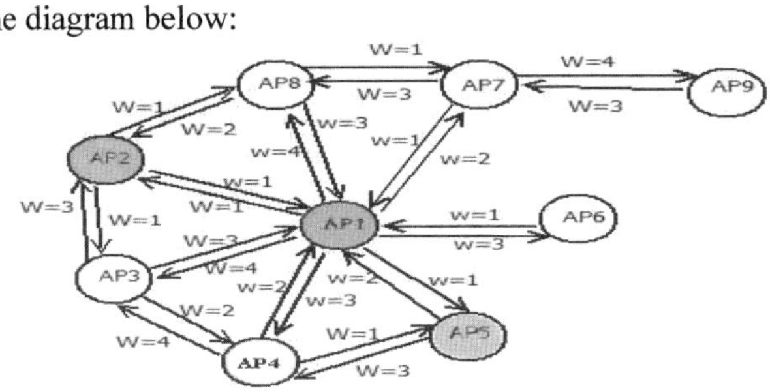

Figure 4. Weight graph (showing pre-authentication for class 1 user).

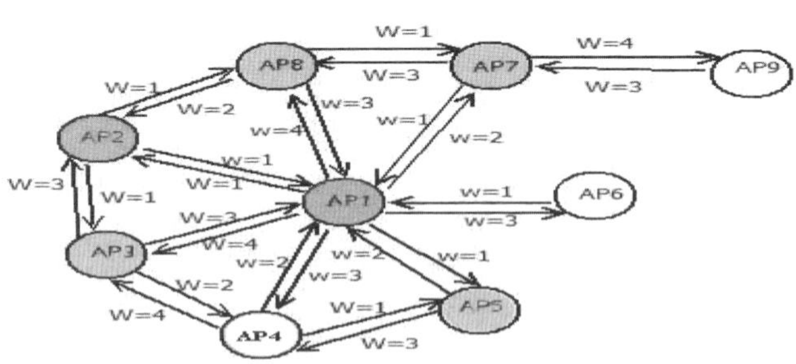

Figure 5. Weight graph (showing pre-authentication for class 2 user).

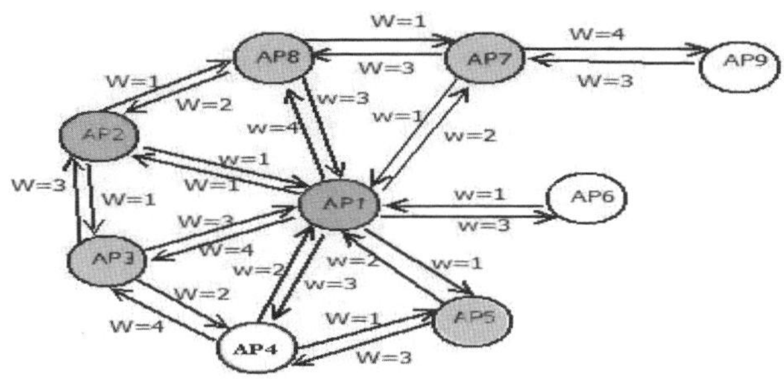

Figure 6. Weight graph (showing pre-authentication for class 3 user).

5. Performance Analysis and Results

Two basic scenario is implemented i.e. the basic full authentication step and the handoff scenario making the use of preauthentication. As the output we basically concentrate on the total time consumed on both the scenario and also the total time required to send the pre-authentication messages to the FHR region.

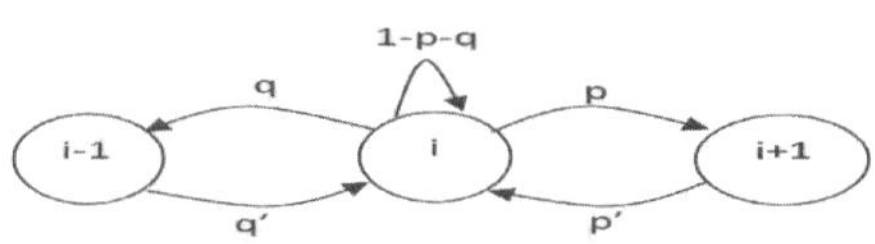

Figure 7. State transition diagram.

5.1. *Probability analysis*

In Figure 7 are state transition diagram the MN move cell(i) to cell(i+1) then probability is p and that MN move cell(i+1) to cell(i) then probability is p'. MN move cell(i) to cell(i-1) then probability is q and MN move cell(i-1) to cell(i) then probability is q' if MN remains same cell then probability is (1-p-q).

Each transition probability, p and q, can be found based on (3). So that we used the stability factor, α. If α ¼ 0 probability the MN hands off to another AP with 1. On the other hand, if α¼1, the MN stays at the current AP with probability 1. P(i, j) is the probability from AP(i) to AP(j) and G is the normalization constant.

$$P(i, j) = \begin{cases} \frac{1}{G} * \frac{1}{w(i, j)} \\ \frac{1}{G} * \alpha \end{cases} \tag{4}$$

$$G = \Sigma\, i = j \left[\frac{1}{w(i, j)} \right] + \alpha \tag{5}$$

For the scenario we implemented the probability calculation is done as follows taking the stability factor as 0:

P(1,1)=0, P(1,2)=1/ 3.66=0.272, P(1,3)=1/(4*3.66)=0.068,

P(1,4)=1/(3*3.66)=0.091, P(1,5)=1/3.66=0.272, P(1,6)=1/(3*3.66)=0.091,

P(1,7) = 1/(2*3.66) = 0.136, P(1,8) = 1/(4*3.66) = 0.068, P(1,9) = 0.

Now for class 1 type of user, the probability of the movement of MS in the preauthenticated AP is = 0.544.

For class 2 type of user, the probability of the movement of MS in the preauthenticated AP is = 0.816.

For class 3 type of user the probability of the movement of MS in the preauthenticated AP is = 0.998.

Hence we can say that for the most sensitive user case, the probability is quiet high for moving in the pre-authenticated AP, and hence the efficiency of handoff is high for them. We can say that for the sensitive user it our algorithm is able to consider the more number of AP for pre-authentication, to avoid suspension of service on the expense of little increase of overhead.

5.2. *Overhead analysis*

Message Overhead analysis: In our scheme, since an 'access reply' message that a MN receives through the authentication procedures contains multiple WEP keys, the length of the message is longer than that of the message used in the general scheme. The length of the message is proportional to the number of pre-authenticated APs calculated according to the FHR selection algorithm.

However, no other information besides the WEP key needs to be duplicated, so the overhead induced by the increased message length is not a critical factor. Furthermore, the number of selected APs can be adjusted according to the users' service class and the number of selected APs is usually quite small due to the users' locality in a local network environment.

Network Overhead analysis: There is a slight network overhead too since the pre authentication message is also send to selected neighboring APs as well. The results in very high efficiency and improvement factor.

The graph is plotted with the number if APs in FHR along X-axis and probability of the MN visiting anyone of the already pre-authenticated AP. Classes represent the weight bound used for ex. Class 1 represents probability of MN going from AP1 to another pre-authenticated AP whose weight exactly equals one, similarly class 2 represents MN from AP1 to another AP with weight of less than or equal to two and so on. We can see that as the weight bound increases the probability factor also increases irrespective of the number of the APs but with the increased number of weight bound we give pre-authentication to various numbers of nodes thus increasing the network congestion. Hence the class of any user should be decide carefully by the subscribers authorizing party.

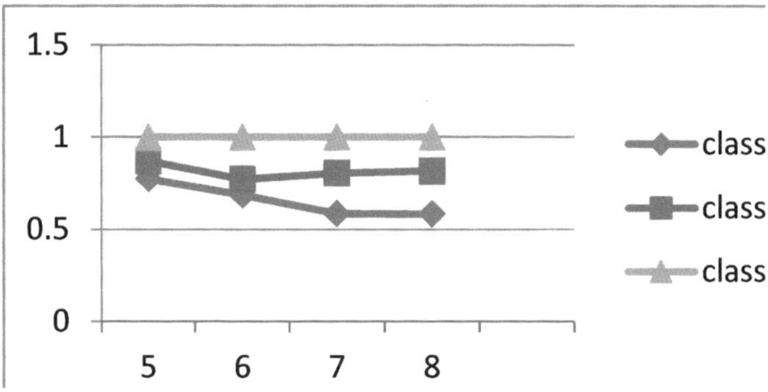

Figure 8. Probability of mobile node visiting a pre-authenticated AP.

6. Conclusion and Future Work

In this paper we reduce the authentication latency up to 41.15% and hence improve the handoff process. For this improvement the mobility pattern prediction is utilized which directs method according to the user behavior in a particular region and hence give an efficient way for the selection of APs for pre-authentication avoiding overhead. Also consideration is given to different kind of user group and there requirements. When the proposed handoff scheme is used, less buffering is required. Therefore, the proposed fast handoff scheme enables public wireless LAN systems to support a variety of applications.

As the part of future work, the prediction of the FHR region can always be improved by using improved techniques of prediction involving the use of AI based approach like using fuzzy logic [9] or using neural networks [8] or other new technologies.

References

1. Pack, S.; Choi, Y., "Fast handoff scheme based on mobility prediction in public wireless LAN systems," Communications, IEE Proceedings, vol. 151, no. 5, pp. 489, 495, 24 Oct. 2004.
2. Mishra, Arunesh, Minho Shin, and William Arbaugh. "An empirical analysis of the IEEE 802.11 MAC layer handoff process." ACM SIGCOMM Computer Communication Review 33, no. 2 (2003): 93–102.

3. Velayos, Héctor and Gunnar Karlsson. Techniques to reduce IEEE 802.11 b MAC layer handover time. Vol. 3. TRITA-IMIT-LCN, 2003.
4. Pack, S.; Jaeyoung Choi; Taekyoung Kwon; Yanghee Choi, "Fast-handoff support in IEEE 802.11 wireless networks," Communications Surveys & Tutorials, IEEE, vol. 9, no. 1, pp. 2, 12, First Quarter 2007.
5. Yaw, Nkansah-Gyekye, "An intelligent vertical handoff decision algorithm in Next Generation Wireless Networks", Computer Science, University of the Western Cape, May 2010.
6. He, Daojing, C. Chi, S. Chan, "A Simple and Robust Vertical Handoff Algorithm for HeterogeneousWireless Mobile Networks," 2010.
7. Koutsorodi, Artemis A., Evgenia F. Adamopoulou, Konstantinos P. Demestichas and Michael E. Theologou, "Terminal management and intelligent access selection in heterogeneous environments." Mobile Networks and Applications 11, no. 6 (2006): 861−871.
8. F. He, F. Wang, "Position aware vertical handoff decision algorithm in heterogeneous wireless networks" Communication Software Center, 2008.
9. J. Capka, R. Boutaba, "Mobility Prediction in Wireless Networks Using Neural Networks by Wireless Networks", Communication Software Center, 2006.
10. X. Shen, J. W. Mark, J. Ye, "User mobility profile prediction: An adaptive fuzzy inference approach" 11-2000, Volume 6, Issue 5, pp. 363−374.
11. Mitton, D., Johns, M., Barkley, S., Nelson, D., Patil, B., Stevens, M., Wolff, B., 'Authentication, authorization, and accounting: protocol evaluation' IETF RFC 3127, June 2001.
12. W. Wanalertlak, B. Lee, C. Yu, M. Kim, S. M. Park, W. T. Kim, "Behavior-based mobility prediction for seamless handoffs in mobile wireless networks," Wireless networks, 15 December 2010-Springer.
13. Rakhi Budhrani, R. Sridaran, "Wireless Local Area Networks: Threats and Their Discovery Using WLANs Scanning Tools," International Journal of Advanced Networking Applications (IJANA), 2014.

Facial Expression Recognition Using Histogram of Oriented Gradients

Jyoti Kumari[1] and R. Rajesh[2]
Department of Computer Science, Central University of South Bihar, Patna, Bihar, India
E-mail: [1]j2kumari@gmail.com, [2]kollamrajeshr@ieee.org
www.cub.ac.in

Most of the human feelings are expressed through face and by seeing one's face, one can easily identify whether he is happy or sad or angry. So for truly knowing the feeling behind words, facial expression must be correctly recognized. This paper gives a brief overview of the facial expression recognition system, discusses few approaches with different strategies, available datasets and the classifiers commonly used. We have used histogram of oriented gradients features for facial expression and the results are promising.

Keywords: Facial Expression Recognition (FER); LBP; HOG; SVM; KNN.

1. Introduction

Facial expression extracts the true or inner emotion which the speaker tries to hide and also find out whether he is happy/sad/angry/disgust/surprise/fear [1][2]. Its applications include human behavior detection, human computer interaction, security, lie detection, pain detection, music system based on mood, automated tutoring, expressing through emoticons etc.

2. Process Discussion

The major steps in any facial expression recognizer (FER) system are preprocessing, face ROI (region of interest) detection, feature extraction, and classifier. The input to the FER system is the image/sequences of images posing from neutral to an expression and back to neutral.

2.1. *Preprocessing*

It is usually done when there is noise or some enhancement is required. Face normalization also comes under this category, if the captured image is non-frontal/rotated to some angle. Gaussian filter, mean filter, contrast changing, edge detection etc. can be used as preprocessing step on requirement.

Pose can be considered changes in terms of angle and camera-person distance and can be dealt with increasing/decreasing resolution. Illumination variation deals with images taken in different lightening conditions and for this, contrast increment/reduction or gradient calculation helps. Occlusion means excluding some features which contributes lesser or even can decrease recognition in the recognition process and sparse representation based classification [3] is robust to occlusion.

2.2. *Face detection*

There are a number of methods available for face detection/extraction but Viola-Jones [4] is the most prominent. It is for the object detection and it can detect different body parts like face, upper body part, head etc. as well as facial components like nose, mouth, eyes etc.

2.3. *Feature extraction*

This is the most important step of the FER system. It says what type of features and in what manner the features are to be extracted. The feature extraction techniques are either based on finding the geometric relationship [5] between facial components/dealing with landmarks of the facial points [6] or applying filter locally or fully [5].

Sometimes either due to fusion of two or more features or bigger image size, the feature vector is so large, it takes more time and storage space for computation. Principal component analysis (PCA) [11], Independent Component Analysis (ICA) [12], and Linear Discriminant Analysis (LDA) [13], are some of the dimensionality reduction mechanisms which are used for feature extraction.

2.3.1. *Action unit (AU) based*

The action unit is the fundamental unit of individual muscle or group of muscles. The FACS (Facial Action Coding System) [6] is the most widely used method. It defines 46 different AU for the non-rigid movements like 6 for chick raiser, 20 for lip stretcher, 26 for jaw drop etc. Other different coding systems are EMFACS [7], MAX [8], AFFEX [9], MPEG-4-SNHC [10].

2.3.2. *Model based*

The active appearance model (AAM) [15][16] and active shape model (ASM) [17][18] are the most widely used model based feature extraction techniques. In ASM, image is represented in terms of set of landmark points. The model is trained for a range of points and the test sample feature points are matched for the best match whereas, the AAM combines statistical model of shape and appearance, i.e. combining ASM model with model built on texture feature or gray level information. By considering the major facial muscle movements to generate the model has been used by Eskil *et al.* [19].

2.3.3. *Texture based and adding temporal information*

The most commonly used texture based feature extraction techniques are Local Binary Patterns (LBP) [20], Local Gradient Code (LGC) [21], Local Ternary Pattern (LTP) [22]. Ji *et al.* used two new textons VTB (Vertical Time Backward) and moments as spatiotemporal descriptors [23]. The other techniques using temporal along with spatial are LBP-TOP (Three Orthogonal Planes) [24][25], LOE-LBP-TOP (Local Oriented Edges) [26]. The other texture based methods combining directional information are Local Directional Pattern (LDP) [27], LDPv [28], Local Directional Number Pattern (LDN) [29], Directional Ternary Pattern (DTP) [30], Gabor filters [14].

2.3.4. *Gradient based*

Now-a-days, gradient has been considered as the powerful mechanism for feature extraction as compared to others. Gradient holds more information rather than the intensity/texture. The most popular is the Histogram of Oriented Gradients (HOG) [31]. Gradient Local Ternary Pattern (GLTP) [32], Local Gradient Pattern [33], Local Gradient Increasing Pattern [34] are the other gradient based techniques for FER.

2.3.5. *Application based*

Some FER systems are specially designed for an application and required real time data. These include spontaneous and posed expression data [35][36], pain detection [37], music system based on mood [38], human computer interaction [39], security [40], driver fatigue detection [41], intelligent tutoring system [42].

2.4. *Classifiers*

Major classifier include but limited to, template matching [43], support vector machine (SVM) [44], nearest neighborhood methods [45], neural networks [46][47] etc.

3. Facial Expression Dataset

The most popular facial expression dataset are provided in this section. JAFFE [47] is the Japanese Female Facial Expression dataset and consists of 10 subjects with seven expression viz. Anger, happy, disgust, fear, neutral, sad, surprise.

CK and CK+ (CK extended) [48] datasets by Cohn-Kanade consists of sequence of images from neutral to peak expression of around 100 university students including female, African-American, Asian. CMU Face image dataset [49] consists of 20 subjects with four expressions namely neutral, sad, happy and angry captured in four different poses viz. left, right, straight, up. Other available datasets are MMI [50], Kaggle [51].

4. Experimental Result with HOG

Histogram of oriented gradient (HOG) [31] is an illumination resistant, gradient based feature descriptor. It divides the image into cells and calculates the magnitude and angular orientations by the help of the gradient filter. The calculated magnitude is then binned depending on the angular orientation cell by cell. The binned values are normalized at the end.

The experiment has been done using JAFFE dataset. For our experiment, 205 images are taken. The original image of size 256 x 256 are divided into 8x8 overlapping (50%) blocks (thus forming 2304 blocks) and HOG features calculated for each block (length of HOG feature for each 8x8 block depends on the number of bins) with various number of bins and are concatenated to form a single feature vector. The experiment has been carried out by leave-one-sample method using KNN(K-Nearest Neighbor) for the classification purpose.

Table 1. Results for FER on JAFFE dataset with different bin size.

No of bins	Feature Length	Recognition Rate
9	20736	87.8049
30	69120	88.2927
24	55296	87.3171
18	41472	88.7805
12	27648	87.8049

5. Conclusion

The application areas of facial expression recognition are increasing and this requires accurate recognition rate on real data. This paper presents a quick survey various feature extraction methods for facial expressions. Experiments are carried out using features extracted from histogram of oriented gradients and the results are promising.

References

1. A. Mehrabian, Communication without words, *Psychological today*, **2**, 53 (1968).
2. P. Ekman and W. V. Friesen, Constants across cultures in the face and emotion, *Journal of personality and social psychology*, **17**, 124 (1971).
3. OuYang, Yan, and Nong Sang, Robust automatic facial expression detection method, *Journal of Software* **8**, 1759 (2013).
4. P Viola, M Jones, Rapid object detection using a boosted cascade of simple features, in *Proc. IEEE Computer Society Conference on Computer Vision and Pattern Recognition (CVPR)*, **1** (2001).
5. Y.-L. Tian, T. Kanade and J. F. Cohn, *Handbook of face recognition*, Ch. Facial Expression Analysis, 487 (London, 2005).
6. P. Ekman and W. Friesen, Facial action coding system: a technique for the measurement of facial movement, *Consulting Psychologists* (San Francisco, 1978).
7. W. V. Friesen and P. Ekman, *Emfacs-7: Emotional facial action coding system*, **2**, 36 (San Francisco, 1983).
8. C. E. Izard, The maximally discriminative facial movement coding system, *Acad. Computing Services and Univ. Media Services*, University of Delaware (1983).
9. C. E. Izard, L. M. Dougherty and E. A. Hembree, A system for identifying affect expressions by holistic judgments (AFFEX), *Instructional Resources Center*, University of Delaware (1983).
10. O. Avaro, A. Eleftheriadis, C. Herpel, G. Rajan and L. Ward, Mpeg-4 systems: overview, *Signal Processing: Image Communication*, **15**, 281 (2000).
11. M. A. Turk and A. P. Pentland, Face recognition using eigenfaces, *Computer Vision and Pattern Recognition*, in *Proc. IEEE Computer Society Conference on CVPR'91*, 586 (1991).

12. M. S. Bartlett, J. R. Movellan and T. J. Sejnowski, Face recognition by independent component analysis, *IEEE Transactions on Neural Networks*, **13**, 1450 (2002).
13. P. N. Belhumeur, J. P. Hespanha and D. Kriegman, Eigenfaces vs. fisherfaces: Recognition using class specific linear projection, *IEEE Transactions on Pattern Analysis and Machine Intelligence*, **19**, 711 (1997).
14. M. J. Lyons, J. Budynek and S. Akamatsu, Automatic classification of single facial images, *IEEE Transactions on Pattern Analysis and Machine Intelligence*, **21**, 1357 (1999).
15. T. F. Cootes, G. J. Edwards and C. J. Taylor, Active appearance models, *IEEE Transactions on pattern analysis and machine intelligence*, **23**, 681 (2001).
16. T. F. Cootes and P. Kittipanya-ngam, Comparing variations on the active appearance model algorithm., *BMVC*, **2**, 837 (2002).
17. E. Setyati, Y. K. Suprapto and M. H. Purnomo, Facial emotional expressions recognition based on active shape model and radial basis function network, *IEEE International Conference on Computational Intelligence for Measurement Systems and Applications (CIMSA)*, 41 (2012).
18. C.-L. Huang and Y.-M. Huang, Facial expression recognition using model-based feature extraction and action parameters classification, *Journal of Visual Communication and Image Representation*, **8**, 278 (1997).
19. M. T. Eskil and K. S. Benli, Facial expression recognition based on anatomy, *Computer Vision and Image Understanding*, **119**, 1 (2014).
20. C. Shan, S. Gong and P. W. McOwan, Facial expression recognition based on local binary patterns: A comprehensive study, *Image and Vision Computing*, **27**, 803 (2009).
21. Y. Tong, R. Chen and Y. Cheng, Facial expression recognition algorithm using lgc based on horizontal and diagonal prior principle, *Optik-International Journal for Light and Electron Optics*, **125**, 4186 (2014).
22. X. Tan and B. Triggs, Enhanced local texture feature sets for face recognition under difficult lighting conditions, *IEEE Transactions on Image Processing*, **19**, 1635 (2010).
23. Y. Ji and K. Idrissi, Automatic facial expression recognition based on spatiotemporal descriptors, *Pattern Recognition Letters*, **33**, 1373 (2012).
24. Y. Guo, Y. Tian, X. Gao and X. Zhang, Micro-expression recognition based on local binary patterns from three orthogonal planes and nearest

neighbor method, *International Joint Conference on Neural Networks (IJCNN)* (2014).

25. G. Zhao and M. Pietik¨ainen, Boosted multi-resolution spatiotemporal descriptors for facial expression recognition, *Pattern recognition letters*, **30**, 1117 (2009).
26. Y. Gizatdinova, V. Surakka, G. Zhao, E. M¨akinen and R. Raisamo, Facial expression classification based on local spatiotemporal edge and texture descriptors, in *Proc. ACM 7th International Conference on Methods and Techniques in Behavioral Research* (2010).
27. T. Jabid, M. H. Kabir and O. Chae, Facial expression recognition using local directional pattern (ldp), *17th IEEE International Conference on in Image Processing (ICIP)*, 1605 (2010).
28. M. H. Kabir, T. Jabid and O. Chae, Local directional pattern variance (ldpv): a robust feature descriptor for facial expression recognition, *Int. Arab J. Inf. Technol.*, **9**, 382 (2012).
29. A. Ramirez Rivera, R. Castillo and O. Chae, Local directional number pattern for face analysis: Face and expression recognition, *IEEE Transactions on Image Processing*, **22**, 1740 (2013).
30. F. Ahmed and M. H. Kabir, Directional ternary pattern (DTP) for facial expression recognition, *IEEE International Conference on Consumer Electronics (ICCE)* (2012).
31. N. Dalal and B. Triggs, Histograms of oriented gradients for human detection, *IEEE Computer Society Conference on Computer Vision and Pattern Recognition*, **1**, 886 (2005).
32. F. Ahmed and E. Hossain, Automated facial expression recognition using gradient-based ternary texture patterns, *Chinese Journal of Engineering*, (2013).
33. M. S. Islam, Local gradient pattern-a novel feature representation for facial expression recognition, *Journal of AI and Data Mining*, **2**, 33 (2014).
34. Z. Lubing and W. Han, Local gradient increasing pattern for facial expression recognition, *19th IEEE International Conference on Image Processing (ICIP)* (2012).
35. H. Dibeklioglu, R. Valenti, A. A. Salah and T. Gevers, Classification of spontaneous and posed smiles, *19th IEEE Conference on Signal Processing and Communications Applications (SIU)* (2011).
36. M. F. Valstar, H. Gunes and M. Pantic, How to distinguish posed from spontaneous smiles using geometric features, in *Proc. ACM 9th international conference on Multimodal interfaces* (2007).

37. A. B. Ashraf, S. Lucey, J. F. Cohn, T. Chen, Z. Ambadar, K. M. Prkachin and P. E. Solomon, The painful face–pain expression recognition using active appearance models, *Image and vision computing*, **27**, 1788 (2009).
38. A. Dureha, An accurate algorithm for generating a music playlist based on facial expressions, *International Journal of Computer Applications*, **100**, (2014).
39. F. Abdat, C. Maaoui and A. Pruski, Human-computer interaction using emotion recognition from facial expression, *IEEE Fifth UKSim European Symposium on Computer Modeling and Simulation (EMS)* (2011).
40. M. A. Butalia, M. Ingle and P. Kulkarni, Facial expression recognition for security, *International Journal of Modern Engineering Research (IJMER)*, **2**, 1449 (2012).
41. Z. Zhang and J. Zhang, A new real-time eye tracking for driver fatigue detection, *in Proc. IEEE 6th International Conference on ITS Telecommunications* (2006).
42. Y.-W. Wu, W. Liu and J.-B. Wang, Application of emotional recognition in intelligent tutoring system, *First IEEE International Workshop on Knowledge Discovery and Data Mining (WKDD)* (2008).
43. Brunelli, R, Template matching techniques in computer vision: Theory and practice (2009).
44. C. Cortes and V. Vapnik, Support-vector networks, *Machine learning*, **20**, 273 (1995).
45. Altman NS, An introduction to kernel and nearest-neighbor nonparametric regression, *The American Statistician*, **46**,175 (1992).
46. F. Rosenblatt, The perceptron: a probabilistic model for information storage and organization in the brain, *Psychological review* (1958).
47. Paul Werbos, Beyond regression: New tools for prediction and analysis in the behavioral sciences (1974).
48. M. J. Lyons, S. Akamatsu, M. Kamachi and J. Gyoba, Coding facial expressions with gabor wavelets, in *Proc. IEEE International Conference on Automatic Face Gesture Recognition and Workshops, FG'98* (1998).
49. T. Kanade, J. F. Cohn and Y. Tian, Comprehensive database for facial expression analysis, in *Proc. Fourth IEEE International Conference on Automatic Face and Gesture Recognition*, 46 (2000).
50. Tom Mitchell, https://archieve.ics.uci.edu/ml/datasets/CMU+Face+Images (1997).
51. M. Pantic, M. F. Valstar, R. Rademaker and L. Maat, Web-based Database for Facial Expression Analysis, in *Proc. IEEE Int'l Conf. Multmedia and Expo (ICME'05)* (Amsterdam, The Netherlands, 2005).

52. Kaggle Facial expression recognition challenge, *https://www.kaggle.com/c/challenges-in-representation-learning-facial-expression-recognition-challenge/data* (2013).

Cloud Computing: Comparative Study Own Server vs Cloud Server

Surendra Kumar Singh
Magadh University Bodhgaya
E-mail: surendra1_singh@rediffmail.com

Cloud computing is internet based network diagram that represent internet, or various parts of it, as schematic clouds. The term, characteristics and services associated with internet based computing is called cloud computing. Characteristics means infrastructures, provisioning, network access and managed metering. The primary business associated model which is employed in software, platform and infrastructure as a service and common deployment model which is deployed by service providers and users to use and maintain the cloud services like private, public, community and hybrid clouds are discussed in this papers. In this paper cloud computing refers to different types of services and applications being delivered in the internet cloud but the fact is, in many cases the devices use to access such services and applications do not require any special applications. That is cloud computing is everywhere. Cloud computing also promise to cut operational and capital costs.

Keywords: Cloud; Operational cost; Capital cost; Hybrid cloud; Managed metering; Deployment model; Network access; Schematic clouds.

1. Introduction

Cloud computing is a technology that uses the internet and central remote servers to maintain data and applications. It allows consumers and businesses to uses application without installation and access their personal files at any computer with internet access. This technology allows for much more efficient computing by centralizing data storage, processing and bandwidth.

This is basically relies on sharing of computing resources to achieve coherence and economies of scale, i.e. Cloud computing is typically defined as a type of computing that relies on sharing computing resources rather than having local servers or personal devices to handle applications. It is just like to utility over a network. The objective of cloud computing is to allow users to take benefit from all of these technologies, without the need for deep knowledge about or expertise with each one of them. The cloud computing main aim to cut costs and helps the users focus on their core business instead of being impeded by IT obstacles. It refers manipulating configuring and accessing the application online. It is also a type of Internet-based computing, where different services such as servers, storage and applications — are delivered to an organization's computers and devices through the Internet.

The cloud computing architecture consist of two components the front end and the back end. The front end of the cloud computing comprises the computer network or client's device. Back end refers to cloud itself which may encompasses computer machines data storage system and servers. A special type of software called middleware is used to allows computer that are connected on the network to communicate with each other.

2. Cloud Computing Model

There are certain services and model working behind the scene making the cloud computing feasible and accessible to end users. Following are the working model for cloud computing.

2.1.1. *Public cloud model*

This deployment model is traditional computing model which is also called external cloud. where resources are dynamically provisioned on fine-grained, self service basis over the internet or via from an offsite third-party provider who bills on a fine-grained basis. This model is useful where several organizations have similar requirements and seek to share infrastructure. Also, it can be economically attractive as the resources (storage, workstations) utilized and shared in the community are already exploited. This is the cloud computing model where service providers make their computing resources available online for the public. It allows the users to access various important resources on cloud, such as Software, Applications or Stored data. One of the prime benefits of using public cloud is that the users are emancipated from performing certain important tasks on their computing machines that they cannot get away with otherwise, these include installation of resources, their configuration and Storage.

2.1.2. *Private cloud model*

A private cloud is created within a business data centre and provides services to internal users. This describes offerings that deploy cloud computing on private network. It consists of applications on virtual machines in a company own sets of hosts. It provide the benefits of utility computing shared hardware costs, the ability to recover from failure and the ability to scale up or down depending upon demand.

2.1.3. *Community cloud model*

If several organization have similar requirements and seek to share infrastructure to realize the benefits of cloud computing, then a community cloud can be established. This is a more expensive option as compared to public cloud as the costs are spread over fewer users as compared to a public cloud. However this option may offer a higher level of privacy, security and policy compliance.

2.1.4. *Hybrid cloud model*

Hybrid cloud means either two separate clouds joined together (public, private) or a combination of virtualized cloud server instances used together with real physical hardware. Thus we can define hybrid cloud is probably the use of physical hardware and virtualized cloud server instances together to provide a single common service. Two clouds that have been joined together are called a combined cloud. Companies can run mission-critical workloads or sensitive applications on the private cloud while using the public cloud for "bursty" workloads that must scaleon-demand. The goal of hybridcloud is to create a unified, automated, scalable environment.

2.2. *Service model*

Service models are the reference model on which cloud computing is based.

2.2.1. *Software as a service (SaaS)*

In this model, a complete application is offered to the customer, as a service on demand. A single instance of the service runs on the cloud & multiple end users are serviced. On the customers' side, there is no need for upfront investment in servers or software licenses, while for the provider, the costs are lowered, since only a single application needs to be hosted & maintained. Now a days SaaS is offered by companies such as Google, Sales force, Microsoft, Zoho, etc. Users can access SaaS applications and services from any location using a computer or mobile device that has Internet access.

2.2.2. *Platform as a service (PaaS)*

A layer of software, or development environment is encapsulated & offered as a service, upon which other higher levels of service can be built. The customer has the freedom to build his own applications, which run on the application servers, such as LAMP platform (Linux, Apache, MySql and PHP), restricted J2EE,

Ruby etc. Google's App Engine, Force.com etc. are some of popular paas example.

2.2.3. *Infrastructure as a service (IaaS)*

IaaS provides basic storage and computing capabilities as standardized services over the network. Servers, storage systems, and made available to handle workloads. The customer would typically deploy his own software on the infrastructure. Some common examples are Amazon GoGrid 3, Tera 3.

2.3. *Cost model*

2.3.1. *CAPEX*

It stands for Capital expenditures. Such expenditures altering the future of the business. A capital expenditure is incurred when a business spends money either to buy fixed assets or to add the value of an existing fixed assets with a useful life extending the taxable year.

2.3.2. *OPEX*

An operating expense, operating expenditure or operational expenditure. Operational expenditure OPEX is an ongoing cost for running a product business or systems. Its counterparts a capital expenditure is the cost of developing or providing non consumable part of the product or system. We can business, like wages, utilities, maintenance, and repairs, fall under the category of **Opex**, or **operational expenditure**. Opex is the money the business spends in order to turn inventory into throughput. Operating expenses also include depreciation of plants and machinery which are used in the production process.

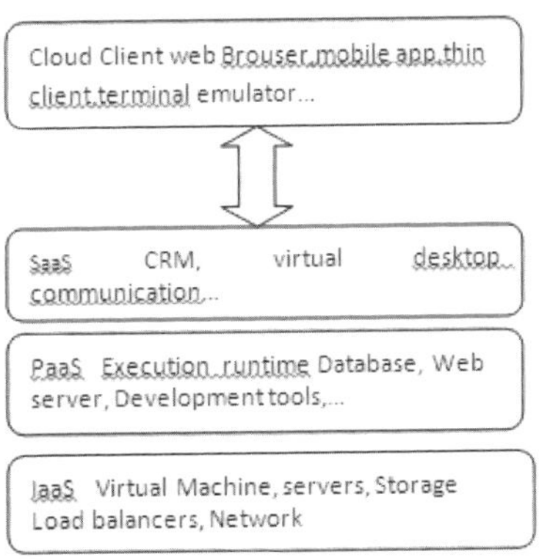

Fig. 1. Service model.

3. Own vs. Cloud Computing

The table below captures some of the differences between the two models of IT. This list is not meant to be exhaustive and is provided to understand the key differences between the two models of IT.

Table 3.1. Key difference.

	Owned and self-managed DC	**Cloud Computing**
Initial / maintenance cost	High Cost of Building, Power Cooling and ongoing maintenance	Low Just the cost of the servers rented. Opex based pricing.
IT Staff	High Requires bigger manpower – across multiple infrastructure skills – networking, firewall etc.	Low Requires only application related management
Reliability	Depends Higher reliability requires extra investment in redundant equipment, power, cooling etc.	24x7 Most of the cloud operators provide 24x7 uptime with high level of SLA guarantee
Efficiency	Difficult to achieve Dedicated environment for one organization	High Virtualized and efficiently built out by specialists
Security	Depends Requires investment in devices, skills and ongoing process optimization	High Typically done for multiple clients, ran with specialists, backed by security certifications
Asset building	Yours to keep Most of the procurement done in this model is on perpetual agreements with hardware or software vendors.	Leased for use The hardware or software that is provided by the cloud provider is only available for use. After the use, it is not owned by the client.
Cost model	Capex + recurring	Opex

Table 3.2. Hardware bill of material (own).

Hardware	Configuration	Budgetary Unit Cost (INR)	Budgetary Total Cost (INR)
Application Server	2 * Quad Core CPU, 64 GB RAM, 3 * 300 GB HDD with RAID 5, Dual Power Supply, 3 year support	5,50,000	5,50,000
Database Server	2 * Quad Core CPU, 64 GB RAM, 5 * 300 GB HDD with RAID 5, Dual Power Supply, 3 year support	6,00,000	6,00,000
Active Directory Server	1 * Quad Core CPU, 32 GB RAM, 2 * 300 GB HDD with RAID 0, Dual Power Supply, 3 year support	4,00,000	4,00,000
Firewall	Standard Firewall with IPS & IDS and VPN connection license.	7,00,000	7,00,000

Table 3.3. Software bill of material (own).

Software	License Detail	Budgetary Unit Cost(INR)	Budgetary Total Cost(INR)
Operating system	Windows 2008 Standard Edition for Active Directory, Application Server and Database server	49,548 per license	1,48,644
	Windows CAL (minimum 25 users)	1,700 per user	42,500
Database Server	2 * Quad Core CPU, 64 GB RAM, 5 * 300 GB HDD with RAID 5, Dual Power Supply, 3 year support	6,00,000	6,00,000
Database	Oracle 11g Standard Edition One – Named User (minimum 100 users)	14,300 per NUP with one year Software update and support	14,30,000
	Oracle 11g Standard Edition One – Processor Based (2 socket)	4,60,000 per processor with one year Software update and support	9,20,000
	Oracle 11g Standard Edition One – Named User (minimum 100 users)	14,300 per NUP with one year Software update and support	14,30,000
	Oracle 11g Standard Edition One – Processor Based (2 socket)	4,60,000 per processor with one year Software update and support	9,20,000
	Oracle 11g Standard Edition – Named User (minimum 100 users)	27,755 per NUP with one year Software update and support	27,75,500
	Oracle 11g Standard Edition – Processor Based (2 socket)	13,87,750 per processor with one year Software update and support	27,75,500
	Oracle 11g Enterprise Edition – Named User (minimum 100 users)	75,335 per NUP with one year Software update and support	75,33,500
	Oracle 11g Enterprise Edition – Processor Based (4 proc)	37,66,750 per processor with one year Software update and support	1,50,67,000
Jasper	Jasper Professional 5.1 (minimum 25 users)	21,000 per user per year	5,25,000
Antivirus	Antivirus clients for each of the servers. (3 servers)	2000 per client	6000

4. Bill of Material - Cloud Server

Table 4.1. H/W bill of material-option 1-Amazon AWS cloud (pay per use model).

Description	Option 1: Hosted on own server		Option 2: Hosted on Amazon AWS Cloud for 1 year (pay per use)	
	Cost (INR)	Remarks	Cost (INR)	Remarks
Hardware	1,000,000	1. Refer to section 3 for details on H/W BoM. 2. No maintenance, setup & support costs included. 3. Hardware cost are indicative and huge margins for vendor negotiation.	2,059,980	1. Refer Section 4 for details on H/W BoM. 2. Includes Amazon AWS Business Support.

Table 4.1. (*Continued*).

Description	Option 1: Hosted on own server		Option 2: Hosted on Amazon AWS Cloud for 1 year (pay per use)	
	Cost (INR)	Remarks	Cost (INR)	Remarks
Software	1,626,000	1. Oracle Standard Edition One Processor based license. 2. Includes first year support costs for Oracle license. 3. Windows OS license cost. 4. Jasper Server license Cost. 5. Rolta as Oracle partner can get a good deal on license cost.	1,626,000	This is at the least same as the cost when hosted on own server.
Total Cost	**2,626,000**		**3,685,980**	

5. Comparative Analysis of Owned vs. Cloud Solution

Description	Option 1: Hosted on own server		Option 2: Hosted on Amazon AWS Cloud for 1 year (pay per use)	
	Cost (INR)	Remarks	Cost (INR)	Remarks
Hardware	1,000,000	1. Refer to section 3 for details on H/W BoM. 2. No maintenance, setup & support costs included. 3. Hardware cost are indicative and huge margins for vendor negotiation.	2,059,980	1. Refer Section 4 for details on H/W BoM. 2. Includes Amazon AWS Business Support.
Software	1,626,000	1. Oracle Standard Edition One Processor based license. 2. Includes first year support costs for Oracle license. 3. Windows OS license cost. 4. Jasper Server license Cost. 5. Rolta as Oracle partner can get a good deal on license cost.	1,626,000	This is at the least same as the cost when hosted on own server.
Total Cost	**2,626,000**		**3,685,980**	

- The pricing has been provided based on the calculator provided on Amazon AWS website. The actual billing may vary depending on the usage of the servers, types of services used and data transfer made during the month.
- Pricing for the Amazon cloud has been provided on the pay-per-use model. This is per hour based billing model.
- Prices are just indicative.

6. Conclusion

Cloud computing is among the most emerging name provides the cost effective and the reliable way to access resources with the help of some emerging tools, techniques and technology. Cloud computing is creating virtualization and thus it result the cost effective round o'clock IT infrastructure availability. It is essential to keep the broadband facility uptodate.

References

1. http://calculator.s3.amazonaws.com/index.html#r=SIN&s=EC2&key=calc-D9B49EAB_096A-4976-8D1A-AA182AE097A3.
2. http://www.oracle.com/us/corporate/pricing/cloud-licencing-070579.pdf
3. "The NIST Definition of Cloud Computing" (http://csrc.nist.gov/publications/nistpubs/800-145/SP800-145.pdf). National Institute of Standards and Technology. Retrieved 24 July 2011.
4. "Know Why Cloud Computing Technology is the New Revolution" (http://www.fonebell.in/cloud- computing-technology-new-revolution/). By Fonebell. Retrieved 8 January 2015.
5. "Where's The Rub: Cloud Computing's Hidden Costs" (http://www.forbes.com/sites/centurylink/2014/02/27/wheres-the-rub-cloud-computings-hidden-costs/). 2014-02-27. Retrieved 2014-07-14.
6. Announcing Amazon Elastic Compute Cloud (Amazon EC2)-beta (https://aws.amazon.com/about-aws/whats-new/2006/08/24/announcing-amazon-elastic-compute-cloud-amazon-ec2---beta/). Amazon.com. 2006-08-24. Retrieved 2014-05-31. www.lcs.mit.edu:8001/~corbato/sjcc62/). *SJCC Proceedings*. MIT. Retrieved 3 July 2012.
7. "Oracle Cloud, Enterprise-Grade Cloud Solutions: SaaS, PaaS, and IaaS" (https://cloud.oracle.com/home). Retrieved (2014).
8. He, Qiang *et al.*, Formulating Cost-Effective Monitoring Strategies for Service-based Systems, (2013).
9. A Self-adaptive hierarchical monitoring mechanism for Clouds (http://dx.doi.org/10.1016/j.jss Elsevier.com. 2011.11.1043)
10. Dario Bruneo, Salvatore Distefano, Francesco Longo, Antonio Puliafito, MarcoScarpa: Workload-Based Software Rejuvenation in Cloud Systems. IEEE Trans. Computers 62(6): 1072-1085 (http://doi.ieeecomputersociety.org/10.1109/TC.2013.30) (2013).
11. Business Adoption of Cloud Computing. Aberdeen Group (2011).

12. Cloud Computing Services Market - Global Industry Size, Market Share, Trends, Analysis And Forecasts, 2012–2018. Transparency Market Research.
13. What Can The Cloud Do To My People. Microsoft.
14. Cloud-Based Disaster Recovery: Ready for the Midmarket. Nethawk - Tech-Tips.
15. Lewis, Grace. *Cloud Computing: Finding the Silver Lining, Not the Silver Bullet*. http://www.sei.cmu.edu/newsitems/cloudcomputing.cfm (2010).

Mobile and GIS Framework for Plantations and Nursery (E-Plantations)

Shailesh Kumar Shrivastava[1] and Dr. S. K. Mahendran[2]
[1]Scientist-E, NIC, Bihar
[2]CS Dept., SVS College, Coimbatore, Tamil Nadu
Amar Nath Pandey and Anant Kumar
Nalanda Open University, Patna, JD Women's College
Patna, Bihar, India

Forests are essential for survival and sustenance of life. Their growth should be optimized so that greater benefits are derived from them. With such a large establishment and geographical base, the monitoring and decision making becomes very critical. The inherent delays hamper the decision process required at a particular time. The increasing area covered by forest plantations creates a demand for trustworthy mechanisms to ensure they are responsibly established and managed. However, most are focused exclusively or prevalently on natural or semi-natural forests, while only a few are specific to planted forests or plantations. The main aim is to assess whether and to what extent planted forests are properly considered within the existing sets of standards/guidelines and to identify areas for improvements, is based on a series of comparative analysis. This paper focus to carry out the full potential of convergence of GIS and Mobile Technology for plantation, with emphasis on technically viable infrastructure solution based on sustainability principles. Integration of GIS and Mobile is being proposed with an objective to enable a single window access to information and services being provided by various formations and to establish a collaborated environment.

Keywords: GIS, e-Governance, Geo-spatial databases, Forest, Plantation, Mobile, Monitoring, Web-GIS, PDAs, GPS, Social Forestry.

1. Introduction

The induction of modern technologies of geospatial tools like Remote Sensing (RS), Geographic Information System (GIS) and Global Positioning System (GPS) have providedvery powerful methods of surveying, identifying, classifying, mapping, monitoring, characterization, and to track changes in the composition, extent, and distribution of several forms of earth resources both renewable and non renewable, living and non-living in nature. Remote Sensing (RS) plays a significant role in providing geo-information in a spatial format and also in determining, enhancing and monitoring the overall capacity of the earth. Satellite observations of land, oceans, atmosphere, and specifically, during natural and human-induced hazards have become crucial for protecting the global environment, reducing disaster losses, and achieving sustainable development. Remote sensing using space-borne sensors is a tool, par excellence, for obtaining repetitive (with a range from minutes to days) and synoptic (with global coverage) observations. GPS provides worldwide positional accurate coordinates, thus, useful to establish geographic location and define the context. A way to acquire recent cost-effective in-situ data, GPS is an important tool for monitoring purpose by acquiring data repetitively about earth features and phenomena. GIS helps in visualization of geospatial data, and the visualization is a convenient and effective way to communicate complex information, e.g. natural resource, and increase our level of understanding about these resources. Besides, GIS also allows data generation, editing, storage, and analysis of spatial data important in planning and decision making. The technology has been an instrumental breakthrough that permits examining natural resources and environment issues in a geographic context. The key point is that GIS provides a means to investigate problems by allowing modeling various phenomena and examining their relationship, e.g. cause and affect relationship, in a place-based context, meaning we can analyze complex, integrated issues from local to global scales.

2. Social Forestry

State of Bihar has a long history of scientific forest management. The forests of Bihar have special significance as they represent a variety of habitats, which are very important from the forestry prospective. Bihar is situated between latitude N to N and longitude E to E. The geographical area of the state is 94,163 Km^2. The forest area of the state is 6,473Km^2 constituting 100 % of the geographical area of the state. With such a large establishment and geographical base, the monitoring and decision making becomes very critical. But, due to lack of adequate information and communication infrastructure, it is challenging to manage and get the day-to-day activity report for the monitoring and decision making. The following broad requirements have come out of the discussions, based on our understanding:

- Capturing the data at source of origin Reliability.
- Easy to maintain.
- Scalability to encompass information of all key stakeholders.
- Secure access and rights based access to stakeholders.
- Standardization and integration of applications & data.
- Business intelligence to analyze and decision making.

ICT deployment is to systematically organize planning implementation and monitoring of forestry and other related operations by systemic collection storage and retrieval of MIS and Geo-spatial data through a computer based communication network. Executing this technology initiatives in an integrated manner where in all its key functions should be carried out through web based workflows which facilitates each role player to log on to the portal and enter his work/data/information and which also facilitates every employee to remain in constant touch with the latest happenings in the department. This working methodology will not only bring transparency responsibility & accountability but will also enhance efficiency.

3. E-Governance Systems

E-governance is the application of Information and communication technologies (ICT) to exchange information between the government and the citizens, government and businesses and between government organizations. E-governance is used to improve the interaction between Government and Citizens, Government and Businesses by the application of some electronic means. It also employs electronic means in internal Government operations to simplify them. E-governance system helps to improve the productivity of Government and helps in decision making. E-governance is the application of Information and communication technologies (ICT) to exchange information between the government and the citizens, government and businesses and between government organizations. E-Governance is all about doing transactions and delivering services online and in a extremely secured manner. Coverage, usability and volumes of transactions being handled are some of the key success factors for any e-governance initiative and govt. agencies are increasingly leveraging the massive advancements in information and communication technology (ICT) to drive these initiatives and reduce the digital divide.

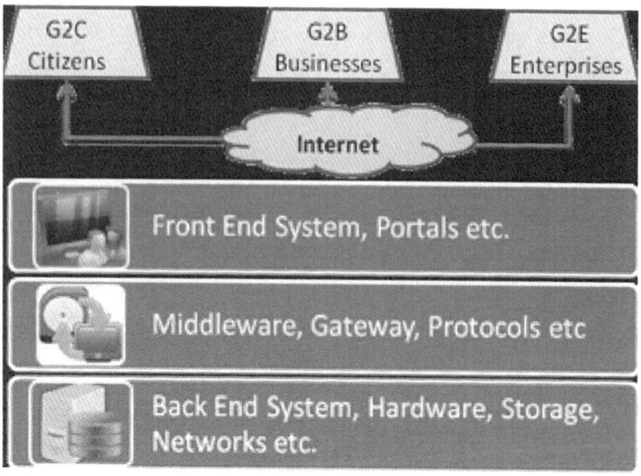

Figure 1. E-governance architecture.

4. Mobile Based E-Plantation Systems

Mobile GIS involves the integration of GIS technology with a variety of other technology which includes Global Positioning Systems (GPS), wireless communication and database technology. It includes accessing, using and maintaining geospatial data directly in the field. The approach is to systematically organize planning implementation and monitoring of forestry and other related operations by systemic collection storage and retrieval of MIS and Geo-spatial data through a computer based communication network. It is proposed to execute its technology initiatives in an integrated manner where in all its key functions should be carried out through web based workflows which facilitates each role player to log on to the departmental portal and enter his work/data/information and which also facilitates every employee to remain in constant touch with the latest happenings in the department. This working methodology will not only bring transparency responsibility & accountability but will also enhance efficiency.

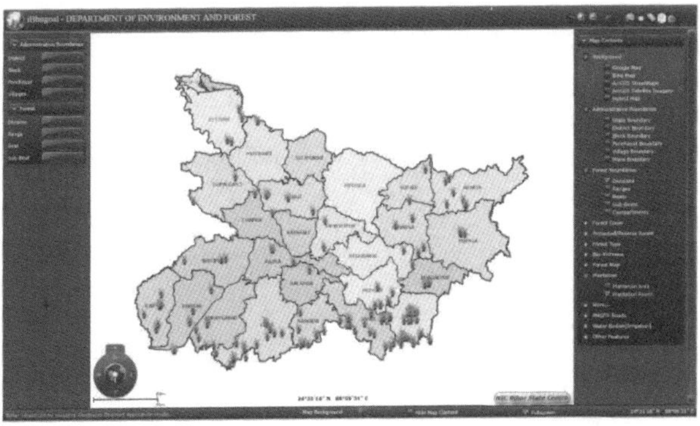

Figure 2. GIS mapping of plantations.

After a thorough assessment of the broad requirements from the department, we have worked out a preliminary solution that is largely focused on creating web based applications framework which will suit the state government's broad vision of information at a mouse click for the stakeholders.

Mobile based forest surveillance & tracking system

Features of the proposed systems include:

- Track real time tracking of forest beat guard,
- Location of the Mobile Device,
- Alert for Fire,
- Tree Cutting activity,
- Forest Encroachment,
- Man – Animal conflict in forest,
- Rescue & Release activity for wild animals,
- Verification of Forest Boundary Pillar,
- Lifecycle Management of Plantation & Nursery details,
- e-Pragati schedule for monitoring of progresses of the Plantation schemes,
- Audit log for the movement of forest personnel.

5. Proposed Architecture

Connected Architecture framework can be proposed with an objective to enable a single window access to information and services being provided by various functionaries of the department and to establish a collaborated environment within the department. Existing employees and other stakeholders will be able to find updated information i.e. guidelines, procedures, policies and contacts etc.

Internal users and departments will have the ability to manage and practice workflow, personalization, content management, and search and business intelligence. Basically, there would be a web-based departmental intranet providing roles based access to its departmental users and restricted connectivity to the public domain information to other users including the citizens.

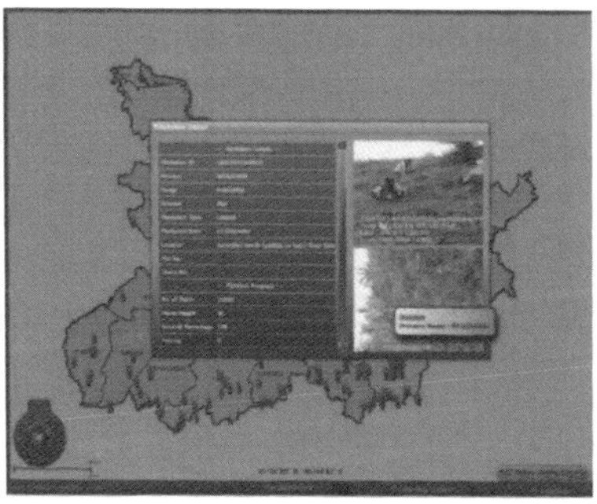

Figure 3. MIS and GIS profile of plantation.

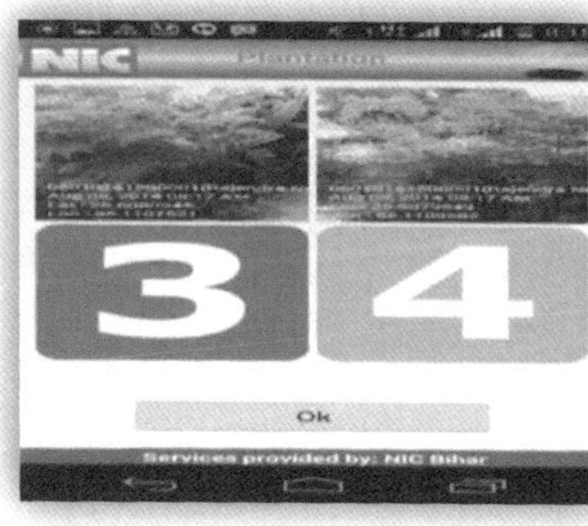

Figure 4. Real-time picture capturing from plantation site.

5.1.1. *Meta data and standard codification*

There is need to prepare the Meta Data repository of all the data items. The ICT Partner can assist the department in creation of the Meta Data repository. Further, in order to have uniformity in data across the department and all its functional units spread across the length and breadth of the State, there is a need to bring about standard codification scheme. The standard codification scheme shall be adopted uniformly and it will be monitored & controlled centrally.

5.1.2. *As-is and to-be data map*

An As-Is and To-Be data map would be created to migrate the existing data. The existing data sets would be mapped to the new data sets and the data types, width and rules shall be ensured for the smooth migration of the data sets.

5.1.3. *Stored procedures and functions*

Appropriate stored procedures and functions shall be created to migrate data from the existing data stores to the new data stores.

5.1.4. *Standardization of classes*

Standard classes shall be created based on the N-tier architecture already prepared. The classes shall be so created to take care of the functionality requirements within the existing applications. The N-tier layered architecture shall be used and each layer shall be based on the classes within that layer. There should be separate layers to handle the errors and exceptions in case of any problem with the applications.

6. Data Capturing and Digitization Strategy

Forests are managed through various kinds of systems which are prescribed by a forestry management experts after carrying out detailed study and taking into consideration all parameters. In recent days use of ICT has become significant for qualitative management of Forests. There are many operations under each system which needs to be monitored with respect to their effect on the forest development and growth so that mid-term corrections could be done based upon the actual field results. The life cycle of forest and plantations need to be indentified and then appropriate technology need to be adopted to ensure proper monitoring and evaluation of the systems. This requires regular collection of huge amount of data annually and its compilation and analysis and report generation. It is a difficult task indeed. The application should facilitate the whole exercise and location wise details along with maps will be possible for decision making.

6.1. *ICT for forest management*

The magnitude and extent of forest and wild life management is large and complex and it involves multi stakeholders breaking all the geographical barriers. Forest and wild life management involves vast geographical areas therefore it is administered by a huge establishment and a large work force. There is widespread public dealing especially with scheduled tribes and backward communities which mostly dwell in and around forests. Since rural life intimately revolves around forests, forestry sector assumes greater significance and sensitivity. National Parks and wild life sanctuaries have different set of management prescriptions which are basically protection and conservation oriented therefore public wild life interface and conflict resolution is a demanding management intervention. Eco-tourism is adding new dimensions to it. Nature and diversity of works executed by the forest department also speaks out the magnitude and complexities involved in the management therefore monitoring and decision making becomes very critical. It is in this context that ICT use is inevitable so that the efficient management of the forests is ensured and quality of goods and services by the forest department to the society is improved.

6.2. *Digitization strategy*

For Conservation and Sustainable Use of Natural Resources, Forest Management Team requires digital maps as it provides ease of working and are handy to carry and helps in geo-positioning. There are many other advantages which need not be discussed here as utility of digital maps is universally known. Everything which is useful in forest management must be captured properly and stored as a digital layer. In the state of Bihar the forest stock maps are partly digitized and it has been observed that there is not much control over the quality of the digitization process. The important features which must be digitized and the various processes involved in digitization are elucidated below:

- Forest Cover, Forest Types, Density Class, Plantations, Pastures, Land use, Water bodies & Drainage, Contours in Watershed areas, Roads etc.
- Administrative Boundaries like Division, Range, Blocks & Beats etc.
- Management Areas, Working Circles, Periodic Blocks, Crops Regeneration Areas, Area rich in non-wood forest products etc.
- Standardization of symbology.

6.3. *Strategy for GIS and mobile enablement of plantations monitoring*

- Creation & integration of **seamless digital data** on the scale 1:15,000 / 1:25,000 scale for entire State ensuring accuracy with respect to Scales and Projections.
- Integrating MIS data available from disparate sources with Geo-spatial data.
- **Standardized Spatial Data Warehouse** which is a collection of key information which may be used to manage data for maximal output and enabling officers to take informed decisions.
- Structured Planning, Management and Analysis using Data Repository for Management Plans & Working Plan Preparation.
- **Standardized Procedures and Workflows** to ensure seamless information flow.
- **Linkages with existing procedures and data sets**.
- Implementing required / mandatory changes (i.e. as per National Working Plan Code etc.) in existing procedures, workflow and responsibilities.
- **Coordination and Networking** amongst Central / State Forests organization for knowledge sharing.

- Need to handle additional technologies like GPS, Web technology etc. at field level.
- **Levels of existing staff experiences** in GIS, cartography, database management and application development and required levels of these skills at field level.
- **Cost & Time estimates** for current manual or automated data handling to implement the functions to be performed by the GIS.
- **Periodical Change Analysis** of Development activities and sustainable use of Natural Resources.
- Controlling **Forest Fires, Encroachments, Poaching and illegal forest fellings.**
- **Preparation of Working Plans** as per **National Working Plan Code** within **stipulated time frames.**

7. GIS and Mobile Solution for Monitoring

- Standardized Forestry Data Model where Data Security, Accuracy, Scalability, Integrity and scale / projection related issues may be addressed with strict QA / QC norms
- Centralized and Standardized Data Warehouse based on proposed National Database Design Code covering Spatial / Non Spatial information required by various levels of organization hierarchy
- User friendly, drop drown lists based digital Forms for Data capturing at field level for improved accuracy and validity with hand held device
- Standardized Modular GIS Applications for Planning, Management, Analyzing Data Repository for GIS based Decision Bihar forest System
- Seamless Integration of MIS and GIS with data validation checks & controls
- Hassle-free Networking amongst Central / State Forests organization for data sharing
- Capacity Building & Development of Expert Level Resources through Structured Trainings at various levels.

8. Process Workflow

The underlined architecture of the solution will be Server and web based which utilizes Central Database Server for ensuring Standardization, Data accuracy, integrity and security. In parallel, Division level officers will use Central Servers in asynchronous mode of data transfer so as they can execute their GIS related activities at local level, compile data offline and Bihar load validated data for approval in Centralized server with restricted access levels.

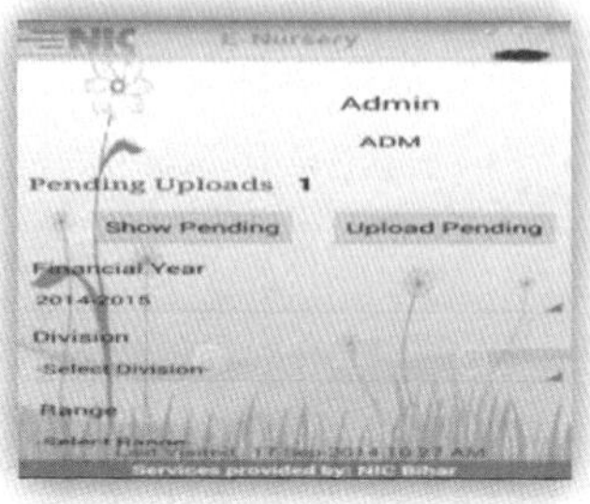

Figure 5. Mobile based data capturing of plantations.

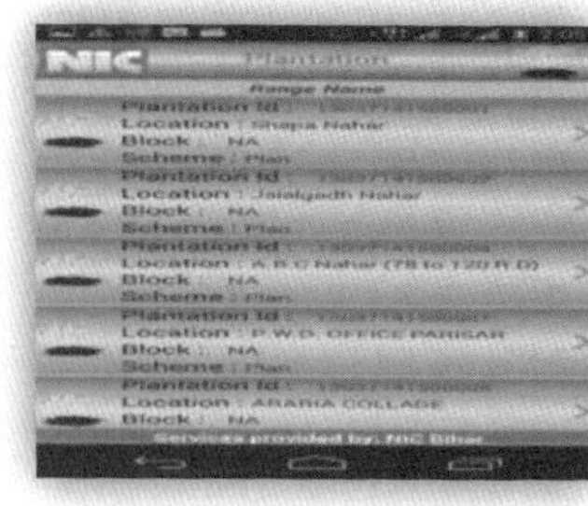

Figure 6. Plantation details with locations.

Using Web browser at Division level, DFOs will be capable of performing structured / non-structured queries required for Working Plan preparation and Forest Management which are not available as per Standard Server Application. Once the data is validated and uploaded, it will be forwarded to respective CFs for further processing and approvals. Circle Officers will use available data for performing structured / unstructured queries and analysis and then consolidated Executive Summaries & reports will be generated and uploaded in the Central Server which provide base for various DSS, MIS, Policies, Projects Monitoring & Evaluation related activities at State level. As per proposed Model, all data will reside on Central Server to maintain uniformity, accuracy and integrity but depending on the levels.

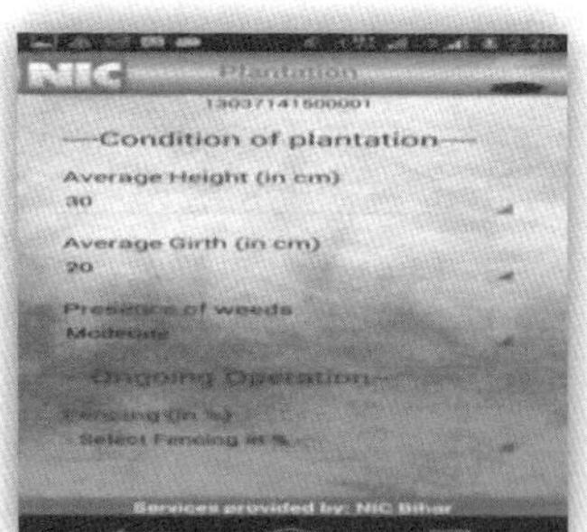

Figure 7. Physical parameters reporting of plantations.

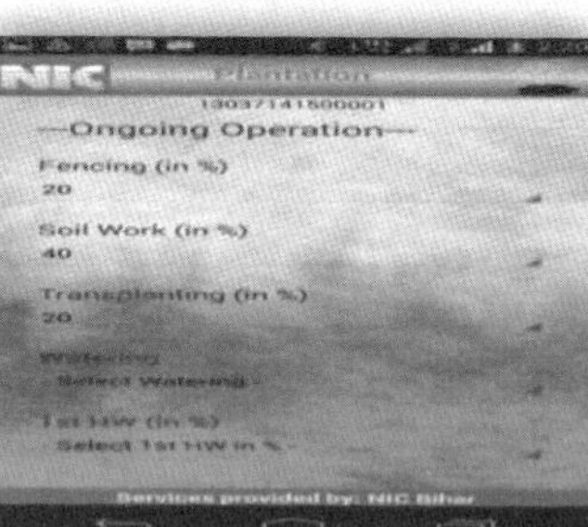

Figure 8. Onsite reporting framework using mobile.

9. Conclusion

"Geographical Information System (GIS)" has emerged as powerful tool which has potential to organize complex spatial environment with tabular relationships. It has been felt to introduce GIS technologies and bring in dynamism and transparency in the nature of working, in order to overcome problem issues related to planning, development monitoring & decision-making. It is largely meant to visualize MIS data spatially through thematic maps for making quicker and better decisions. e-Plantation Software provides features of Registering Farmers, Allocating Scheme to the farmer, Maintaining Stock of Plants both for departmental use and Sale, Approval, Inspection details, Fund transfer to farmers, planning for next year requirements etc. This can be milestone in bringing transparency through e-governance applications.

Acknowledgments

Authors would like to acknowledge ShVivek Kr. Singh, IAS, Principal Secretary, Environment and Forest, Govt. of Bihar helping to conceptualize the paper, Dr. Vandana Sharma, DDG and Head RS & GIS Division, NIC-HQ for technical inputs, Sh. Rajesh Kumar Singh, State Informatics Officer, NIC, Bihar for guidance and support. Authors also like to acknowledge kind support of Vice Chancellor, Nalanda Open University and HOD, JD Women's Collge, Patna.

References

1. Kumar, R. Siva, P. S. Acharya, D. Dutta, M. Prithviraj, N. Mendiratta and B. Singh, 2002, Natural Resources Data Management System (NRDMS- A Suite of Technologies for Local Level Planning, Map Asia Conference Proceeding, Aug. 2002, Bangkok, Thailand.
2. Ghosh, S. K. (2005) GIS Based Modeling for Rural Infrastructure Development, National Seminar on GIS Application in Rural Development, organized by Geomatics Cell, National Institute of Rural Development, Hyderabad September 26–28, 2005, pp. 273–287.
3. GIS Modeling for Rural Roads – I GeoApproach Project study of MP Govt. PMGSY project.
4. http://darpg.gov.in/darpgwebsite_cms/document/file/iBhugoal_case.pdf
5. https://negp.gov.in/pdfs/c22.pdf
6. iGeoForest Project of NIC, Madhya Pradesh.
7. http://proceedings.esri.com/library/userconf/proc04/docs/pap2101.pdf
8. https://mpforest.gov.in/pdf/MPFDICTINITIATIVES.pdf

9. http://www.researchgate.net/profile/Karel_Charvat2/publication/236213975_Mobile_forest_management/links/00b4951705a772aea7000000.pdf

Internet Traffic Classification: A Survey

Gargi Srivastava
Computer Science and Engineering Department, NIT Patna,
Patna, Bihar800005, India
E-mail: gargisri68@gmail.com
www.nitp.ac.in

M. P. Singh
Computer Science and Engineering Department, NIT Patna,
Patna, Bihar 800005, India
E-mail: mps@nitp.ac.in

Prabhat Kumar
Computer Science and Engineering Department, NIT Patna,
Patna, Bihar 800005, India
E-mail: prabhat@nitp.ac.in

J. P. Singh
Computer Science and Engineering Department, NIT Patna,
Patna, Bihar 800005, India
E-mail: jps@nitp.ac.in

Internet traffic analysis is one of the popular research interest area because of its benefits for many applications primarily to the task of internet traffic classification. Many work in this field have progressed in the direction of classification and also for traffic prediction. This paper presents survey on traffic classification and prediction. The paper also mentions the importance of decision trees in the field of traffic classification.

Keywords: Traffic analysis; Traffic classification; Traffic prediction; Decision trees.

1. Introduction

Traffic analysis is important to understand the behavior of networks and resource usage. The need for traffic analysis also arises for QoS assessment and effective troubleshooting. For maintaining security in network, traffic analysis is inevitable.

Demands to improve the traffic classification speed and accuracy ushers the way for a lot of research in this field.

Nguyen and Armitage [25] clearly state in their work that nowadays it is impossible to classify network packets on the basis of port number and payload. Statistical features of network packets are now used together with machine learning approaches to classify the internet traffic.

A lot of research work is going on in the field of network traffic prediction that bears a lot of similarity to road traffic prediction. The results of prediction can be used to avoid congestion at peak time and also properly manage the traffic flowing through the network.

Recent work focuses on machine learning techniques. Decision trees and its different versions are used for traffic classification owing to their simple and comprehensive nature.

There are some recent issues in the field of traffic classification [1]. The continued increase in the number of different Internet applications and the growing motivation to mask them to avoid blocking is a major hurdle in traffic classification. Large portion of network traffic still remains to be unclassified by all techniques. UDP traffic is ignored and many traffic traces do not include bidirectional flows. Researchers use different levels of granularity for defining traffic flows and classes which makes different methods difficult to compare. Protocol encapsulation and multiple service channels make the situation worse.

Organization of this paper is as follows. Section 2 deals with nature of traffic and solutions for its analysis. Section 3 discusses traffic prediction. Section 4 describes the importance of feature selection for classification. And finally Section 5 explains congestion avoidance as an application of traffic classification.

2. Nature of Traffic and its Analysis

This section explains the nature of traffic as generated by a single source and it also presents various methodologies for analyzing the internet traffic.

2.1. *Nature of traffic*

The work in [11] considers traffic generated by a single source. Authors study the statistical properties of traffic and analyze the correlation structure that exist among the network packets. The goal is to single out the most suitable distribution that fits to the interarrival time, packet size and byte count processes.

Gomes *et al.* [11] collect and analyze traces containing traffic from various classes and then fit several known distributions like pareto, log-normal, weibull, rayleigh and gamma to experimental distributions.

Leland *et al.* [28] claim that the packet or byte count in Ethernet LAN is self-similar i.e. the network traffic behaves same when viewed at different degrees of magnification, or different scales on single dimension. Willinger *et al.* [27] claim that session interarrival time is independent and identically distributed that is, it follows the Poisson process. Packet-byte count shows the phenomenon of long-range dependence. Cao *et al.* [10] find that the marginal distribution of the interarrival times was matching Weibull distribution best.

The research in [11] totally focuses on traffic generated by an individual source. But this cannot give us a prediction about the network traffic generated in volume by different users in a network or across several networks. Moreover, the paper takes into consideration only unidirectional flows whereas bidirectional flows are completely neglected for determining the final results.

2.2. *Streaming solutions for analysis of network traffic*

The paper [9] states that traffic measurements were performed by configuring conservative sampling factors at the routers with very limited local storage. Iterative measurement paradigm has the following advantages:

- Interesting traffic patterns could be detected or learned on the fly.
- Online analysis of collected information.
- Iterative evolution of subsequent rules for finer traffic inspection.
- Reducing redundant measurements.
- More accurate response to user query.

The paper [8] describes “Multi-Resolution Tiling (MRT)” in which the rule sets are configured by performing iterative analysis over collected data.

This paper proposes three traffic streaming algorithms:

- Equilibrium rollback (ER).
- Flow momentum (FM).
- Directed momentum (DM).

These algorithms overcome the offline nature of traditional measurements.

2.3. *Content-addressable memory architecture*

Banerjee *et al.* [23] discuss about TCAM architecture, update algorithms and TCAM lookup mechanism. They introduce an advanced incremental update scheme that reduces TCAM writes by up to 52% and reduces search latencies by 50%. This paper discusses about a packet classifier dual TCAM architecture: ITCAM (Interior TCAM) & LTCAM (Leaf TCAM) with simple SRAM. The main attributes for optimization are:

- TCAM space

- Power consumption
- Updating mechanism
- Lookup performance.

Work in [12] alleviate issues related to port range expansion and updates to packet classifiers. The new Packet Classifier-Dual TCAM Architecture+ (PC-DUOS+) deals with such problem as it requires less additional hardware, no re-balancing and moreover the TCAMs are never locked out so no delay.

PC-DUOS+ has better improvement in lookup time.

This method increases the classification speed at the routers, primarily for deciding what action is to be taken for a given packet that satisfies a particular criteria.

3. Traffic Prediction

There are two categories of network traffic prediction models based on the characteristics of the network traffic:

(i) Single forecast model: It uses a mathematical model to predict network traffic. Few examples are: autoregressive model, gray model, self-similarity model and chaos model. But these models suffer from lack of accuracy.

(ii) Combination forecast model: This predicts by combining several single models. Examples include Poisson neural network combined model, ARIMA (Auto Regressive Integrated Moving Average) and SNF (Neuro Fuzzy) combined model, Covariant and artificial neural network combined model, QPSO (Quantum behaved Particle Swarm Optimization) and BPNN (Back propagation neural network) combined model and wavelet analysis and AR-LSSVM (Auto Regressive-Least Square Support Vector Machine) combined model. These suffer from error in predicting dynamic real-time update network.

Guang [22] proposes a network traffic prediction model based on alpha Trous wavelet analysis and Hopfield neural network.

Dong *et al.* propose another combination forecast model in [5]. They combine the statistical and heuristic model to predict flow rate.

Dong *et al.* [5] propose a hybrid support vector machine for regression for flow rate prediction using both statistical and heuristic models. The paper describes the following three hybrid models:

(i) ARIMA model and the Elman NN model with temporal contributions (Elman_time)

(ii) ARIMA model and the Elman NN model with spatial contributions (Elman_space)

(iii) ARIMA model and the Elman NN model with temporal and spatial contributions (Elman_all).

The results show 9.04% accuracy improvement and reliable flow rate predictions under both low- and high-flow traffic conditions.

ARIMA models provide a responsive but time-delayed representation of the fluctuation pattern, resulting in higher prediction errors. Also there are large variances in prediction results of Elman NN models. They also lack consistency and robustness.

Hybrid predictors have a more rapid and synchronized response behavior.

ARIMA+Elman_allhas the best performance and Elman_space+ARIMA has the worst performance among the three hybrid models. For individual models, Elman_all performs best; Elman_time performs better than ARIMA which in turn is better than Elman_space.

3.1. *Analyzing complexity of traffic flow*

Xiang *et al.* [29] analyze complexity of traffic flow time series generated by Nagel Shreckenberg (NS) model.

Costa *et al.* [15] propose multiscale entropy. NS model is a one-dimensional traffic model based on cellular automata.

The parameters for NS model are:

- Circle lane length L ≈ data size
- Vehicle density ≈ number of users in the network.

The results claim the following:

- Randomization probability and vehicle density affect the complexity of time headway.
- On large time scales, with same randomization probabilities, the complexity is at a stable and high level in higher vehicle density scenarios.
- On increasing time scale, with same randomization probabilities thecomplexity decreases in lower vehicle density scenarios.

Models present to predict network traffic may prove very useful for real-time bandwidth management. Predicting the nature of traffic generated by a certain organization at different times of the day or year can help concerned authorities to react to congestion in a proactive manner rather than a reactive manner. This may also help in the cost pricing of the bandwidth utilized by the organization.

4. Feature Selection for Traffic Classification

Feature selection is a very important step that is performed before classification. It plays an important role in improving the performance of the classifier.

4.1. *Multi-class imbalance*

The paper [3] states that if 248 statistical features are used to characterize network traffic flows, the computation cost of classifier will be very high. There are three categories of feature selection methods:

(i) Filter methods: It selects "k" attributes having higher correlation with output class. RELIEF and FOCUS are the most commonly used filter techniques. Williams *et al.* [18] find that filter feature selection methods can greatly improve computational performance and do not degrade classification accuracy significantly.
(ii) Wrapper methods: It is a search strategy which finds an optimal subset by adding/deleting features from the input feature set.
(iii) Embedded Selection Methods: They carry out feature subset selection by inducing logical descriptions. ID3(Iterative Dichotomizer 3), C4.5 and CART(Classification And Regression Tree) are few examples.

En-Najjary *et al.* [24] build logistic regression model to handle the class imbalance problem through transferring the multi-class classification into two-class classification.

Qiong and Zhen [21] investigate ability of three methods:

(i) Random under-sampling
(ii) Random over-sampling
(iii) Cost-sensitive learning.

Zhen and Qiong[14] use information theory to assess how much one feature is biased towards one class. In this paper, authors propose a new feature selection method (BFS) to alleviate the multi-class imbalance problem. The paper compares BFS with fast correlation-based filter (FCBF) and Fullset using Naive Bayes.

4.2. *Decision tree for feature selection*

In [18] Nigel *et al.* undertake two feature-set reduction algorithms and five classifiers and measure the performance on the basis of classification speed and build up time.

The feature-set reduction algorithms are:
(i) Consistency based feature selection
(ii) Correlation based feature selection.

The classifiers are:
(i) BayesNet
(ii) Naive Bayes using dicretisation
(iii) Naive Bayes using kernel density estimation (NBK)
(iv) Naive Bayes Tree
(v) C4.5.

The paper lists twenty two features of network traffic flow considered to be the full feature set.

This paper concludes that C4.5 is the fastest algorithm when using any of the feature sets. NBK builds the fastest classifier.

Feature reduction significantly reduces feature space with a minimal impact on accuracy.

Grabczewski and Jankowski [13] propose two different algorithms, SSV1 and SSV2, to augment the efficiency of feature selection Decision tree algorithms have an inherent feature to estimate a suitability of features for separation of objects representing different classes.

Ratanamahatana and Gunopulos [4] describe a Selective Bayesian Classifier (SBC). It uses only those features that C4.5 uses in its decision tree. Authors also test Augmented Bayesian Classifier (ABC).

C4.5 constructs decision trees.

Boosting on NB classifier fails on average. Langley and Sage [20] use wrapper approach to select relevant features. Cardie [7] uses decision trees together with nearest neighbor. In the paper [17] Kubat *et al.* use decision tree in feature selection for NB classifier. In ABC, correlation arcs between attributes augment NB. Feature selection improves NB.

Mark and Oded [16] describe and evaluate an information-theoretic algorithm for data-driven induction of classification models based on a minimal subset of available features.

A recent improvement over C4.5 is the C5.0 [26], [19] decision tree algorithm.

Compared to all feature reduction algorithms decision trees have shown more promising results than any other methods like filter feature selection and wrapper methods. They are simple, comprehensive and concise. The time complexity of the decision tree is also less.

5. Congestion Avoidance

There are various mechanisms to manage the bandwidth:
- Traffic shaping
- Scheduling algorithms
- Congestion avoidance
- Bandwidth reservation protocols/algorithms
- Traffic classification.

Congestion avoidance techniques monitor network traffic loads in an effort to anticipate and avoid congestion at common network bottlenecks. Cisco [6] lists the following congestion avoidance mechanisms:
- Tail drop
- Weighted random early detection (WRED) and distributed WRED
- DiffServ Compliant WRED.

Flow-based WRED forces WRED to afford grater fairness to all flows on an interface in regard to how packets are dropped.

6. Conclusion

This paper reviews various work done in different streams of internet traffic analysis. The nature of traffic is self-similar and there are many streaming solutions available for improving the packet classification accuracy through software as well as hardware. Prominent work has also been done in the field of traffic prediction. Selecting a feature subset out of the full feature set has been shown to improve accuracy and speed of classification. Bandwidth management is a very lucrative field that requires traffic classification as its primary task.

References

1. Alberto Dainotti and Antonio Pescape, Issues and Future Directions in Traffic Classification, *IEEE Network*, 35 (2012).
2. Alisha Cecil, A Summary of Network Traffic Monitoring and Analysis Techniques, http://www.cse.wustl.edu/~jain/cse567-06/ftp/net_monitoring /index.html
3. Andrew Moore, Denis Zuev and Michael Crogan, Discriminators for use in flow-based classification, 1 (August 2005).
4. Chotirat Ann Ratanamahatana, Dimitrios Gunopulos, Scaling up the Naive Bayesian Classifier: Using Decision Trees for Feature Selection, *Proceedings of Workshop on Data Cleaning and Preprocessing*, 1, (2002).

5. Chunjiao Dong, Stephen H. Richards, Qingfang Yang, Chunfu Shao, Combining the Statistical Model and Heuristic Model to Predict Flow Rate, *Journal of Transportation Engineering*, **1** (2014).
6. Congestion Avoidance Overview, *Cisco IOS Quality of Service Solutions Configuration Guide*, Release **12.2**, 175.
7. Claire Cardie, Using decision trees to improve case-based learning, *ICML*, 25 (1993).
8. Faisal Khan, Nicholas Hosein, Soheil Ghiasi, Chen-Nee Chuah, Puneet Sharma, Streaming Solutions for Fine-Grained Network Traffic Measurements and Analysis, *IEEE/ACM Transactions On Networking*, Vol. **22**, 377 (2014).
9. Jianning Mai, Chen-Nee Chuah, Ashwin Sridharan, Tao Ye and Hui Zang, Is sampled data sufficient for anomaly detection, in *Proc. 6th ACM SIGCOMM IMC*, 165 (2006).
10. Jin Cao, William S. Cleveland, Dong Lin, Don X. Sun, Internet traffic tends toward poisson and independent as the load increases, *Nonlinear Estimation and Classification*, C. Holmes, D. Denison, M. Hansen, B. Yu, B. Mallick, 83 (2002).
11. Joao V. P. Gomes, Pedro R. M. Inacio, Branka Lakic, Mario M. Freire, Henrique J. A. DA Silva, Paulo P. Monteiro, Source Traffic Analysis, *ACM Transactions on Multimedia Computing, Communications and Applications*, **6**, 21 (2010).
12. Karthik Lakshminarayan, Anand Rangarajan and SrinivasanVenkatachary, Algorithms for advanced packet classification with ternary CAMs, *Proceedings of the ACM SIGCOMM Internet Measurement Workshop*, 193 (2005).
13. Krzysztof Grabczewski, Norbert Jankowski, Feature Selection with Decision Tree Criterion, *Fifth International Conference on Hybrid Intelligent Systems*, 1 (2005).
14. Liu Zhen, Liu Qiong, A New Feature Selection Method for Internet Traffic Classification Using ML, *Physics Procedia* **33**, 1338 (2012).
15. Madalena Costa, Ary L. Goldberger, C. K. Peng, Multiscale entropy analysis of biological signals, *Physical Review E*, **71**, 1, (Feb. 2005).
16. Mark Last, Oded Maimon, A Compact and Accurate Model for Classification,*IEEE Transactions on Knowledge and Data Engineering*, **16**, No. 2, 203, (2004).
17. Miroslav Kubat, Doris Flotzinger, Gert Pfurtscheller, Discovering patterns in EEG-signals: Comparative study of a few methods, European Conference on Machine Learning, *LNCS* Vol. **667**, 366 (1993).

18. Nigel Williams, Sebastian Zander, Grenville Armitage, A Preliminary Performance Comparison of Five Machine Learning Algorithms for Practical IP Traffic Flow Classification, *ACM SIGCOMM Computer Communication Review*, **36**, 5, 7, (October 2006).
19. Nilima Patil, Rekha Lathi, Vidya Chitre, Comparison of C5.0 & CART Classification algorithms using pruning technique, *International Journal of Engineering Research & Technology*, Vol. **1** 1, (June 2012).
20. Pat Langley and Stephanie Sage, Induction of selective Bayesian classifiers, *Proceedings of the Tenth Conference on Uncertainty in Artificial Intelligence*, 399 (1994).
21. Qiong Liu, Zhen Liu, A comparison of improving multi-class imbalance for internet traffic classification, *Information System Frontiers*, 509 (2014).
22. Sun Guang, Network Traffic Prediction based on the Wavelet Analysis and Hopfield Neural Network, *International Journal of Future Computer and Communication*, **2**, 101 (April 2013).
23. Tania Banerjee, Sartaj Sahni, Gunasekaran Seetharaman, PC-DUOS+: A TCAM Architecture for Packet Classifiers, *IEEE Transactions On Computers*, **63**, 6, 1526 (June 2014).
24. Taoufik En-Najjary, Guillauume Urvoy-Keller, Marcin Pietrzyk, Jean-Laurent Costeux, Application-based feature selection for internet traffic classification, *Proceedings of 22nd International Teletraffic Congress*, 1 (2010).
25. Thuy T. T. Nguyen and Grenville Armitage, A Survey of Techniques for Internet Traffic Classification using Machine Learning, *IEEE Communications Surveys & Tutorials*, **10**, 56, (2008).
26. Tomasz Bujlow, Tahir Riaz, Jens Myrup Pederson, A method for classification of network traffic based on C5.0 Machine Learning Algorithm, *Workshop on Computing, Networking and Communications*, 237 (2012).
27. Walter Willinger, Vern Paxson, Murad S. Taqqu, Self-similarity and heavy tails: Structural modeling of network traffic, *A Practical Guide to Heavy Tails: Statistical Techniques and Applications*, R. J. Adler, R. E. Feldman, M. S Taqqu, 27 (1998).
28. Will E. Leland, Murad S. Taqqu, Walter Willinger, Daniel V. Wilson, On the self-similar nature of Ethernet traffic (extended version), *IEEE/ACM Transactions on Networking*, **2**, 203 (1994).
29. Zhengtao Xiang, Yujin Li, Li Xiong, Using Multiscale Entropy Method to Analyze the Complexity of Traffic Flow, *Second International Conference on Business Computing and Global Informatization*, 785 (2012).

Comprehensive Study of Search Engine

Sarowar Kumar[1], Kumar Abhishek[2], Abhay Kumar and M. P. Singh
Dept. of Computer Sc. & Engg., NIT Patna, India,
E-mail: [1]sarowarkumar@gmail.com, [2]kumar.abhishek@nitp.ac.in

Search Engines are integral part of web today. This paper attempts to give overview of available search engine, with main focus on Google and VisiNav which are using different techniques for searching the web data.

Keywords: Search Engine; Semantic Web; Page Rank; Sub topical Graph.

1. Introduction

Searching is one of the fundamental problems in the area of Computer Science. There are many techniques which are suggested for searching a given data may be from file, database or web.

A part of searching process is a Filtering process. As most of the information present in the World Wide Web is not relevant for the result which is expected by user. World Wide Web just dumps the information in front of user. Here comes the role of Search engines as filters. Search engines will structure and organize relevant search results and filter out unnecessary results. As information present in web is huge, even such filter with 10% gain is considered as efficient filter [1].

Another important concept related to web is indexing. As indexes are used in books to access information faster and in effective way, they are also required in web for faster information retrieval. Indexes reduce the time and effort required for searching in the World Wide Web. Search Engines are required to do indexing on web pages for faster retrieval of results. It will reduce their effort and time to search web documents.

2. History of Search Engines

This section talks about some milestones in the journey of search engines. The first search engine developed was Archie. It was developed by Alan Emtage, a student at McGill University in Montreal in 1990. It was a database of web filenames. Those filenames were matched with user queries [2].

Lycos was one of the milestones in search engine history. It was developed in 1994. It provided ranked search along with prefix matching in the query [2][3].

Altavista was the first search engine which was having unlimited bandwidth and natural language queries [2].

Yahoo the acronym for Yet Another Hierarchical Officious Oracle is the milestone in the development of search engines. David Filo and Jerry Yang PhD students of Stanford University started it to track their personal interests. Today yahoo is one of the prime search engines [2] [4].

Google revolutionalized the concept of search and search engines. Google was also started in Stanford University by Larry Page and Sergey Brin. Today the situation is search engines means Google [2][5].

Apart from that MSN, Ask.com etc. are also some of the common search engines.

3. Components of Search Engine

The following diagram depicts some peculiar components of search engines.

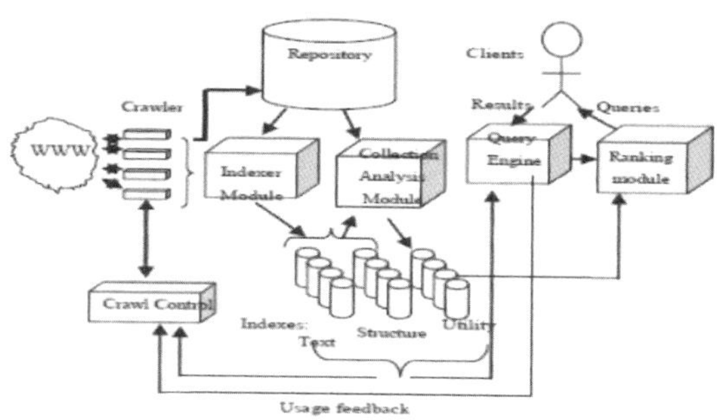

Fig. 1. General components of search engines [6].

Search Engines use the components like crawlers to search the entire web space. Results returned by crawlers will be stored in Repository. To retrieve the results from repository more efficiently, a special module called Indexer is used by search engines. As the results returned by crawlers can be very huge, Indexer module helps to search them faster. Indexer module uses index to search from Repository. Collection analysis module will analyze results from the repository based on user query and relevance of the result for the query.

From the client's perspective, Query engine and Ranking module are the most important parts. Query fired by client will be given to Query module. Based on the query submitted to Query module and the results returned,

Ranking module ranks the results and gives most frequently used pages first. For this purpose, it can use page rank algorithm.

Crawl control is the module responsible for controlling crawlers. Crawlers can give many results. To restrict the number of pages retrieved or to categorize the results obtained etc. Crawl control module is used.

More or less every search engine has these modules as their fundamental parts. Google also uses these kinds of modules for searching and uses page rank algorithm to prioritize the pages. VisiNav also has these basic modules but it uses a different technology for searching and for displaying the results to the client [6].

4. Comparison Of search Engines

This section compares results Obtained by some search engines with that obtained by Google.

4.1. *Comparison of Google and Yahoo*

Yahoo is better than many search engines but it cannot efficiently determine if the link is natural citation or not. Whereas Google is better in determining whether the link is true editorial citation or not. Even Google has aggressive filters which will eliminate many similar search results [11].

It is clearly observed that Google results are based on popularity of search results and content of information available.

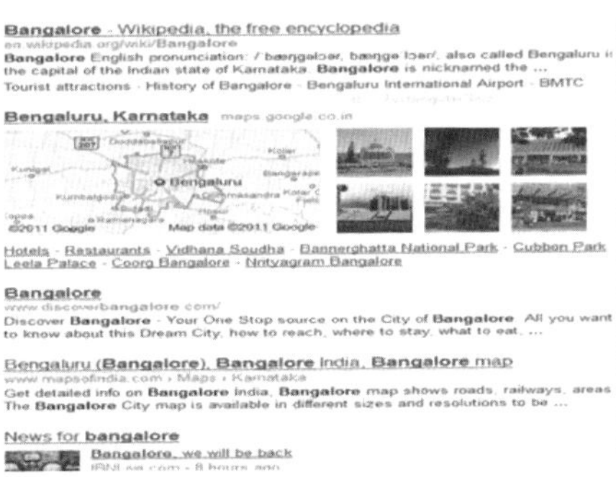

Fig. 2. Search results for keyword 'Bangalore' in Google [7].

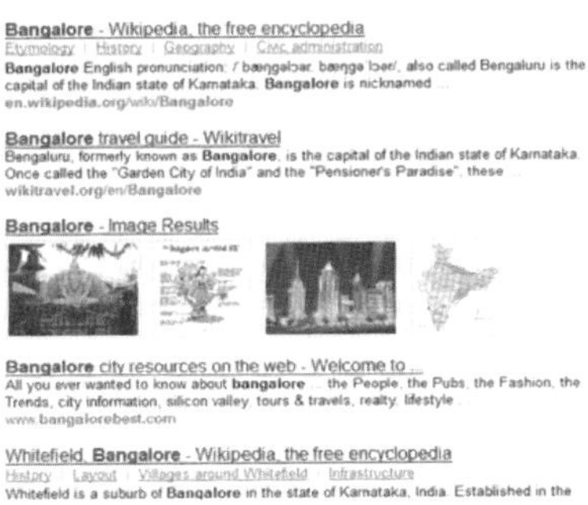

Fig. 3. Search results for keyword 'Bangalore' in Yahoo [12].

Fig. 4. Search results which are related to keyword 'Bangalore' in Google [7].

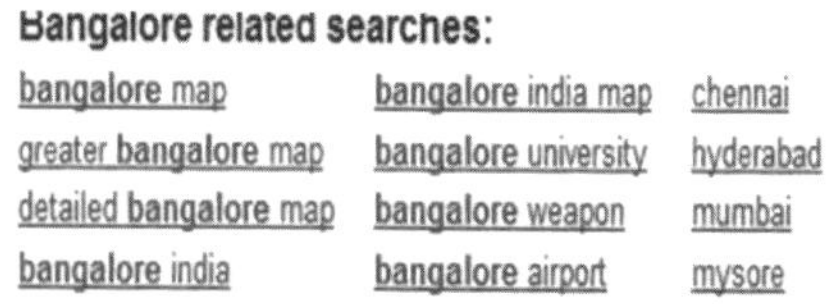

Fig. 5. Search results which are related to keyword 'Bangalore' in Yahoo [12].

4.2. *Comparison of Ask.com and Google*

Ask.com performs search by identifying topics and then finding experts on those topics and then finding the popularity of those topics in those experts similar to Page rank algorithm used by Google [14].

Fig. 6. Search results for keyword 'Bangalore' in ask.com [13]

Fig .7. Search results which are related to the keyword 'Bangalore' in ask.com [13]

As seen from Fig. 6, ask.com puts all its advertisements in search results main page whereas Google puts all advertisements in right side pane of search results.

Ask.com also gives user option 'smart answer' which enables user drill down operation to be performed on first page itself. Results obtained by ask.com are almost equivalent to Google in certain aspects like search, map results etc. [14].

4.3. *Comparison of Google and Bing*

Bing also produces almost comparable results as that of Google as shown in Fig. 8 and Fig. 9.

Fig. 8. Search results for keyword 'Bangalore' in Bing.

Fig. 9. Search results related to 'Bangalore' in Bing.

In Bing, related search results appear on the left side pane and sponsored results appear as a part of search results [15]. Even results obtained are different as compared to Google as seen from Fig. 4 and Fig. 9.

There is option to navigate through results based on region. So it allows drill down operation with region [15].

5. Working of Google

In recent years, Google is considered to be the most widely used and most famous search engine. The reason for this success may be the quality of search results one can get by firing a query. The basis for this lies in the Page Rank algorithm used by Google to rank the web page documents. This algorithm is devised by Google founders Lawrence Page and Sergey Brin. Over the years this algorithm has undergone many changes but still the fundamental working principle remains the same. This section discusses some aspects of Page Rank algorithm.

There are many methods of ranking a web page. One method can be length of contents of a web page. Another method talks about popularity of web page which is decided by number of incoming links to a page. Drawback with this approach is there may be some deceptive web pages having many incoming links to it but the content of those pages may not be of much value to the user.

Page Rank is a concept introduced to improve the quality of search results produced. Rank of a page should not only depend on the number of incoming pages but it should depend on the number of high quality incoming pages.

The Page Rank of a page A is given by:

$$PR(A) = (1-d) + d\,(PR(T1)/C(T1) + \ldots + PR(Tn)/C(Tn)) \quad \ldots \quad \text{Eq. (1)}$$

Where PR(A)= Page Rank of A, PR(Ti)= page Rank of Ti which ranks to page A, C(Ti)= Number of outgoing links on page Ti, d= damping factor. Typical value is between 0-1. Usually the value is generally set to 0.85.

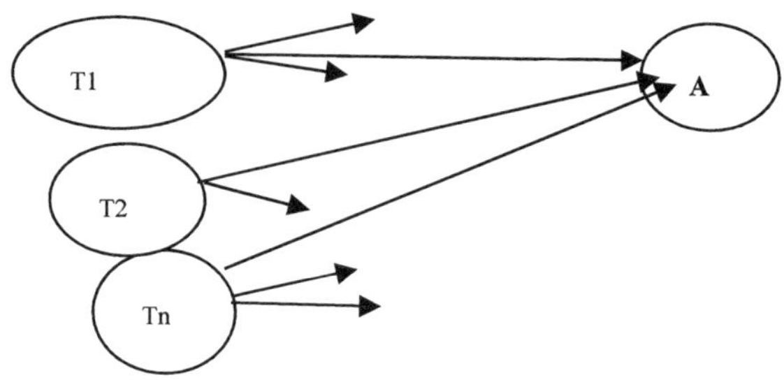

Fig. 10. Page Rank Algorithm.

So page rank definition given above is a recursive definition for page rank of page.

Page rank can also be calculated using different notation as follows:

$$PR(A) = (1\text{-}d) / N + d\,(PR(T1)/C(T1) + \ldots + PR(Tn)/C(Tn)) \ldots \quad \text{Eq. (2)}$$

Where N= Total number of pages in entire web.

A page's page rank can also be written as page rank= 0.15+0.85*(a "share" of page rank of all pages linked to that page)... Eq. (3)

Share= linking page's page rank divided by number of outbound links on the page.

For example, Consider Fig. 11. Which is a site having three pages A, B and C.

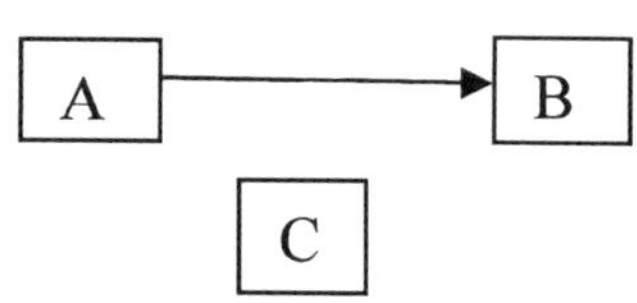

Fig. 11. A simple example for explaining page rank.

Assume that each page is allocated a page rank PR1 initially. After linking page A to page B in the same site, if we run the calculation as per Eq (3),

Page rank of A= 0.15

Page rank of B= 1

Page rank of C= 0.15

Here page A has voted for page B as a result Page B's page rank is increased.

After 100 iterations, the results will be

Page rank of A=0.15

Page rank of B= 0.2775

Page rank of C= 0.15

And total page rank in the site is 0.15+0.2775+0.15=0.5775. [16].

Because of the large size of web, Google search engine uses approximate iterative computation process for page rank. In this method, each page is assigned a small initial page rank value and then real page ranks are calculated based on the page rank equation. Generally around 100 iterations are required for getting good approximation of entire web.

The ranking of web pages in Google was originally dependent on following factors.

1. Page specific factors which include body text, contents of title tag and url etc.
2. Anchor text of incoming links.
3. Page Rank

There is rule called page-rank 1 rule which helps to improve efficiency of page rank algorithm. Web pages are important if other important pages point to them. To have a higher rank, it is not required to have many incoming links but links coming from small number of high importance pages is sufficient.

5.1. *Some factors affecting page rank*

5.1.1. *Dangling links*

Some pages don't have any outgoing links. Therefore its page rank cannot be propagated to other pages. So links to those pages are called dangling links.

One approach of preventing dangling links is not to include them in database until the page rank is computed. PDF files are good examples of dangling links. But links to PDF files don't drain out the page rank for sites having links to them.

5.1.2. *Effect of number of pages*

Combined page rank of all pages in the web equals total number of pages in the web. So addition of web page increases page rank of all the pages by one. It also has impact on Page rank of actual web sites.

Addition of web pages to existing sites reduces page rank for already existing pages of site. So page rank algorithm seems to be biased towards web sites with small number of pages. It is also possible to increase the page rank of existing pages by addition of new pages. So care must be taken to distribute small page rank to the additional pages.

5.1.3. *Some other factors influencing page rank*

1. Link visibility
2. Position of link within a document
3. Distance between web pages
4. Importance of linking page
5. Maintenance of the web page

5.2. *Modified page rank algorithm*

To include all the factors discussed here, page rank algorithm is modified. So, different evaluation is made for links within a document. The document structure is also having some importance in finding page rank of web pages. The original page rank algorithm is gradually modified to give more efficient results. Therefore results given by Google are by far considered to be more efficient than other search engines [7][8].

6. VisiNav (Visual Data Search Navigation)

VisiNav is a search engine which works on different concepts and model than traditional search engines including Google. This section tries to answer "Why VisiNav?" if Google is there as a powerful search engine. This section also describes components and concepts involved in developing VisiNav.

Most of the search engines today including those mentioned here use keyword search technique for searching the web to get the results. So the principle involved here is keyword search on hypertext document. The search engine returns links to documents or results and not the actual data. The query structure is very simple. The query here consists of small keywords. These search engines cannot handle complex queries.

Semantic web is a technology which allows integration of data from large datasets and expressing them as objects, their attributes and links to other objects. VisiNav is based on this technology. It answers the question of how the users can express complex queries.

The best approach for allowing complex queries is incremental construction of queries by users using primitive operations on them. The first step is keyword search and then followed by specified set of operations on the results obtained. This is used in VisiNav hence the name Visual Data Search Navigation. Users can navigate through results based on primitive operations provided by search engine so he/she can get more constrained results and chances of getting null results are reduced.

Dataset for VisiNav is collected from web. VisiNav allows for

a. searching objects.
b. teleporting between objects.
c. faceted browsing for incrementally restrict the result.
d. set based link navigation for navigating between objects.
e. visualization of query results in graph, table, map, timeline view format [9].

6.1. *Operations in VisiNav [10]*

i) Keyword search: To begin with, user can start simple keyword search on objects of interest. There is no synonym search or natural language processing stuff in this part of search.
ii) Object Navigation: This is similar to hypertext navigation. This operation will return single node or set of nodes.
iii) Facet selection: Facet is a combination of Object property and literal value for object. This allows user to restrict the query result.
iv) Path traversal: Apart from object navigation, navigation along object properties to get result set is also possible using this operation.
v) Projection: This operation is done to select only dataset properties for display.
vi) Sorting: The sorting of result set is possible along one or more criterion.

6.2. *System architecture*

The system architecture of VisiNav is based on MVC. The controller is implemented in Servlet. It will receive queries from user and retrieves topical subgraphs. The Model classes parse the statements given by topical subgraphs into Java objects. In the view, results are given to web browser.

Topical subgraphs contain all the information required to display result tree. They also contain information for further navigation. It will first look for nodes directly matching the searching criterion. These set of nodes is then expanded using outgoing links upto defined threshold. Similarly incoming links are also considered for expanding nodde set.

The results are displayed in three ways:

a. results display in web browser: The web application gives result in XML, and then XSL and CSS is applied to it to give it to user via browser.

b. XML based data export: The web application returns result in XML and gives back to user in same format.
c. Text based data export: The web application returns result in plain text and gives back to user in same format.

In semantic web environment, objects are expressed in common data format called RDF. So to display results in different formats to user, additional plug – ins are required [9][10].

7. Result Comparison of Google and VisiNav

Table 1 gives comparison of results obtained in Google and VisiNav.

Table 1. Comparison of results from Google and VisiNav.

Query	Google results	VisiNav results
1. Isaac Newton	1. Related to Isaac Newton and his life.	1. Related to Isaac Newton 2. Related to scientist as a profession 3. Related to Physics 4. Related to calculus
2. Pete Sampras	1. Related to Pete Sampras	1. Related to Pete Sampras
3. Football	1. Related to football 2. Related to FIFA 3. Football Matches 4. Football images	1. Related to football 2. Related to College football teams 3. Related to football team 4. Related to football helmet.
4. Sachin Tendulkar	1. Related to Sachin Tendulkar	---

As seen from results, in some cases VisiNav gives specific results as well as results related to query fired. So user can navigate through different object properties like football, football teams and football helmet. Even it has foaf (Friends of a friend).e.g. if search is performed on Tim Berners-Lee then it will give people known by Tim Berners-Lee. It will also create profile of him. If a search is performed on sir Isaac Newton, apart from search results of Isaac Newton, it will give results of related results like Calculus, Physics etc. But in some cases VisiNav will rely on Google to give results like in case of Sachin Tendulkar.

By comparing Fig. 12 and Fig. 13, we can clearly see the difference of results obtained by VisiNav and Google.

VisiNav classifies the search results obtained. For example fourth result shown in Fig. 13 is classified by VisiNav as document, post and wiki article. Similarly fifth result is classified as Male person, human and former agent. VisiNav also shows domains related to Issac Newton like scientist showing its classification as profession, Physics as a scientific discipline etc. This indicates

VisiNav can link Issac Newton to these fields for which in Google you may have to type Issac Newton+ scientist or Issac Newton+ Physics.

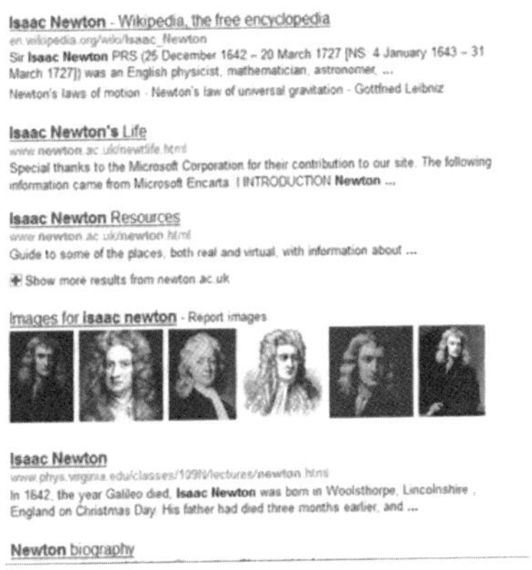

Fig. 12. Search results obtained for keyword 'Issac Newton' in Google [7]

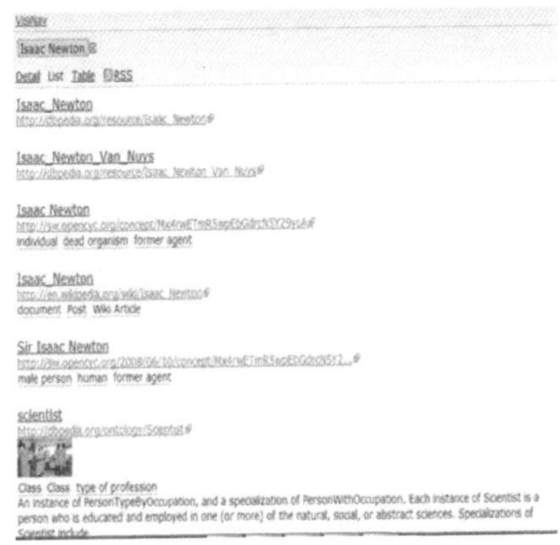

Fig. 13. Search results for keyword 'Issac Newton' in VisiNav in List view [9].

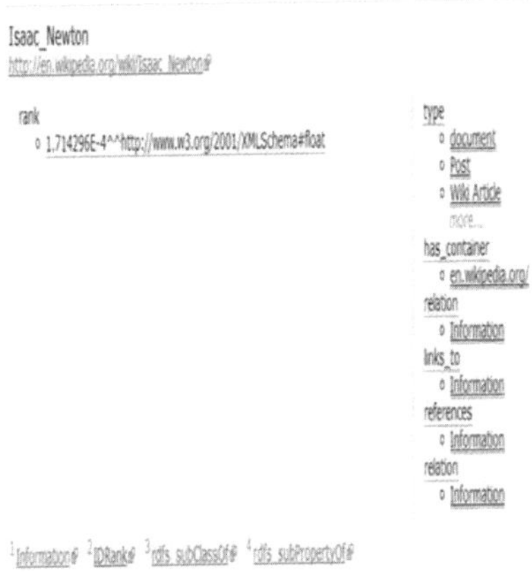

Fig. 14. Detailed view for fourth result in Fig. 13 [9].

	rank
Isaac_Newton	1.714296E-4

Fig. 15. Table view for fourth result in Fig. 13 [9].

Detailed view of search result in VisiNav shows rank of page in left pane as shown in Fig. 14. In the right pane it will show type of article, container of the page, the page is having links to what type of resources, references etc.

Table view of result will give rank of page in tabular format as shown in Fig. 15. User can also download this view in .CSV format.

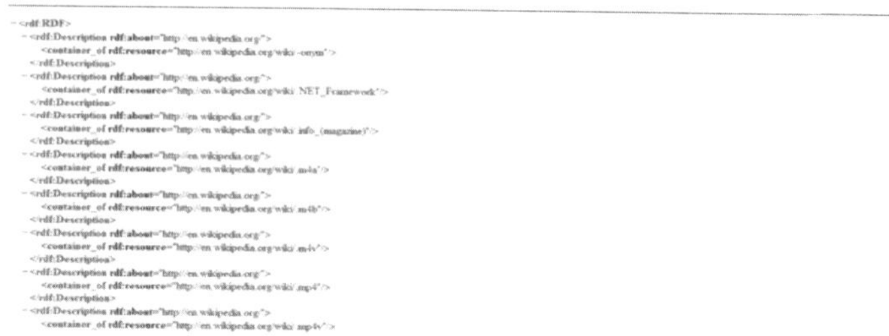

Fig. 16. RDF view of fourth result in Fig. 13 [9].

User can get RDF for search result in RDF view of search Result as shown in Fig. 16.

As VisiNav provides many options to view and navigate through results, it is called visual Data Search and Navigation (VisiNav). Clearly results obtained by VisiNav are different and allows many options than Google. But still VisiNav is not full proof search engine like Google as google gives exhaustive search results filtered using well tested filters.

8. Conclusion

Although lot of improvement needs to be made in VisiNav, it provides strong foundation for improved search based on semantic web methodology. It definitely improves user experience in query construction and retrieving results. Google by far is best search engine as far as result retrieval is concerned. So this paper attempts to address working, concepts and results given by both search engines.

References

1. Kumar Abhishek, Saurabh Kulkarni, Vipin Kumar, N. Archana, P. Kumar, “A Review on Personalized Information Recommendation System Using Collaborative Filtering”, *IJCSIT*, **2**, 2011, 1272.
2. *http://www.searchenginehistory.com/.*
3. *http://www.lycos.com/.*
4. *http://docs.yahoo.com/info/misc/history.html.*
5. *http://www.google.com/about/corporate/company/history.html.*
6. Divakar Yadav, A. K. Sharma, J. P. Gupta, Users search trends on www and their analysis, *Proceedings of the First International Conference on Intelligent Interactive Technologies and Multimedia*, 59 (2010).
7. *http://www.google.com.*
8. *http://pr.efactory.de/e-pagerank-algorithm.shtml.*

9. *http://sw.deri.org/2009/01/visinav.*
10. Andreas harth, VisiNav: Visual Web Data Search And navigation, *Web Semantics: Science, Services and Agents on the World Wide Web*, 8, 348 (2010).
11. *http://www.seobook.com/relevancy/.*
12. *http://www.yahoo.com.*
13. *http://www.ask.com/.*
14. *http://www.readwriteweb.com/archives/ask_what_differentiates_them_from_google.php*
15. *http://www.bing.com.*
16. *http://www.webworkshop.net/pagerank.html.*

A Survey on Social Networks: Issues and Attacks

Anubha Maurya
Computer Science and Engineering, National Institute of Technology Patna
Patna, Bihar-80001, India
E-mail: anubhamaurya@nitp.ac.in
www.nitp.ac.in

M. P. Singh
Computer Science and Engineering, National Institute of Technology Patna
Patna, Bihar-80001, India
E-mail: mps@nitp.ac.in

The importance of social network is increasing day by day as it is becoming a popular and convenient medium for communication between the social users. It enables user to share their profile data, ideas, videos and any content they have with them. It gives public developer opportunity to use it and extend applications of social network service. The people associated with social network are not much aware about the security threats and risk involved with sharing their data. They trust the network and easily avail their data to them. They reduce the original privacy alerts for their convenience and hence make it easier for attackers to attack. The users awareness about these risks and challenges will protect the loss of private and personal information. This survey paper examines the issues related to the social networks and several possible ways for an attacker to attack.

Keywords: Social Network; Security Issues; Privacy Issues; Trust Issues; Social Attacks.

1. Introduction

In recent years, social network usage has increased sharply as these networks facilitate communication and become a spot of virtual meeting. Social networks like Twitter, Facebook, Google+, LinkedIn play an important role in people's daily life. Through these social networking sites users communicate to known as well as unknown person. To unknown person first a trust is build and then a relationship is formed. Social networking sites give numerous applications to interact and build relationship with friends and third party as well [1]. The third party communication can be done in two ways:

(i) Application accessed directly by the user;
(ii) An application launched in collaboration with the social network service provider.

In both ways sensitive data is revealed and used to jeopardize the user. In this survey, the focus is on the security and privacy aspects that can arise due to misuse of published information by the user. It aims to give a collaborative view of social networking, related issues and attacks possible on these social networks.

Organization of this paper is as follows: Section 2 discusses what social networking is and classification of users different types of data present in social network. Section 3 describes the disclosure of huge amount of data that give rise to many social issues such as security, privacy and trust. Section 4 reviews the exploitation possible on social network due to availability of user's personal information. Section 5 discuss about the attacks possible on Social Network because of leakage of information and invaders action. And finally, Section 6 concludes the survey.

2. Social Networking

To build connection with friends, relatives, co-workers and family, users use social network. Social network can be defined as the place where users can interact with others by creating their own profile. The profile generally includes the information in the form of name, age, contact information, videos, blogs, interest, group information, photos and many other attributes [2]. Social Networking provides services to the users which reflects the social network relations among the users. These services have accumulated several resources to provide a better service to the users so that they can get easily what they want. It also provides the service of instant messaging where user can write messages to their friends. Social network according to [3] is the place where users have personal space for them, have social connection facility, means of communication to stay connected with others and the social graph which shows how one is related to the other. All this functions give availability of data. Hence the user data present on network is comprised of three types of information [1] as shown in Fig. 1:

i. Identity data
ii. Social graph data
iii. Content data

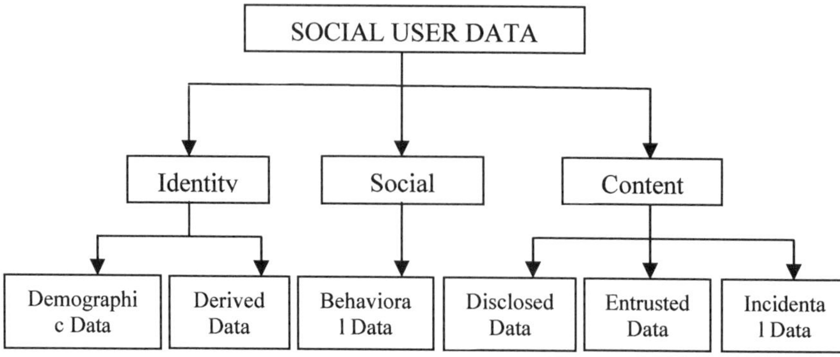

Fig. 1. Types of user data in social network.

2.1. *Identity data*

It is the data related to user identity. It informs outer world about the user, who the user is. It includes the profile data, user's identity information and the privacy policy. It can be further classified as: Demographic Data and Derived Data.

Demographic data is the data which users provide to networking sites for use to refer them such as name, date of birth, credit card number [4]. Derived data is the data that is derived from the information that is being provided by the user to sites like age etc.

2.2. *Social graph data*

It is the data which helps user and person associated to user, know the social world around them. It tells the relationship between the users like friendship connections in form of friends, family, co-workers. Social graph helps to collect behavioral data in form of links users create with others and activities performed between them [5].

2.3. *Content data*

It is the data that represents what user owns and have in social world. It includes user's photo, messages, status updates and videos. It can be further classified as Disclosed data, Entrusted data and Incidental Data.

Disclosed data is in the form of users update on their social webpage. Whatever they update, they have access control over it. Entrusted data is the data that user post on other's page like comments, likes, photos etc. This data once posted, user do not have access control over it. Incidental Data is the data that other people post on user's social webpage. In this data type also user cannot control others from posting [5].

3. Issues Related to Social Networking

Different social networking sites give different type of accessing capability to users for information present over network. Some information can be made private and some information is always public. Therefore controlling access to the user data is a difficult task and even user sometimes does not have control on it [2]. In [6], notion of privacy bits concept is introduced. In this privacy control bits are divided into bare minimum and supremum class. The access of data will depend on these bits and user grants permission for this. But still, illegal access and disclosure of user's data leads to issues in social network in form of security issues, privacy issues and trust related issues [7].

3.1. *Security issues*

Security is the confidentiality, integrity and availability of data. It includes both the physical access and logical security issues. The data stored should be safe from unauthorized access and should be reliable and accurate. Recent concern in field of security for social network user is the harvesting of information by both social networking sites and third party companies. Mobile applications for social networking are increasing and give an opportunity to be more close to our social world. But at the same time it increases the security threats for collection of users data, such as, location data and users habits can easily be identified. This helps third parties for advertisement [8] and misuse of this information.

3.2. *Privacy issues*

Privacy says right to be let alone along with right to be left in secret. The personal information available on social websites is included in privacy threats as it needs to be kept secret [9]. Illegal disclosure and misuse of this information can lead damage to user's lives. User's personal space privacy and communication privacy are the main issues as crawling the search engine the profile information from some Social networking sites get available whether users know it or not [3]. In [10], the authors describe about the attack using learning algorithm on the available data of user to predict the private information about them. Privacy basically deals with 'which information' of user is 'going where' as their identity, location or the activity list.

3.3. *Trust issues*

Trust is defined as the estimate how much a user trusts its neighbor user. It also assures that the function, process or human will behave in the expected manner in future. Building trust depends upon friend's recommendation and initial trust of user on the trustee. Trust is a critical factor in sharing and creating information among communities [11]. Sometimes even the service provider are unaware of what data the third party requires for running the application and on trust basis users all data gets revealed [12]. Issue arises how to distinguish between the newcomer (the real person) and the misbehaver on basis of trust value. The social network rely on the trust value to create new relationship.

4. Possible Exploits On Social Networking Sites

The information of users present over social network is vast in amount and online social network service providers store this information for their use [12]. Major online platforms are open Application Programming Interface (API) and allow third party to directly interact with them. On behalf of the users, these API's provide mechanism to read write or update users data by third party application [13]. Sometimes to provide richer content these networking sites share this information to third party sites without informing the users. Such abundant information availability gives rise to attack. Group of users having same social interests develop trust relationship. They try to access same resource at a time where issue of collision arises and attacks occur. Another way to collect a user information given by [14] is that, if a user have multiple accounts on different sites then relative information can be gathered from there about their accounts. A table of relevant information is created by collection distinct information about the victim from different sites.

The users are generally not aware of the security risks associated with the use of social networking sites. They easily trust the social network content [4], [11] and give their information to them. Leakage of information, the information that is explicitly not available but rather inferred from relationship about the user is an issue which threats the user privacy and security. Here is the list of few possible threats to user account: malicious propagation, spamming of messages, privacy violation, disk compromise, context-aware spam, social phishing and identity anonymity. In Section V the attacks are elaborated in detail.

5. Attacks on Social Networking Sites

Users irresponsible behavior and lack of security awareness gives attackers the space to misuse the social networking sites. They gain the sensitive information from the user's account and misuse it [9], [15]. Only leakage of information is not the source of availability of information to the outside world. But the attack techniques can also be used to acquire information of the social users.

In [9] authors have classified the attacks as direct attack techniques and malicious attacks. Direct attack includes DDoS attack and Sybil attack.

The *DDoS (Distributed Denial of Service) attack* delivers the huge amount of request for a web resource simultaneously so as that web resource becomes unavailable to users'.

The *Sybil attack* is serious threat to privacy of users and generally occur in peer to peer networks. In this attack fake profile of the user is created by the attacker to create relationship with the friends of user and gain sensitive information about them [16].

Malicious Attack defined as malware, is a software that is designed to collect users credential data (private data) from their computers by interrupting computer operations. It propagates itself from one user to another user (person in his friend list) in social network because of the property of frequent interaction among them [8], [12].

In [17], an attack known as *friend – in – the – middle attack* is explained which occurs at network layer. It captures the communication link by hijacking the session cookies between users and social networking providers and impersonates the user and interacts with social network without proper authorization.

A special attack in [13], on the use of extension for same application most of the time is discussed. Attackers in the hope of increasing the frequency of their decision use the extension for same application.

Socio-technical attacks are used to acquire information that are not meant to be announced in public. It includes stealing of account password, financial details and used to discredit person or company [15].

Social graphs are now days more prone to attacks in form of *Identity theft and social link forging*. Identity theft is cloning the identity of a user for profit from their reputation. Social link forging is to create relationship with a person to influence and win their trust them by fooling them with a fake relationship with reputed person [3].

In [18], *neighborhood attack* is described as the attack on concept that if attacker knows the friends of the victim and their relationship, then identification of the victim can be easily done. Social graph is used to explore this relationship.

Cross-site scripting attack (XSS) hits the existing vulnerabilities to expand its effect in social network. XSS is a worm which propagates in the network on the basis of three factors [19]:

i. User behavior of visiting others profile,
ii. Structure of community i.e. how much clustered it is and
iii. Community size.

Clickjacking attack is the attack in which users click a link for some intended purpose but it gets forwarded to some other site created by the attacker and by this they post spam messages in the user profile on their behalf [8].

Infilteration Attack is the one in which the victim is caught by sending a friend request from a fake account. As the victim accepts the friend request by default his all details by passing the privacy policy are shared with the attacker [20].

Social phishing attack is the one in which the fraud acts as trusted third party to the user. They acquire sensitive and personal information of the user on the basis of their past history and preferences of websites they use [21].

URL shortening [8] is the process used when there is a link to be posted on site and its URL length is very long and making it difficult to post in a small space. But this can become a security risk when the user don't know that at which link he is being forwarded after clicking this URL. Before it can be realized the user is now under attack of malware.

De-Anonymization Attack is the attack in which victim's real identity is captured. Users aware of risks associated with revealing personal information provides pseudonyms for their identity. But attackers by using cookies, network topology and user's group information uncover the real identity of users [8], [12].

6. Conclusion

Social Networking Sites are playing an important role in the Internet era. Users spend more time on social network than on any other activity. Social network has both advantage and disadvantage associated with it. This paper analyzes different aspects of social networks for identifying the dark side i.e. issues related to the network. Since users are not afraid of providing their personal information on these networks such as real name, date of birth, location but they don't want it to be published in public. But service providers for better service give this information to third party. This way they are publishing usersinformation in public without their permission and gives possibility of attacks on users profile and related data. Hence as much as possible, the service provider should take in account the security and privacy breaches associated with the users details.

References

1. Moo Nam Ko, Gorrell P. Cheek, Mohamed Shehab, Ravi Sandhu, Social-Networks Connect Services, *IEEE Computer Society*, 37 (August 2010).
2. Mohamed Shehab, Anna Squicciarini, Gail-Joon Ahn, Irini Kokkinou, Access Control for Online Social Networks Third party Applications, *Computers and Security*, **31**, 897 (2012).
3. Chi Zhang and Jinyuan Sun, Xiaoyan Zhu, Yuguang Fang, Privacy and Security for Online Social Networks: Challenges and Opportunities, *Network, IEEE*, **24**, 13 (2010).
4. Ai Ho, Abdou Maiga, Esma Aimeur, Privacy Protection Issues in Social Networking Sites, *IEEE/ACS International Conference on, Computer Systems and Applications, (AICCSA 2009)*, 271 (10–13 May 2009).
5. Bruce Schneier, A Taxonomy of Social Networking Data, *IEEE Computer And Relaibility Socities*, (2010).
6. B. Krishnamurthy, C. E.Wills, Characterizing Privacy in Online Social Networks, *Workshop on Online Social Networks (WOS, SIGCOMM 2008), ACM*, pp. 37–42, (2008).
7. S. T. R. Rajam, S. B. R. Kumar, Security and Privacy issues in Social Network Services: An Overview, *International Journal of Innovative Research in Electrical, Electronics, Instrumentation and Control Engineering 2014, IJIREEICE*, **2**, (2014).
8. Michael Fire, Roy Goldschmidt, and Yuval Elovici, Online Social Networks: Threats and Solutions, *IEEE Communication Surveys & Tutorials* **16**, 2019 (2014).
9. Xi Chen, K. Michael, Privacy Issues and Solutions in Social Network Sites, *Technology and Society Magazine, IEEE*, **31**, 43 (2012).
10. Raymond Heatherly, Murat Kantarcioglu and Bhavani Thuraisingham, Preventing Private Information Inference Attacks on Social Networks, *IEEE Transactions On Knowledge And Data Engineering*, **25**, 1849 (2013).
11. C. Vorakulpipat, A. Marks, Y. Rezgui, S. Siwamogsatham, Security and privacy issues in Social Networking sites from user's viewpoint, *Technology Management in the Energy Smart World (PICMET), 2011 Proceedings of PICMET '11*, 1, (2011).

12. H. Gao, J. Hu, T. Huang, J. Wang, Y. Chen, Security Issues in Online Social Networks, *Internet Computing, IEEE*, **15**, 56 (2011).
13. Mohammed Shehab, Said Marouf, Recommendation Models for Open Authorization, *IEEE Transactions on Dependable and Secure Computing*, **9** 583 (2012).
14. Danesh Irani, Steve Webb, Calton Pu and Kang Li, Modeling Unintended Personal-Information Leakage from Multiple Online Social Networks, *Internet Computing, IEEE*, **15**, 13 (2011).
15. J. Nagy, P. Pecho, *Third International Conference on* Social Networks Security, *Emerging Security Information, Systems and Technologies, SECURWARE '09*, IEEE, 321, (2009).
16. L. A. Cutillo, R. Molva, M. Onen, Analysis of Privacy in Online Social Networks from the Graph Theory Perspective, *Global Telecommunications Conference (GLOBECOM 2011), IEEE*, 1, (2011).
17. Markus Huber, Martin Mulazzani, Gerhard Kitzler, Sigrun Goluch, Edgar Weippl, Friend-in-the-Middle Attacks Exploiting Social Networking Sites for Spam, *IEEE Internet Computing*, 28 (2011).
18. Bin Zhou, Jian Pei, Preserving Privacy in Social Networks Against Neighborhood Attacks, *Data Engineering, 2008, ICDE 2008, IEEE 24th International Conference on*, pp. 506–515, 7 (2008).
19. Mohammad Reza Faghani, UyenTrang Nguyen, A Study of XSS Worm Propagation and Detection Mechanisms in Online Social Networks, *IEEE Transactions On Information Forensics And Security*, **8**, 11, 1815 (2013).
20. Bogdan Carbunar, Mahmudur Rahman, Niki Pissinou, Athanasios V. Vasilakos, A Survey of Privacy Vulnerabilities and Defenses in GeoSocial Networks, *IEEE Communications Magazine*, pp. 114 (2013).
21. P. Joshi, Kuo, C.-C. J., Security and privacy in online social networks: A survey, *2011 IEEE International Conference on Multimedia and Expo (ICME)*, 1 (2011).

Reduced Rule for Banknote Genuinity

Chhotu Kumar[†] and Anil Kumar Dudyala

Dept. of Computer Science, NIT, Patna, Patna, India

E-mail: [†]chhotu.k31@gmail.com, ak@nitp.ac.in

This paper shows the reduced rule can be nearly efficient to without pruned decision tree rules for checking the genuinity of banknotes (currency). Pruning is required to minimize the decision tree, its additional nodes and rules, which is irrelevant to reduce the error-rate for classification. Pruning reduces the complexity and easy to understand. In this paper Pruned decision tree rules decides the classification of banknotes whether it is genuine or fake with a good accuracy. Sensitivity, accuracy, specificity are the metrics applied to check the performance of all experiments. All the conducted experiments have been done using cart. The experimental results obtained by pruned decision tree rules and without decision tree rules are compared also. We also put the extracted rule by which it perfectly explains that how it comes to a better prediction accuracy. Its major aim to show by decision tree and its extracted rule how fake notes and genuine notes have been classified.

Keywords: Banknote Genuinity; CART; Decision Tree; Pruned Decision tree rules; Without pruned decision tree rules.

1. Introduction

Authenticating banknotes (currency) is a hot topic in financial market about its genuinity and avoiding fraud during marketing exchange and in doing transactions. The accuracy of the tool for predicting banknotes genuinity is much needed and important because many of the miscreants try to induce fake notes in the financial market using advanced copy machine or scanner. Financial market, banks, country, people are affected by the fake notes. It is difficult to identify which note (currency) is genuine and which one is fake especially when the notes are in a Lot and have similar features. Hence there is a need for a system which allows the user to identify whether the given set of notes are genuine or not based on their features. This need gave us a scope to propose an efficient system for banknote authentication using pruned decision tree rules. Pruned decision tree rules is an optimal way to predicting banknotes genuinity in recent. Pruning is the process for reducing the decision tree from the bottom of the tree by removing some of the subtrees, which is not important in prediction.

Pruning should be used for introducing simplicity provide accuracy remain nearly same. The tree generated by using without pruned decision tree rules may contain over-fitting and their result to predict the test data may not always right. So, the decision tree should be pruned to reduce the over-fitting. Here, pruned decision tree learning algorithm such as CART [6] and pruned rules generated by it performs optimally and nearly same to a good classifier. We also put the extracted rule by which it perfectly explains that how it comes to a better prediction accuracy. Its major aim toshow by decision tree and its extracted rule how fake notes and genuine notes classified.

The remaining sections are arranged as follows. Section 2 deals with literature survey corresponding to Reduced Rule For Banknote Genuinity used in the similar field. Section 3 gives the dataset description. Methodology used for problem solving is discussed in Section 4. Various machine learning techniques applied for solving this problem are explained briefly in Section 5. Result and discussion are dealt in Section 6. Last section discusses about conclusion and future work.

2. Literature Survey

Preserving genuinity of higher denomination printed banknotes is one of the critical issues. It has major role in financial activities of every country. Aoba *et al.* used three layered perceptron and RBF networks for Euro banknote recognition. They have used various Euro denominations with various RBF kernels [1]. Zhou *et al.* has done the performance evaluation of corporate bankruptcy prediction models on imbalanced dataset by applying sampling methods [2]. Yu *et al.* applied Leave-One-Out incremental extreme machine learning for bankruptcy prediction and has given specific financial indicators [3]. Guangli *et al.* has applied decision tree and logistic regression for forecasting credit card churners. They concluded that LR performs better than decision tree for the real Chinese bank dataset [4].

Kawamura *et al.* applied decision tree to predict whether or not to cure a new patient who have HCV (Hepatitis C Virus) using a decision tree model working by evaluation of SNPs (Single nucleotide polymorphism) in a genome-wide association study [8]. Lee *et al.* applied decision tree to predict people's locations with the help of observing people's activity. They have applied nearly 6000 datasets and done experiments on WEKA tool [9]. Son *et al.* applied decision tree to design a model to predict characteristics and knowledge associated with congestive heart failure for early diagnosis [12]. Wang *et al.*

applied ID3 to form a decision tree based on information gain and mined valuables purchasing associations [13].

Zhang *et al.* applied logistic regression and artificial neural network for bankruptcy prediction. They have shown that neural networks perform significantly better than logistic regression models [10]. Sun *et al.* applied support vector machine for financial distress prediction (FDP) for studying the performance of the SVM when used individually and in ensemble mode. The outcome of the above paper indicates that SVM ensemble is significantly superior to individual SVM classifier [11]. Kim *et al.* applied both ensemble methods bagging and boosting for improving the performance of traditional neural network in bankruptcy prediction and concluded that the ensemble bagged and boosted neural networks performance is better than traditional neural networks [15].

Danenas *et al.* applied linear SVM for credit risk evaluation to show that it performs better than logistic regression and RBF network in terms of accuracy and identification of each class [16]. Boyacioglu *et al.* applied MLP, self-organizing map (SOM), learning vector quantization (LVQ), competitive learning, multivariate discriminant, k-means cluster and logistic regression for predicting bank financial failures and concluded that MLP and LVQ performs well in terms of accuracy [17].

Azad *et al.* analysed with three approaches for extracting decision rules. They extracted directly from rows, using paths and attributes [19]. Trabelsi *et al.* applied pruning on belief decision tree (BDT). They found interesting results of pruned BDT compared to without pruned BDT [20]. Baesens *et al.* applied neural network to evaluate credit risk. They applied Neurorule, Trepan, Nefclass on three datasets. They concluded Neurorule and Trepan perform better to C4.5 algorithm and logistic regression classifier. They described how decision trees applied for extracting rules [21]. Nikita *et al.* applied various pruning techniques like post-pruning and pre-pruning. They conduct experiments on two datasets (diabetes and glass) using weka. They concluded size of decision tree reduces when minobj increases accuracy also increases [22].

3. Dataset Description

The dataset used for carrying out the experiments is taken from UCI machine learning repository [5]. The dataset is owned by Volker Lohweg (University of Applied Sciences) and donated by Helene Darksen (University of Applied Sciences). The dataset has five attributes out of which four attributes are real valued attributes and one attribute is target attribute. There are total 1372 instances. The dataset contains balanced ratio of both classes which is 55:45 (genuine: counterfeit). The target class contains two values namely 0 & 1, where 0 represents genuine banknotes and 1 represents fake banknotes. The description of the attributes is as follows.

Table 1. Dataset description.

Attribute Name	Value Type	Description
Variance	Double	It is a measure of the 'spread' of a distribution about its average value.
Skewness	Double	Skewness tells about the direction of variation of the lack of symmetry.
Kurtosis	Double	Kurtosis is a parameter that describes the peakedness of distribution.
Entropy	Double	Image entropy is the amount of information which must be coded for, by a compression algorithm.
Class	Nominal	Class contains two values namely 0 and 1 where 0 represents genuine banknotes and 1 represents fake banknotes.

4. Problem Solving Methodology & Experimental Setup

The experimental models have been setup using two methods, namely hold-out and 10 fold cross validation. Hold-out method is one in which the data set is separated into two subsets (say 70:30 ratio) called the training set and the test set respectively. The training set is used to train the classifier while the testing set is used to estimate the error rate of the trained classifier. Ten fold cross validation is another method in which the dataset is partitioned into 10 subsets of equal size. The algorithm is trained using nine subsets and is tested with remaining one subset. In the next iteration the subset which is used for testing is mutually substituted with one of the nine training subsets. This process is repeated until all the subsets have been trained. CART [6] is used for making the experimental results.

As in this paper, all experiments have been done for prediction on categorical variable, so here CART [6] builds the classification tree only.

CART [6] develops binary tree where each parent node has only two child nodes, one is left child node and other one is right child node. According to definition, the leaf nodes have a less percentage of impurity in comparison to root node.

From the structural view of decision tree, we have extracted decision tree rules. The decision tree rules can be analyzed as conditional decision like if-else statements.

4.1. *Performance measure*

Following metrics have been used for measuring the performance of the proposed system.

Sensitivity is the measure of how much ratio of target class is correctly classified by the particular classifier. It is given as

$$\textbf{Sensitivity} = tp / (tp + fn) \quad (1)$$

Specificity is the measure of how much ratio of non-target class is correctly classified by the particular classifier. It is given as

$$\textbf{Specificity} = tn / (tn + fp) \quad (2)$$

Accuracy is the measure of how much ratio of whole dataset (including both target and non-target class) is correctly classified by the particular classifier. It is given as

$$\textbf{Accuracy} = (tp + tn) / (tp + tn + fp + fn) \quad (3)$$

Where,

tp= true positive (target class correctly classified)
tn= true negative (target class wrongly classified)
fp= false positive (non-target class correctly classified)
fn= false negative (non-target class wrongly classified)
All these metrics are given in percentages in the result tables.

5. Machine Learning Techniques

Various machine learning techniques used in this paper are as follows

5.1. *Decision tree (DT)*

A decision tree is a classifier expressed as a recursive partition of the instance space. The decision tree consists of nodes that form a rooted tree, meaning it is a directed tree with a node called "root" that has no incoming edges. All other nodes have exactly one incoming edge. A node with outgoing edges is called an internal or test node. All other nodes are called leaves (also known as terminal or decision nodes). In a decision tree, each internal node splits the instance space into two or more sub-spaces according to a certain discrete function of the input attributes values. In the simplest and most frequent case, each test considers a single attribute, such that the instance space is partitioned according to the attribute's value. In the case of numeric attributes, the condition refers to a range.

Each leaf is assigned to one class representing the most appropriate target value. Alternatively, the leaf may hold a probability vector indicating the probability of the target attribute having a certain value. Instances are classified by navigating them from the root of the tree down to a leaf, according to the outcome of the tests along the path [4].

6. Result & Discussion

The experiments yielded huge sets of result as an effect of lot of parameters tuning. Only the top performing set of the pruned decision tree rules and without pruned decision tree rules using both methodology is displayed as a result in Table 2 and Table 3. The top performing techniques are represented in bold and italic font in the table.

Table 2. Result of without pruned decision tree rules.

Methodology	Accuracy	Sensitvity	Secificity
Hold-Out	99.12	100	98.93
Cross-Validation	98.82	98.36	98.52

Table 2 summarizes the results of pruned decision tree rules used under hold-out method and cross validation method. It is quite evident from the above table that hold-out method in pruned decision tree has topped over all with respect to sensitivity, accuracy and specificity.

As any predictor's goal is to get high sensitivity, decision tree fulfills this need for our problem earlier. Hence the rules yielded by without pruned and pruned decision tree are extracted and represented in human understandable format shown respectively at appendix A and appendix B. We also present the pruned decision tree in appendix C which shows how the classification of fake notes and genuine notes have been done. Thus obtained rules are coded into a program to check their accuracy. This program is tested with hold-out data and tenfold cross-validation data in both pruned and without pruned decision tree rules. The top performing set in pruned decision tree rules are represented in bold and italic font in the Table 3. The rules can be used for identifying the fake notes with the given parameters.

Table 3. Result of pruned decision tree rules.

Methodology	Accuracy	Sensitvity	Secificity
Hold-Out	94.27	83.78	92.26
Cross-Validation	94.09	84.26	91.95

Table 3 summarizes the results of without pruned decision tree rules used under hold-out method and cross validation method. It is quite evident from the above table that hold-out method in without pruned decision tree has topped with respect to accuracy and specificity while cross validation has topped with respect to sensitivity.

Table 4. Result of pruned decision tree rules using coding.

Methodology	Accuracy	Sensitvity	Secificity
Hold-Out(70:30)	100	99.8361	99.9271
Cross-Validation	98.275	98.526	98.394

Table 4 summarizes the result of using coding on extracted pruned decision tree rules for both hold-out and ten-fold cross validation. We can see that these rules have given good result for both the methods. It can be observed from the result that the sensitivity obtained is 100 percent for hold-out data and 98.526 percent for the 10-fold cross validation data which is quite good. Hence, these rules can be used confidently for identifying the fake notes with given the bank note features.

7. Conclusion & Future Work

We have used pruned and without pruned decision tree rules for banknotes authentication in two methodologies to find optimal and less complex classifier. Exhaustive experiments have been conducted and the best performance result of each methodology is presented in the results table. It has been found that overall, hold-out method has given the competently best result in both pruned and without pruned decision tree rules. We have used coding also on extracted pruned decision tree rules, thus the result given by rules have quite good sensitivity of 99.8361 percent in hold-out method, while 98.275 percent in cross-validation method. The rules given by both pruned and without pruned decision tree can be used to find out whether the given note is fake or not just by applying them. In future, this work can be extended by reducing the rules obtained with the help of pruning along with not compromise with accuracy, sensitivity and specificity and an ensemble method can be developed for bank note authentication.

References

1. M. Aoba, T. Kikuchi, Y. Takefuzi, Euro banknote recognition system using a three layered perceptron and rbf networks, IPSJ Transactions on mathematical modelling and its applications, vol. 44 (May 2003) No. SIG 7 (TOM 8). Machine learning and cybernetics, 2260–2264.

2. Ligang Zhou, Performance of corporate bankruptcy prediction models on imbalanced dataset: The effect of sampling methods, Knowledge-Based Systems 41 (2013) 16–25.
3. Qi Yu, Yoan Miche, Eric Severin, Amaury Lendasse, Bankruptcy prediction using extreme learning machine and financial expertise, Neuro-computing 128 (2014) 296–302.
4. Guangli Nie, Wei Rowe, Lingling Zhang, Yingjie Tian, Yong Shi, Credit card churn forecasting by logistic regression and decision tree, Expert System with Application 38 (2011) 15273–15285.
5. https://archive.ics.uci.edu/ml/machine-learning data bases/00267/data_ banknote_authentication.txt accessed on (27/09/2014).
6. http://www.salford-systems.com/products/cart#download/win32exe accessed on (15/08/2014).
7. Laura Garach Morcilloa, Francisco Javier Calvo Poyoa, Griselda López Maldonado, Using Decision Trees for Comparing Different Consistency Models, Procedia - Social and Behavioral Sciences 160 (2014) 332–341.
8. Yoshihiro Kawamura, Shigeru Takasaki, Masashi Mizokami, Using decision tree learning to predict the responsiveness of hepatitis C patient to drug treatment, FEBS Open Bio 2 (2012) 98–102.
9. Jae Sung Lee, Eun Sung Lee, Exploring the Usefulness of a Decision Tree in Predicting People's Locations, Procedia - Social and Behavioral Sciences 140 (2014) 447–451.
10. Guoqiang Zhang, Michael Y. Hu, B. Eddy Patuwo, Daniel C. Indro, Artificial neural networks in bankruptcy prediction: General framework and cross-validation analysis, European Journal of Operation Research 116 (1999) 16–32.
11. Jie Sun, Hui Li, Financial distress prediction using support vector machines: Ensenble vs. individual, Applied Soft Computing 12 (2012) 2254–2265.
12. Chang-Sik Son, Yoon-Nyun Kima, Hyung-Seop Kimb, Hyoung-Seob Park, Min-Soo Kim, Decision-making model for early diagnosis of congestive heart failure using rough set and decision tree approaches, Journal of Biomedical Informatics 45 (2012) 999–1008.
13. Wang Xiaohu, Wang Lele, Li Nianfeng, An Application of Decision Tree Based on ID3, Physics Procedia 25 (2012) 1017–1021.
14. Chiara Gigliaranoa, Silvia Figini, Pietro Muliere, Making classifier performance comparisons when ROC curves intersect, Computational Statistics and Data Analysis 77 (2014) 300–312.

15. Myoung-Jong Kim, Dae-Ki Kang, Ensemble with neural networks for bankruptcy prediction, Expert Systems with Applications 37 (2010) 3373–3379.
16. Paulius Danenas, Gintautas Garsva, Selection of Support Vector Machines based classifiers for credit risk domain, Expert Systems with Applications 42 (2015) 3194–3204.
17. Melek Acar Boyacioglu, Yakup Kara, Omer Kaan Baykan, Predicting bank financial failures using neural networks, support vector machines and multivariate statistical methods: A comparative analysis in the sample of savings deposit insurance fund (SDIF) transferred banks in Turkey, Expert Systems with Applications 36 (2009) 3355–3366.
18. Seim Timung, Tapas K. Mandal, Prediction of flow pattern of gas–liquid flow through circular microchannel using probabilistic neural network, Applied Soft Computing 13 (2013) 1674–1685.
19. Mohammad Azada, Beata Zieloskoa, Mikhail Moshkova, Igor Chikalov, Decision Rules, Trees and Tests for Tables with Many-Valued Decisions Comparative Study, Procedia Computer Science 22 (2013) 87−94.
20. Salsabil Trabelsi, Zied Elouedi, Khaled Mellouli, Pruning belief decision tree methods in averaging and conjunctive approaches, International Journal of Approximate Reasoning 46 (2007) 568−595.
21. Bart Baesens, Rudy Setiono, Christophe Mues, Jan Vanthienen, Using Neural Network Rule Extraction and Decision Tables for Credit-Risk Evaluation, Management Science © 2003 INFORMS Vol. 49, No. 3, March (2003) 312−329.
22. Nikita Patel, Saurabh Upadhyay, Study of Various Decision Tree Pruning Methods with their Empirical Comparison in WEKA, International Journal of Computer Applications (0975-8887) Volume 60 No.12, December 2012.

Appendix A.

SL No.	Rules	Outcome
1	if (a <= 0.3202 && b <= 7.563 && a <= -0.2573 && c <= 6.8536)	Fake
2	if (a <= 0.3202 && b <= 7.563 && a <= -0.2573 && c > 6.8536 && b <= -4.7997)	Fake
3	if (a <= 0.3202 && b <= 7.563 && a <= -0.2573 && c > 6.8536 && b > -4.7997)	Genuine
4	if (a <= 0.3202 && b <= 7.563 && a > -0.2573 &&	Fake

	c <= 1.9045 && b <= 5.5762)	
5	if (a <= 0.3202 && b <= 7.563 && a > -0.2573 && c <= 1.9045 && b > 5.5762)	Genuine
6	if (a <= 0.3202 && b <= 7.563 && a > -0.2573 && c > 1.9045)	Genuine
7	if (a <= 0.3202 && b > 7.563 && a <= -4.726)	Fake
8	if (a <= 0.3202 && b > 7.563 && a > -4.726)	Genuine
9	if (a > 0.3202 && c <= -4.3706 && b <= 7.1918)	Fake
10	if (a > 0.3202 && c <= -4.3706 && b > 7.1918)	Genuine
11	if (a > 0.3202 && c > -4.3706 && a > 1.5922 && a > 2.03660)	Genuine
12	if (a > 0.3202 && c > -4.3706 && a > 1.5922 && a <= 2.0366 && c <= -2.6483)	Fake
13	if (a > 0.3202 && c > -4.3706 && a > 1.5922 && a <= 2.0366 && c > -2.6483)	Genuine
14	if (a > 0.3202 && c > -4.3706 && a <= 1.5922 && c <= -2.2859 && b <= 3.6512)	Fake
15	if (a > 0.3202 && c > -4.3706 && a <= 1.5922 && c <= -2.2859 && b > 3.6512)	Genuine
16	if (a > 0.3202 && c > -4.3706 && a <= 1.5922 && c > -2.2859 && d <= 0.2707)	Genuine
17	if (a > 0.3202 && c > -4.3706 && a <= 1.5922 && c > -2.2859 && d > 0.2707 && c <= 1.9194)	Fake
18	if (a > 0.3202 && c > -4.3706 && a <= 1.5922 && c > -2.2859 && d > 0.2707 && c > 1.9194)	Genuine

Where ‘a’ represents variance, ‘b’ represents skewness, ‘c’ represents kurtosis and ‘d’ represents entropy.

Appendix B.

SL NO.	Rules	Outcome
1	if (a <= 0.32 && b <= 7.57)	Fake
2	if (a <= 0.32 && b > 7.57 && a <= -4.73)	Fake
3	if (a <= 0.32 && b > 7.57 && a > -4.73)	Genuine
4	if (a > 0.32 && c > -4.39)	Genuine
5	if (a > 0.32 && c <= -4.39)	Fake

Where ‘a’ represents variance, ‘b’ represents skewness, ‘c’ represents kurtosis and ‘d’ represents entropy.

Appendix C.

Extracted decision tree using CART.

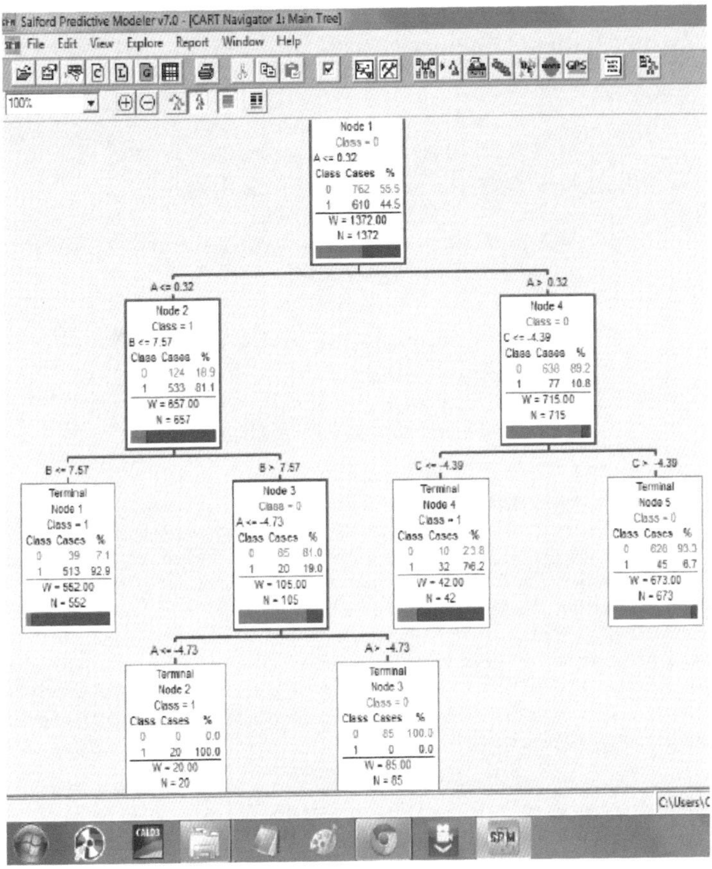

Fig. 1. Extracted decision tree using CART.

A Study on Medical Diagnosis Based on Inter Valued Fuzzy Cluster Analysis

Bhagwan Sahay Meena
Dept. of Computer Science, Assam University Silchar, Silchar, Assam, India
E-mail: bhagwan.sahay.meena@aus.ac.in
www.aus.in

Sharmila Bhattacharjee
Dept. of Computer Science, Assam University, Silchar, Silchar, Assam, India
E-mail: bhattacharjeesharmila2@gmail.com

Inspire of using standardization efforts, medical diagnosis is still considered an art. This status is owed due to the fact that medical diagnosis required a proficiency in coping with uncertainty simply that is not found in today's computing machinery. The diagnostic decision in medicine is frequently encountered with uncertainties. Modeling of this uncertainties in the process of diagnosis of disease under fuzzy environment is an important subject. Various efforts have been made to model then uncertainties in this area through fuzzy sets and its generalizations. The interval-valued fuzzy set is referred to as an i-v fuzzy set. This study is to propose a comparative study for medical diagnosis based on cluster analysis by taking the lower and upper bounds from a given matrix of patients, symptoms, diseases relation.

Keywords: Fuzzy set, fuzzy membership function, fuzzy reference function, membership value matrix, inter value (i-v), Inter value fuzzy matrix (IVFM).

1. Introduction

The basic idea of i-v fuzzy set, and i-v fuzzy relations, the Sanchez's approach [1,2] for S = a set of symptoms, D = a set of diagnosis, P = a set of patients. 'Medical knowledge' is an i-v fuzzy relation R from S to D (i.e. on S x D) expressing relationship between symptoms and diagnoses.

In the present paper, it discuss i-v fuzzy medical diagnoses based on i-v composite fuzzy relations and they correspond to three stages:

- determination of symptoms,
- determination of a medical knowledge (based on i-v fuzzy relations),
- determination of diagnosis (based on i-v composite fuzzy relations);

i.e. all in the sense of interval-valued degree of membership functions.

It is known that the fuzzy relational equation xA = b, where A is a fuzzy matrix and b is a fuzzy vector is consistent, then it has a unique maximum solution and a finite number of minimum solutions.

A study on fuzzy relational equations was introduced by Sanchez [6,7]. Recently the concept of Interval valued fuzzy matrices as a generalization of fuzzy matrix was introduced and developed by extending the max-min operations on fuzzy algebra F = [0, 1], for elements a, b ∈ F.

A new algorithm is proposed to solve the fuzzy relation equation P ∘ Q = R with max–min composition and max–product composition. Here [6,7], it has represented an IVFM A = (aij) = ([aijL , aijU]) where each aij is a subinterval of interval [0, 1], as the Interval matrix A = [AL, AU] whose ijth entry is the interval [aijL, aijU]. Hence the lower limit AL = (aijL) and upper limit AU = (aijU) are fuzzy matrices such that AL ≤ AU. Here, AL and AU are lower and upper limit of matrices of Inter value fuzzy metrics [3,4]. In this study, It has covered two algorithm to solve the Interval valued fuzzy relational equations as a generalization of fuzzy relational equations.

2. Preliminaries

Here we give some basic preliminaries which are basically used in this paper.

Definition 2.1. Let X be a set. A mapping A :X→[0, 1] is called a fuzzy subset of X.

Definition 2.2. Let *X, Y* be two sets. A fuzzy relation R from X to Y is a fuzzy subset of X × Y, characterized by its membership function R: X×Y→[0, 1], A fuzzy relation R from X to Y will be referred to as R (X→Y).

Definition 2.3. Let X be a set. A mapping ~ A : X →D[0, 1] is called an interval-valued fuzzy set of X, where D[0, 1] denotes the subintervals of [0, 1], and AL and AU are two fuzzy sets of X.

Definitions 2.4. For a pair of fuzzy matrices A =(L_{ij}) and B = (U_{ij}) in F_{mn} such that A ≤ B, let us define the interval matrix denoted as [A, B], whose ij^{th} entry is the interval with lower limit L_{ij} and upper limit U_{ij}, is [L_{ij}, U_{ij}].

In scrupulous for A = B, IVFM [A, B] reduces to A ∈B_{mn}.

For A = (a_{ij}) = ([a_{ij}L ,a_{ij} U]) ∈(IVFM)$_{mn}$.

Let us define AL = (a_{ij}L) and AU = (a_{ij} U) clearly AL and AU belongs to F_{mn} such that AL ≤ AU. Therefore an IVFM A can be written as A = [AL, AU].

For A = (a_{ij})m×n = ([a_{ij}L , a_{ij}U]) and B = (b_{ij})n×p = ([b_{ij}L, b_{ij} U]), their product denoted as AB = (C_{ij}) = $\sum_{k=1}^{n}(a_{ik}.b_{kj})$ (1)
where i = 1,2,…,m and j = 1,2,…,p.

$$AB = \left[\sum_{k=1}^{n}(a_{ik}L.b_{kj}L), \sum_{k=1}^{n}(a_{ik}U.b_{kj}U)\right]$$

If A = [AL, AU] and B = [BL, BU] then AB = [ALBL, AUBU].

For $A = (a_{ij})$, $B = (b_{ij}) \in (IVFM)_{mn}$

$A \leq B$ if and only if $a_{ij}L \leq b_{ij}L$ and $a_{ij}U \leq b_{ij}U$ for all i =1,...m and j = 1,2,…n

In particular if $a_{ij}L = a_{ij}U$ and $b_{ij}L = b_{ij}U$ then reduces to the standard max-min composition of fuzzy matrices.

3. Matrix in Medical Diagnosis

In a given sample of medical diagnosis, suppose S is a set of symptoms, D is a set disease and P is a set of patients. Now, S denote the crisp universal set of all symptoms, D be the crisp universal set of all diseases, and P be the crisp universal set of all patients. Let us define a fuzzy relation Rs on the set P x S in which membership grades Rs(p, s) (where p € P, s € S) indicate the degree to which the symptom s is present in patient p. Let us further define a fuzzy relation Ro on the universal set S x D, where Ro(s, d) is (s€ S, d € D) indicates the frequency of occurrence of symptom s with disease d.

Let Re also be a fuzzy relation on the same universal set, where Rc (s, d) corresponds to the degree to which symptom s confirms the presence of disease d .Suppose S is a set of symptoms of certain disease, d is a set of disease and P is a set of patients, construct a relational matrix S over d. Two relation matrices R1 and R2 are implemented by using inter valued-fuzzy matrix. R1 is called symptom-disease matrix and R2 is called non symptom disease matrix.

Another relational matrix Q is created which is called patient-symptom matrix. After computing more relations are calculated T_{1L}, T_{2L}, T_{1U}, T_{2U}, T_{3L}, T_{4L}. Finally, after computing S_{T1} and S_{T2} we will get the result.

3.1. *Algorithm*

Step (1): Input the interval valued fuzzy matrices (F, D) and $(F, D)^C$ where S is a set of symptoms S, and D is the set of diseases. Also write the medical knowledge matrix $R1$ and $R2$ representing the relation matrices of the IVFM (F, D) and $(F, D)^C$ respectively.

$$R2 = 1 - R1 = [1 - R_{IU}, 1 - R_{IL}]$$

Step (2): Input the IVFM ($F1$, S) over the set P of patients and write its relation matrix Q.

Step (3): Compute the relation matrices

- $T_{1L} = Q_L.R_{IL}$ and $T_{1U} = Q_U.R_{IU}$
- $T_{2L} = Q_L.R_{2L}$ and $T_{2U} = Q_U.R_{2U}$ then we get,
- $T_{3L} = Q_L.(J - R_{1L})$ and $T_{3L} = Q_U.(J - R_{1U})$
- $T_{4L} = Q_L.(J - R_{2L})$ and $T_{4L} = Q_U.(J - R_{2U})$

Step (4): Compute S_{T1L}, S_{T1U}, S_{T2L} and S_{T2U}.

Step (5): Compute the diagnosis scores S_{T1} and S_{T2}

- S_{T1}= max {T1(pi,dj),T3(pi,dj)} for i=1,2,3 j=1,2
- S_{T2}=max {T2(pi,dj),T4(pi,dj)} for i=1,2,3 j=1,2

Step (6): Find S_K = max{S_{T1}(pi, dj) − S_{T2}(pi, dj)} then we conclude that the patient pi is suffering from the disease d_k.

Step (7): If S_K has more than one value then go to step one and repeat the process by reassessing the symptoms for the patient.

4. Case Study 1

Step 1:

Let us consider there are three patient's P={Ramesh, Ganesh, Dinesh} in a hospital with symptoms S={temperature, headache, cough and vomiting} problem. Let the possible diseases relating to the above symptoms be Maleria and Dengue. We consider the set S = {s1, s2, s3, s4} as universal set, where *s*1, *s*2, *s*3 and *s*4 represent the symptoms temperature, headache, cough and vomiting respectively and the set D = {d1, d2} where *d*1 and *d*2 represent the parameters malaria and Dengue respectively. Suppose that F (*d*1) = [< *s*1, [.5, 4] >, < *s*2, [.1, .4] >, < *s*3, [.3, .3] >, < *s*4, [.2, .5] >] and F (*d*2) = [< *s*1, [.6, .5] >, < *s*2, [.2, .5] >, < *s*3, [.6, .4] >, < *s*4, [.3, .6] >]. The Interval valued fuzzy matrix (F, D) is a parameterized family [F(*d*1), F(*d*2)] of all interval valued fuzzy matrix over the set S and are determined from expert medical documentation. This interval valued fuzzy matrix (F, D) and its complement $(F,D)^C$ are represented by two relation matrices *R*1 and *R*2 called symptom - disease matrix and non symptom disease matrix respectively given by

$$R1 = \begin{array}{c} \\ S1 \\ S2 \\ S3 \\ S4 \end{array} \begin{array}{c} \begin{array}{cc} d1 & d2 \end{array} \\ \begin{pmatrix} [0.5, 0.4] & [0.6, 0.5] \\ [0.1, 0.4] & [0.2, 0.5] \\ [0.3, 0.3] & [0.6, 0.4] \\ [0.2, 0.5] & [0.3, 0.6] \end{pmatrix} \end{array} \qquad R2 = \begin{array}{c} \\ S1 \\ S2 \\ S3 \\ S4 \end{array} \begin{array}{c} \begin{array}{cc} d1 & d2 \end{array} \\ \begin{pmatrix} [0.6, 0.5] & [0.5, 0.4] \\ [0.6, 0.9] & [0.5, 0.8] \\ [0.7, 0.7] & [0.6, 0.4] \\ [0.5, 0.8] & [0.4, 0.7] \end{pmatrix} \end{array}$$

Now, we calculate the value of R_{1L}, R_{1U}, R_{2L}, R_{2U}.

$$R_{1L} = \begin{array}{c} \\ S1 \\ S2 \\ S3 \\ S4 \end{array} \begin{array}{c} \begin{array}{cc} d1 & d2 \end{array} \\ \begin{pmatrix} 0.5 & 0.6 \\ 0.1 & 0.2 \\ 0.3 & 0.6 \\ 0.2 & 0.3 \end{pmatrix} \end{array} \qquad R_{1U} = \begin{array}{c} \\ S1 \\ S2 \\ S3 \\ S4 \end{array} \begin{array}{c} \begin{array}{cc} d1 & d2 \end{array} \\ \begin{pmatrix} 0.4 & 0.5 \\ 0.4 & 0.5 \\ 0.3 & 0.4 \\ 0.5 & 0.6 \end{pmatrix} \end{array}$$

$$R_{2L} = \begin{array}{c} \\ S1 \\ S2 \\ S3 \\ S4 \end{array} \begin{array}{c} \begin{array}{cc} d1 & d2 \end{array} \\ \begin{pmatrix} 0.6 & 0.5 \\ 0.6 & 0.5 \\ 0.7 & 0.6 \\ 0.5 & 0.4 \end{pmatrix} \end{array} \qquad R_{2U} = \begin{array}{c} \\ S1 \\ S2 \\ S3 \\ S4 \end{array} \begin{array}{c} \begin{array}{cc} d1 & d2 \end{array} \\ \begin{pmatrix} 0.5 & 0.4 \\ 0.9 & 0.8 \\ 0.7 & 0.4 \\ 0.8 & 0.7 \end{pmatrix} \end{array}$$

Step 2:

Again we take P = {P1, P2, P3} as the universal set where $P1$, $P2$ and $P3$ represent patients respectively and S = {s1, s2, s3, s4} as the set of parameters suppose that,

$F1(s1) = [< P1, [.4, .3] >, < P2, [.3, .3] >, < P3, [.4, .2] >]$
$F2(s2) = [< P1, [.5, .4] >, < P2, [.2, .3] >, < P3, [.4, .2] >]$
$F3(s3) = [< P1, [.5, .4] >, < P2, [.4, .4] >, < P3, [.6, .3] >]$ and
$F4(s4) = [< P1, [.6, .5] >, < P2, [.3, .4] >, < P3, [.6, .3] >]$

The Interval valued fuzzy matrix (F1, S) is another parameterized family of all interval valued fuzzy matrices and gives a collections of approximate description of the patient-symptoms in the hospital. This interval valued fuzzy matrix (F1, S) represents a relation matrix Q called patient-symptom matrix given by

$$Q = \begin{array}{c} \\ P1 \\ P2 \\ P3 \end{array} \begin{array}{c} \begin{array}{cccc} S1 & S2 & S3 & S4 \end{array} \\ \begin{pmatrix} [.4, .3] & [.5, .4] & [.5, .4] & [.6, .5] \\ [.3, .3] & [.2, .3] & [.4, .4] & [.3, .4] \\ [.4, .2] & [.4, .2] & [.6, .3] & [.6, .3] \end{pmatrix} \end{array}$$

Now, We calculate,Q=[Q_L, Q_U]

$$Q_L = \begin{array}{c} \\ P1 \\ P2 \\ P3 \end{array} \begin{array}{c} \begin{array}{cccc} S1 & S2 & S3 & S4 \end{array} \\ \begin{pmatrix} .4 & .5 & .5 & .6 \\ .3 & .2 & .4 & .3 \\ .4 & .4 & .6 & .6 \end{pmatrix} \end{array} \qquad Q_U = \begin{array}{c} \\ P1 \\ P2 \\ P3 \end{array} \begin{array}{c} \begin{array}{cccc} S1 & S2 & S3 & S4 \end{array} \\ \begin{pmatrix} .3 & .4 & .4 & .5 \\ .3 & .3 & .4 & .4 \\ .2 & .2 & .3 & .3 \end{pmatrix} \end{array}$$

Step 3:

Now combine the relation matrices R_{1L}, R_{1U} and R_{2L}, R_{2U} with Q_L and Q_U we get the matrices $T1 = [T_{1L}, T_{1U}]$ and $T2 = [T_{2L}, T_{2U}]$.

$$T_{1L} = Q_L.R_{1L} = \begin{array}{c} \\ P1 \\ P2 \\ P3 \end{array} \begin{array}{c} \begin{array}{cc} d1 & d2 \end{array} \\ \begin{pmatrix} 0.5 & 0.8 \\ 0.3 & 0.5 \\ 0.5 & 0.8 \end{pmatrix} \end{array} \qquad T_{1U} = Q_U.R_{1U} = \begin{array}{c} \\ P1 \\ P2 \\ P3 \end{array} \begin{array}{c} \begin{array}{cc} d1 & d2 \end{array} \\ \begin{pmatrix} 0.7 & 0.8 \\ 0.6 & 0.7 \\ 0.4 & 0.5 \end{pmatrix} \end{array}$$

Similarly the value of $T_{2L}[Q_L.R_{1L}]$, $T_{2U}[Q_U.R_{2U}]$ could be calculated.

Now here, it will calculate the value of T_{3L}, T_{4L}. According to the below given relation between Q and R1and R2

$T_{3L} = Q_L.(J - R_{1L}), T_{3U} = Q_U (J - R_{1U})$, where value of J set to 1

$T_{4L} = Q_L. (J - R_{2L}), T_{4U} = Q_U (J - R_{2U})$

$$T_{4L} = Q_L.(J-R_{2L}) = \begin{array}{c} \\ P1 \\ P2 \\ P3 \end{array}\begin{array}{c} \begin{array}{cc} d1 & d2 \end{array} \\ \begin{pmatrix} 0.8 & 1.0 \\ 0.5 & 0.6 \\ 0.8 & 1.0 \end{pmatrix} \end{array}$$

$$T_{4U} = Q_U.(J-R_{2U}) = \begin{array}{c} \\ P1 \\ P2 \\ P3 \end{array}\begin{array}{c} \begin{array}{cc} d1 & d2 \end{array} \\ \begin{pmatrix} 0.4 & 0.6 \\ 0.4 & 0.6 \\ 0.3 & 0.4 \end{pmatrix} \end{array}$$

Step 4:

Now, Compute S_{T1L}, S_{T1U}, S_{T2L} and S_{T2L} from the given relation -

$S_{T1L}=\max(T_{1L},T_{4L})$

$S_{T1U}=\max(T_{1U},T_{4U})$

$S_{T2L}=\max(T_{2L},T_{3L})$

$S_{T1U}=\max(T_{2U},T_{3U})$

$$S_{T1L} = \max(T_{1L}, T_{4L}) = \begin{array}{c} \\ P1 \\ P2 \\ P3 \end{array}\begin{array}{c} \begin{array}{cc} d1 & d2 \end{array} \\ \begin{pmatrix} 0.8 & 1.0 \\ 0.5 & 0.6 \\ 0.8 & 1.0 \end{pmatrix} \end{array}$$

$$S_{T1U} = \max(T_{1U}, T_{4U}) = \begin{array}{c} \\ P1 \\ P2 \\ P3 \end{array}\begin{array}{c} \begin{array}{cc} d1 & d2 \end{array} \\ \begin{pmatrix} 0.7 & 0.8 \\ 0.6 & 0.7 \\ 0.4 & 0.5 \end{pmatrix} \end{array}$$

Similarly the value of $S_{T2L}=\max(T_{2L},T_{3L})$, $S_{T1U}=\max(T_{2U},T_{3U})$ could be calculated.

Step 5:

We calculate the diagnosis score S_{T1} and S_{T2} from the below given relation

$S_{T1} = \max [S_{T1U}(pi, dj), S_{T2L}(pi, dj)]$

$S_{T2} = \max [S_{T1L}(pi, dj), S_{T2U}(pi, dj)]$

Now, we have the difference for patient and the diseases by IVFM Method.

Step 6:

Finally calculate the value of $S = S_{T1} - S_{T2}$

Table 1. Maximum score for relation patient over diseases.

$S_{T1}-S_{T2}$	D1	D2
P1	0.2900	0.1700
P2	-0.1700	-0.1000
P3	0.6600	0.1400

We conclude the patient *P3* and P1, both are suffering from the disease D1 and patient's *P2* do not have any above mention diseases.

5. Case Study 2

Step 1:

In medical diagnosis, the fuzzy cluster analysis can also be used. This approach basically perform clustering algorithms on the set of patients, symptoms and diseases. The degree of membership can be define on the basisof severity of symptoms. The similarity is computed between symptoms of the patients that could be known and symptoms of a patient possessing the symptoms pattern for each possible disease. The cluster of the patients can be form whose symptoms are similar and the diagnostic candidate can be determine on the basis of patient's degree of membership which is greatest.

Let us take a patient Ramesh who displays the symptoms S = { temperature, headache, cough and vomiting }, it can be denoted by S1, S2, S3, S4 at the level of strictness given by fuzzy set

$Ax = .2/S1 + .7/S2 + .5/S3 + .7/S4$

where Ax =[0.1] it show the membership in fuzzy set considered patient x and defined in the set S= {S1,S2,S3, S4}, its indicate the strictness level of the symptoms Si for the patient. Now it determine a diagnosis among two possible diseases d1, d2 (malaria, dengue). Now on the basis of medical knowledge, first of all it determine the matrices R1 between symptoms

S= {S1,S2,S3,S4} and diseases D= { d1,d2}

$$R1 = \begin{array}{c} \\ S1 \\ S2 \\ S3 \\ S4 \end{array} \begin{array}{c} \begin{array}{cc} d1 & d2 \end{array} \\ \begin{pmatrix} [0.5, 0.4] & [0.6, 0.5] \\ [0.1, 0.4] & [0.2, 0.5] \\ [0.3, 0.3] & [0.6, 0.4] \\ [0.2, 0.5] & [0.3, 0.6] \end{pmatrix} \end{array}$$

Now, it make the separate two cluster B1 and B2 for both diseases d1 and d2 respectively, taking the lower bounds and upper bounds of symptoms of diseases.

$$B1 = \begin{matrix} & D_{1L} & D_{1U} \\ S1 & 0.5, & 0.4 \\ S2 & 0.1, & 0.4 \\ S3 & 0.3, & 0.3 \\ S4 & 0.2, & 0.5 \end{matrix} \qquad B2 = \begin{matrix} & D_{2L} & D_{2U} \\ S1 & 0.6, & 0.5 \\ S2 & 0.2, & 0.5 \\ S3 & 0.6, & 0.4 \\ S4 & 0.3, & 0.6 \end{matrix}$$

Step 2:

Now it define the fuzzy relation W, which is the weight of symptoms and diseases that specify the importance of each symptoms Si in the diagnosis for each diseases, which is given by

$$W = \begin{matrix} & S1 & S2 & S3 & S4 \\ d1 & 0.4 & 0.5 & 0.7 & 0.9 \\ d2 & 0.8 & 0.6 & 0.1 & 0.3 \end{matrix}$$

Step 3:

In process to diagnose patient x, it use a clustering technique to determine which diagnostic cluster (B1,B2) the patient is most similar. To compute the similarity, it use distance measure based on minkowski distance and it given by–

$$D=[\Sigma | w(si,dj)(Bjl. (si)- Ax(si)) |^{p} + \Sigma | w(si,dj)(Bju. (si) - Ax (si)) |^{p}]^{1/p} \qquad (2)$$

Where, $I\,l = \{ Ax (S\,i\,) < (Bjl. (S\,i\,) \}$

$I\,u = \{ Ax (S\,i\,) < (Bju. (S\,i\,) \}$

Step 4:

By the given formula and the above given W, B1, B2 and Ax, it calculate the value of possible diseases d1 and d2 are 0.50 and 0.38.

So, It concludes that the patient *Ramesh* is suffering from **malaria**.

Now finally it conclude that the first case study is more reliable then the second case study.

References

1. E. Sanchez, Solutions in composite fuzzy relation equations, *Fuzzy Automata and Decision Processes* (Elsevier/North-Holland, New York, 1977).

2. E. Sanchez, Inverses of fuzzy relations: Application to possibility distributions and medical diagnosis, *Fuzzy Sets and Systems* 2 (1979) 75–86.
3. B. Chetia and P. K. Das, An Application of Interval valued fuzzy soft set in medical diagnosis, *Int. J. contempt. math.,* sciences (2010).
4. S. K. Biswas and A. R. Roy, An Application of intuitionist fuzzy sets in medical diagnosis, *Fuzzy Sets and Systems* (2001).
5. E. Sanchez, Inverse of fuzzy relations, application to possibility distributions and medical diagnosis, *Fuzzy sets and Systems* (1979).
6. A. K. Shyamal and A. Pal, Interval valued fuzzy matrices, *Journal of Fuzzy Mathematics* (2006).
7. X. Zhou and P. Angelov, Real-time joint landmark recognition and classifier generation by an evolving fuzzy system. *Proceedings of IEEE world congress on computational intelligence* 2006 6314–21.
8. C. Xydeas, P. Angelov, S Chiao and M Reoullas, Advances in classification of EEG signals via evolving fuzzy classifiers and dependant multiple HMMs, *Computers in Biology and Medicine* 2005; 36(10): 1064–83.
9. S. Lekkas and L. Mikhailov, Breast cancer diagnosis based on evolvable fuzzy classifiers and feature selection In: T. Allen and R. Ellis, M. Petridis, *Proceedings of the 28th international conference on innovative techniques and applications of artificial intelligence*. Cambridge, UK: Springer; 2008.
10. R. Holt and N Hanley, Essential endocrinology and diabetes. Malden, USA: *Blackwell Publishing*, 2006.